WITHDRAWN

Slurry Walls

McGRAW-HILL SERIES IN MODERN STRUCTURES:
Systems and Management
Thomas C. Kavanagh, Consulting Editor

Baker, Kovalevsky, and Rish *Structural Analysis of Shells*
Coombs and Palmer *Construction Accounting and Financial Management*
Desai and Christian *Numerical Methods in Geotechnical Engineering*
Dubin and Long *Energy Conservation Standards*
Forsyth *Unified Design of Reinforced Concrete Members*
Foster *Construction Estimates from Take-Off to Bid*
Gill *Systems Management Techniques for Builders and Contractors*
Johnson *Deterioration, Maintenance and Repair of Structures*
Johnson and Kavanagh *The Design of Foundations for Buildings*
Kavanagh, Muller, and O'Brien *Construction Management*
Kreider and Kreith *Solar Heating and Cooling*
Krishna *Cable-Suspended Roofs*
O'Brien *Value Analysis in Design and Construction*
Oppenheimer *Directing Construction for Profit*
Parker *Planning and Estimating Dam Construction*
Parker *Planning and Estimating Underground Construction*
Pulver *Construction Estimates and Costs*
Ramaswamy *Design and Construction of Concrete Shell Roofs*
Schwartz *Civil Engineering for the Plant Engineer*
Tschebotarioff *Foundations, Retaining and Earth Structures*
Tsytovich *The Mechanics of Frozen Ground*
Walker, Walker, and Rohdenburg *Legal Pitfalls in Architecture, Engineering, and Building Construction*
Woodward, Gardner, and Greer *Drilled Pier Foundations*
Xanthakos *Slurry Walls*
Yang *Design of Functional Pavements*
Yu *Cold-formed Steel Structures*

Slurry Walls

Petros P. Xanthakos
Consulting Engineer

McGRAW-HILL BOOK COMPANY
New York St. Louis San Francisco Auckland Bogotá
Düsseldorf Johannesburg London Madrid Mexico
Montreal New Delhi Panama Paris São Paulo
Singapore Sydney Tokyo Toronto

Library of Congress Cataloging in Publication Data

Xanthakos, Petros P
  Slurry walls.

  Includes bibliographical references and index.
  1. Slurry trench construction.   2. Concrete walls.
3. Foundations.   I. Title.
TA775.X36   624'.1834   79-10095
ISBN 0-07-072215-3

Copyright © 1979 by McGraw-Hill, Inc. All rights reserved.
Printed in the United States of America. No part of this
publication may be reproduced, stored in a retrieval system,
or transmitted, in any form or by any means, electronic,
mechanical, photocopying, recording, or otherwise, without
the prior written permission of the publisher.

1234567890   KPKP   7865432109

The editors for this book were Jeremy Robinson and Elizabeth P. Richardson,
the designer was Naomi Auerbach, and the production supervisor
was Teresa F. Leaden. It was set in Bruce Old Style
by Black Dot, Inc.

Printed and bound by The Kingsport Press.

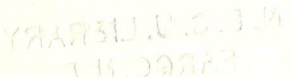

**With affection and love
I dedicate this book
to my daughter Eleni.**

Contents

Foreword by Ralph B. Peck   xiii
Preface   xv

1. State of the Art   1

1-1   Origin and First Uses of Slurries   1
1-2   Present Applications   2
1-3   A Brief Historical Review   4
1-4   First Experimental Work   11
1-5   Comparative Review of Ground-Support Systems   12
       REFERENCES   17

2. Stability of Trenches and Cuts   18

2-1   Stability of Unsupported Trenches   18
2-2   Stability of Slurry-filled Trenches in Clay   20
2-3   Stability of Slurry-filled Trenches in Sand   24

Contents

2-4	Effect of Gel Strength of Slurry on Trench Stability	27
2-5	Special Considerations for Excavations in Sand	29
2-6	Changes in the Properties of Adjacent Soil	35
2-7	Special Considerations for Excavations in Clay	40
2-8	Stability of Trenches with Horizontal Curvature	44
2-9	Stability under External Concentrated Loads	47
2-10	Stability under Dynamic Loading	47
2-11	The Cylindrical-Surface Method	52
	REVIEW PROBLEMS	54
	REFERENCES	56

3. Soil-Slurry Interaction   58

3-1	Formation of Seal	58
3-2	Thixotropic Gelation	64
3-3	Stability against Sloughing and Peel-off	67
3-4	Stagnation Gradient	71
3-5	Effect of Penetration on Soil	76
3-6	Permeability of Gravel Bed Penetrated with Slurry	78
3-7	Blowout Gradient	82
3-8	Changes in Slurry Density	84
3-9	Filter Cake	85
3-10	Electroosmotic Phenomena	86
3-11	Fluctuations in Slurry Level and Water Table	88
	REVIEW PROBLEMS	90
	REFERENCES	91

4. Technology, Preparation, Uses, and Control of Slurries   93

4-1	Introduction to Slurry Systems	93
4-2	Fundamentals of Clay Colloidal Systems	97
4-3	Applications of Stability Controls	104
4-4	Properties of Slurries	107
4-5	Measurement of Flow and Physical Properties	112
4-6	Materials Used in Slurries	117
4-7	Functions of Slurries	126
4-8	Control Limits	127
4-9	Effect of Soil Conditions on Control Limits	134
4-10	Proportioning the Slurry	135
4-11	Preparation and Control of Slurries	136
4-12	Slurry Disposal and Salvage	141
4-13	Bleeding of Slurries	144
	REVIEW PROBLEMS	146
	REFERENCES	148

5. Excavation and Equipment   150

5-1	Introduction to Excavating Systems	150
5-2	Factors Affecting the Selection of Equipment	153
5-3	Excavation for Slurry-Trench Cutoffs	155
5-4	Excavation with Bucket Scraper	160
5-5	Basic Types of Clamshells	162
5-6	Tools and Equipment for Hard Ground and Boulders	168

5-7	Excavation with Clamshells and Percussive Tools	171
5-8	Excavation with Rotary Drilling Equipment	177
	REFERENCES	193

6. Slurry-Trench-Cutoff Walls — 194

6-1	Introduction to Cutoff Systems	194
6-2	Earth Backfill	196
6-3	Construction of Earth Cutoffs	200
6-4	Design of Earth Cutoffs	206
6-5	Blowout Tests	208
6-6	A Case Study of Earth Backfill	211
6-7	Efficiency of Trench Cutoffs	215
	REVIEW PROBLEMS	219
	REFERENCES	221

7. Semirigid and Rigid Cutoff Walls — 222

7-1	Clay-Cement Mixes	222
7-2	Cement-Bentonite Cutoffs (Solidified Walls)	227
7-3	Injected Screens (Impervious Walls) and Vibrated Membranes	231
7-4	Similarities between Grouts and Slurries	236
7-5	Plastic-Concrete Cutoffs	237
7-6	Rigid Cutoff Walls	244
7-7	High-Resistance Noncorrosive Cutoffs	247
	REVIEW PROBLEMS	253
	REFERENCES	253

8. Concrete Technology and Construction — 255

Part A: Technology — 255

8-1	Practical Requirements Affecting Mix Design	255
8-2	Proportioning Concrete Mixes	257
8-3	Concrete Placement	260
8-4	Flow Motion of Tremied Concrete	264
8-5	Strength of Concrete	271
8-6	Bond Strength and Bond Stress	278

Part B: Construction — 286

8-7	Assembly and Details of Reinforcement	286
8-8	Common Types of Construction Joints	290
8-9	Special Construction Joints	297
8-10	Efficiency of Construction Joints	302
8-11	Structural Connections	306
	REVIEW PROBLEMS	311
	REFERENCES	312

9. Construction Fundamentals of Diaphragm Walls — 313

9-1	Site Inspection and Preparation	313
9-2	Panel Dimensions	315

x Contents

9-3	Panel Sequence and Arrangement	319
9-4	Guide Walls	320
9-5	Construction Accuracy and Tolerance	323
9-6	Watertightness of Diaphragm Walls	328
9-7	Problems Caused by Underground Structures and Utilities	331
9-8	Problems Caused by Boulders and Obstructions	333
	REFERENCES	335

10. Various Wall Systems — 336

10-1	Prefabricated Diaphragm Walls	336
10-2	Bored-Pile Walls	345
10-3	Composite Walls	353
10-4	Circular and Polygonal Enclosures	359
10-5	Posttensioned Diaphragm Walls	363
10-6	Buttressed Walls, Cells, and Arched Structures	367
	REVIEW PROBLEMS	373
	REFERENCES	374

11. Load-Bearing Elements and Foundations — 375

11-1	The Use of Slurries in Large-Diameter Piles	375
11-2	The Use of Slurries in Prismatic and Linear Elements	377
11-3	Usual Defects and Repairs	380
11-4	The Transfer of Load by Base Bearing and Shaft Resistance	381
11-5	Basic Concepts of Load Transfer	384
11-6	Tests on Bored Piles Installed by Slurry Displacement	386
11-7	Tests on Skin Friction and Wall Adhesion	392
11-8	Tests on Diaphragm-Wall Panels	396
11-9	Guidelines for the Selection of Load-bearing Elements	400
11-10	Design for Vertical Loads; Prismatic or Circular Elements	404
	REVIEW PROBLEMS	408
	REFERENCES	410

12. Uses and Applications — 412

12-1	Underpinning of Structures	413
12-2	Diaphragm Walls with Tiebacks	414
12-3	Diaphragm Walls on Stilts, Piles, Subpiers, and Rock	415
12-4	Prefounded Columns	418
12-5	Strip Panels	419
12-6	Protection of the Base of an Excavation	422
12-7	The Downward Construction Method	425
12-8	Use of Prefabricated Panels	427
12-9	Construction from a Lower Level	429
12-10	Applications for Subway Construction	431
12-11	Traffic Underpasses and Depressed Roadways	449
12-12	Underground Construction for Buildings	456
12-13	Special Problems of Deep Vertical Cuts	462

12-14	Utility Tunnels	464
12-15	Underground Parking	468
12-16	Industrial and Service Installations	469
12-17	Waterfront Facilities	473
12-18	Special Examples	476
	REFERENCES	484

13. Topics Relevant to Analysis and Design — 485

13-1	Settlement Detrimental to Surroundings	486
13-2	Ground Movement and Settlement Due to Excavation	487
13-3	Heave in Narrow Excavations Supported by Diaphragm Walls	490
13-4	Diaphragm Walls in Lieu of Underpinning	495
13-5	Stability of Circular, Polygonal, and Arched Structures	497
13-6	Failure Criteria for Unreinforced Steel-and-Concrete Panels	502
13-7	Miscellaneous Problems Related to Construction Conditions	505
13-8	Posttensioned Diaphragm Walls	508
13-9	Loads Acting on Gound Support	515
13-10	Experimental Comparison of the Stability of Rigid and Flexible Walls	520
13-11	Ground Movement and Settlement in Excavations Supported by Diaphragm Walls	525
13-12	General Stability of Linear Walls	561
13-13	Lateral Earth Stresses	563
	REVIEW PROBLEMS	605
	REFERENCES	608

Index — 611

Foreword

Everyone interested in slurry walls, from the most experienced specialty contractor to the engineer or architect considering the use of a slurry wall on a specfic job, stands to benefit from this book.

The techniques and practical applications of slurry-wall construction were developed largely by competitive, private contractors in several parts of the world. As a result, the literature contains more than its share of promotional material, and the claims for the various systems often reflect commercial as well as engineering considerations. Early development, largely in Italy and France, was favored by the European practice of choosing among contractor-designed alternatives. This practice gave full rein to ingenuity on the part of the contractors, but it probably inhibited adoption of the systems in countries where design is customarily done by engineers on behalf of the owners. Hence, in many countries, even though slurry-wall construction has become more usual, it retains an element of mystery.

Readers will discover that the author has assembled, evaluated, and organized present knowledge about the properties and function of slurries, about equipment and its use, about concrete and its placement under slurry, about structural requirements and design of the walls, and about the advantages and limitations of slurry-wall construction in practice. They will find, probably to their surprise, that the information, gathered from many disciplines and specialties and from many parts of the world, adds up to a coherent and rational whole that justifies the recognition of slurry walls as a respected engineering solution to many problems in underground construction. On the basis of the comprehensive treatment of the subject, designers will be able to make a sound decision as to whether slurry-wall construction should be considered for the project at hand. If they decide in the affirmative, they will have a background against which to judge the qualifications, the proposed methods, and the performance of the contractor.

The need for such a book has been pressing, and the author has filled it fully and elegantly.

Ralph B. Peck

Preface

This book is intended to set forth the technology of slurry walls and related construction, including current practices and applications. Although a great number of references are listed, the text hardly constitutes a survey of existing literature. Instead, considerable effort has been made to organize existing knowledge and develop it into a form that can be readily used by practicing engineers, architects, contractors, students, and owners of underground installations. Since its initial inception, this type of work has interacted with many engineering disciplines and produced new and fast-developing ideas. The result has been an impressive expansion of the construction market, increased uses, and new applications. In spite of the interest aroused and the attention attracted, this technique is still a new field of endeavor and warrants a complete understanding of its advantages as well as its limitations.

The experience gained thus far suggests that slurry walls and related elements are likely to remain more an art than a science, like any other foundation system, because of the many uncertainties associated with underground construction and soils engineering. However, the book places the emphasis equally on art and science, pointing out, where possible, the limitations of our present knowledge and understanding of the subject.

The material has been organized in much the same sequence in which it was initially introduced in 1974 (in the author's "Underground Construction in Fluid Trenches," Engineering Colleges, University of Illinois, Chicago Circle). Thus it is compatible with the normal construction procedures and follows the events that ordinarily lead to the finished structure. The book begins with a general background and historical notes. Next, it deals with the conditions of trench support and with the soil-slurry interaction which is credited with the stability mechanism. The technology of slurries continues the subject and completes three chapters of immediate geotechnical interest.

The excavation phase is an integral part of the overall process and requires elegant procedures as well as sophisticated equipment, discussed in Chap. 5. With this chapter the technical background is sufficient for a study of cutoff walls, dealt with in Chaps. 6 and 7. Concrete technology and details are treated in Chap. 8 on a provisional basis, as it is evident that the state of the art is advancing fast and that new scientific data will be forthcoming. This subject is further supplemented by the construction fundamentals of diaphragm walls, discussed in Chap. 9. Various wall systems are presented in Chap. 10; besides linear diaphragm walls they include prefabricated sections, bored-pile walls, composite walls, posttensioned diaphragm walls, buttressed walls, and arches. The load-carrying capacity and the construction fundamentals of foundation elements built by the slurry process are the subject of Chap. 11.

Uses and applications are discussed in Chap. 12, with special emphasis on urban situations. Chapter 13 completes the subject with a balance of geotechnical and structural approaches relevant to analytical and design problems. Throughout the text the terms *diaphragm walls* and *slurry walls* are used interchangeably and are intended to mean the same thing, i.e., an underground concrete wall inserted by the slurry-trench process.

Practical problems and examples of reasonable simplicity but sometimes of complex origin have been inserted where possible to demonstrate the validity and the limitations of the technique. Review problems have been included at the end of most chapters, particularly those with a theoretical background, so that sufficient material is available with the text to cover a course in geotechnical, structural, and construction engineering at the graduate level.

My gratitude is extended to Dr. Ralph B. Peck, who graciously read the complete text and provided the foreword. During the many years the book was under preparation, its chapters were written several times. My sincere thanks are due my wife, who typed all the material at least twice until a satisfactory final manuscript was obtained and who assisted in proofreading.

Petros P. Xanthakos

Slurry Walls

Chapter One
STATE OF THE ART

1-1 ORIGIN AND FIRST USES OF SLURRIES

Using slurries for drilling and trenching is a comparatively recent development which appears to have originated independently of any related work in the grouting field. It is conceivable that the method was suggested to oil engineers for well drilling as early as 1914. According to the record, a common problem in cable-tool drilling could arise whenever the hole became filled with a thick mud of water and cuttings from the excavation. When this mud contained too much solid material, it could interfere with the movement of equipment and the drilling process. However, with deeper holes it became evident that the muddy slurry filled a useful function in balancing inward artesian water and gas pressure and supporting the face of the hole in unstable formations.

Although the supporting action and the stabilizing effect of slurries must have been known since 1900, the first publication on the subject did not appear until 1913. The technology of drilling fluids dates back to 1921, but significant progress in the preparation and control of slurries probably did not occur until the early 1940s. Bentonite was first introduced in slurry systems in 1929. In 1927 Peterfi and Freundlich were the first to investigate and define the all important isothermal transformation of colloidal suspensions termed *thixotropy*. In 1931 the Stormer viscometer was applied to the measurement of slurry properties, and in the same year Marsh invented the funnel viscometer still in broad use today.

Drilled Concrete Piers Stabilizing muds have been used since the inception of rotary drilling. The slurry was initially used in the hole to remove cuttings by reverse circulation, and soon it was recognized as a prime factor in preventing sloughing, peel-off, and face collapse. The technique was adapted to deep foundation drilling from oil-well practice but required suitable materials and special slurry-control agents.

Contiguous Drilled-Pier Walls A row of drilled piers (also called *bored piles*) can be used as ground support. Walls of this type did not appear until the early 1950s, or more than 30 years after the introduction of bored piles. A probable explanation for this delay is that urban renewal did not start until after the second World War. At present bored-pile walls are used in lieu of steel sheeting and soldier piles with lagging.

Slurry-Trench-Cutoff Walls The first slurry-trench-cutoff wall was probably built at Terminal Island near Long Beach, California, in 1948 (Sherard, 1969; Leps, 1973). It was 45 ft deep and consisted of earth backfill. The trench was excavated with a bucket, and its face was protected by a water-clay slurry upgraded with bentonite.

Recent techniques and materials used in cutoff walls encompass a wide range. Examples are the plastic concrete cutoffs adapted for deformability and strength, the solidified walls allowing considerable flexibility by transforming the initial bentonite slurry into a permanent construction material; and the injected-grout screens whereby an impermeable membrane is obtained by injecting cement-bentonite grout into a preformed slot.

The Continuous Diaphragm Wall This structure, formed and cast in a slurry trench, was first conceived by Veder in 1938 and can be looked upon as a combined development from two related systems; the mud-filled borehole and the continuous bored-pile wall. The complexity of the operation was somewhat simplified with the use of conventional grabs for direct linear excavation while the slurry was only used for face support. Although test concrete panels were inserted in linear trenches in the 1940s, the first underground reinforced-concrete walls were not attempted until the 1950s. The construction sequence for a continuous diaphragm wall using modern methods and equipment is shown in Fig. 1-1 and is self-explanatory.

1-2 PRESENT APPLICATIONS

A 1977 assessment of the total area of diaphragm walls built throughout the world is in excess of 10 million square meters, excluding cutoff walls, bored piles, and similar installations. The main uses are as retaining structures and

State of the Art 3

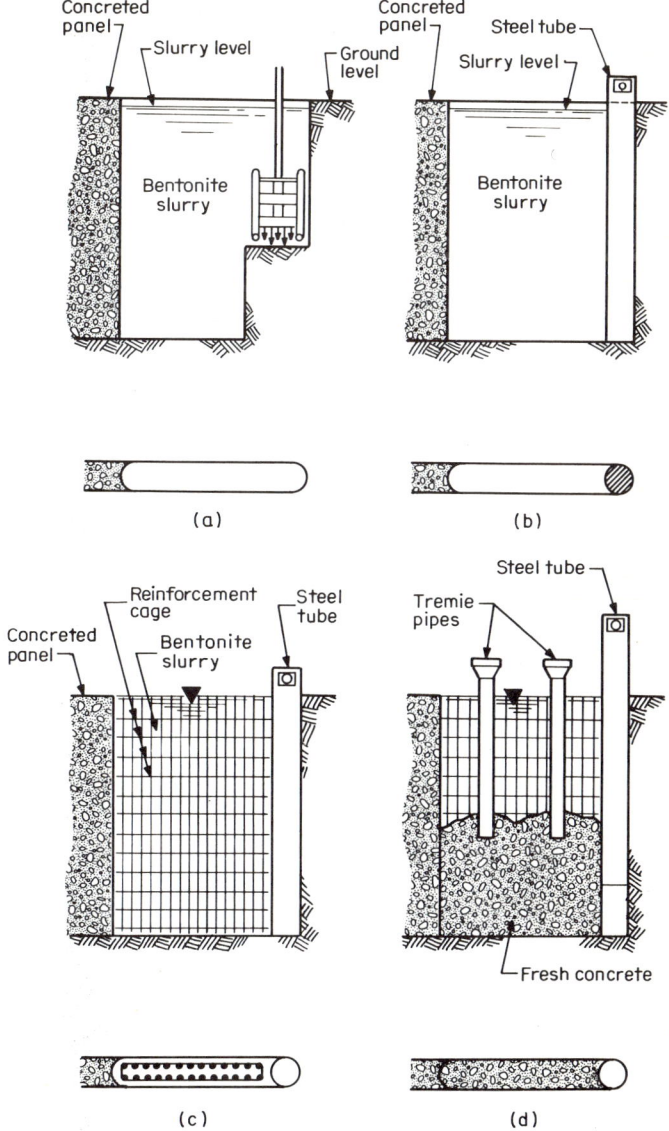

Fig. 1-1 Typical construction sequence of a diaphragm wall, executed in four stages: (a) excavation, (b) insertion of steel tubing, (c) placement of reinforcement cage, (d) concrete placement.

load-bearing elements for deep building basements, traffic underpasses, underground mass-transit stations, and cut-and-cover tunnels, parking garages, underground industrial facilities, docks and waterfront installations, waterworks, and miscellaneous foundations.

In general, three factors have contributed to the expansion of the market.

The first is the commercial availability of bentonite. This colloid is prepared from montmorillonite clays and is easily refined and converted into sodium form for engineering uses. Second, the method accommodates difficult situations and conditions, including troublesome soils, and it can to a great extent remedy the problems of underground construction in urban areas. A third favorable factor is the resolution of certain practical and technical matters, including the improvement of excavating techniques, the development at the site of mud plants for processing slurries, including mixing and desanding facilities, and the control of concrete quality and strength.

The use of slurries is also evident in special applications. Examples are (1) the slip-caisson construction, where the slurry is used merely to reduce the frictional resistance between soil and a foundation caisson so that the latter can be lowered without driving or pounding (Lorenz, 1963); (2) slurry trenches intercepting vibrations and shock waves, thus protecting sensitive buildings from dynamic loads (Xanthakos, 1974a); (3) slurry face shields, which enable the cutterhead of a tunneling machine to operate in a slurry medium; and (4) the horizontal penetration of large pipes and tubes, where the slurry reduces skin friction thereby acting as a lubricant.

1-3 A BRIEF HISTORICAL REVIEW

At first attention was directed to the excavation process, and a common impediment to slurry trenching was the necessity of passing through very pervious formations and sealing them off. Contractors showed a tendency to compromise between boring and earthwork equipment since this was thought to lead to cost savings. Economy presumably ought to be the primary reason for using slurry trenches in preference to other techniques and types of foundations. Thus, when applications were limited in both volume and variety, cost factors dictated diversified equipment that could be adapted to broader output. Extension of the market allowed some redemption of the corresponding investments, although there was still no distinction between this type of work and other forms of excavation.

In the course of time it became apparent that economic considerations alone should not dictate the principal characteristics of the technique. This attitude helped establish the boundaries of the system and made better engineering judgment possible. Thus, attention was directed also to the construction accuracy of the finished structure, the details of construction joints, preparation and control of slurries, concrete quality control, and the all-important matter of soil-structure interaction, leading eventually to the adaptation of diaphragm walls as permanent structures.

Statistical Data and Regional Developments

Applications in Europe On a volume and application basis Europe has led the diaphragm-wall market, beginning with Italy. A summary of regional construction in Italy is given by Sadleir and Dominioni (1963), Baldovin and Berra (1969), and ICOS (1968). One of the most significant and difficult projects ever to be undertaken is the proposed aquifer recharge to lift Venice and stop the damage caused by excessive groundwater extraction and the resulting sinking of foundations. This scheme foresees the construction of a perimeter diaphragm wall 120 m (390 ft) deep to enclose and isolate the ground formation beneath the city, to be followed by the installation of recharge wells to pump water and thus restore the original groundwater pressure balance beneath Venice (Braun, 1975).

In France the linear diaphragm wall probably evolved from the bored-pile wall and led to such innovative concepts as precast diaphragm-wall panels and solidified cutoff walls executed without construction joints.

In England the technique was first introduced with the construction of the Hyde Park underpass in London. Recent examples are the Piccadilly line extension to Heathrow Airport, a portion of which was constructed by bored piles and continuous diaphragm walls; the anchorage structures for the Humber suspension bridge; and the German Embassy in London, built with posttensioned diaphragm walls.

Belgium is among the leading European countries that have extensively adapted the diaphragm-wall technique. One of the largest and most recent projects is the Brussels metro, where diaphragm walls provided the permanent structure for the transit line. During this construction it was possible to use 12-m-long panels (about 40 ft) without problems.

In Germany and Switzerland the method is basically adapted to city construction, and notable examples are posttensioned diaphragm walls. In Poland the Institute of Building Technology was conducting investigations on the properties of thixotropic suspensions as early as 1952. In Norway the first applications were probably in conjunction with the construction of the Oslo subway in soft clay, but notable technical contributions have been made by the Norwegian Geotechnical Institute. In Spain diaphragm walls have been used in building basements, waterfront facilities, and subway construction.

Applications in Japan and the Far East In Japan the technique was introduced in 1959, and construction has now reached an average yearly volume in excess of 200,000 m^2 (Ikuta, 1974). This activity is largely due to an increased demand for underground space but also reflects current progress in the state of the art. Bentonite does not appear to be as plentiful as it presumably is in North Africa, Europe, and the United States, and this shortage has been compensated for by the production of new colloid materials that also make further recycling of slurries possible.

A 1974 breakdown of all applications in Japan is given in Fig. 1-2. Although diaphragm walls appear to be permanent structures more in buildings and less in other types of underground construction, this trend has been rapidly changing. Thus, as of 1977 there are no restrictions in the use of

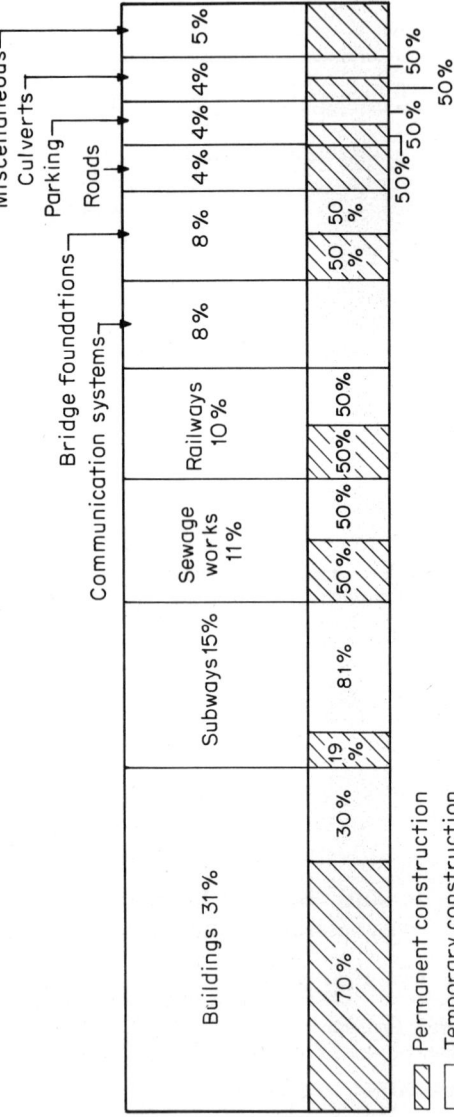

Fig. 1-2 Breakdown of diaphragm wall applications in Japan. (*From Ikuta, 1974.*)

diaphragm walls, and current Japanese building codes and standards include provisions for their design and construction.

Elsewhere in the Far East diaphragm walls have been used in Hong Kong (mainly in subway construction) and in Taipei (for deep building basements).

Applications in North and South America The technique was introduced in Canada in 1957, and since then it has followed a growth rate similar to that in Europe and Japan. In South America, diaphragm walls are currently used for subway work in Rio de Janeiro and Sao Paulo, Brazil, and in Caracas, Venezuela.

In the United States diaphragm-wall construction was initiated in 1962. Sherard (1969) and Kapp (1969) have compiled statistical surveys and have given the first account of this work. A more recent assessment is Xanthakos (1974b). The activity appears to be concentrated largely in metropolitan areas, but a great many applications are evident with the use of cutoff walls for groundwater control. Examples from the various regions are the World Trade Center in New York (Fig. 1-3), the Water Tower Place in Chicago (Fig. 1-4), the Huxtable Pumping Station in Arkansas (Fig. 1-5), the Cobian Plaza in Puerto Rico (Fig. 1-6), sections of the New York subway (Fig. 1-7), the construction shafts for the Chicago underflow system (Fig. 1-8), and the Sixty

Fig. 1-3 Diaphragm walls for the World Trade Center, New York (*ICOS.*)

Fig. 1-4 Diaphragm walls for the Water Tower Place, Chicago. (*Tone Boring Co.*)

Fig. 1-5 Cutoff walls for the Huxtable Pumping Station, Arkansas. (*ICOS.*)

Fig. 1-6 Diaphragm walls for the Cobian Plaza in Puerto Rico. (*ICOS.*)

Fig. 1-7 Diaphragm walls for the New York City subway.

Fig. 1-8 Construction shafts for the Chicago underflow system. (*Sumitomo America.*)

State Street Tower in Boston (Fig. 1-9). Notable examples of the early applications are the walls for the stations of the Bay Area Rapid Transit (BART) project.

Although it is difficult to assign growth rates to regional construction, it is evident that certain factors have prompted the use of the technique in most parts of the country. Some regions have responded more favorably than others, and in some instances the choice is influenced by nontechnical considerations. Thus, in many areas this type of construction has become standard practice, whereas it is still unknown in some others. As of 1977, diaphragm walls were contemplated in major urban projects, including Boston and Chicago.

Fig. 1-9 Diaphragm walls for the Sixty State Street Tower, Boston. (*Franki.*)

A national conference was held at the University of Illinois Chicago Circle in the spring of 1974, focusing on underground construction in slurry trenches. An interim report (Xanthakos, 1974a) dealing with the fundamentals of the technique distributed at the conference provided the basis for this book.

1-4 FIRST EXPERIMENTAL WORK

Valuable contributions to the state of the art made between 1950 and 1960 involved laboratory research conducted by independent investigators, various institutions, and a number of specialist firms. Credit must be given to Veder (1953, 1954, 1963) and Lorenz (1950, 1953, 1954, 1961, 1963) for their investigations of the behavior of bentonite slurries and of the various aspects of trench support.

Following laboratory tests, Veder continued the investigation with full-scale field models, building short-length diaphragm-wall panels in a series of circular 0.6-m-diameter elements. The stability of these panels confirmed that equilibrium was obtained between soil and groundwater on one side and bentonite slurry on the other. Thus the first experimental panels consisted of circular concrete elements built to overlap each other, which were later converted into straight walls when it became possible to excavate linear trenches.

Veder (1963) presented the following conclusions about the supporting action of bentonite slurries:

1. When a trench is excavated, a bentonite cake is formed at the interface. The cake is generally impermeable and adheres fully to the soil, thus allowing the slurry to exert its full hydrostatic thrust against the face.
2. The formation of the cake is influenced by the degree of stability of the slurry as it penetrates into the soil pores, its ability to gel in these voids, and probably by certain electroosmotic phenomena.
3. The bentonite cake thus constitutes a kind of plaster on the face, linking the individual grains which constitute the soil and keeping them in their relative position.
4. In percussion boring and to some extent in rotary drilling there is some compaction of soil ahead of the tool, and this tends to increase the stability of the face. Furthermore, the kinematic action of the excavating tool favors soil impregnation by bentonite in the immediate vicinity to form the cake.

This supporting action was credited for the stability of trench excavations and made it possible to reach considerable depths. Veder also suggested that increasing the excavation depth should not materially affect the balance of lateral forces and that the most difficult conditions usually are encountered at the beginning of the operation close to the ground surface. These points are discussed in detail in subsequent sections.

1-5 COMPARATIVE REVIEW OF GROUND-SUPPORT SYSTEMS

It is neither unusual nor surprising that extension of the diaphragm-wall market has been followed by a trend for the average prices to drop. Baldovin and Berra (1969) compiled a cost summary showing that average prices in Italy fell by as much as 50 percent between 1954 and 1968, in spite of rising construction costs. In France, at least until 1969, the cost of diaphragm walls dropped by 5 percent every year. Although this trend is not likely to continue under the current inflationary pressures, it reflects greater competition between bidders due to greater availability of specialty contractors, continuing progress and improvement of construction techniques, and increased construction volumes through market extension allowing better recovery of investments and lower mobilization costs.

Table 1-1 demonstrates how the use of diaphragm walls may influence the cost of underground construction, showing an average cost breakdown for cut-and-cover subway construction. In this case the construction involves a temporary support and a permanent structure, and evidently substantial savings may be possible if these two items are combined into one system. The cumulative percentage shows also that items presumably having a major effect on disruption and public inconvenience have only a minor effect on cost. For example, street-decking, utility-adjustment, and surface-restoration costs can be changed significantly without affecting the overall cost materially.

TABLE 1-1 Cut-and-Cover Cost Breakdown for Subway Construction

Item	Percentage of total cost
Mobilization*	4–10
Decking	2–8
Excavation and ground support	30–35
Permanent structure and underpinning	30–40
Backfill	3–6
Utilities (maintenance and relocation)	5–15
Restoration	2

*Includes traffic maintenance.

The direct cost of ground support varies according to the method used. For average conditions soldier piles with lagging usually provide the least expensive system, followed by sheet piling, diaphragm walls, contiguous bored-pile walls, and ground treatment. However, this comparison can be misleading unless it takes into consideration the many factors that determine the overall cost. For example, a system that alleviates the need for underpinning and groundwater control can result in an overall economy although the ground-support system itself is not necessarily the most economical.

The following sections provide a comparative review of ground-support systems.

Soldier Piles with Lagging

Soldier piles with wood lagging represent a time-tested ground support which generally is economically attractive. The system is adapted where ground movement can be tolerated and groundwater control is not required. Certain problems will arise, however, where it is necessary to underpin existing foundations or where the excavation involves water-bearing ground. Ground loss can occur as the excavation is carried down and the lagging is installed, especially in the more granular soils. Both surface water and groundwater leaking through the lagging will carry the fine fraction and thus cause subsidence.

Structurally, this support is flexible, particularly below excavation level, and offers little resistance to ground movement in this region. Since the wall is not watertight, the excavation must depend on well points or on deep-well dewatering. Because there is no seal, there is some danger of piping. On the other hand, if the dewatering is extensive, settlement can occur as result of lowering the water table.

The installation is more economical if the piles can be withdrawn and reused. If they must be left in place, the choice is more attractive if they can become part of the permanent structure. The method is uniquely suited to situations involving utilities.

Sheet-Pile Walls

Steel sheeting eliminates many of the disadvantages of soldier piles with lagging. A seal is obtained at the base of the excavation if the sheeting is driven in interlock, and this generally ensures adequate control of groundwater. There is more resistance to ground movement, especially below excavation level, but because of its inherent flexibility sheet piling is more suitable for relatively shallow excavations or where some soil deformation can be tolerated. The system should not be used where ground movement must be rigidly controlled.

Where the ground is hard or contains boulders and other obstructions, it may be quite difficult and even impossible to drive sheet piling. Alternatively sheet-pile driving can have depth limitations. In congested urban sites sheet-pile installations may not be possible because of the danger of cutting utilities. Noise and vibrations from driving operations commonly are objectionable. Further disadvantages are the headroom requirements, which sometimes necessitate the removal of aerial obstacles for positioning and driving the piles, and problems inherent in the transportation and delivery of long pile sections in built-up areas.

Sheet-pile walls are more expensive than soldier piles with lagging unless they can be pulled out and reused.

Ground Treatment

Freezing Freezing can be used practically in any type of soil which has pore water. Almost invariably, however, it has been used below the groundwater table, i.e., in fully saturated soils. In some countries, including the United States, freezing is regarded as a last-resort remedial measure employed where other techniques are not feasible. This is probably due to the relative high cost and the specialized nature of the application. Because of the rather limited uses, experience with freezing is confined to a few specialist contractors, and such important details as instrumentation, freezing plant, and freeze elements still remain a proprietary matter. Published records usually omit useful cost data, soil properties of frozen ground, and other details necessary to evaluate the technical merits and the relative cost of the technique.

A major advantage of freezing is its independence of the soil permeability, and in fact successful applications have been reported in soils of low permeability such as fine sands and silts. It is possible to evaluate the strength-deformation characteristics of frozen soil in terms of temperature, the time necessary to achieve freezing, and the predicted ground movement. However, the time required to achieve freezing can increase drastically due to such factors as groundwater flow and the presence of organic materials. Furthermore, it is difficult to control freezing temperatures closely due to uncertainties in heat loss and deviations in the assumed soil thermal properties. The strength of frozen soil is materially influenced by changes in temperature; hence a difference of only a few degrees can alter the behavior of frozen ground and require extra refrigeration capacity which will increase costs accordingly. A slight but persistent seepage can result in progressive and hard-to-stop thawing of the frozen section and thus require additional freezing units with a corresponding increase in the cost.

Grouting In the last few years considerable progress has been made, particularly abroad, in grouting. In addition to being used for granular soils to reduce permeability and improve strength, consolidation grouting is used to increase the load-carrying capacity of plastic soils.

The cost of materials still continues to be a serious factor in the total cost of the application. A 1975 cost breakdown of grouting gives a range from $15 for a cement-bentonite grout to as high as $150 for sodium silicate gels per cubic yard of grouted soil (Tallard, 1975). Accordingly, the ability to modify grout materials in the field to accommodate variable in situ conditions can result in considerable savings. Grouting techniques, on the other hand, still remain to considerable extent a proprietary matter, and despite the published technical data they are unfamiliar to most engineers.

Although the technique still is not a standard construction practice, especially in the United States, it has received immediate applications abroad to remedy the problems caused by the concentration of utilities at crossings. Thus, tunneling in a grouted area between two open cuts is becoming more common.

Shotcrete Although shotcrete applications do not constitute direct ground treatment, they are commonly employed to control ground behavior in underground openings. Despite the fact that in many instances shotcrete has been misused, in general it has given satisfactory performance and has been economical.

Shotcrete has proved particularly useful as a remedial support in many situations. Alternatively, it can be used in underground openings for initial ground control, for supplemental control, or as permanent lining. The ground conditions in which shotcrete is appropriate must be carefully evaluated, together with the economic considerations and the compatibility of the application with the methods of excavation. In this respect shotcrete is uniquely effective and useful as secondary lining in cut-and-cover tunnels where the initial support system consists of diaphragm walls.

Bored-Pile Walls

Contiguous or secant piles built in prebored holes stabilized with bentonite slurry can be adopted in the most unfavorable and adverse ground conditions, and this is possible with a higher factor of safety than in diaphragm walls. In the event of slurry loss through open ground there is no immediate risk of collapse. On the other hand, the bearing capacity of the wall is greater due to the greater contact surface with the ground for the same volume of concrete. Because of the flexibility inherent in the installation bored-pile walls, like soldier piles with lagging, are quite suitable at sites involving utilities.

Among the disadvantages are certain depth limitations imposed by resistance to the sinking and raising of the guide tube, a tendency to introduce misalignment in hard or inclined stratifications, and difficulties in ensuring watertightness if the piles are not secant.

Diaphragm Walls

Diaphragm-wall construction in general provides maximum economy where it can transplant the temporary sheeting and the permanent ground support or where the excavation involves underpinning or groundwater control and the diaphragm wall can remedy this problem.

Regarding the direct cost of the installation, the ability to produce a low-cost structure will generally depend on many factors. Thus, besides the physical dimensions and the configuration of the wall, its cost may be influenced by (1) the required embedment below excavation level for stability or seepage control; (2) the nature of ground to be excavated and the presence of boulders or other hard obstructions; (3) the associated stability requirements; and (4) such site factors as mud disposal, availability of utilities, time restrictions, and availability of working space.

Advantages It is possible to produce time savings and thus satisfy the goal of rapid completion. This is true, for example, with the under-the-roof method or by the downward construction process, whereby the excavation of the basement and the erection of the superstructure can proceed simultaneously.

The method can be adopted under unfavorable soil and hydrologic conditions and where other techniques may have limitations. The walls can be constructed to considerable depths ahead of the main excavation and act as structural underpinning to adjacent structures.

The installation is completed essentially free from noise and vibrations and can accommodate the requirements of dry excavation. Maximum usage of the site is possible by building the walls along the site boundary, and panels can actually be constructed against existing buildings under minimum clearance. The wall layout is not sensitive to irregular and difficult sites.

Finally ground movement and settlement can be eliminated or sufficiently controlled with bracing such as cross walls at the bottom and the permanent floors elsewhere.

Disadvantages Site congestion seriously impedes traffic, access to adjoining property, and other activities. In order to lessen this drawback it may be necessary to (1) operate on one side only; (2) restrict panel size, mud plant, and equipment; and (3) operate during certain hours only.

The finished product is influenced by the type of soil, and the surface finish may be rough and thus require further treatment. Obstructions may cause concrete blisters and protrusions that must be broken out, and in sloped hard formations there is a danger of deflection in the excavation with a corresponding loss of verticality. Likewise, the overall matter of quality assurance depends entirely on processes, methods, and operations carried out at the site; hence it requires special field inspection and supervision. Furthermore, the accuracy of the excavation to a certain extent depends on the skills and experience of the operator. Finally, the disposal of used slurries in urban areas may pose special problems.

REFERENCES

Baldovin, G., and E. Berra, 1969: In Situ Diaphragm Walls, *Proc. 7th Int. Conf. Soil Mech. Found. Eng., Mexico City, Spec. Sess.* 14, 15, pp. 86-90.
Braun, W., 1975: Aquifer Recharge to Lift Venice, *Underground Serv.*, vol. 1, no. 3 (Foundation Publications, Ltd., London).
ICOS, 1968: "Underground Works," ICOS, Milan.
Ikuta, Y., 1974: Diaphragm Walling in Japan, *Ground Eng.*, **7** (5): 39-44.
Kapp, M. S., 1969: The Application of Cast-in-Situ Diaphragm Walls in the United States, *Proc. 7th Int. Conf. Soil Mech. Found. Eng., Mexico City, Spec. Sess.* 14, 15, pp. 97-99.
Leps, T. M., 1973: Early United States experience with slurry-trench construction, personal communication.
Lorenz, H., 1950: *Bautechnik*, **27:**313.
———, 1953: *Bautechnik*, **30:**232.
———, 1954: *Beautechnik*, **34:**250.
———, 1961: *Bauingenieur*, **36:**5.
———, 1963: "Utilization of a Thixotropic Fluid in Trench Cutting and the Sinking of Caissons," Butterworths, London, pp. 202-205.
Sadleir, W. A., and G. C. Dominioni, 1963: "Underground Structural Concrete Walls," Butterworths, London, p. 189.
Sherard, J. L., 1969: Statistical Survey of Diaphragm Wall Applications, *Proc. 7th Int. Conf. Soil Mech. Found. Eng., Mexico City, Spec. Sess.* 14, 15, p. 96.
Tallard G., 1975: Dewatering and Grouting as Supplementary Ground Engineering Techniques, *U.S. Dept. Transp. Chicago Urban Transp. Dist., Proc. Underground Constr. Prob., Tech. Solutions.*
Veder, C., 1953: Procedure for the Construction of Impermeable Diaphragms at Great Depths by Way of Thixotropic Seals, *Proc. 3d Int. Conf. Soil Mech. Found. Eng., Zurich.*
———, 1954: Procedure for Constructing Impermeable Diaphragms, *Cement Mag.* no. 8.
———, 1963: "Excavation of Trenches in the Presence of Bentonite Suspensions for the Construction of Impermeable and Load Bearing Diaphragms," Butterworths, London, pp. 181-188.
Xanthakos, P. P., 1974a: "Underground Construction in Fluid Trenches," Colleges of Engineering, University of Illinois, Chicago.
———, 1974b: Diaphragm Wall Construction and Slurry Trench Applications in U.S.A., *Ground Eng.*, **7** (5): 31-33.

Chapter Two

STABILITY OF TRENCHES AND CUTS

2-1 STABILITY OF UNSUPPORTED TRENCHES

Under certain conditions trenches and cuts can be excavated without lateral support. For such cases the excavation is carried out without the use and protection of slurries, either partially or to its full depth if the soil is stable. The resulting advantages are mainly the elimination of mud plants and facilities for the preparation, supply, and disposal of slurries and the ability to concrete the panel in the dry.

A vertical cut in cohesionless soil is feasible only if the face is protected. An inclined cut in clean dry sand generally is stable regardless of height as long as the angle i which its slope makes with the horizontal is smaller than the angle of friction ϕ of the sand in a loose state. The factor of safety for such a slope usually is defined as the ratio $(\tan \phi)/(\tan i)$. The only unknown factor is an appropriate value for the angle ϕ, which usually can be estimated with sufficient accuracy.

A trench in cohesive soil can stay open without slurry support provided the site is underlain by firm materials. This problem can be analyzed by noting that saturated clays, when loaded under conditions causing no changes in water content, behave with respect to the applied stress at failure as purely cohesive materials; i.e., the angle of shear resistance is zero. The stability is then investigated on the basis of limit theory. Terzaghi and Peck (1948) have presented the analysis illustrated in Fig. 2-1. Slip failure occurs along plane surface BC making an angle with the horizontal, as shown in Fig. 2-1a. The following expressions are easily derived:

$$z_0 = \frac{2c}{\gamma} \tan\left(45 + \frac{\phi}{2}\right) \tag{2-1}$$

and
$$H_{cr} = 2z_0 = \frac{4c}{\gamma} \tan\left(45 + \frac{\phi}{2}\right) \tag{2-2}$$

where z_0 = depth at which horizontal stress is zero
γ = unit weight or bulk density of soil
c = cohesion factor
H_{cr} = maximum height for which cut is stable with factor of safety of 1
ϕ = angle of shear resistance

The stress diagram corresponding to this condition is shown in Fig. 2-1b. The assumption of $\phi = 0$ reduces the foregoing expressions to

$$z_0 = \frac{2c}{\gamma} \tag{2-1a}$$

and
$$H_{cr} = 2z_0 = \frac{4c}{\gamma} \tag{2-2a}$$

If a uniform surcharge acts on the surface, these relations are easily modified to include its effect. The critical height is now

$$H_{cr} = \frac{4c - 2q_s}{\gamma} \tag{2-2b}$$

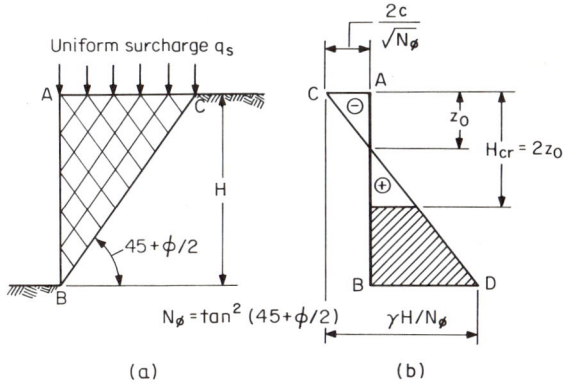

Fig. 2-1 Conditons of a cut in cohesive soil at slip failure: (*a*) section through cut; (*b*) horizontal earth-stress diagram.

At any depth less than z_0 the soil is in tension; hence tension cracks tend to develop near the surface, associated with the horizontal stretching which is inherent in the active state of stress. In general, if tension cracks propagate to a depth z_t, the maximum unsupported height must be adjusted so that $H_{cr} = 4c/\gamma - z_t$. Terzaghi (1941) and Terzaghi and Peck (1967) have estimated that tension cracks may extend to a depth about one-half of the total height of the cut. However, in most practical problems tension cracks are disregarded below the depth z_0 since in this region the soil is in compression. For this condition the critical height is

$$H_{cr} = \frac{2c}{\gamma} = z_0 \tag{2-3}$$

i.e., the maximum height of unsupported soil is also the assumed depth of tension cracks.

From the foregoing it is evident that the development of tensile stresses affects trench stability, and in some cases it may be quite unsafe to assume that soil can withstand horizontal tension. However, where the water table is at or near the ground surface, it is unlikely that tension cracks actually will develop; if they do, they will extend to a depth less than z_0. The presence of a uniform surcharge q_s results in compressive stresses which tend to close the tension cracks. If q_s is larger than $2c$, tension cracks can be ignored, but in this case H_{cr} is negative; i.e., the trench cannot be excavated. A special solution is needed if q_s is less than $2c$.

In some instances failure is temporarily retarded by factors contributing to the tensile strength of the soil, particularly in the upper crust of the ground. An example is the frost condition in cold climates. Failures of this nature are not uncommon (Tschebotarioff, 1973).

On the basis of the $\phi = 0$ assumption the corresponding theoretical failure plane is inclined at 45° to the horizontal. Analysis based on circular failure arcs leads to similar expressions except that the numerical coefficient 3.85 replaces the coefficient 4 in the right-hand term of Eq. (2-2a). This corresponds to the stability number estimated by Taylor (1948).

2-2 STABILITY OF SLURRY-FILLED TRENCHES IN CLAY

Nash and Jones (1963) have considered the stability of trenches on the basis of limit theory. The supporting element in this case is the hydrostatic force exerted by the slurry upon the face of the excavation. This action generally is facilitated by a watertight membrane built at the slurry-earth interface according to a mechanism discussed in subsequent sections.

The forces for simple wedge failure are shown in Fig. 2-2. Figure 2-2b is for cohesive materials ($\phi = 0$), whereas Fig. 2-2c corresponds to purely cohesionless sand ($c = 0$).

Stability of Trenches and Cuts 21

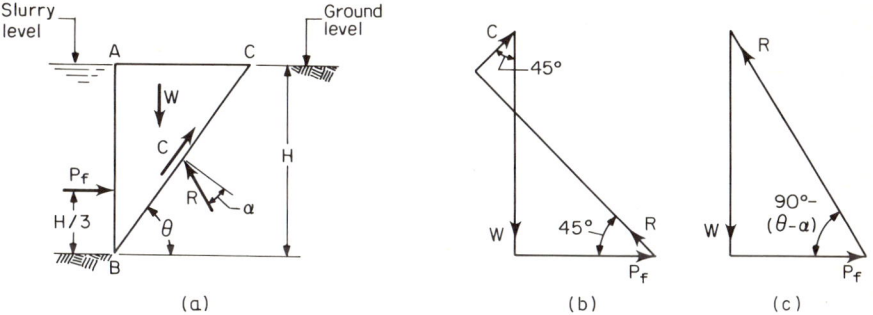

Fig. 2-2 Stability of a slurry trench: (*a*) section through trench; (*b*) force polygon for purely cohesive soil ($\phi = 0$); (*c*) force triangle for purely cohesionless sand ($c = 0$).

The active wedge *ABC* tends to slide along plane *BC* under its own weight, but this tendency is resisted by the shear force *C* and by the hydrostatic thrust P_f of the slurry. The condition $\phi = 0$ yields $\alpha = 0$ and $\theta = 45°$. When all the forces in the polygon of Fig. 2-2*b* are expressed in terms of H, c, γ, and γ_f, equilibrium leads to

$$H_{\text{cr}} = \frac{4c}{\gamma - \gamma_f} \tag{2-4}$$

If a uniform surcharge q_s acts on the surface, the critical height is

$$H_{\text{cr}} = \frac{4c - 2q_s}{\gamma - \gamma_f} \tag{2-5}$$

Experience shows that Eqs. (2-4) amd (2-5) give satisfactory results for the conditions under which they are applied, namely (1) the trench is relatively long compared with its depth; (2) the cohesion factor *c* is representative for the depth over which the analysis is considered; and (3) there is no fluid loss to the ground through the interface. The thrust of the slurry also reduces the tendency of the soil to develop tension cracks and fissures in the upper part (Xanthakos, 1974; Tschebotarioff, 1967). It is also evident that for the same soil parameters the critical height depends on the slurry density or unit weight, so that the heavier the slurry the greater this height. However, as shown in Chap. 4, there are limitations on the maximum practicable weight of slurry.

Validity of the $\phi = 0$ Concept

The $\phi = 0$ analysis for trench-stability problems in cohesive soils has considerable practical significance and is valid where the soil is stressed under undrained conditions, i.e., where there is no change in the water content. Examples are saturated soils where a cut or a trench is excavated quickly, with or without a retaining structure. Thus, the $\phi = 0$ concept, also known as *total-stress analysis*, is an *end-of-construction* method, but stability at the end of excavation clearly is no guarantee that the trench will remain stable forever.

For ordinary construction in slurry trenches the excavation is temporary, and the trench usually remains open only for a few days merely to allow the placement of concrete or other backfill material. This time is short compared with the time required for excess pore pressures within the soil to dissipate. In this case it is appropriate to use the $\phi = 0$ concept and undrained loading for stability problems. This leads to an overall simplification and allows the use of Eqs. (2-4) and (2-5), in which the factor c is replaced by the undrained shear strength s_u, generally quoted as one-half of the unconfined compressive strength.

After an excavation if the trench is intended to remain open for a long time, certain changes are likely to occur. The soil will begin to swell and this is accompanied by changes in the pore pressure and the effective stress (Lambe and Whitman, 1969). This means that with time the soil gradually will become weaker and will therefore exert a greater horizontal thrust when the wedge reaches the active state. For this condition an effective stress analysis is more appropriate.

The two methods, total- and effective-stress analysis, can be related by comparing the factor of safety with plastic failure, i.e., limiting equilibrium (Bishop and Bjerrum, 1960). For a soil entering this state the factor of safety should be 1 regardless of the method used in the analysis. In this respect both methods agree in providing a factor of safety of 1 for a soil on the verge of collapse caused by a change of stress under undrained conditions. Although this factor is the same, the theoretical location of the failure surface is dependent upon the value of ϕ, and the closer this value approaches the true angle of shear resistance the more realistic the position of the failure surface in the calculations will be. However, the total-stress analysis does not establish this surface since $\phi = 0$ always implies a critical plane inclined at 45° with the horizontal.

This discrepancy is rectified by noting that in a triaxial test it is not necessary to know the exact location of the failure plane. The $\phi = 0$ analysis is valid as long as failure is assumed to occur when the maximum shear stress reaches the maximum shear existing at failure in a corresponding triaxial test. Thus it is possible to analyze the stability of a trench without really knowing the exact inclination of the failure plane. The choice of method for short-term problems in saturated soils therefore becomes a matter of convenience, and the $\phi = 0$ concept is preferred for its simplicity.

Situations where the $\phi = 0$ Analysis Is Not Valid

For conditions other than undrained shear the $\phi = 0$ method can lead to unrealistic predictions. The concept similarly loses its meaning whenever field measurements of pore pressure are used as control. A total-stress analysis with $\phi \neq 0$ or where the angle of consolidated undrained shear is used should be avoided because of the difficulty of determining the physical significance of the factor of safety thus obtained (Bishop and Bjerrum, 1960).

The $\phi = 0$ concept clearly does not apply to partially saturated soils. In this case an approximate expression of undrained strength in terms of the two parameters c and ϕ is possible, but because of the many factors influencing them results from such analyses must be interpreted with caution.

Excavation of trenches in stiff fissured and weathered clays can give rise to special problems. Stress reduction caused by unloading causes the fissures to open up and provide weak zones for the sliding surface to follow. This problem is more serious where the excavation is made without slurry.

If the end-of-construction condition is not the prevailing criterion, the $\phi = 0$ method should not be used. For prolonged construction operations allowance must be made for the rapid adjustment of the pore pressure to the equilibrium condition. Typical examples of overestimating the factor of safety are found in case histories. A circular slip failure occurred during excavation of an open cut in Chicago for the Congress Expressway in 1952. According to the $\phi = 0$ analysis, the estimated factor of safety was about 1.11, but this value was reduced to 0.90 after allowing for the effect of progressive failure, rate of testing, and some cracks and joints due to desiccation in the upper crust of clay. An overestimation of the factor of safety by as much as 80 percent, reported by Skempton and LaRochelle (1965) in connection with the Bradwell deep excavation in overconsolidated London clay, led to slide failure. This failure was explained by the combination of the opening of fissures caused by stress release and change in pore pressure due to the high bulk permeability, despite the short period of excavation.

Although the foregoing examples involved inclined rather than vertical slopes, problems of a similar nature cannot be precluded in slurry-trench excavations, mainly because of the similarity in the mechanism of slip failure (see also Sec. 3-13).

Choice of Parameters Discrepancies between field and laboratory strength are principal causes of errors in stability problems. Certain factors can cause the sample strength as measured in the laboratory to be different from the in situ strength. Procedures regarding sampling, sample disturbance, and rate of shear are known, among others, to influence the strength of a soil. Major variations in the strength of soft clays also can be caused by strength anisotropy and strain-rate effects (Ladd 1967; Ladd and Foott, 1974). Despite the circumstantial evidence indicating that these variations frequently tend to compensate for each other, none of their effects are explicitly included in the present procedures. The existence of such uncontrollable factors can thus make the $\phi = 0$ analysis either unsafe or too conservative under certain conditions.

2-3 STABILITY OF SLURRY-FILLED TRENCHES IN SAND

Trenches in Dry Sand

The method of analysis presented by Nash and Jones (1963) is extended to dry cohesionless sand as shown in Fig. 2-2a and c. The active wedge ABC is prevented from sliding by the frictional resistance along plane BC and by the thrust of the slurry P_f. From the equilibrium of forces

$$\tan \alpha = \frac{\gamma - \gamma_f}{2\sqrt{\gamma \gamma_f}} \qquad (2\text{-}6)$$

The factor of safety F against sliding is defined as the ratio $(\tan \phi)/(\tan \alpha)$, from which it follows that

$$F = \frac{2\sqrt{\gamma \gamma_f} \tan \phi}{\gamma - \gamma_f} \qquad (2\text{-}7)$$

It is seem from Eq. (2-7) that for simple wedge failure the stability of trenches in dry sand is independent of the height of the excavation and dependent only upon the properties of the soil and the weight of slurry. Again, stability is improved with a heavier slurry for the same soil parameters and conditions.

Trenches in Sand with Water

Where the free water level is near the ground surface, it is quite difficult to maintain trench stability. This may require one or all of the following measures: (1) lower the natural water table somewhat in the vicinity of the excavation; (2) raise the slurry level in the trench; (3) use a heavier slurry; and (4) resort to other factors for stability.

When the soil consists predominantly of coarse sand and gravel, some penetration of the adjacent zone by slurry will undoubtedly occur, generally above the natural water table, until the formation of the impermeable barrier at the interface restricts this flow. Disregarding the associated effects, one carries out the analysis of stability assuming drained conditions whereby the soil stress is an effective stress and the pore pressure is static water pressure. For a water table at the ground surface this requires

$$\tfrac{1}{2}\gamma_f H^2 = \tfrac{1}{2}\gamma' H^2 K_\alpha + \tfrac{1}{2}\gamma_w H^2$$

where γ' is the effective (buoyant) weight of soil. This relation yields

$$K_\alpha = \tan^2\left(45 - \frac{\phi'}{2}\right) = \frac{\gamma_f - \gamma_w}{\gamma'}$$

Setting $\gamma_f - \gamma_w = \gamma_f'$ and solving for $\tan \phi'$ under an appropriate factor of safety F gives

$$F = \frac{2\sqrt{\gamma' \gamma_f'} \tan \phi'}{\gamma' - \gamma_f'} \qquad (2\text{-}8)$$

which is also derived directly from Eq. (2-7) by substituting γ' and γ_f' for γ and γ_f, respectively, i.e., effective weights for total weights. An interesting relation is

$$\gamma_f = K_\alpha \gamma' + \gamma_w \qquad (2\text{-}9)$$

which states that for stability the unit weight of slurry must equal the active effective weight of soil plus the weight of water.

Variations in the Slurry and Groundwater Level

The general case of stability with arbitrary groundwater and slurry level is shown in Fig. 2-3. The groundwater lever is at distance mH and the slurry level at distance nH, respectively, from the bottom of the trench. The analysis

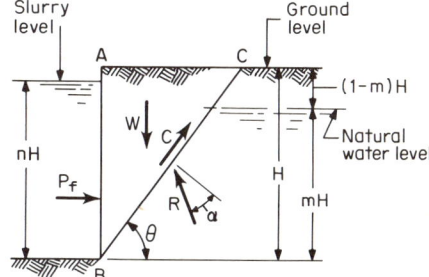

Fig. 2-3 Stability of a trench for arbitrary slurry and natural water level.

can be carried out taking total earth stress above the water table and effective earth stress below the water table. This condition leads to

$$\gamma_f = \frac{\gamma(1 - m^2)K_\alpha + \gamma' m^2 K_\alpha + \gamma_w m^2}{n^2} \qquad (2\text{-}10)$$

which for $m = n = 1$ reduces to Eq. (2-9). In the foregoing expression the angle ϕ is taken the same for both dry and submerged soil, which generally is a satisfactory approximation. Also saturation of the soil with water above the free level is likely to increase the assumed dry weight. Probable saturation capillary heads (which determine the highest elevation above the free water surface at which complete saturation exists) may vary from a few inches for coarse gravel to almost 6 ft for silty sand (Lane and Washburn, 1946).

Equation (2-10) shows the sensitivity of a trench in sand to fluctuations in the natural water and slurry level. This becomes apparent by varying m and n and then solving for γ_f. Thus, problems can conceivably arise where slurry trenches are excavated near large bodies of natural water. In this case it is essential to determine how consistently the groundwater table at the site responds to fluctuations in the level of the nearby waterway.

A useful example of trench failure in water-bearing ground has been reported by Morgenstern and Amir-Tahmasseb (1965). The failure occurred during the construction of a cutoff diaphragm wall for the protection of the main power station at Pierre-Bénite, France, located on the banks of the Rhône River. A section of the cutoff is shown in Fig. 2-4, together with the site and ground conditions. As shown, the natural ground level was raised by 3.5 m to provide a construction level higher than the highest anticipated river level. The initial slurry density was about 1.025 t/m³ (64 lb/ft³) but was raised during excavation to about 1.20 t/m³ (75 lb/ft³) due to the retention of soil from the excavation.

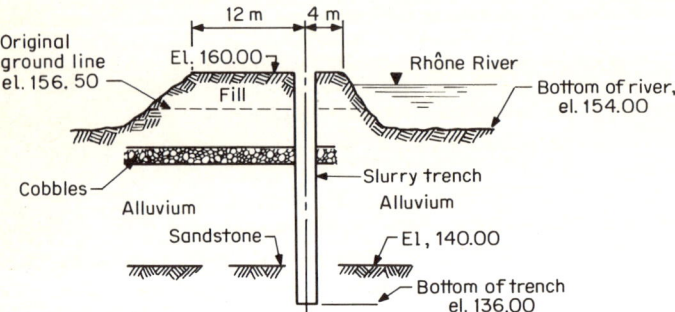

Fig. 2-4 Pierre-Bénite site. Typical cross section adjacent to river.

The slips occurred in the fill portion of the trench following unexpected floods during heavy rainfalls in March and April of 1963, which raised the groundwater level in the vicinity of the excavation almost to the top of the fill. The slips extended the entire length of the panel.

The results of the investigation are shown in Fig. 2-5 for two different values of the angle of friction, namely 30 and 35°. When the tip of the slips is taken at elevation 156.50, the top of the fill at 160.00, and the slurry level 15 cm from the top, n has a value of 0.96. Three values of m are used, namely 0.87, 0.93, and 1.00. It is evident that for a given shear strength of the soil the required slurry density to maintain stability is quite sensitive to fluctua-

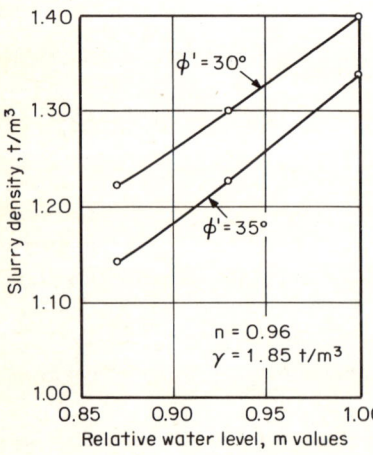

Fig. 2-5 Relation of slurry density (measured) to relative water level. Slips at Pierre-Bénite. (*From Morgenstern and Amir-Tahmasseb, 1965.*)

tions in the groundwater level. Using slurry densities from 1.20 to 1.25 t/m³ (75 to 78 lb/ft³), one can see that the corresponding groundwater level at the time of slip is very near the top of the trench. This comparison thus corroborates the criteria of failure discussed in the foregoing analysis.

2-4 EFFECT OF GEL STRENGTH OF SLURRY ON TRENCH STABILITY

As shown in subsequent chapters, in certain applications the slurry in the trench may possess a certain shear (gel) strength. An example is the solidified cutoff wall. In general, a small shear (for instance 10 to 15 lb/ft²) can help stability appreciably. Such a system will tend to undergo a physical deformation as the earth material tends to slide into the trench and displace the slurry, but this tendency is resisted by the shear strength of the slurry (Morgenstern, 1963). The problem is equivalent to the compression of a perfectly plastic material between two rough rigid plates, analyzed by Prandtl (1923) and extended by Bishop (1952) to the computation of active stress of a puddle core in earth dams. The slurry-trench stability problem basically is the opposite of that of the earth stress induced in the puddle core.

When we consider an infinitesimal element in the mass of slurry acted upon as shown in Fig. 2-6, equilibrium exists if the following expressions are satisfied:

$$\frac{\partial \sigma_x}{\partial x} + \frac{\partial \tau_{xy}}{\partial y} = 0 \qquad \frac{\partial \sigma_y}{\partial y} + \frac{\partial \tau_{xy}}{\partial x} - \gamma_f = 0 \qquad (2\text{-}11)$$

where γ_f is the slurry density. Plastic flow occurs when the stresses satisfy the yield criterion

$$(\sigma_x - \sigma_y)^2 + 4\tau_{xy}^2 = 4\tau_f^2 \qquad (2\text{-}12)$$

in which τ_f is the shear strength of the slurry, also called *gel strength*. Subject to the appropriate boundary conditions, the solution of Eqs. (2-11) and

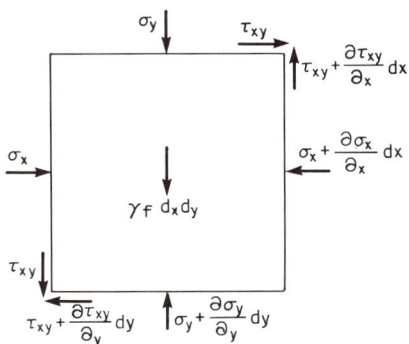

Fig. 2-6 Stress diagram for an element of gelled slurry.

(2-12) is a statically determinate problem. For a trench width $2a$ it can be shown that the horizontal stress is

$$\sigma_x = \gamma_f y + \frac{\tau_f y}{a} + \tfrac{1}{2}\pi\tau_f \qquad (2\text{-}13)$$

The total resistance of the slurry is therefore

$$P_f = \int_0^H \sigma_x \, dy$$

or

$$P_f = \tfrac{1}{2}\gamma_f H^2 + \frac{\tau_f H^2}{2a} + \tfrac{1}{2}\pi\tau_f H \qquad (2\text{-}14)$$

where H is the height of the trench. If the shear strength is neglected in Eq. (2-14), the total resistance is reduced to

$$P_f = \tfrac{1}{2}\gamma_f H^2$$

Hence the cohesion factor in Eq. (2-14) due to the shear strength τ_f is

$$C_f = \frac{\tau_f H^2}{2a} + \tfrac{1}{2}\pi\tau_f H \qquad (2\text{-}14a)$$

Equations (2-14) and (2-14a) are quite general and can be used for stability analyses in any type of soil. For example, in dry cohesionless sand a trench is stable if

$$\tfrac{1}{2}\gamma H^2 K_\alpha = \tfrac{1}{2}\gamma_f H^2 + \frac{\tau_f H^2}{2a} + \tfrac{1}{2}\pi\tau_f H$$

which gives

$$H_{\text{cr}} = \frac{\pi\tau_f}{\gamma K_\alpha - \gamma_f - \tau_f/a} \qquad (2\text{-}15)$$

Likewise for clay with $\phi = 0$ and cohesion c, the critical height is

$$H_{\text{cr}} = \frac{4c + \pi\tau_f}{\gamma - \gamma_f - \tau_f/a} \qquad (2\text{-}16)$$

It is evident from Eqs. (2-15) and (2-16) that the existence of a small shear strength (10 to 20 lb/ft²) can aid stability appreciably and that relatively narrow trenches are stable with a greater factor of safety than relatively wide openings for the same value of τ_f.

The effect of shear strength τ_f on stability is more pronounced if τ_f exceeds 10 lb/ft². Although it is possible to prepare slurries that will develop this gel strength, in most practical situations this is neither desired nor practicable (see also Chap. 4). It follows, therefore, that the advantages of the foregoing analysis can be utilized in certain applications only, e.g., the construction of prefabricated walls and solidified cutoffs, but very seldom in the usual slurry-trench excavations.

2-5 SPECIAL CONSIDERATIONS FOR EXCAVATIONS IN SAND

In the analytical expressions derived thus far the geometry and shape of the excavation were ignored, and the trench was considered infinitely long. Slip failure was assumed to occur when the total thrust from the active wedge just exceeds the total slurry pressure. This condition is reached when the area of the stress diagrams shown on the right of Fig. 2-7a to c equals the area of the slurry pressure diagram shown on the left.

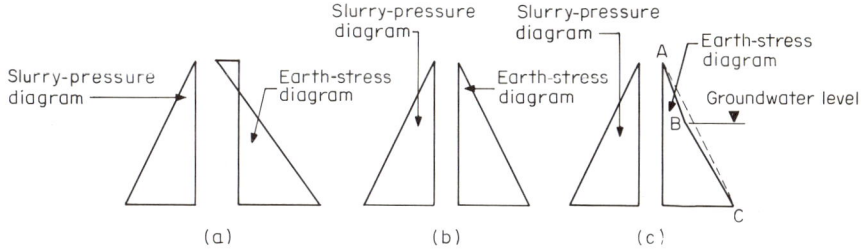

Fig. 2-7 Stress diagrams for slurry trenches: (a) trench in clay; (b) trench in dry sand or in fully submerged sand; (c) trench in sand with water table below the ground surface.

The stability condition can also be expressed in terms of unit stress, and a derivative of this principle is Eq. (2-9). For this case it is necessary to equate the lateral stress of the soil, including water, to the lateral pressure of the slurry at any given depth. This concept is not convenient for clays but applies readily to sands. The two methods, total thrust or unit stress, apparently are equivalent for the case shown in Fig. 2-7b but are not comparable for the case shown in Fig. 2-7c. For multilayered soil, stability is better analyzed by the unit-stress method.

Equal-Stress Method

Figure 2-8 shows the section of a trench. The notation is slightly different from the symbols previously used. At depth h_x stability is expressed as

$$p_f = p_w + p_a \qquad (2\text{-}17)$$

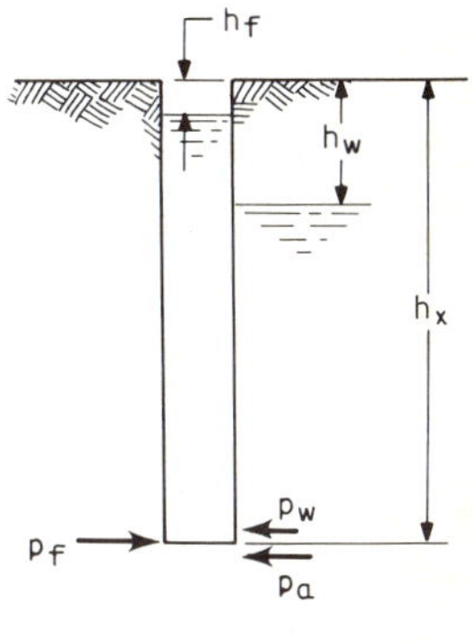

Fig. 2-8 Trench in sand with variable slurry and groundwater level.

where p_f = hydrostatic pressure of slurry
p_w = pressure of groundwater (static)
p_α = horizontal active earth stress

The following relations hold:

$$p_f = \gamma_f(h_x - h_f) \quad \text{and} \quad p_w = \gamma_w(h_x - h_w)$$

$$\text{if } h_x \leq h_w \quad \text{then} \quad p_\alpha = \gamma h_x K_\alpha$$

$$\text{if } h_x > h_w \quad \text{then} \quad p_\alpha = [\gamma h_w + \gamma'(h_x - h_w)]K_\alpha$$

and if $h_f = h_w = 0$, Eq. (2-17) readily yields Eq. (2-9).

Arching Effect of Short Trenches in Sand

In relatively short trenches the arching effect is developed when the part of the soil along the excavation tends to yield, but this tendency is resisted at the ends of the panel, which form the boundary lines between yielding and stationary soil. In this case the arching effect may be considerable and must therefore be included in the analysis. Kowalewski (1964) has suggested that in a trench of finite dimensions arching occurs under conditions analogous to those which determine the vertical stresses acting on the crown of a tunnel.

Figure 2-9a shows a short panel where the coulomb edge is assumed to have the shape of a segment of the vertical parabolic cylinder $ABC - A'B'C'$. Sliding occurs on plane $A'B'C'$ inclined at an angle θ to the horizontal. Piaskowski and Kowalewski (1965) have proposed nomographs similar to the group plotted in Fig. 2-9b. These curves express the conditions of equilibrium for soil of a given strength and density. It is evident that K_α decreases with increasing h_x/l ratio; hence the total earth thrust per unit length is less for a short trench than for a long one, the depth h_x being the same. It should be noted that h_w defines the groundwater table as shown in Fig. 2-8.

Alternatively, for short trenches the coefficient K_α can be adjusted using an analysis similar to that proposed by Terzaghi (1943) for the investigation of the stress condition in a small fissure. The application of this principle is exemplified by applying a reduction factor A to the active-stress coefficient K_α (Huder, 1972) before entering Eq. (2-17). The value of A is given by

$$A = \frac{1 - \exp(-2nK_\alpha \tan \phi)}{2nK_\alpha \tan \phi} \quad (2\text{-}18)$$

so that the active stress p_α is now calculated from

$$p_\alpha = [\gamma h_w + \gamma'(h_x - h_w)]AK_\alpha \quad (2\text{-}19)$$

In Eq. (2-18) n is the ratio of depth to length; that is, $n = h_x/l$. Values of A for the most ususal cases can be found graphically with the help of Fig. 2-10.

It is interesting to compare results obtained from these two methods. For example, using $\phi = 35°$ and $h_x/l = 3$, we estimate that $K_\alpha \tan \phi = 0.19$. From

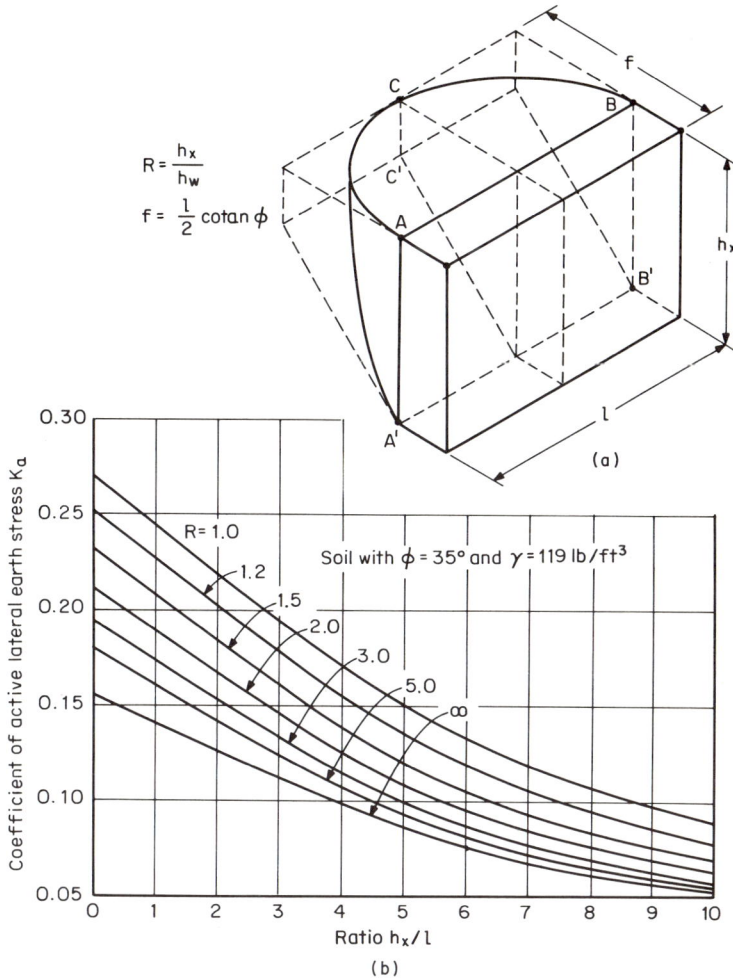

Fig. 2-9 Variation of active earth-stress coefficient with trench dimensions. (*From Piaskowski and Kowalewski, 1965.*)

the graph of Fig. 2-10 the reduction factor A is about 0.61. The factor applied to the coefficient K_α from the Piaskowski solution is obtained as the ratio $0.19/0.27 = 0.70$, which is in good agreement with the Huder method considering the different assumptions involved. The reduction factor should also be applied to surcharge loads.

Circular Excavations in Sand

Circular cuts in sand are approximately analyzed following the same procedure, now taking the diameter of the cut as the dimension l. Results thus obtained are not entirely accurate but are conservative for the general application.

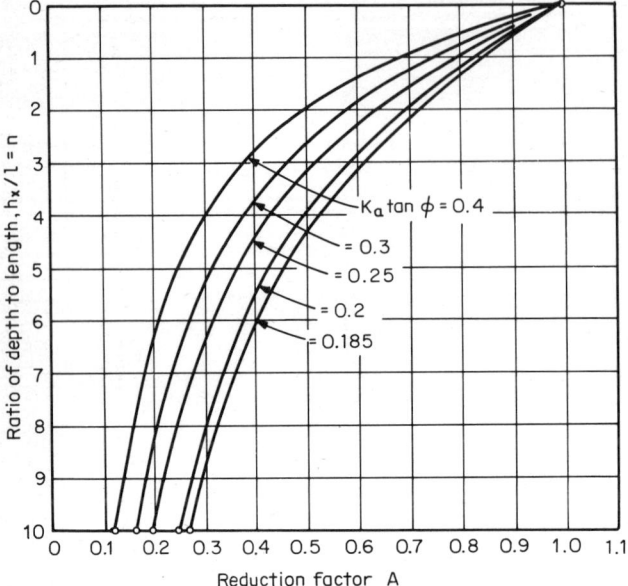

Fig. 2-10 Reduction factor A for lateral earth stress in trenches. Idealized soil without cohesion. (*From Huder, 1972.*)

Effect of Groundwater Level on Stability

The sensitivity of trench stability to the groundwater level demonstrated by Eq. (2-10) is also shown by the graphs of Fig. 2-11. The variation of γ_f (unit weight of slurry) is plotted vs. the groundwater level. The symbols correspond to the notation of Fig. 2-8.

The following conclusions are significant: (1) if h_w is less than a certain value h_0, γ_f decreases with increasing critical depth h_c, but this behavior is reversed when h_w is greater than h_0; (2) the asymptote of the function $\gamma_f = f(h_c)$ in either case is a line parallel to the h_c axis at a distance h_0; and (3) for a given soil and a value of h_f there is a certain value of h_0 for which γ_f is the same for any value of h_c.

Stability of Multilayered Soil

The equal-stress method becomes very convenient where the soil profile is not uniform, particularly in multilayered soil. If the profile consists of n different layers above the water table and m different layers below the level, the stability can be expressed in terms of the slurry unit weight as

$$\gamma_f = \frac{K_\alpha}{h_x - h_f}\left(\sum_{i=1}^{n} \Delta h_i\, \gamma + \sum_{i=1}^{m} \Delta h_i\, \gamma'\right) + \frac{\gamma_w(h_x - h_w)}{h_x - h_f} \qquad (2\text{-}20)$$

where h_x, h_f, and h_w correspond to the notation of Fig. 2-8 and h_i is the thickness of each individual layer. Where ϕ is not approximately the same for

Stability of Trenches and Cuts

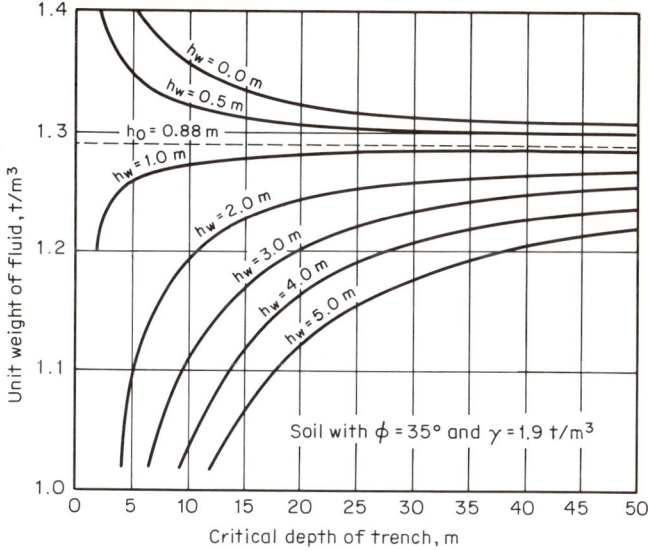

Fig. 2-11 Unit weight of fluid vs. critical depth of trench. (*From Piaskowski and Kowalewski, 1965.*)

all layers, each factor in Eq. (2-20) should be multiplied by the corresponding value of K_α.

The factor of safety is estimated from the expression

$$f = \frac{\gamma_{fa}(h_x - h_f) - \gamma_w(h_x - h_w)}{\gamma_f(h_x - h_f) - \gamma_w(h_x - h_w)} \quad (2\text{-}21)$$

where γ_{fa} is the actual unit weight of the slurry and γ_f is the slurry weight required for a factor of safety of 1.

Ground Settlement

Figure 2-12 shows the variation of ground settlement caused by slurry-trench excavation as a function of the factor of safety. The data are taken from tests

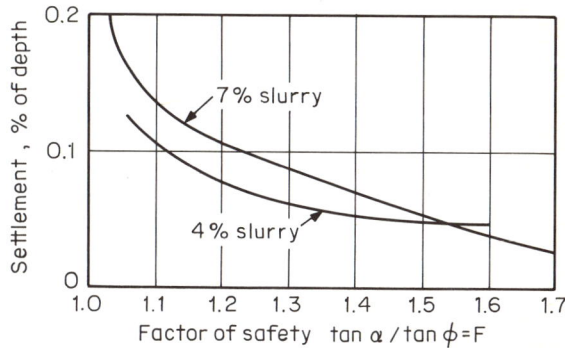

Fig. 2-12 Percentage of settlement in sand vs. factor of safety. (*From Elson, 1968.*)

carried out by Elson (1968). The factor of safety is expressed as the ratio (tan ϕ)/(tan α), and tan α is estimated from Eq. (2-6). Although the results cannot be considered general, it is evident that when F is greater than 1.5, the probable settlement is less than 0.05 percent of the trench height. This settlement is assumed to extend across the theoretical sliding wedge, but in reality it is greater near the face.

Further Analysis of Arching

The arching effect of short panels has also been investigated by Schneebeli (1964). This analysis is based on the theory of lateral earth stresses for silos (Caquet and Kérisel, 1956), from which an analogous reduction is obtained for the active thrust. For a panel of length l excavated in cohesionless soil, Schneebeli has derived the following expression for the horizontal stress σ_z at depth z:

$$\sigma_z = \frac{\gamma l}{N_\phi \sin 2\phi} [1 - \exp(-n \sin 2\phi)] \qquad (2\text{-}22)$$

where $\qquad n = \dfrac{z}{l} =$ ratio of depth z to length l

and $\qquad N_\phi = \dfrac{1 + \sin \phi}{1 - \sin \phi}$

which in soil mechanics is called the *flow factor*. When the depth z becomes very great compared with the length l, the horizontal stress approaches the asymptotic value

$$\sigma_{max} = \frac{\gamma l}{N_\phi \sin 2\phi} \qquad (2\text{-}23)$$

Thus, according to the Schneebeli theory the horizontal earth stress attains a maximum value and remains constant thereafter, whereas according to the Huder theory, the reduction factor A approaches a limiting value but the horizontal stress still continues to increase with depth.

It is apparent that arching is caused by the lateral movement of the soil toward the trench just before yielding and is associated with a vertical shortening. Frictional resistance along the boundary edges where the sliding soil meets stationary soil opposes subsidence and thus supports part of the weight of the sliding soil. Just before failure, the principal vertical stress in the lower part of the wedge gets smaller and less than the weight of the soil above it.

In the arching effect the final depth is not necessarily the most critical depth. This would depend on the details of the problem and also upon whether the reduction in the earth stress is according to Eq. (2-18) or (2-22).

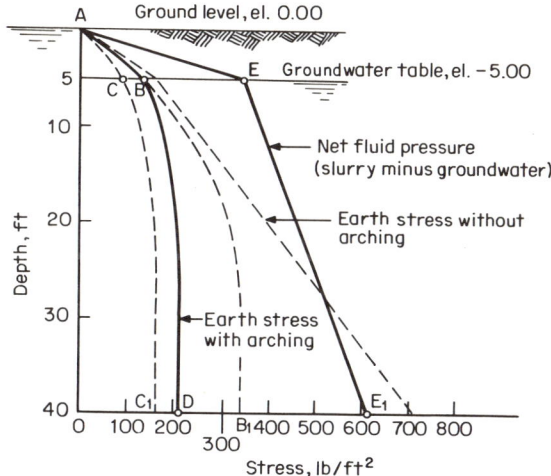

Fig. 2-13 Application of arching effect to trench-stability problems. Earth stresses according to the Schneebeli theory.

Figure 2-13 illustrates how the Schneebeli method can be applied to trench-stability problems. This example is for a soil with $\phi = 35°$, $\gamma = 120$ lb/ft³, and groundwater table at elevation -5.00. The slurry weight is 70 lb/ft³. Line AEE_1 represents the net hydrostatic pressure, i.e., slurry minus groundwater pressure. Without arching the lateral earth stress is seen to exceed the net hydrostatic pressure at a depth of about 28 ft. Next the arching effect is considered for a panel 10 ft long, the value of σ_{max} is computed from Eq. (2-23) for both the dry and submerged weight, and the curves ABB_1 and ACC_1 are drawn accordingly. Beginning at point B, the curve BD is drawn parallel to curve CC_1 ($CB = C_1D$). Evidently the curve ABD is the lateral-earth-stress diagram adjusted for the arching effect. The trench now is stable, and, furthermore, the difference between the net fluid pressure and earth stress increases with depth.

2-6 CHANGES IN THE PROPERTIES OF ADJACENT SOIL

In the preceding sections the analysis generally ignores the interaction between the soil and the slurry, and solutions are obtained assuming that the slurry is separated from the adjacent soil by an impermeable layer at the interface. This assumption is quite appropriate for soils of low permeability, but appreciable penetration by slurry may occur into the pores of coarse granular soils and gravels until a seal is formed and stops further penetration. This problem is discussed in Chap. 3. Factors relevant to stability are (1) strength changes of the soil caused by slurry impregnation and (2) a slurry thrust less than the theoretical if the filter cake of the interface is permeable, resulting in a flow of slurry toward the ground.

Gelation in the Soil Adjacent to Excavation

The presence of voids generally causes the slurry to flow into the soil adjacent to an excavation. The extent of this penetration depends mainly on the pore size, the differential hydrostatic head, and the shear strength of the slurry. Gelled slurry in the pores can impart a certain shear strength to the soil. Elson (1968) has investigated this problem disregarding any changes in the friction angle.

Figure 2-14 shows the stress diagrams for the zone impregnated with slurry. Diagram ABC expresses the stress without filter cake, and ACD corresponds to the stress condition with filter cake. The symbols τ_f and r are the gel strength of the slurry and the mean pore radius of the soil, respectively. From the diagram ABC, C_a is estimated as

$$C_a = \frac{2\tau_f l^2}{r \cos \theta} \tan \phi \qquad (2\text{-}24)$$

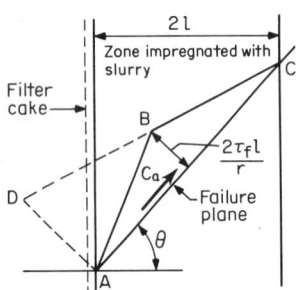

Fig. 2-14 Stress diagram in zone of soil impregnated with slurry; ABC = without filter cake; DAC = with filter cake.

If the filter cake is formed at the left boundary, C_a is found from the stress diagram ACD as

$$C_a = \frac{4\tau_f l^2}{r \cos \theta} \tan \phi \qquad (2\text{-}25)$$

This is the apparent increase in the shear strength of the soil. The horizontal component is

$$C_h = 2C_a \cos\left(45 + \frac{\phi}{2}\right)$$

or

$$C_h = \frac{8\tau_f l^2}{r} \tan \phi \qquad (2\text{-}26)$$

which is included in the analysis as a net reduction in the active thrust of the soil. Since the difficulty in selecting the parameters of Eq. (2-26) makes the analysis very approximate for practical use, this theory is useful mainly in obtaining comparative solutions.

Stability without Filter Cake

The importance of the impermeable layer (also called filter cake) at the interface can be appreciated if the stability is considered without the filter cake. Two significant things may happen: (1) the actual thrust exerted by the slurry upon the face may be much less than the theoretical and (2) the slurry in the soil pores may gel without imparting an increase of strength to the soil. If the fluid in its final position is diluted, its effect on the contact soil surfaces may be a reduction of the friction angle, giving a weaker soil.

The stability of trenches without filter cake has been studied by Müller-Kirchenbauer (1972), who includes a practical and simple method for determining the so-called *stagnation gradient* i_0. For a slurry having a differential head h and penetrating into the soil for a distance l, as shown in Fig. 2-15c, the stagnation gradient is defined as the ratio h/l.

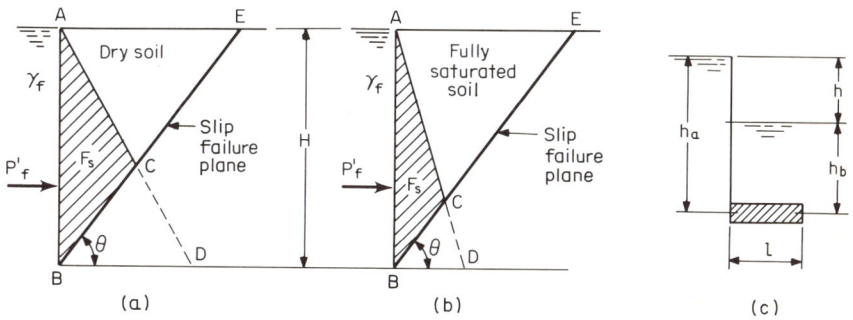

Fig. 2-15 Penetration of granular soil by fluid and application of slurry pressure.

Figure 2-15a and b shows how the hydrostatic thrust of the slurry should be applied when there is no filter cake. In this case the actual thrust exerted by the slurry is

$$P'_f = V i_0 \gamma_{fp} \tag{2-27}$$

where P'_f = actual thrust of slurry without filter cake
V = volume of zone within active wedge penetrated by slurry (ABC)
γ_{fp} = unit weight of slurry in penetrated zone, usually = γ_f

If F_s is the area of triangle ABC, then

$$P'_f = F_s i_0 \gamma_f \tag{2-28}$$

which generally is much smaller than the hydrostatic force P_f computed according to the impermeable-membrane theory. A comparison of P'_f and P_f can be made by noting that

$$\frac{P'_f}{P_f} = \frac{\text{triangle } ABC}{\text{triangle } ABD}$$

Likewise, a smaller factor of safety can result from a reduced friction angle (see also Sec. 3-5), since this means a weaker soil and hence more active thrust. In some instances this reduction has been as much as 5° (Müller-Kirchenbauer, 1972) and is attributed to changes in the mechanism of friction caused by diluted slurry in the soil pores. This effect should be taken into account by considering the portion ABC of the active wedge ABE as being resisted along the slip plane BC according to the reduced friction angle.

Stability with Permeable Filter Cake

It is not uncommon to carry out an excavation under the protection of a relatively permeable filter cake. This situation arises, as mentioned, in trench excavations for the construction of a solidified cutoff wall or a prefabricated diaphragm wall. On the other hand, even if a stable slurry has been provided initially, it is conceivable that contamination and other adverse effects might alter the colloidal stability during excavation, resulting in a face protected by a relatively permeable filter cake. In this case, the actual thrust mobilized by the slurry is between P'_f and P_f.

Figure 2-16 shows the formation of such a barrier by filtering a slurry containing a certain proportion of both colloid and noncolloid solids, i.e., bentonite and some aggregate from the excavation. For a volume element dV and a corresponding penetration dl the following notation is used:

n_1 = void volume of soil in situ = porosity ratio
n_2 = void volume of filter cake = porosity ratio
i_{01} = stagnation gradient of slurry related to soil
i_{02} = stagnation gradient of slurry related to filter cake
dV, V = volume of slurry element, including all solids
dV_s, V_s = volume of all solids (colloid and aggregate) in element
γ_s = unit weight of aggregate (soil)
γ_f = unit weight of slurry without aggregate
γ_{fs} = unit weight of slurry with aggregate
z = depth of trench
h = head of slurry, generally $< z$
h_1 = fraction of h transmitted to soil region
h_2 = fraction of h transmitted to filter-cake region
m = concentration factor = V_s/V (not bentonite concentration)

When we consider the surface area of a single volume element, it is evident that for $a = 1$

Stability of Trenches and Cuts 39

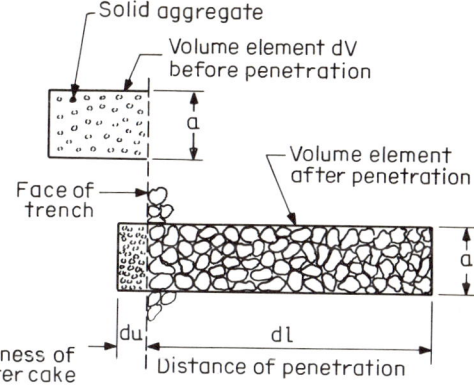

Fig. 2-16 Formation of surface filtrate by penetration of suspension containing solid aggregate.

$$dV_s = m\, dV = (dl\, n_1 + du)\, m$$

and also

$$dV_s = du\,(1 - n_2)$$

When combined, these relations yield

$$\frac{du}{dl} = \frac{mn_1}{1 - n_2 - m} = \lambda \qquad (2\text{-}29)$$

The following are true by definition of the stagnation gradient:

$$\frac{dh_2}{du} = i_{02} \quad \text{and} \quad \frac{dh_1}{dl} = i_{01}$$

Combined with Eq. (2-29), they give

$$\frac{dh_2}{dh_1} = \frac{h_2}{h_1} = \lambda\,\frac{i_{02}}{i_{01}} \qquad (2\text{-}30)$$

where

$$h_1 + h_2 = h \qquad (2\text{-}31)$$

If the stagnation gradients are known, the fraction heads and the dimensions l and u can be estimated from the foregoing expressions. For a slurry that has a unit weight γ_{fs} rather than γ_f, the head at depth z is adjusted accordingly, so that

$$h_2 = \frac{1}{\gamma_f}\int_0^z \gamma_{fs}\, dz \qquad (2\text{-}32)$$

Now the fraction of the slurry pressure p'_{f1} and p'_{f2} transmitted to the soil and the filter cake respectively is, from Eq. (2-28),

$$p'_{f1} = i_{01}\gamma_f l \qquad (2\text{-}33a)$$

and

$$p'_{f2} = i_{02}\gamma_f u \qquad (2\text{-}33b)$$

and also

$$p'_{f1} + p'_{f2} = p'_{fs} = h_2\gamma_{fs} \qquad (2\text{-}34)$$

The estimation of stagnation gradients requires a series of simple tests, described in detail in Chap. 3. The same chapter also includes examples and problems involving stability analyses with a permeable filter cake.

2-7 SPECIAL CONSIDERATIONS FOR EXCAVATIONS IN CLAY

The arching effect shown for trenches in sand can also be appreciable for trenches in clay. For very shallow cuts of any planar shape, including rectangular and circular configurations, the methods of analysis discussed in Sec. 2-2 generally are applicable. As the depth increases and exceeds the length, the stability can be significantly improved by introducing the resistance of the soil on vertical planes through the edges of the cut. This problem has been studied by Schneebeli (1964) and Meyerhof (1972), among others.

Stability of Circular Cuts

In a circular cut hoop stresses act normal to the radial planes. Near the face they are roughly equal to the vertical stresses, but away from the face both hoop and radial stresses approach the earth stress at rest.

Shallow Circular Cuts Shallow circular cuts are defined as having a depth-to-diameter ratio less than about 12. Meyerhof (1972) has derived an approximate solution beginning from the expression

$$p_z = (\gamma - \gamma_f)z - 2c = (\gamma' - \gamma'_f)z - 2c \tag{2-35}$$

which generally is true for long excavations. In this case p_z is the net active pressure for a completely saturated soil at depth z. For a shallow circular cut the coefficient 2 in Eq. (2-35) is replaced by the factor K such that

$$K = 2\left[\ln\left(\frac{2d}{b} + 1\right) + 1\right] \tag{2-36}$$

in which d is the depth of the cut and b is the diameter (or width) of the cut. Likewise Eq. (2-4) is rewritten by replacing total weights by effective weights and the factor $4c$ by $2Kc$, so that

$$H_{cr} = \frac{2Kc}{\gamma' - \gamma'_f} \tag{2-37}$$

which shows that for shallow circular cuts the stability number $2K$ increases with increasing depth-to-diameter ratio d/b. The K function of Eq. (2-36) is plotted in the upper curve of Fig. 2-17, which is for a length-to-width ratio of 1 and therefore applies to circular shapes.

The inclination and shape of the K curve indicate that although the factor K increases with increasing depth-to-diameter ratio, this increase is not linear. This means that the factor of safety H_{cr}/H_{actual} decreases with increas-

Stability of Trenches and Cuts   41

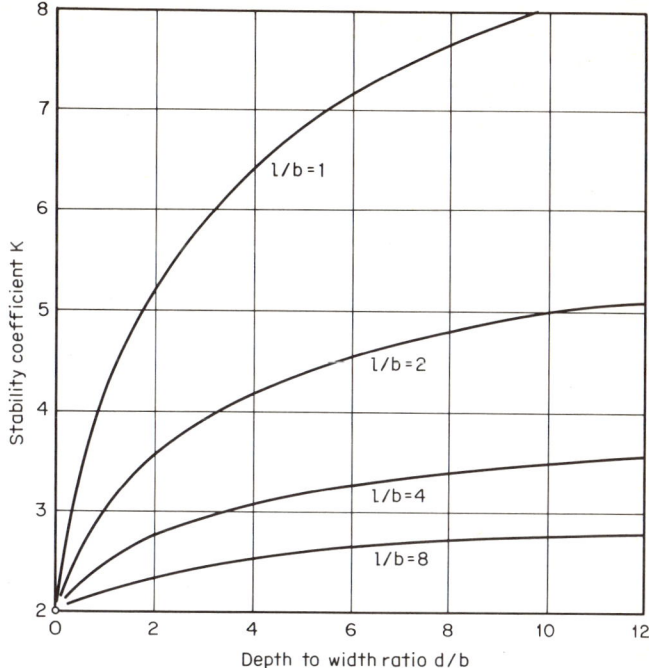

Fig. 2-17 Chart for stability coefficients of earth stresses. Rectangular cuts in clay; l = length, b = width, and d = depth. (*From Meyerhof, 1972.*)

ing depth, and therefore the stability should be checked for the maximum excavation depth.

Deep Circular Cuts When the depth-to-diameter ratio exceeds about 12, the earth stress and stability of circular or square cuts approaches the bearing-capacity problem of a deep vertical strip foundation with an overburden stress equal to the earth stress at rest. It can be shown that the critical height is now (Meyerhof, 1972)

$$H_{cr} = \frac{Nc}{K_0\gamma' - \gamma'_f} \tag{2-38}$$

where K_0 is the coefficient of earth stress at rest and N is a bearing-capacity factor for deep strip foundations. Meyerhof (1951) has shown that for rigid-plastic materials $N = 2\pi + 2 = 8.28$, which establishes an upper limit for both the coefficient K and the stability number N in saturated clays.

If the soil is assumed to be elastic, a solution of stability for deep circular cuts is obtained by the analytical modeling of a thick cylinder under an external pressure equal to the earth stress at rest. For elastoplastic material the expression for N is

$$N = \ln \frac{E_i}{3c} + 1 \qquad (2\text{-}39)$$

where E_i is the initial tangent modulus of the clay (Bishop et al., 1945).

Accurate estimations of the undrained modulus are quite difficult, because of its sensitivity to stress level and the dependence upon such factors as the rate of loading, time of consolidation, sampling disturbance, and others. For saturated clays where the ratio E_i/c is between 5000 and 2000, the value of N in Eq. (2-39) ranges from 6 to 7.5. If the value of E_i/c is not known, N usually is taken as 8 for most applications.

Since the analysis of deep cuts does not include the lateral shear resistance at the bottom of the excavation, the results are conservative and on the safe side. If this resistance is included, the corresponding bearing-capacity factor has a value of 9.34 for a rigid-plastic material (Meyerhof, 1951). The effect of base shear resistance is therefore an increase by 1 in the value of K and N for deep circular or square cuts, or about 12 percent increase in the critical height.

Arching Effect on Short Rectangular Trenches

An approximate but conservative analysis is obtained for a rectangular trench of length l, width b, and depth d by taking the earth stress along the perimeter of the two end zones of width b and length $b/2$ the same as for a square or circular cut. The earth stress for the remaining length $l - b$ is the same as for a long trench. In this case the coefficient K in Eq. (2-36) increases from a value of 2 for $d/b = 0$ to a maximum

$$K = 2\left(1 + \frac{3b}{l}\right) \qquad (2\text{-}40)$$

for fairly great depths. Likewise the stability factor N in Eq. (2-38) increases from a value of 4 for $d/b = 0$ to a maximum

$$N = 4\left(1 + \frac{b}{l}\right) \qquad (2\text{-}41)$$

for fairly great depths. For intermediate depths both K and N are estimated by interpolating between these limits, as suggested by Hansen (1961). For convenience the results are plotted in Figs. 2-17 and 2-18 ignoring base shear, from which K and N can be found graphically.

The foregoing analyses are readily extended to include a uniform surcharge q_s and stratified soil conditions. In this case, plastic failure requires

$$p_t - p_f = Nc \qquad (2\text{-}42)$$

where p_t is the maximum total horizontal thrust from the soil including any surcharge and p_f is the total slurry pressure.

Stability of Trenches and Cuts 43

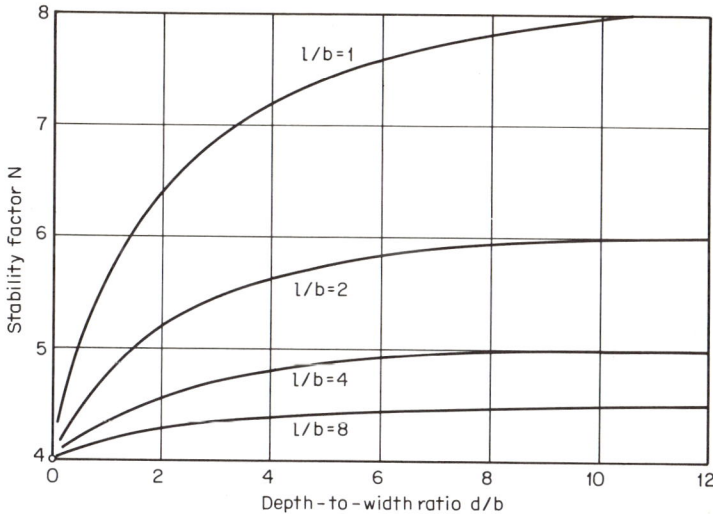

Fig. 2-18 Chart for stability factors. Rectangular cuts in clay; l = length, b = width, and d = depth. (*From Meyerhof, 1972.*)

Lateral Earth Displacement

During an excavation the lateral stress at rest is replaced by the slurry pressure, and this change of stress generally causes the soil to undergo some deformation. The soil movement can be estimated by considering the stress-strain relationship and assuming that the soil is homogeneous and isotropic with a geostatic stress.

For a deep circular cut the radial deformation at depth z is (Timoshenko and Goodier, 1951)

$$\Delta = \frac{(1 + \mu)p_z b}{2E_i} \quad (2\text{-}43)$$

in which p_z is the net lateral stress at depth z, μ is Poisson's ratio, and the other parameters are as before. For saturated clay $\mu = 0.5$, and for a slurry-filled cut $p_z = (K_0\gamma' - \gamma'_f)z$. If p_z is positive, the movement of the face is toward the excavation and vice versa. Substituting the value of p_z in Eq. (2-43) gives

$$\Delta = 0.75(K_0\gamma' - \gamma'_f)\frac{2b}{E_i} \quad (2\text{-}44)$$

Likewise the lateral displacement at the center of the long side of a deep rectangular cut of length l is

$$\Delta = 0.75(K_0\gamma' - \gamma'_f)\frac{2l}{E_i} \quad (2\text{-}45)$$

2-8 STABILITY OF TRENCHES WITH HORIZONTAL CURVATURE

Figure 2-19 shows two similar excavations with curvature in the horizontal plane; Fig. 2-19a is a circular cut, and Fig. 2-19b is a circular trench. For the latter the stability of the excavation depends upon the stability of both the concave and the convex faces. As a first approximation the stability of the outside (concave) face is roughly equivalent to the stability of the circular cut if changes in the bottom of the hole are ignored, although inward movement of this face is less in the circular trench than in the circular cut.

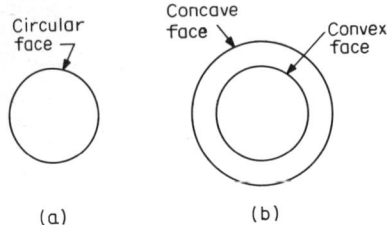

Fig. 2-19 Excavations with horizontal curvature: (a) circular cut; (b) circular trench.

The convex face in Fig. 2-19b undergoes different stretching and deformation just before failure, and it can logically be inferred that its stability is more critical than the stability of the concave face.

An analysis of stability for circular trenches (Lorente De No, 1969) includes both convex and concave vertical slopes. Solutions are provided by the use of finite-difference methods applied to lumped-parameter models. The soil is considered homogeneous and isotropic, and additionally the elastic medium is assumed frictionless. The analysis considers excavations in the dry, i.e., without slurry support. This means the removal of soil mass, which changes the geometry of the model, so that a unique solution is not always assured. The problem is avoided by simulating the increase in the depth of the excavation by an increase in the weight of material. The excavation is thus assumed to have been carried out in a weightless medium to its final depth, and thereafter the overburden is increased until failure occurs.

First yielding is developed at the bottom of the model, and failure is determined by the difference of the principal stresses. This difference increases with depth except when the coefficient K_0 has a value of 1. However, the analysis presented by Lorente De No is based on a value of μ (Poisson's ratio) of 0.25, so that the value of K_0 derived from the relation $K_0 = \mu/(1 - \mu)$ is not 1.

Results of Analysis

Figure 2-20 shows the deformation of the model at failure. Figure 2-20a is for a concave slope and therefore it represents a conventional circular cut. A strip excavation (infinite rectangular cut) is shown in Fig. 2-20b. It is evident that both the vertical settlement at the ground surface and the inward movement

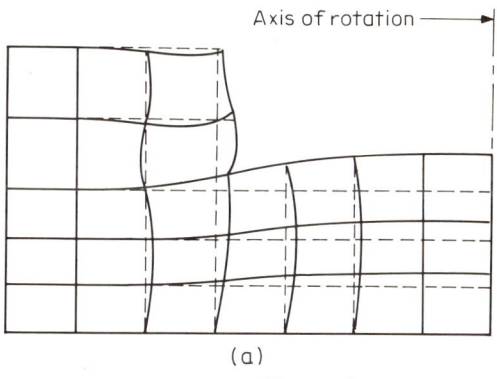

Fig. 2-20 Deformation of soil medium due to excavation: (a) circular excavation; (b) long rectangular cut. (*From Lorente De No, 1969.*)

of the face are somewhat less for the concave excavation than for the long strip. Figure 2-21 shows the deformation of compound excavation, consisting of concave and convex vertical slopes. Although these analyses have been carried out separately, when the two sketches are superimposed, they provide the deformation of the soil in a circular trench, showing that for the same curvature and depth the concave slope causes less surface settlement and inward movement than the corresponding convex slope.

In Fig. 2-22 is plotted the factor $f = \gamma H/c$ vs. the ratio $r = R/H$ of the circular trench. R is the radius of horizontal curvature, and H is the height of the excavation. The upper curve is for concave slopes and the lower for convex. The lower limit for the concave and the upper limit for the convex slope is obtained when $r \to \infty$ (linear trenches) and is represented by the asymptote for $\gamma H/c = 4$. The factor f actually is equivalent to the stability number.

As a first approximation, f can be obtained from the following expressions:

Concave slopes:
$$f = 4 + \frac{2}{3r + 1} \tag{2-46}$$

Convex slopes:
$$f = 4 - \frac{2}{3r - 1} \tag{2-47}$$

which apply to dry excavations. Apparently there is no distinction between relatively shallow and very deep excavations. The factor f reaches a maximum of about 6 for concave slopes of great depth ($r = 0$) or for circular cuts with very small horizontal radius. This generally is more conservative than the results obtained by the bearing-capacity theory, but it is difficult to compare the two methods since they are based on different assumptions.

Convex vertical slopes appear to approach the asymptotic line for values of r close to unity, which establishes the lower limit of f at about 3. In fact, Eq. (2-47) gives better results for $r \geq 1$. For $r < 1$, plastic flow reaches the axis of symmetry before it is developed at the surface, creating a tension zone around this axis and causing failure by the combined action of shear and tension cracks. This type of failure should not be investigated as part of this analysis since tension cracks would cause problems with respect to discontinuities and geometric changes.

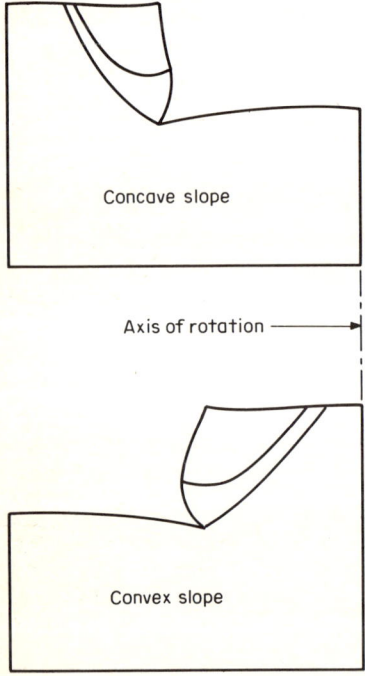

Fig. 2-21 Typical plastic zones at failure for concave and convex slopes. (*From Lorente De No, 1969.*)

On the basis of the foregoing theory the analysis of slurry-filled circular trenches can be approximated by estimating f from Eqs. (2-46) and (2-47) for the concave and convex face, respectively, and then entering Eq. (2-4) or (2-5), replacing the coefficient 4 by f. In the lower zone of r the results are likely to be misleading or too conservative for the convex face, but this solution is admissible under an appropriate factor of safety.

2-9 STABILITY UNDER EXTERNAL CONCENTRATED LOADS

In practical situations trenches must remain stable under the effect of concentrated loads, e.g., existing foundations, underground structures, heavy moving loads, and miscellaneous loads incidental to the construction. Stresses induced by these loads are often estimated from the principles of elastic theories, which as a rule are based on the assumption that stress is proportional to strain. Most of the classical solutions also assume that soil is homogeneous (its properties are constant from point to point) and isotropic (its properties are the same in each direction through a point). Since soil never exactly satisfies and often violates these assumptions, the results often constitute a crude approximation.

In practice, the effect of certain concentrated loads such as foundations is compensated for by limiting the panel length. This helps stability in two ways: (1) arching takes effect within the panel with a redistribution of stress to all four sides, and (2) the foundation usually possesses sufficient rigidity longitudinally to ensure that if the soil loses some of its strength, the structure will tend to ignore this localized loss and adjust the load transfer by overload-

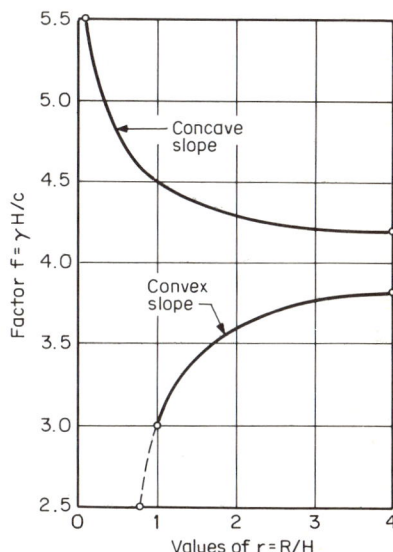

Fig. 2-22 Chart for factor $f = \gamma H/c$, concave and convex slopes. Variable values of $r = R/H$, where R = radius of horizontal curvature and H = height of excavation.

ing the soil beyond the boundaries of the panel. This solution has been tried empirically and found to work well.

2-10 STABILITY UNDER DYNAMIC LOADING

Dynamic loading is induced during an earthquake or from the application of a vibratory excitation, e.g., the passage of a railroad loading. The effect upon stability is partly an increase in the shear stress, under which sliding occurs, and partly a decrease or loss of strength during cyclic loading. The analysis usually involves a modification of the limiting-equilibrium method to include a dynamic factor. For seismic forces this factor is related in some way to the acceleration of the earth (Newmark, 1964), and numerically the seismic coefficient ranges from 0.10 to 0.25.

Seed (1966) has proposed a method of analysis assuming some motion in the underlying earth and regarding the excavation as an elastic deformable body with damping. This was necessitated because growing evidence shows that the soil strength mobilized during earthquakes may be quite different from that determined under static loading conditions and is a function of the entire time history of the stresses developed during the earthquake. Thus, in addition to an estimation of the average seismic coefficient, laboratory strength tests are performed using cyclic loading, and the observed maximum strain provides an indication of the safety of the excavation during an earthquake.

Effect of Ground Vibrations

If saturated cohesionless soil is subjected to ground vibrations, the result is an increase in the pore-water pressure due to movement of water from voids as the soil is compacted and densified. The corresponding decrease in effective stress may turn the sand into a *quick*, or liquefied, condition for which the soil strength is practically zero; thus no load can be supported. This situation arises primarily in loose silts and sands and can eventually lead to trench collapse.

Vibrations can also cause problems in clay deposits, making an excavation unstable, but this condition is more critical in highly sensitive clays. On the other hand, if soft clays contain sand lenses, their liquefaction may contribute significantly to the development of a slide.

In general, ground motion produces a corresponding acceleration that can considerably alter the state of stress in the active wedge. For a potential circular-arc failure ground acceleration causes an active moment about the center of the arc which adds to the overturning moment of the sliding wedge. These phenomena are intensified in earthquakes and large-scale explosions or blasting operations. More frequently, however, it is desired to estimate the effect of moving loads alongside an excavation or the effect of vibrations from construction operations.

Essentially, the property of ground-wave motion that governs effects on trench stability is the energy released to the active wedge. This can be represented by the amplitude of the motion produced, the frequency of this motion, the acceleration which results from combining the amplitude and frequency, the force which it delivers to the wedge, or the energy itself, defined in terms of the velocity of the motion. All these concepts are subject to direct measurement, and various combinations of them have been used to correlate the observed results. It should be noted that an important factor is the total energy, which is also governed by the duration of motion.

Effect of Moving Loads and Construction Equipment

Probably the best way to assess this situation is by means of full-scale field tests, in which appropriate dynamic measurements can be taken. Such a test panel is shown in Fig. 2-23, excavated adjacent to a railroad track at a distance of only 3.8 m (12.5 ft) from the track center. Various measurements were taken to study (1) the effect on stability of freight trains and diesel locomotives running alongside and (2) the effect of vibrations caused by excavating equipment such as drills and chisels. The results are summarized in the following statements (Saito et al, 1974).

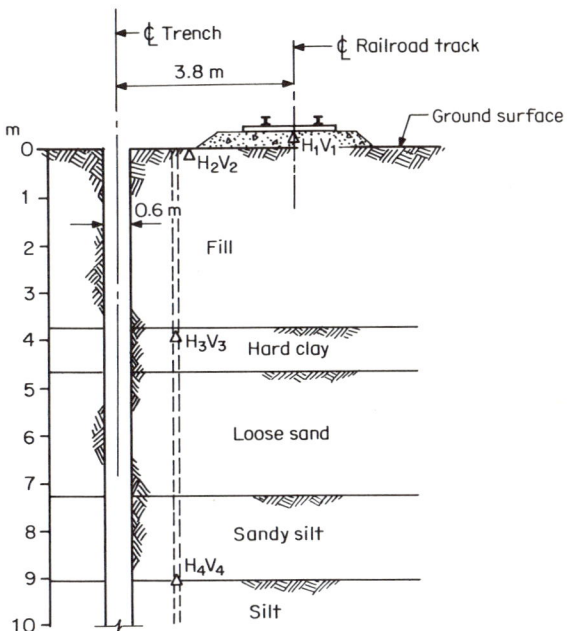

Fig. 2-23 Test panel for the study of the effect of dynamic loads; V = vertical component of vibration, H = horizontal component of vibration. (*From Saito et al., 1974.*)

Dynamic Measurements Upon the passage of trains the vibrations induced in the ground just beneath the ballast are significantly reduced at the hard-clay layer, but little reduction is observed from the hard clay down to the loose sand and silt since this soil is conducive to vibrations. The vibration pattern is shown in Fig. 2-24 for the various construction stages, i.e., before excavation, when excavation reached 6.6 and 20 m, and after completion of the concrete wall.

Response spectra are shown in Fig. 2-25. Figure 2-25a is for the vertical and horizontal component of ground vibrations just beneath the ballast before excavation. The period for V_1 is between 0.01 and 0.3 s. The period for H_1 has

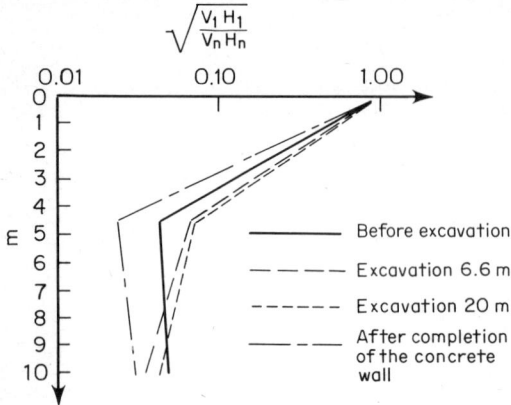

Fig. 2-24 Decrease of the intensity of vibration with distance and through different soil layers. (*From Saito et al., 1974.*)

some peaks between 0.08 and 0.1 s. Figure 2-25b to d shows response spectra for V_3 at various stages of construction. Acceleration varies with excavation, and its level increases as the excavation becomes deeper but is reduced after the completion of the concrete wall, as shown in Fig. 2-25d. It is possible to assess these effects by comparing the relative acceleration values. Thus, if the relative acceleration level 0.6 m from the face is taken as 1, during construction it changes as follows: 0.94 for excavation depth of 3 m, 1.25 for excavation depth of 8 m, 1.24 for excavation depth of 40 m, and 0.46 after completion of the concrete wall.

The dynamic effect of construction equipment, e.g., a chisel used to break hard obstacles, is the impact imparted to the ground. For a chisel maximum acceleration occurs just below the operating level and generally is greater than that transmitted to the trench by adjacent railroad movements. Furthermore, a chisel affects both faces of the trench equally, whereas vibrations from moving trains affect only the adjacent face and are not transmitted across the trench.

Strain Measurements Strains give an indication of the stability of the trench during vibrations. An estimation of strains is possible if the acceleration, the average period, and the velocity of vibrations are known. The passage of trains produces strains of the order of 5.0×10^{-5} for H_3 and 1.0×10^{-4} for H_4. Strains caused by chiseling are 2.7×10^{-5} for H_3 and 9.4×10^{-4} for H_4. All these strain levels are well below the dynamic yielding strain and the liquefaction limit of loose sand and silt.

Pore-Water Pressure Measurements It generally is difficult to obtain measurements of dynamic pore-water pressures. For the test panel of Fig. 2-23 pore-water pressure measurements were possible for both dynamic and static conditions. Pore-water pressure changes are shown in Fig. 2-26, associated with the passage of trains. The pore-water pressure begins to increase just as soon as the train begins to pass and reaches a maximum when the vibration level becomes maximum. The component due to vibrations is about 0.01 kg/cm² (20.5 lb/ft²), which is fairly small. This is just an approximate

Stability of Trenches and Cuts 51

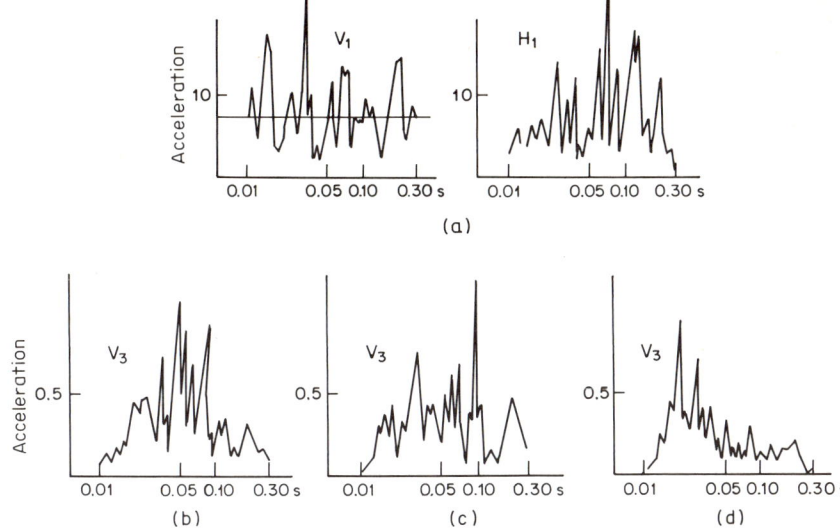

Fig. 2-25 Response spectra; V = vertical component of ground vibrations; H = horizontal component of ground vibrations: (a) pattern of vibrations induced in the ground just beneath the ballast; patterns of vibration of V_3 (b) before excavation, (c) for excavation at 40 m, and (d) after completion of the wall. (*From Saito et al., 1974.*)

estimation; vibration strains are too small for pore-water pressure instruments to trace short-term pressure changes with better accuracy.

From Fig. 2-26 the maximum measured pore-water pressure is 0.05 kg/cm² (102 lb/ft²). For a static train loading KS-18 (Japanese standard) the corresponding earth pressure is 0.076 kg/cm² according to the Westergaard distribution theory, from which it follows that the pore-water pressure coefficient is about 0.66. Pore-water pressure changes due to chiseling are greatest when the chisel operates near the instrument. For this case the maximum measured pore-water pressure is 0.056 kg/cm² (115 lb/ft²).

The foregoing results lead to the conclusion that the dynamic effect of conventional moving loads on the stability of trenches in loose or soft soil is

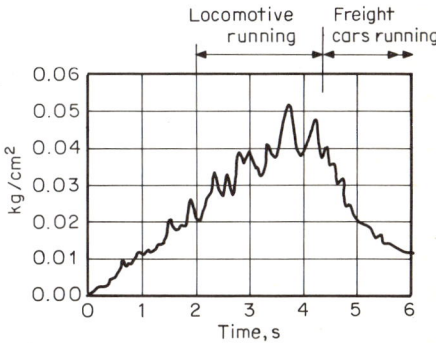

Fig. 2-26 Variations in the pore-water pressure due to the passage of trains. Point 8 m deep when excavation reached a depth of 6.6 m. (*From Saito et al., 1974.*)

greater near the upper part of the excavation and gradually diminishes probably below 4 m (about 13 ft). The effect of chiseling is more serious near the operating level but diminishes away from that level, so that it does not affect the stability of the trench as a whole although it may result in some localized sloughing.

2-11 THE CYLINDRICAL-SURFACE METHOD

According to this method of analysis, a rotational type of failure occurs along a slip surface approximating circular arcs. The rotational character of some slides in deep excavations and particularly in soft or loose soil has been established beyond doubt. In this case the analysis involves a determination of the circle (its center and radius) which best represents the surface along which sliding occurs. This critical circle fulfills the condition that the ratio of the sliding moment and the resisting moment is a maximum. The investigation is therefore within the category of maximum and minimum problems exemplified by the Coulomb theory and involves possible combinations of the cohesion and the friction angle which might give the minimum factor of safety along potential cylindrical failure surfaces. Nonetheless, slight variations in the geometry of the assumed circle are likely to have a minor effect on the exact stability; errors in the assumed panel boundaries, the stress distribution, and the soil parameters are far more critical.

All potential circles have one common point, which is the intersection of the vertical face and the bottom line of the trench, and they are therefore toe circles. Once the radius and the center are determined, the analysis is carried out as shown in Fig. 2-27. The loads or forces acting within the active wedge are determined, and the distance from their line of action to the center of the circle is computed; these may include the total or effective weight W or W' of the soil mass, the hydrostatic force P_w due to groundwater, and externally applied loads such as the concentrated load P or the uniform surcharge q_s. The summation of their moments about the center of rotation O is the overturning moment, which tends to cause a slide along curved surface BC. The resisting moment M_r is provided by (1) the shear resistance along the failure surface BC represented by the resultant friction R_f and the resultant cohesion R_c, (2) the hydrostatic thrust P_f of the slurry, and (3) the shear resistance, both friction and cohesion, mobilized along the end sides of the cylindrical section ABC. The summation of the moments of these forces about O gives the resisting moment M_r. Thus the factor of safety along the curved surface BC is

$$F = \frac{M_r}{M_0} \tag{2-48}$$

It is evident that the length of the cylindrical section must be known before the analysis, and this length is quite relevant to the factor of safety with which stability is maintained. Since the end returns of the section give the

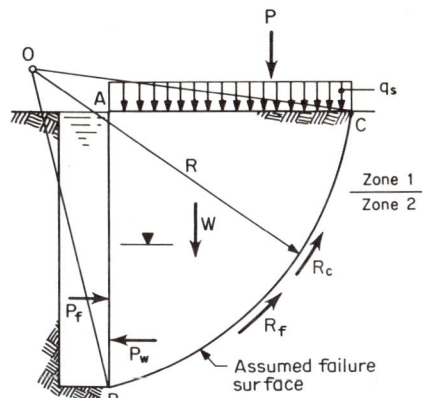

Fig. 2-27 Analysis of trench stability according to the cylindrical-surface method.

same shear resistance regardless of length, it follows that relatively short panels are safer than relatively long trenches.

In practice, the panel length used in the analysis should take into consideration the actual data. If an excavation is carried out in a series of panels under external loads producing a uniform surcharge, it is quite logical to use the same panel length in the analysis of stability. If, according to some elastic theory, the external loads have a distribution which is less than the actual panel length, two analyses may be necessary, one using a length equal to the spread of the concentrated load, the other using the actual panel length and distributing the load over it. Without this double analysis it is not always obvious which case is more critical.

For continuous trenches, the longer the length the less significant the effect of the end sides, and in many instances it may be ignored. If concentrated loads act on continuous trenches, a special solution will be needed, but satisfactory results are obtained if the spread of the load is taken as the finite panel.

A concentrated load acting on the surface of the sliding wedge or in the interior mass makes the same contribution to the overturning moment, and this depends only on the distance from the center of rotation O. The greater this distance the greater the overturning moment. Computer programming usually is necessary to estimate each factor entering Eq. (2-48). Variation in the data of each case is not a problem since each corresponding factor can be integrated separately, e.g., the change in shear strength from zone 1 to zone 2 shown in Fig. 2-27. Estimation of the moments in terms of known parameters is given by Uchida et al. (1973).

Suggested Reading

The foregoing analysis is a partial case of slope-stability problems which are dependent upon the slope angle β and a depth factor n_d. This factor is the ratio of the depth at which firm soil exists to the depth of excavation. According to Taylor (1948), failure of all slopes rising at an angle of more than 53° occurs along a toe circle regardless of the value of n_d. Circular slides and the pertiment theory, including the friction-circle method and graphical solutions, have also been studied by Petterson (1916), Terzaghi and Peck (1948), Bishop (1954), Morgenstern and Price (1965, 1967), Skempton (1948), Tschebotarioff (1973), and many others.

There is some indication that the failure surface is a derivative of the family of circles and approaches a cycloid or a similar curve, particularly in the case of vertical slopes. In fact, a cycloid is the path of the fastest descent of a body moving under the influence of gravity and without friction from one point to another at a lower level but not directly below it. In reality, it is very difficult to detect the difference between circular and cycloidal arcs since the observed slip failures appear very much the same in the field. In this respect the concept of circular-arc failure has received more acceptance primarily because of its relative simplicity and because it is more manageable by mathematical treatment. Ellis (1973) has introduced the cycloidal-arc method in the analysis of shallow ditches.

The method is quite convenient and allows a direct estimation of the arching effect. It applies equally to cohesionless and cohesive soil and also to soil that has both friction and cohesion, which is frequently the case. One disadvantage is that computations are laborious and require computer programming. However, computerized solutions are quite useful since they show directly the effect of varying the parameters of the analysis and thereby provide a better understanding of the causes of failure.

REVIEW PROBLEMS

2-1 A slurry-filled trench is excavated in clay. The undrained shear strength is 2000 lb/ft², and the unit weight of the soil can be taken as 120 lb/ft³. The slurry weighs 70 lb/ft³. For a factor of safety of 1.5 estimate the maximum height first without external loads and then for a uniform surcharge of 250 lb/ft². What is the maximum height without slurry? State your assumptions.

2-2 A long trench is excavated in dense, well-graded dry sand having a friction angle of 40° and a unit weight of 120 lb/ft³. Estimate the factor of safety if the slurry weighs 70 lb/ft³. Repeat using Eq. (2-9), without groundwater, and explain the difference if any.

2-3 Repeat Prob. 2-2 taking the natural water table at the surface and determine the slurry weight for a factor of safety of 1.33.

2-4 Using the data of Prob. 2-2 and again taking the natural water table at the surface, estimate the minimum slurry weight for a factor of safety of 1 according to Eqs. (2-8) and (2-9).

2-5 Derive the expression of stability for a trench in clay when the ground surface slopes upward at an angle i to the horizontal. Disregard surcharge and assume $\phi = 0$.

2-6 (*a*) Analyze a long trench in dry sand to include a uniform surcharge. Extend the analysis to sand with water; (*b*) derive Eq. (2-10).

2-7 Discuss the effect of capillary action for the general case of trench stability in clay.

2-8 A 50-ft-deep trench is excavated in clay using a 70 lb/ft³ slurry. The soil has an undrained shear strength of 250 lb/ft² and a unit weight of 110 lb/ft³ for the first 20 ft. Below that the undrained shear is 500 lb/ft² and the unit weight 120 lb/ft³. Determine whether the trench is stable and estimate the factor of safety.

2-9 Analyze the stability of a trench in sand with water assuming that the slurry has an appreciable shear strength. Derive an expression for the critical height.

2-10 For a long trench in sand h_x = 70 ft, h_w = 7 ft, h_f = 0, according to Fig. 2-8, γ = 120 lb/ft³, and ϕ = 40°. Estimate the minimum slurry weight necessary for stability using Eqs. (2-10) and (2-17). Explain why the answers are different.

2-11 Estimate the factor of safety for Prob. 2-10 if the actual weight of slurry is 75 lb/ft³. Compare the different values.

2-12 A long trench is excavated in sand with the groundwater and the slurry level at the surface. In this case ϕ = 40° and γ = 120 lb/ft³, and the trench height is 60 ft. Determine the factor of safety if the slurry weighs 80 lb/ft³ using Eq. (2-21). Compare the answer with that of Prob. 2-3.

2-13 A 3-ft-wide trench is excavated in 10-ft panels. The site is underlain with clay, s_u = 500 lb/ft², K_0 = 0.6, γ = 115 lb/ft³, and E_i = 4000 lb/in². The slurry weighs 70 lb/ft³. Estimate the critical height using (a) conventional methods and (b) considering the arching effect. What is the probable maximum soil displacement? Assume fully saturated soil.

2-14 A circular shaft is excavated in clay with γ = 110 lb/ft³, K_0 = 0.6, E_i/c = 1500, and average cohesion 300 lb/ft². The shaft has a diameter of 10 ft and is 120 ft deep. Estimate the minimum slurry weight necessary for stability assuming elastoplastic soil.

2-15 A circular cut 20 ft in diameter and 40 ft deep is excavated in clay. It will be filled with an oil-base fluid weighing 55 lb/ft³. Determine the minimum shear strength of soil required for stability. Assume γ = 110 lb/ft³.

2-16 A trench 3 ft wide and 30 ft deep in excavated adjacent to a manufacturing plant to intercept vibrations from a nearby subway construction. The trench is filled with a thick slurry, but it is estimated that the total earth thrust exceeds the slurry pressure by 3400 lb per foot of trench. Is it possible to maintain stability, and if so, what is the minimum shear strength of slurry?

2-17 A trench is excavated in dry sand, the weight of which varies according to the relation γ = 90 + 0.0006σ_v, where σ_v is the vertical geostatic stress. The angle of friction also varies, but for practical purposes it can be taken as 35°. Determine the minimum slurry weight necessary to just maintain stability.

2-18 Analyze the stability of a slurry-filled trench assuming that tension cracks can develop. The soil is stiff clay.

2-19 A 40-ft-deep trench is excavated in sand with ϕ = 35°, γ = 120 lb/ft³, and γ_f = 70 lb/ft³. Determine whether the excavation is safe. If not, estimate how much the slurry level should be raised. Assume that the natural water table is at the ground surface.

2-20 For Prob. 2-19 the engineer decides that because it is not feasible to raise the construction level, it is better to use 10-ft panels and also to lower the groundwater table somewhat. Provide a solution using the Schneebeli method for a factor of safety of 1.5.

2-21 A 3-ft-wide circular trench has an inside radius 60 ft. It is excavated in clay with γ = 110 lb/ft³ and s_u = 400 lb/ft². If a slurry weighing 70 lb/ft³ is used, estimate the maximum height for a factor of safety of 1.

2-22 Determine the critical height of a vertical slope in clay without slurry support assuming that tension cracks extend to about one-half the height.

2-23 Discuss in a general manner how you would analyze stability to include the dynamic effect of pore water.

2-24 Analyze the stability of a trench making an angle of 10° with the vertical. Investigate both cohesive and cohesionless soil.

2-25 Derive Eqs. (2-4), (2-5), and (2-6). Make reference to the diagrams of Fig. 2-2.

2-26 Trenches are often excavated using the two-stage panel process as shown in Fig. 5-21. Analyze the stability of the earth column between pilot holes.

REFERENCES

Bishop, A. W., 1952: The Stability of Earth Dams, Ph.D. Thesis, University of London.

———, 1954: The Use of the Slip Circle in the Stability Analysis of Slopes, *Geotechnique*, **5:** 7-17.

——— and L. Bjerrum, 1960: The Relevance of the Triaxial Test to the Solution of Stability Problems, *Proc. ASCE Res. Conf. Shear Strength Cohesive Soils, Boulder, Colo.*, pp. 437-501.

Bishop, R. F., R. Hill, and N. F. Mott, 1945: The Theory of Indentation and Hardness Test, *Proc. Phys. Soc.*, **57:** 147.

Bowles, J. E., 1968: "Foundation Analysis and Design," McGraw-Hill, New York.

Caquet, A., and J. Kérisel, 1956: "Traité de mécanique des sols," 3d ed., Gauthier-Villars, Paris.

Ellis, H. B., 1973: Use of Cycloidal Arcs for Estimating Ditch Safety, *J. ASCE Soil Mech. Found. Div.*, **99**(SM2): 165-179.

Elson, W. K., 1968: An Experimental Investigation of the Stability of Slurry Trenches, *Geotechnique*, **18**(37): 39-49.

Hansen, J. B., 1961: The Ultimate Resistance of Rigid Piles against Transverse Forces, *Dan. Geot. Inst. Bull.* 12, p. 5.

Huder, J., 1972: Stability of Bentonite Slurry Trenches with Some Experience in Swiss Practice, *Proc. 5th Eur. Conf. Soil Mech. Found. Eng., Madrid*, vol. 1, pp. 517-522.

ICOS, 1968: Trench Excavation Close to Heavy Footing, Miscellaneous Works, Milan, Italy.

Kowalewski, Z., 1964: "Parcie czynne gruntu w wycopie a skończoney dlugosci," Blul. Inf. Naun.-Techn., Warsaw.

Ladd, C. C., 1967: "Strength and Compressibility of Saturated Clays," class notes for Pan-Am soils course, Universidad Catolica Andress, Belto, Caracas.

——— and R. Foott, 1974: New Design Procedure for Stability of Soft Clays, *J. ASCE Geotechn. Div.*, **100**(GT7): 763-786.

Lambe, T. W., and R. V. Whitman, 1969: "Soil Mechanics," M.I.T.-Wiley, New York.

Lane, K. S., and D. E. Washburn, 1946: Capillarity Tests by Capillarimeter and by Soil Filled Tubes, *Proc. Highw. Res. Board*.

Lorente De No, C., 1969: Stability of Slopes with Curvature in Plane View, *Proc. 7th Int. Conf. Soil Mech. Found. Eng., Mexico City*, pp. 635-638.

Meyerhof, G. G., 1951: The Ultimate Bearing Capacity of Foundations, *Geotechnique*, **2:** 301.

———, 1972: Stability of Slurry Trench Cuts in Saturated Clay, *Proc. Conf. Performance Earth Earth-supported Struct., Purdue Univ.*, vol. 1, pt. 2, 1451-1466.

Morgenstern, N. R., 1963: Comments, *Symp. Grouts Drilling Muds*, Butterworths, London.

——— and I. Amir-Tahmasseb, 1965: The Stability of Slurry Trench in Cohesionless Soils, *Geotechnique*, **6**(4): 387-395.

——— and V. E. Price, 1965: "The Analysis of Stability of General Slip Surfaces," *Geotechnique*, **15:** 79-93.

——— and ———, 1967: A Numerical Method for Solving the Equations of Stability of General Slip Surfaces, *Comp. J.*, **9:** 388-393.

Müller-Kirchenbauer, H., 1972: Stability of Slurry Trenches, *Proc. 5th Eur. Conf. Soil Mech. Found. Eng., Madrid*, vol. 1, pp. 543-553.

Nash, J. K. T. L., and G. K. Jones, 1963: "The Support of Trenches Using Fluid Mud," Butterworths London.

Newmark, N.M., 1942: Influence Charts for Computations of Stresses in Elastic Foundations, *Univ. Ill. Bull.* 338.

———, 1964: Effects of Earthquakes on Dams and Embankments, *Geotechnique*, **15:** 139-160.

Petterson, K. E., 1916: Kajraseti Goteborg des Mars 1916, *Tek. Tidskr.*, **46:** 289.

Piaskowski A., and Z. Kowalewski, 1965: Applications of Thixotropic Clay Suspensions for Stability of Vertical Sides of Deep Trenches without Strutting, *Proc. 6th Int. Conf. Soil Mech. Found. Eng., Montreal*, vol. 2, pp. 526-529.

Prandtl, L., 1923: Compression of Plastic Blocks between Rough Rigid Plates, *Z. Angew. Math. Mech.*, **3:** 401.

Saito, J., Y. Goto, and H. Sato, 1974: Stability of Trench against Dynamic Loads during Slurry Excavation *Kajima Corp. Spec. Bull.*, Tokyo.

Schneebeli, G., 1964: "Le Stabilité des tranchées profondes forées en presence de boue," *Houille Blanche*, **19**(7): 815-820.

Seed, H. B., 1966: Soil Stability Problems Caused by Earthquakes, *Univ. Calif., Berkeley, Soil Mech. Bitum. Mat. Lab. Res.*

Skempton, A. W., 1948: The $\phi = 0$ Analysis of Stability and Its Theoretical Basis, *Proc. 2d Int. Conf. Soil Mech. Found. Eng., Rotterdam*, vol. 1, pp. 72-78.

——— and H. Q. Golder, 1948: Practical Examples of the $\phi = 0$ Analysis of Stability of Clays, *Proc. 2d Int. Conf. Soil Mech. Found. Eng., Rotterdam*, vol. 2, pp. 63-70.

——— and P. LaRochelle, 1965: The Bradwell Slip: A Short Term Failure in London Clay, *Geotechnique*, **15:** 222-242.

Taylor, D. W., 1948: "Fundamentals of Soil Mechanics," Wiley, New York.

Terzaghi, K., 1941: General Wedge Theory of Earth Pressure, *Trans. ASCE*, pp. 68-97.

———, 1943: "Theoretical Soil Mechanics," Wiley, New York.

——— and R. B. Peck, 1948, 1967: "Soil Mechanics in Engineering Practice," Wiley, New York.

Timoshenko, S., and J. N. Goodier, 1951: "Theory of Elasticity," McGraw-Hill, New York.

Tschebotarioff, G. P., 1967: *Proc. 3d Pan-Am. Conf. Soil Mech. Found. Eng., Caracas*, Gen. Rep. Div. 4, pp. 301-322.

———, 1973: "Foundations, Retaining and Earth Structures," McGraw-Hill, New York.

Uchida, H., K. Tanaka, and H. Mizutani, 1973: On the Stability of Vertical Sides in a Slurry Trench, *Kajima Corp. Spec. Bull.*, Tokyo.

Xanthakos, P. P., 1974: "Underground Construction in Fluid Trenches," Colleges of Engineering, University of Illinois, Chicago.

Chapter Three

SOIL-SLURRY INTERACTION

The concepts discussed in this chapter have an immediate geotechnical interest and therefore appeal directly to students and researchers of soil mechanics. However, there are many practical problems resulting from the way slurries interact with soil, and unless they are properly understood and controlled in the field, they may lead to such unexpected and dangerous situations as high slurry loss or even loss of stability. In this context engineers and contractors are encouraged to become familiar with the contents of this chapter, especially the last sections.

3-1 FORMATION OF SEAL

As already mentioned, the analysis of stability almost invariably assumes the formation of an impermeable barrier at the interface in a manner that enables the slurry to exert its full hydrostatic thrust. Moreover, the barrier prevents slurry loss toward the ground. Since plain water alone in the slurry does not provide this function, even though its thrust may otherwise be sufficient to prevent sliding, it is necessary to use a colloid type of material in the slurry. The deposition of its particles in the soil pores causes the formation of a seal by *thixotropy*, discussed in some detail in this chapter and in Chap. 4.

Figures 3-1 and 3-2 show how granular soil and bentonite interact when they are in contact. Two compartments of a glass tank are filled with sand and a suspension of bentonite, respectively. The separating panel is then gradually lifted, allowing the suspension to penetrate the sand, as shown in Fig. 3-1*a*, in which the penetrated zone is shown dark. Although this pattern of penetration is not necessarily typical, it confirms an important fact: a seal is formed within the soil and restricts further flow of slurry toward the ground. Figure 3-1*b* shows what happens if the slurry is allowed to escape, i.e., if its level drops. The adjacent soil now exhibits a certain cohesion which allows it to stand almost vertically. The tree-shaped triangular intrusion shown in Fig. 3-2 occurs when an excavation is carried out in coarse material and is exemplified by a horizontal spread which increases with increasing pressure, particularly in soil of low water content.

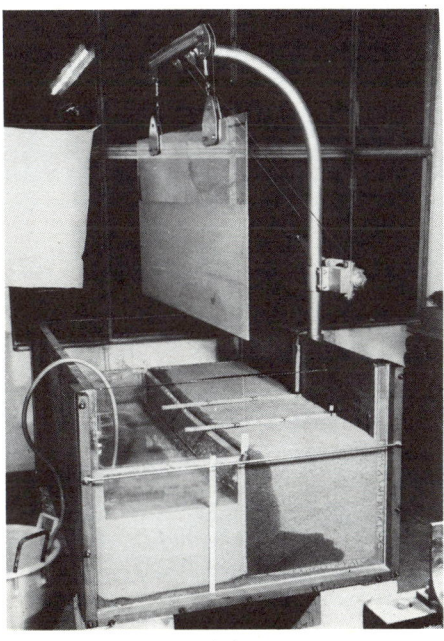
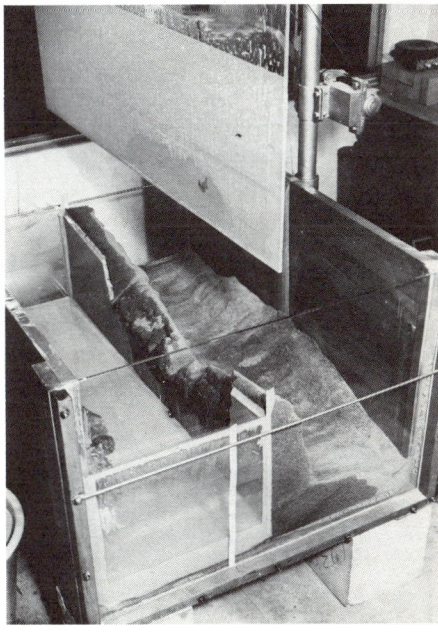

(a)   (b)

Fig. 3-1 Test showing the interaction of bentonite slurry with fine granular soil: (*a*) the granular soil is penetrated by the bentonite slurry, as shown by the dark area; (*b*) the soil face is relatively stable in spite of the lowering of the bentonite level. (*ICOS.*)

In water-bearing formations any flow likely to occur until the barrier separates the two media again will be toward the ground as long as the slurry is heavier than plain water and its level stands higher than the natural water table, giving a pressure differential.

Filter Cake and Penetration

Figure 3-3 shows schematically how a colloidal slurry penetrates into a porous granular medium. In Fig. 3-3*a* the slurry enters into the pores due to some

60 Slurry Walls

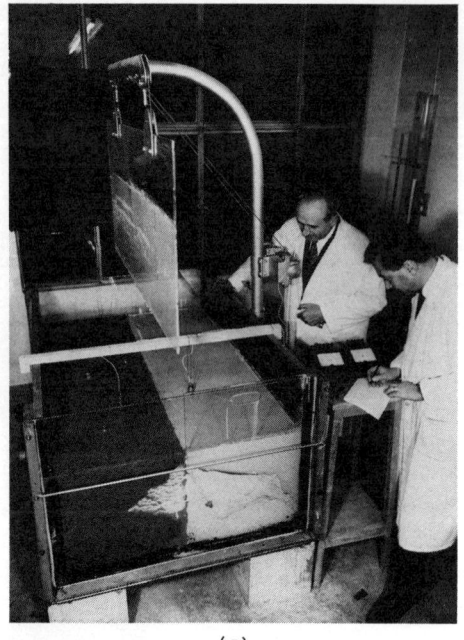

Fig. 3-2 Penetration of bentonite slurry into a granular medium: (*a*) general view of the glass tank; (*b*) tree-shaped triangular intrusion. (*ICOS.*)

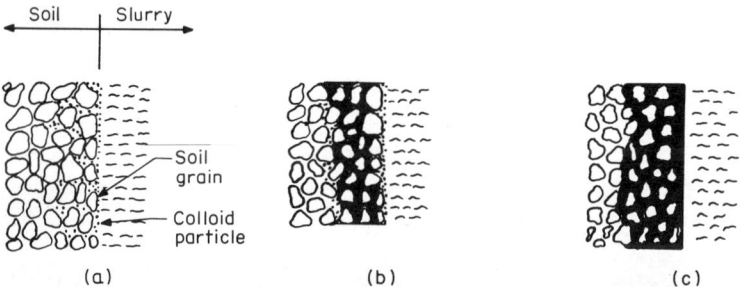

Fig. 3-3 Formation of filter cake: (*a*) deposition of colloid fraction in the soil voids; (*b*) filtration of slurry; (*c*) formation of impermeable film along the face.

pressure difference, and during this process groups of solid particles begin to occupy void space between soil grains. In Fig. 3-3*b* the slurry continues its filtration as more solid particles accumulate in the pores, and the system now forms a tightly packed zone of gelled material, commonly called *filter cake*. Very soon, usually in a few seconds, the cake is covered by a thin layer of bentonite particles, also called *protective film*, as shown in Fig. 3-3*c*, and at this stage the barrier is fairly impermeable and offers complete resistance to further penetration.

It is seen that the filter cake actually is obtained through the filtration of slurry; hence its formation is influenced by the degree of permeability of the soil. Every soil and every rock contains continuous voids, but for filtration to occur these voids must be of a certain size. Thus, the filter cake is not formed in soils where the permeability is close to zero.

If we consider a tube of diameter $2r$ and length l as an idealized representation of the soil pores, and if we assume that the tube is filled with slurry, it is possible to relate the pressure difference Δp necessary to displace the slurry and produce flow to the shear strength τ_f of the slurry as it gels in the pores. The slurry is on the verge of shear failure if

$$2\pi r l \tau_f = \pi r^2 \Delta p$$

which gives

$$\Delta p = \frac{2\tau_f l}{r} \tag{3-1}$$

This is easily modified to include the so-called *critical hydraulic gradient*. The slurry therefore penetrates the soil to different distances l, which depend upon the pore size of the material, the properties of the slurry (mainly its shear strength), and the pressure difference existing between slurry and ground water.

Seal Mechanism and Slurry Loss

The formation of a seal can occur according to several distinct mechanisms under varying soil conditions, encompassing the entire range from very fine-grained soils to open gravel. Basically, a seal is obtained according to three modes.

Surface Filtration This occurs when a filter cake is formed, as hydrated bentonite particles are brought together in the soil pores. The result is a dense packing of material allowing only limited penetration of the soil by the slurry. Since no material is absolutely watertight, during and after the formation of the filter cake water from the slurry continues to percolate through it toward the soil. Water lost in this way is referred to as *filtered water*. Water loss during surface filtration is well defined and consists of two different parts: (1) initial slurry loss, which extends over the period required for the formation of the cake, and (2) steady slurry loss, which obeys the flow

$$v = mT^{-1/2}$$

where v = flow rate
T = time after filter-cake formation
m = const

According to Hutchinson et al. (1974), slurry loss during and after filter-cake formation can be represented by the curve of Fig. 3-4.

Deep Filtration This mechanism takes effect in medium to coarse soil, and in this case the penetration may extend from a few inches to several feet.

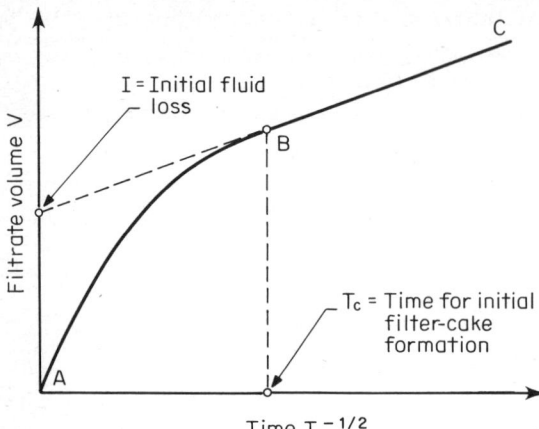

Fig. 3-4 Fluid loss during and after filter-cake formation. (*From Hutchinson et al., 1974.*)

Eventually, dense packing of colloid particles in the zone just adjacent to the face provides a filter cake and stops further penetration.

The interception and deposition of colloid particles in the soil pores near the interface results in the dilution of any slurry escaping to the ground. Accordingly, the impregnating fluid is a waterlike liquid which has lost a considerable portion of its colloid fraction, and this is true for both surface and deep filtration.

Rheological Blocking In this case the slurry flows directly into porous formations, such as gravel, until it is restrained by its own shear strength. The rheological behavior of slurry under flow conditions can be described by the relation between the shear stress and the resulting rate of shear. This mechanism takes effect within fairly large pores and is confirmed when bentonite is found in formations many feet away from the face. The process is characterized by gradual gelation of the slurry in the zone of penetration, the extent of which is governed by Eq. (3-1). The initial slurry loss is erratic.

Of these distinct mechanisms surface filtration is the preferred way of obtaining a seal because it minimizes slurry loss and does not change the soil characteristics.

Slurry Loss Figure 3-5 shows initial slurry loss vs. bentonite concentration. The curve discloses that there is a cutoff concentration which for the type of bentonite used (Berkbent) is between 4 and 4.5 percent. Below that, the initial slurry loss increases very sharply even in soil of very low permeability such as 5×10^{-3} cm/s (Hutchinson et al., 1974). It is therefore necessary to select a minimum bentonite concentration according to this behavior.

The presence of a small quantity of fine sand in the slurry may alter the sealing mechanism in open ground ($k = 1$ cm/s) from rheological blocking to deep or even surface filtration, as shown in Fig. 3-6, with the subsequent

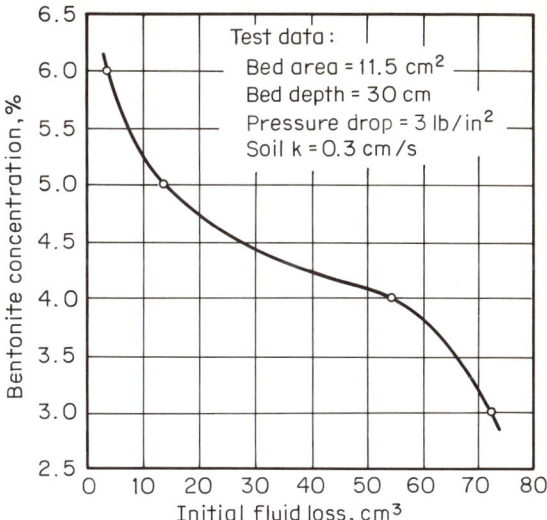

Fig. 3-5 Initial fluid loss vs. bentonite concentration. *(From Hutchinson et al., 1974.)*

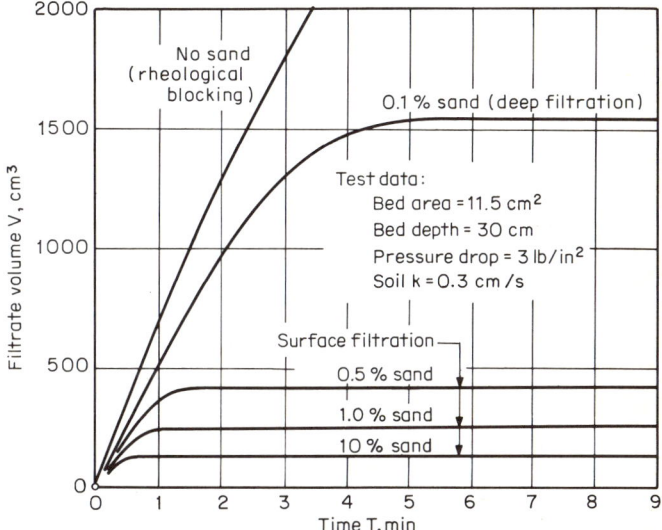

Fig. 3-6 Effect of sand on filtration of 5% Berkbent fluid through fine gravel bed. *(From Hutchinson et al., 1974.)*

dramatic reduction in the initial slurry loss demonstrated in Fig. 3-7. Although there always is some suspended fine materials in the slurry as a by-product of the excavation, for trenching in open ground a small fraction of fine granular soil in the initial slurry is beneficial to the sealing process.

The gradual loss of water rather than slurry which accompanies surface or

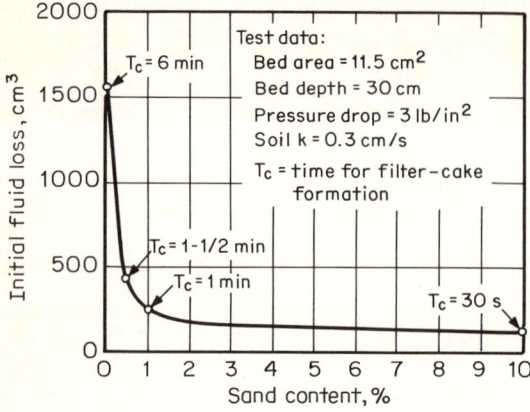

Fig. 3-7 Effect of sand content on initial fluid loss through fine gravel bed. (*From Hutchinson et al., 1974.*)

deep filtration, although small, must sometimes be considered since it may increase the initial colloid concentration. In this case, it is possible to predict the subsequent changes in the flow properties of the slurry, and, as shown in other chapters, a colloid concentration higher than necessary can cause a variety of problems. On the other hand, filter cakes and protective films generally are sufficiently compressible to reduce rather than increase leakage even if the pressure differential increases. The rate of slurry loss is not necessarily pressure/sensitive but depends also on many other factors (see also Chap 4), although this is a convenient assumption when investigating the process of penetration.

3-2 THIXOTROPIC GELATION

The importance of the sealing process became apparent as early field techniques were improved, particularly with regard to the methods of excavation. The existence of a low-permeability barrier at the interface was confirmed in the early stages as attention was focused on the behavior and control of slurries. For practical field controls significant improvements have been possible through the use of thixotropic materials in slurry systems. In general terms thixotropic behavior is based on the initial nonequilibrium of interparticle forces due to remolding or disturbance and the effects on subsequent structure changes within a slurry system left undisturbed. Thixotropy is of quite general occurrence in all fine-grained materials and consists of a reversible process of softening caused by disturbance, followed by a time-dependent return to the original state.

Thixotropy of Slurries

To the slurry specialist thixotropy means slurries (as opposed to "concentrated" systems, which are solids of high water content) and gel setting times of the order of minutes or, at most, hours. For slurries that gel when allowed to stand thixotropy results in an isothermal, reversible, time-dependent sequence occurring under conditions of constant composition and volume whereby the slurry stiffens to a gel while at rest and liquefies upon stirring or remolding. As shown in Chap. 4, the interparticle-force mechanism and principles of aggregation and dispersion are quite important. Freundlich (1935) was the first to show the dependence of thixotropic behavior on the balance of forces acting between particles. The electrolyte concentration is also important, and an increase in electrolyte content usually leads to a decrease in setting time for a thixotropic gel.

The tendency of particles of the same nature to adhere upon contact is relevant to thixotropy. All very fine particles exhibit a lack of balance of forces at their surface. These forces are satisfied either by contact of particles or by adsorption of ions from an adjacent phase. This explains thixotropy on the basis of an attraction-repulsion force balance illustrated in Fig. 3-8. The ordinates to the curves represent the energy (positive if repulsion and negative if attraction) necessary to bring the particles from an infinite spacing to a given spacing along the horizontal axis. Curve A represents a stable suspension which exhibits neither flocculation nor thixotropy because the energy barrier prevents close approach to particles. Curve B is for particles which spontaneously agglomerate and settle out of suspension. The energy minimum indicated by curve C represents a system of particles in a thixotropic gel. Any movement, e.g., shaking or shearing, tending to change the particle spacing causes an increase in the energy of repulsion and leads to a more liquid state (see also Chap. 4).

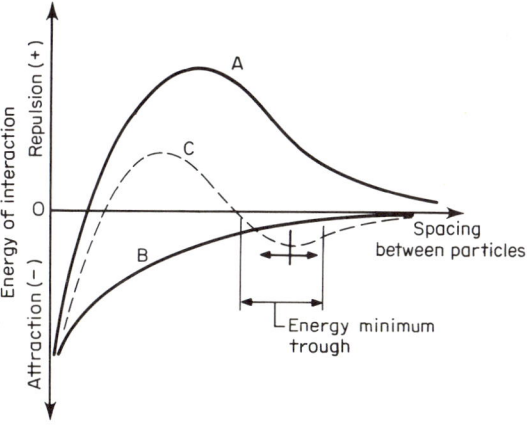

Fig. 3-8 Energy-distance curves for dilute suspensions of dispersed, flocculated, and thixotropic materials. Curve A: close particle approach prevented by energy barrier; particles disperse. Curve B: energy decreases as particles approach; particles flocculate. Curve C: behavior of thixotropic material; spacing of particles at rest. Remolding moves particles away, and energy increase causes the strength to decrease. (*From Mitchell, 1960.*)

There is evidence (Marshall, 1949; Van Olphen, 1956) that at times two separate zones of thixotropic behavior exist for the same material. For example, reversible hardening-liquefaction effects may first occur at very low salt concentrations, but then higher salt content may lead to a stable system. Further increase in salt content causes the thixotropic characteristics to reappear. These phenomena are quite important in the control of slurries and are discussed at some length in other sections.

Kruyt (1952) has shown that thixotropy occurs preferentially, and perhaps almost exclusively, in systems with elongated particles such as clays. It must

66   Slurry Walls

also be noted that thixotropic suspensions of bentonite have been observed in which the particles come to rest although there is no material contact between them. On the other hand, a suspension of graphite in mineral oil is nonconducting in a liquid state but conducting after gelation, indicating that particles now must be in contact and form a continuous network.

Reversibility

Laboratory and field tests show that the sol-gel transformation of thixotropic suspensions of the type used as slurries is completely reversible (Mitchell, 1960). When shaking or stirring of the suspension stops, the material gels but the gel strength varies with many factors. When stirring starts again, the material liquefies. This reversibility may continue indefinitely.

Reversibility may also occur in more concentrated systems (as used here, concentrated means solids of very high water content), but in this case it may be limited. Figure 3-9 shows results of reversibility studies on silty clay soil from Vicksburg, Mississippi (LL = 39, PL = 25). The aged strength of this

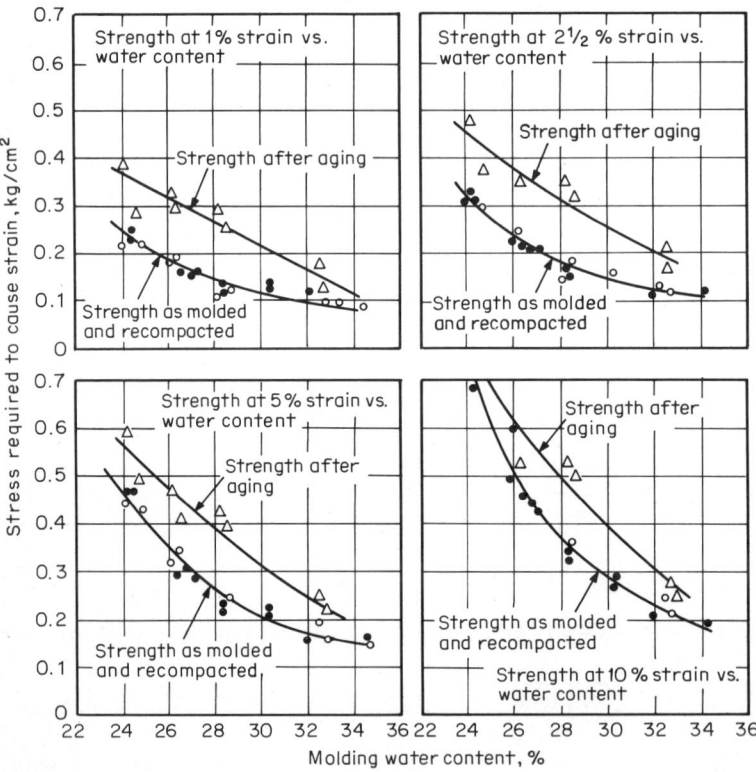

Fig. 3-9 Comparison of the as-compacted and the aged-remolded-recompacted strength of Vicksburg silty clay: ● = samples tested immediately after compaction; ○ = samples aged 6 days after compaction, remolded, recompacted, and then tested; △ = samples aged 6 days after compaction. (*From Mitchell, 1960.*)

compacted soil after 6 days' storage varies from 1.1 to 1.8 times the original as-molded strength, depending on the water content and the strain at which the strength is measured. Results are shown for different values of stress necessary to cause four different strains, namely 1, $2\frac{1}{2}$, 5 and 10 percent, as a function of molding water content for samples tested immediately after compaction and for samples aged 6 days, remolded, and compacted. The data for the as-compacted and the remolded aged samples show no significant variation, leading to the conclusion that the stiffening process is reversible.

The role of water appears to be of primary importance in determining the status of interparticle forces. If a material can pass from a state of flocculation to a state of dispersion when placed in water, thixotropic effects will be noted. In fact for concentrated (solid) systems thixotropic effects have been found to increase with increased water content.

A practical aspect of reversibility in slurries is the linkage between colloid particles to form a structure which within a short time (usually seconds) causes the system to gel in the soil pores. For colloid concentrations which are practical and attainable in slurry-trench excavations gelation leads rapidly to the buildup of the filter cake and the deposition of the protective film at the interface. Although theoretical thixotropic strength-change relations are not available to permit direct correlation of thixotropic effects with design problems, the behavior of thixotropic slurries can be reasonably predicted by means of simple tests before and during construction. These tests are described in Chap. 4. Thus, although a comprehensive theory regarding thixotropy and its causes is not yet available, its practical aspects are fairly well understood and adequately controlled.

3-3 STABILITY AGAINST SLOUGHING AND PEEL-OFF

Peel-off Theory

Besides providing a seal, the filter cake has also a plastering effect on the interface so that individual soil particles and grains are compelled to remain in the earth structure. This is particularly important in cohesionless soil, where kinetic and rolling friction are the only forces to keep outermost particles at the exposed face from collapsing due to their own weight or under the action of tools. If enough soil grains can break away, the face will eventually collapse. Sloughing and peel-off are such examples. Thus, soil particles at the interface must be supported by the shear strength of the filter cake as the excavation proceeds.

If the average soil grain is represented by a sphere of diameter D submerged in a slurry of shear strength τ_f, motion is produced in any direction if the applied force exceeds the total shear resistance on the surface of the sphere. According to Weiss (1967), the shear resistance is

$$T = \frac{\pi D^2 \tau_f}{4} \qquad (3\text{-}2)$$

The sphere is prevented from downward movement if T exceeds the buoyant weight of the particle with respect to the slurry. This weight is

$$W' = \frac{\pi D^3}{6} (\gamma - \gamma_f)$$

and the requirement $W' = T$ yields

$$D_V = \frac{3}{2} \frac{\tau_f}{\gamma - \gamma_f} \qquad (3\text{-}3)$$

in which γ is the density of the grain (without voids) and D_V is a notation used to distinguish the vertical movement of the particle. Equation (3-3) is for particles submerged in slurry but also applies to the stability of overhanging boundaries and to irregular portions of the face.

However, Eq. (3-3) does not relate the stability of particles with respect to movement toward the face. Resistance to such movement depends on the orientation of grains in the earth structure, the inclination of the bearing area for a given particle, and the shear effect of the deposited filter cake along the boundary separation zone.

Figure 3-10 shows a regular array of grain particles idealized as a group of spheres of uniform diameter. As shown in Fig. 3-10b, β is the angle of inclination of the face from the vertical, and α is the angle of the common tangent to two contiguous spheres with the vertical to the face. Outward movement of a sphere will begin parallel to the common tangent and will occur under an angle $\psi = \alpha - \beta$, as shown. Such movement is caused by the component F of the submerged weight W'. For the outermost particles the motion may be conceived of as rolling rather than sliding, so that it is independent of friction.

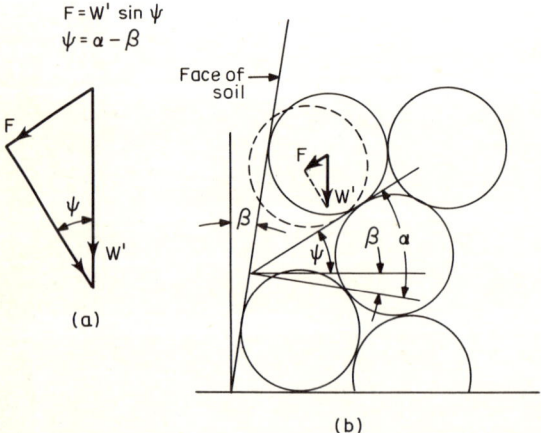

Fig. 3-10 Mechanism of peel-off failure for single soil grain idealized as a sphere.

Then the force F is resisted by the force T. Obviously

$$F = W' \sin \psi = \frac{\pi D^3}{6} (\gamma - \gamma_f) \sin \psi$$

and the requirement $T = F$ gives

$$D = \frac{3}{2} \frac{\tau_f}{(\gamma - \gamma_f) \sin \psi} \qquad (3\text{-}4)$$

For a regular arrangement of spheres, the angle α is constant within the soil. Examples of constant angles α are shown Fig. 3-11a and b for loose and dense array of spheres, respectively. If $\alpha = \beta$, $\psi = 0$ and $F = 0$, so that movement of a single sphere is not possible. The angles α and β should not be correlated to the friction angle or the angle of repose since they express the orientation of particles rather than interparticle friction. The condition $\alpha = \beta$ requires the simple cubic packing of spheres shown in Fig. 3-11c. Müller-Kirchenbauer (1972) has suggested that the average angle α varies from 35 to 40°, but this is based on limited data.

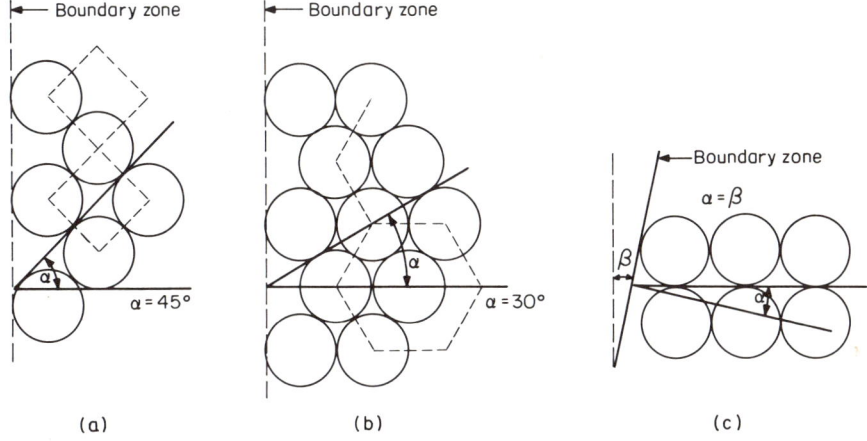

Fig. 3-11 Tangential angle α for regular arrangements of spheres.

From Eqs. (3-3) and (3-4) it is seen that the ratio D/D_V is a function of the angle $\psi = \alpha - \beta$. For a value of α of 37.5°, the variation of the ratio D/D_V is shown in Fig. 3-12 as a function of the inclination angle $90° - \beta$.

Stability against sloughing and peel-off is thus seen to depend on the ability of soil grains to remain stacked in their initial position. Any tendency for the outermost grains to move must be resisted by the shear strength of the filter cake. The most frequent disturbance of the filter cake occurs near the ground surface, and this may indicate that peel-off will probably be observed in the upper part of the trench not only because of this disturbance but also due to

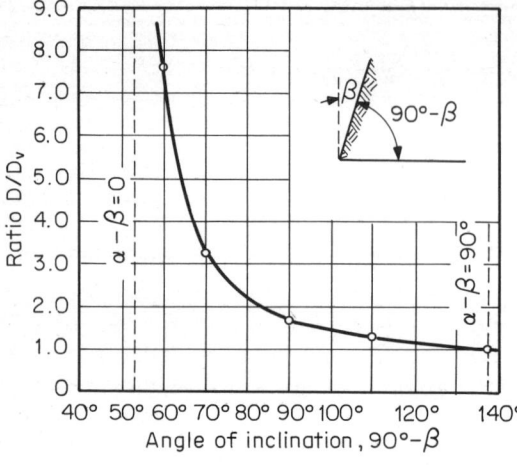

Fig. 3-12 Dependence of ratio D/D_V upon the angle of inclination $90° - \beta$ for a tangential angle α of 37.5°.

the generally loose array of soil particles in the top layers. Nonetheless, experience shows that peel-off can occur anywhere in the trench.

Peel-off Tests

Figure 3-13 shows a relatively simple device used to test the stability of a soil sample against peel-off. A container is filled with the sample, which is extracted and placed so that the vertical face becomes horizontal inside the container. The sample may be saturated before starting the test. A bentonite suspension is poured into the container until it is filled to the top, and at the same time the bottom valve is open to allow all but a small quantity of water to be discharged. After any air bubbles have been let out, the valves are turned

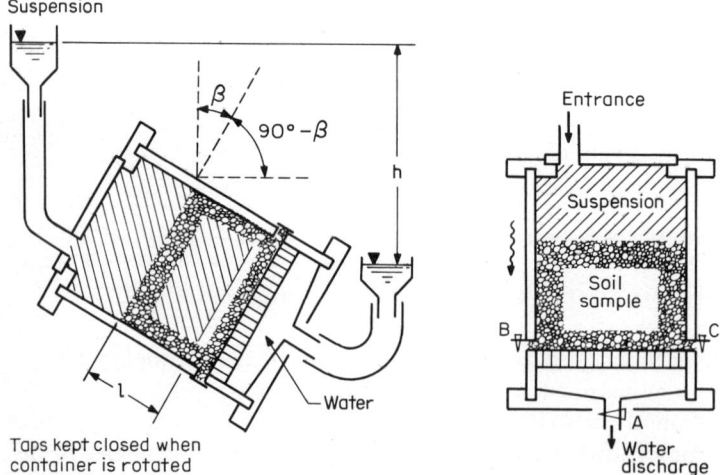

Fig. 3-13 Peel-off test to investigate stability of individual grains at various angles of inclination.

off and the container is rotated as shown to cause peel-off from the interface of the sample. The latter may remain stable when it becomes vertical, in which case the rotation continues until it completes 145° from the initial position. If no particles fall off, this means that the sample is stable against peel-off. Sample disturbance and volume distortions in the conduct of the test give rise to uncertainties in the interpretation of results. Nonetheless, the test is very simple and useful and can be carried out several times using bentonite suspensions of varying concentration and flow properties.

As shown in Fig. 3-13, a small quantity of water remains inside the container according to the pressure head h. The ratio h/l is the stagnation gradient defined in Chap. 2. Besides the direct information which is obtained, this procedure can be used to investigate the relationship of the flow properties of the slurry to relevant soil properties, e.g., grain size and shape, angle α, and shear strength. The test may also indicate the influence of pore size on the extent of penetration and the ability of slurry to form the filter cake.

3-4 STAGNATION GRADIENT

A very useful and relatively simple test can be carried out to determine the stagnation gradient and thus permit estimation of the actual thrust that might be exerted by the slurry against the face. By definition, the stagnation gradient is the ratio h/l of the pressure head to the penetration distance. For this analysis it is convenient to take the boundaries as shown in Fig. 2-16. The notation is the same as that used in Sec. 2-6 for stability with permeable filter cake.

Rewriting Eq. (3-1) and setting $\Delta p = h\gamma_f$, we obtain

$$\frac{h}{l} = \frac{2\tau_f}{r} \frac{1}{\gamma_f}$$

or

$$i_0 = \frac{2\tau_f}{r} \frac{1}{\gamma_f} \tag{3-5}$$

which relates the stagnation gradient to the parameters τ_f and r. However, since accurate determination of the average pore diameter $2r$ and the in situ gel strength τ_f of the slurry is quite difficult, the use of Eq. (3-5) often leads to unsafe predictions. Thus, in addition to the standard viscosity and filtration tests important information is obtained from stagnation-gradient tests. Since the purpose is to provide the relationship $i_{01} = f(i_{02})$, at least two tests are required, one using pure slurry and the second using slurry aggregated with solid matter from the excavation. For all cases m is the total solid ratio by volume.

Figure 3-14 shows the apparatus used for the stagnation test. It consists essentially of a sufficiently long pipe with an upper chamber separated from the pipe by means of a tap and a disk. The pipe is filled with a soil sample in a saturated state. The tap is then closed, and the chamber is filled with a colloi-

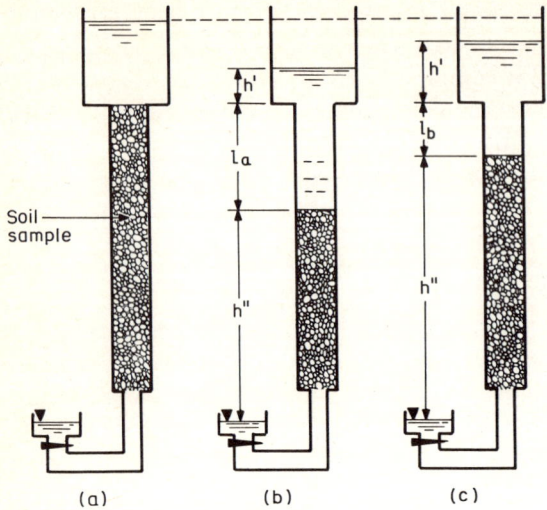

Fig. 3-14 Stagnation gradient test for the determination of the values i_0 and i_{01}/i_{02}: (a) situation before starting test; (b) penetration without filtrate; (c) penetration with filtrate.

dal suspension. When the tap is opened and the disk removed, the suspension penetrates the sample, as shown in Fig. 3-14b. At this stage the penetration occurs without filter cake, and the fluid loss is the initial loss. By definition the stagnation gradient is (assuming the wall of the pipe to be of frictionless material)

$$i_{01} = \frac{h_a}{l_a} = \frac{h' + l_a + h''(\gamma_w/\gamma_f)}{l_a} \tag{3-6}$$

which actually gives a numerical estimation of i_{01} since all the parameters in the right-hand term of Eq. (3-6) are measured directly. Figure 3-14c shows the penetration for a relatively impermeable filter cake. This penetration is now l_b and generally is less than l_a. Apparently

$$h_b = h' + l_b + h'' \frac{\gamma_w}{\gamma_f} \tag{3-7}$$

With a value of i_{01} computed from Eq. (3-6) and a value of h_b computed from Eq. (3-7), the fraction of slurry pressure transmitted to the filter cake is, from Eq. (2-34),

$$p'_{f2} = p'_f - p'_{f1} = h_b\gamma_f - i_{01}\gamma_f l_b \tag{3-8}$$

and the thickness of the filter cake is calculated from Eq. (2-29) as

$$u = \lambda l_b \tag{3-9}$$

which requires that λ be known. For all practical purposes it can be assumed that $n_1 = n_2$, in which case λ can be computed from Eq. (2-29) or with the help of Fig. 3-15. The stagnation gradient i_{02} corresponding to a filter-cake thickness u is

$$i_{02} = \frac{h_b - i_{01}l_b}{u} \tag{3-10}$$

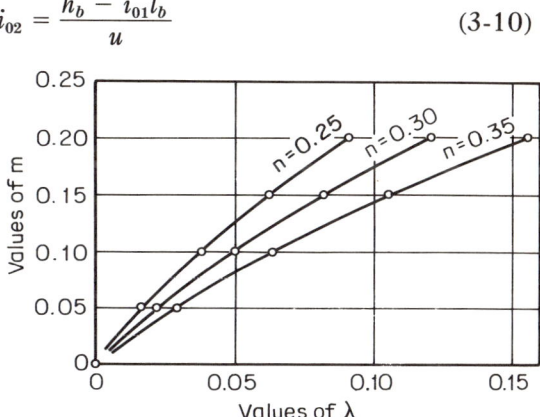

Fig. 3-15 Relation between λ and the concentration m of solid matter in suspension (for $n_1 = n_2$).

If the test is carried out using a slurry aggregated with soil from the excavation, the head h_b is adjusted, noting that just above the top of the pipe the slurry weight is γ_{fs} but just below the boundary the weight is γ_f since most soil particles are intercepted there. Hence the head h_b is

$$h_b = h' \frac{\gamma_{fs}}{\gamma_f} + l_b + h'' \frac{\gamma_w}{\gamma_f} \tag{3-11}$$

If the filter cake has zero permeability, l_b is also zero and h_b is the actual pressure head. Then from Eq. (3-8) the fraction pressure transmitted to the filter cake is the total hydrostatic pressure of the slurry.

It is not necessary to consider the thickness of the filter cake in order to estimate the fraction pressure exerted on it. Indeed, Eq. (3-8) is rewritten as

$$\frac{p'_{f2}}{p'_f} = 1 - \frac{i_{01}l_b}{h_b} \tag{3-8a}$$

which is generally valid for a thin filter cake. In practice, however, any excess material deposited on the face is scraped off by the up-and-down action of the excavating tools, and a reasonable u_{max} is about 1 cm (or $\frac{3}{8}$ in). The fraction pressure head transmitted to the region beyond the filter cake is therefore increased according to

$$p'_{f1} = h\gamma_{fs} - i_{02}\gamma_f u_{max} \tag{3-12}$$

Figure 3-16 shows schematically the penetration for a filter cake of increasing thickness $u_2 = \lambda l_b$ and a filter cake of increasing thickness to u_{max}, which remains constant thereafter. It can be shown that

$$z_u = \frac{\gamma_f}{\gamma_{fs}} u_{max} \left(i_{02} + \frac{1}{\lambda} i_{01} \right) \tag{3-13}$$

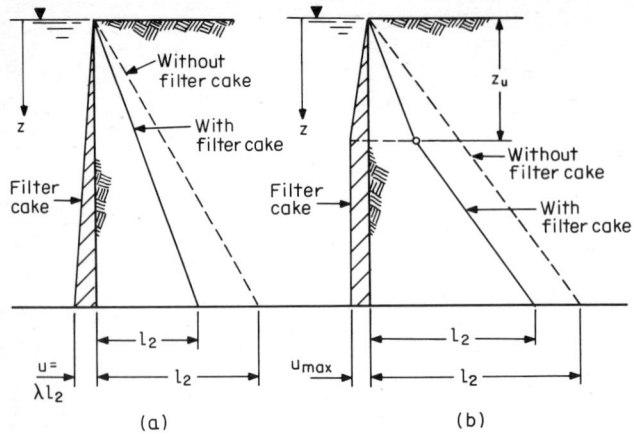

Fig. 3-16 Zones and shape of penetration: (*a*) filter cake of increasing thickness; (*b*) filter cake of limited thickness.

Comments on Stagnation-Gradient Tests This test is a time-dependent process; i.e., the results depend on the time at which measurements are made. Furthermore, the test accommodates situations exemplified by rheological blocking or deep filtration, and its main purpose is to disclose the probable penetration at the site and the reduction in the effective thrust of the slurry caused by a relatively permeable filter cake. It is not a permeability test. On the other hand, the test is useful if high-quality undisturbed soil samples can be obtained, especially from considerable depths. Where this is not possible, suitable samples may be prepared in the laboratory with the same gradation as the soil in situ and compacted to a similar density. The length of the sample is quite important since it should not be completely penetrated by the slurry in the pipe; hence some previous experience with the entire test usually is necessary and useful.

Example 3-1 The analysis of stability for a trench in sand is investigated assuming that the fluid penetrates through a relatively permeable filter cake of limited thickness, as shown in Fig. 3-16*b*. The soil has the following properties:

$$\phi = 30° \qquad \gamma = 120 \text{ lb/ft}^3$$
$$n_1 = 30\% \qquad \gamma_s = 150 \text{ lb/ft}^3$$

The bentonite concentration in the fluid gives $\gamma_f = 66$ lb/ft³, which is raised by mixing with soil aggregate to $\gamma_{fs} = 75$ lb/ft³. The concentration factor is $m = 0.10$ by volume. The porosity of the filter cake is taken the same as n_1. The trench has an actual depth of 15 ft. On the basis of the impermeable-membrane theory, the analysis of stability merely requires reference to Eq. (2-7), by which the trench is found stable with a factor of safety $F = 1.9$ regardless of depth. The safety factor is estimated for $\gamma_f = 66$ lb/ft³.

Now the analysis is considered for a relatively permeable cake. Stagnation-gradient tests give $i_{01} = 2$ and $i_{02} = 10$. The maximum thickness of filter cake $u_{\max}$ is taken as 0.4 in. The analysis is carried out according to the following steps.

Soil-Slurry Interaction

STEP 1. Compute λ either from Fig. 3-15 or from Eq. (2-29). For $m = 0.10$ and $n = 0.30$, find $\lambda = 0.05$.

STEP 2. Calculate z_u from Eq. (3-13)

$$z_u = \tfrac{66}{75}(0.4)\left(10 + \frac{2}{0.05}\right) = 17.5 \text{ in}$$

STEP 3. Referring to Fig. E3-1, compute the penetration at depth of 15 ft

$$l = \tfrac{15}{2}(1.20) = 9 \text{ ft} = 108 \text{ in}$$

(Note that 1.20 is the specific gravity corresponding to $\gamma_{fs} = 75 \text{ lb/ft}^3$.)

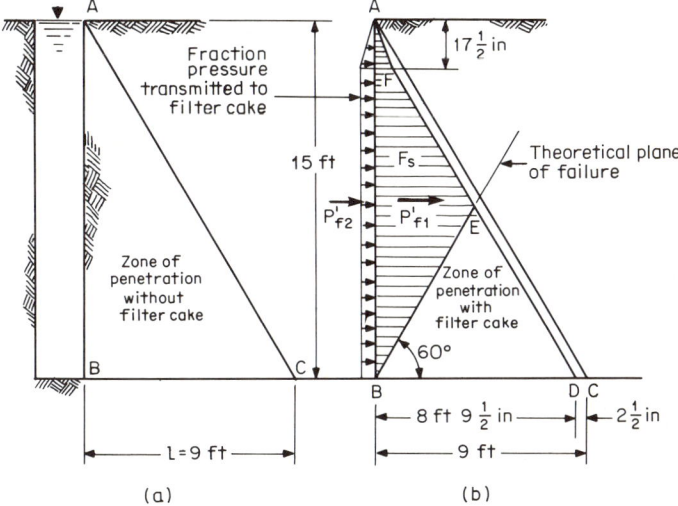

Fig. E3-1 Stability analysis of trench under permeable filter cake.

STEP 4. At depth of 17.5 in find the penetration with and without filter cake:

$$l = \begin{cases} \dfrac{u}{\lambda} = \dfrac{0.4}{0.05} = 8 \text{ in} & \text{with filter cake} \\[2mm] \dfrac{17.5}{2}(1.20) = 10.5 \text{ in} & \text{without filter cake} \end{cases}$$

Note that the difference $10.5 - 8 = 2.5$ in is the distance DC in Fig. E3-1b.

STEP 5. Calculate the area of penetration $ABEF = F_2$ and compute P'_{f1}. This area is approximately 32.5 ft². Then

$$P'_{f1} = 2(32.5)(66) = 4290 \text{ lb/ft of trench}$$

STEP 6. Estimate the fraction force P'_{f2}. First find

$$p'_{f2} = i_{02}\gamma_{fs}u = 10(75)(\tfrac{0.4}{12}) = 25 \text{ lb/ft}^2$$

at depth of 17.5 in and constant thereafter. Hence

$$P'_{f2} = 25(14.27) = 360 \text{ lb/ft}$$

The total force mobilized against the face is thus 4650 lb/ft compared with the available hydrostatic thrust of 8450 lb/ft. Since the actual active thrust exerted by the soil is 4500 lb/ft, it is evident that the factor of safety is now reduced to 1.03 and the trench is almost on the verge of failure.

3-5 EFFECT OF PENETRATION ON SOIL

Effect on Clay

Excavations in clay generally are exemplified by surface filtration. Filtration is not likely to occur in soil of very low permeability, and the protective cake will not be formed. However, the low permeability of the soil is sufficient to keep slurry loss to a minimum, and in many instances a trench in clay can be excavated using only plain water as slurry merely to exert the thrust against the face.

When a cut is open, the lateral stress along the vertical face is reduced from the value at rest to a value close to zero, and along this section the pore pressure is negative, i.e., in tension. As the trench is filled with slurry, the latter acts against the face and the two pressures tend to equalize. This means that the soil absorbs water near the face, and this is accompanied by a corresponding loss of strength. The reduction in soil strength is observed very near the face, where there is a marked increase in water content. The softening soon dissipates and usually does not extend more than 1 in from the face.

The softening process is shown by the results of Table 3-1, representing tests made on $1\frac{1}{2}$-in-diameter samples of London clay, LL = 76 and PL = 28. Before testing, the samples were immersed in a 4% bentonite suspension prepared from tap water.

Similar results were obtained for $1\frac{1}{2}$-in-diameter samples tested in 6 and 8% bentonite slurries. The loss of strength was shown to be considerably less for

TABLE 3-1 Results of Triaxial Tests on Clay Samples (from Nash and Jones, 1963)

TEST DATA: London clay, $1\frac{1}{2}$ in diameter, LL = 76 and PL = 28, immersed in 4% bentonite mixed in tap water

Depth of sample	Natural water content $w, \%$	Density, lb/ft³	Water content after immersion, %	Confining stress, lb/in²	Undrained shear strength s_u, lb/ft²	Immersion time, h
13 ft 6 in–14 ft 8 in	32	118	....	10	1100	0
			38.9	10	130	4
			45.0	10	90	8
	34.4	115	....	20	770	0
			43.8	20	130	4
			43.2	20	90	8
25 ft–25 ft 10 in	29.4	121	....	30	2150	0
			35.6	30	50	4
				30	70	8

4-in-diameter samples, where the water content was markedly increased at the edges only. For these samples the initial water content of 27.1 percent was increased to 28.8 percent at the center, to 29.1 percent 1 in from the edge, and to 40 percent $\frac{1}{4}$ in from the face (Nash and Jones, 1963). The samples were immersed in a 6% bentonite slurry for 24 h under a confining stress of 30 lb/in², and the results were similar to those observed when clay was immersed in plain water.

The softening very near the face is not only a laboratory phenomenon but is also observed in actual construction. As long as it does not extend more than a few inches, it has negligible effect, if any, upon the stability of the trench. On the other hand, a small permeability in a clay matrix causes colloid-sized particles to be deposited at the interface, restricting further penetration. Impermeable membranes at the junction are confirmed by direct observations. Nash and Jones (1963) have reported measured permeability coefficients of the order of 2.3×10^{-9} cm/s for films in contact with London clay.

Effect on Granular Soil

Figure 3-17 shows results of tests on samples of coarse sand and sandy gravel. In the natural state and water content this soil had an angle of friction of about 35°. When the samples were saturated with slurry, the friction angle was reduced to 30° (Müller-Kirchenbauer, 1972). Similar results have been reported by Farkas (1971).

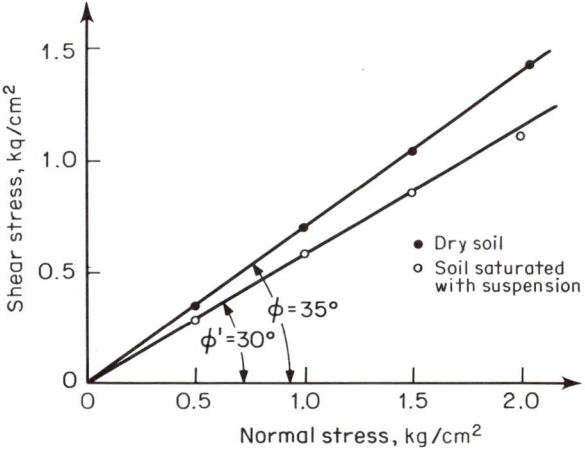

Fig. 3-17 Reduction in the angle of friction due to saturation with bentonite suspension. (*From Müller-Kirchenbauer, 1972.*)

If this situation could arise in the field, sliding along the failure surface in the zone impregnated with slurry should be considered for the reduced angle of friction. It is not always clear under what conditions to expect these changes in the properties of granular soil, except for the following comments:

1. The interaction of granular soil and slurry is either deep filtration or

rheological blocking. Unless the ground is an open bed, the former process is likely to dominate during excavation. The slurry is filtered through the cake, so that only diluted liquid and sometimes plain water penetrate the soil pores. The gel strength of this liquid is negligible for all practical purposes.

2. The surfaces of soil particles generally are contaminated with various ions and possibly other materials, and this might be expected to contribute to the frictional resistance. When slurry penetrates soil, these contaminants are largely squeezed out from between the actual points of contact, although a small quantity of the initial material still remains to influence the shear strength at the contact junction. A colloidal suspension is very slippery, even when diluted. Such a coating provides a thin lubricating layer and may act as boundary lubricant along the contact surface.

3. For the more contaminated surfaces water increases the friction, i.e., acts as an antilubricant. The water will disrupt any contaminating layer, reduce its effectiveness as a lubricant, and thereby increase interparticle friction. However, for clean surfaces water may not have the same effect. It might appear that if a thin coating of diluted bentonite slurry covers a soil particle, water will tend to disrupt it, but it is not known at what point this action will take effect.

4. As the surfaces of particles become rougher, the effects of the cleaning process on friction are reduced and very rough surfaces have the same degree of friction independent of surface cleanliness. This means that the ability of a colloidal slurry to lubricate the particle surface is reduced as the surface roughness is increased; hence more reduction in interparticle friction must be expected for particles of relatively smooth surface.

3-6 PERMEABILITY OF GRAVEL BED PENETRATED WITH SLURRY

The mechanism of rheological blocking and the associated process can be disclosed in laboratory tests (Jefferis, 1972). A thixotropic slurry penetrating a gravel bed will eventually seal it through a process of gelation provided the voids are not very large. The seal, however, may not be formed for some time until a sufficient penetration occurs, resulting in a corresponding slurry loss toward the ground. Sometimes this may lead to a loss of stability because of a drop in the hydrostatic head as slurry flows toward the ground. The resulting effects warrant the complete understanding of the problem.

Piezometer Tests

Neither Eq. (3-1) nor the stagnation tests explain the mechanism of rheological blocking. Within a continuous gravel bed the penetration is stopped either when colloid particles accumulate in the voids near the face, much as they do in fine material, or by letting the slurry flow as a fluid under the applied hydraulic gradient until its gel strength restricts further flow and forms a continuous seal downstream. Permeability reduction by filter cake means a high rate of flow, initially, which is reduced as soon as the filter cake is formed, as shown in Fig. 3-4. Thus, in the initial stages the permeability of the face decreases with time, whereas if the slurry sets in the voids inside the gravel bed, the permeability is independent of time.

These two mechanisms are revealed in piezometer tests (Jefferis, 1972). If in these tests the slurry remains fluid (rheological blocking), the effective radius of the piezometer increases during inflow to the sample as a void is formed around the tip. During outflow from the sample the slurry moves toward the piezometer to form a filter cake. Thus the measured permeability will increase with inflow and decrease with outflow.

The apparatus used for this test (Fig. 3-18) consists of a tank fitted with porous tiles across two opposite faces and connected to a reservoir of water. The piezometer has a diameter of $\frac{3}{4}$ in (1.9 cm). Flow rates are measured with a horizontal flow gauge. Initially the tank is partially filled with a slurry, and then gravel is gradually introduced and carefully tamped down. The pressure in the piezometer is subsequently raised, and the corresponding inflow rate is measured over a certain period. The application of pressure is continued in increments until the sample fails, presumably by cracking, although the failure surface is not identified due to the weakness of the material.

Fig. 3-18 Apparatus used to conduct piezometer tests of bentonite slurry in continuous gravel band. (*From Jefferis. 1972.*)

Results of Piezometer Tests Figures 3-19 and 3-20 show results of tests carried out by Jefferis (1972). The flow rate is plotted vs. the head. It is evident that for both cases the flow rate is independent of the direction of flow and dependent only upon the applied head; i.e., the inflow and outflow permeabilities of the slurry-gravel system are the same. The measured permeabilities are shown in Table 3-2.

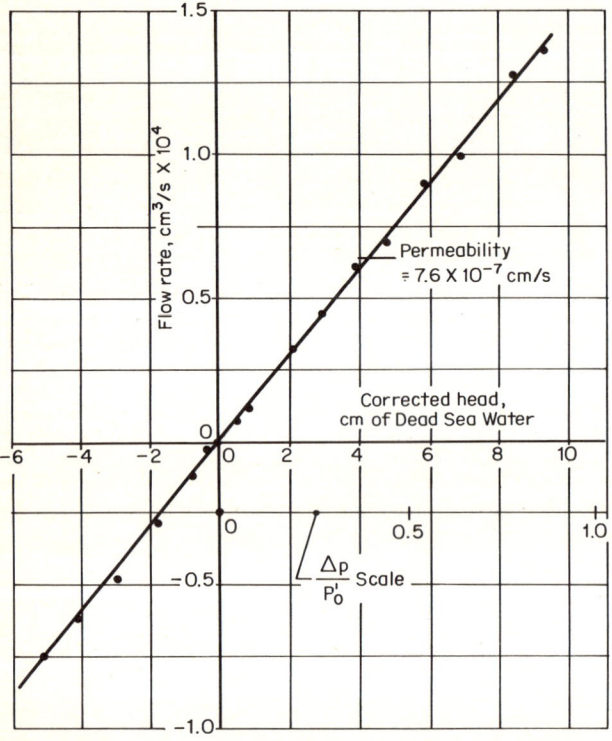

Fig. 3-19 Results of piezometer tests with dry-clay slurry in passing $\frac{3}{4}$-in gravel. (*From Jefferis, 1972.*)

For the passing $\frac{3}{4}$-in sample the value of p'_0 is calculated for homogeneous material. Figure 3-19 includes a scale for $\Delta p/p'_0$, and evidently the sample failed when this ratio reached 1, which is the value predicted for incompressible material by Bjerrum and Nash (1972). However, for the $\frac{3}{4}$-in no. 7 (ASTM) gravel no estimate of p'_0 could be made as the gravel formed a continuous structure with the slurry supported in the voids; hence this sample was not tested to failure.

The conclusion from these tests is that rheological blocking is the mechanism of slurry-impregnated gravel beds. The slurry seals the gravel by forming a gel structure in the voids, and this blocking results in a significant reduction in the permeability of the system. The tests give no indication of the blowout gradient.

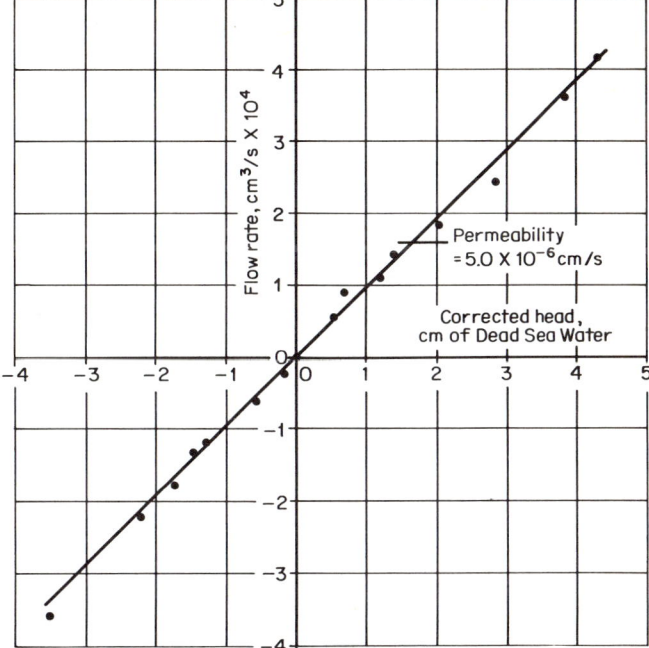

Fig. 3-20 Results of piezometer tests with dry-clay slurry in $\frac{3}{4}$-in no. 7 (ASTM) gravel. (*From Jefferis, 1972.*)

TABLE 3-2 Permeabilities of Gravel Band (from Jefferis, 1972)

Grading	k, cm/s	
	Slurry with gravel	Gravel without slurry
Passing $\frac{3}{4}$ in	7.6×10^{-7}	3×10^{-4}
$\frac{3}{4}$-in no. 7 (ASTM)	5.0×10^{-6}	5.0

Effect of Clay on the Permeability of Gravel

A relatively small quantity of clay in a soil can dominate its permeability (Nash, 1972). If the fine fraction of the soil includes true clay particles, the specific surface (total surface area of particles per unit volume of solids) is far greater than the value given for equivalent cubes or spheres. For instance, a montmorillonite clay broken down to nearly unit-cell size in flat plates might have a theoretical specific surface of about 800 m²/g (Grim, 1953).

To demonstrate the effect of clay on the permeability of gravel Nash (1972) carried out the following simple test. Three samples of gravel having permeabilities of 3.4×10^{-4}, 9.5×10^{-1}, and 4.2×10^{-3} cm/s were gradually mixed with clay in increasing quantities. As long as the percentage of clay in the gravel was relatively low, there was no significant change in the permeability of the system. When the clay ratio approached 20 percent by volume, the permeability dropped rapidly to about 5×10^{-8} cm/s for all three samples.

For these effects to occur, the clay must be uniformly distributed throughout the gravel bed. For the permeability to undergo substantial reduction, it is only necessary to block the pore spaces of the gravel. Until there is sufficient clay to do this, the permeability remains high. With a clay ratio from 15 to 25 percent of the total volume, the gravel attains the characteristics of a well-packed system, which can be looked upon as inert plums in clay matrix having the permeability of the clay itself. This is very useful in designing earth backfills for cutoff walls.

3-7 BLOWOUT GRADIENT

The single-tube equilibrium expressed by Eq. (3-1) implies that failure can occur by forcing the slurry out of the gravel voids when the applied hydraulic gradient delivers a force exceeding the gel strength of the slurry. The analysis of the blowout or failure gradient may follow the same principles which govern the flow of fluids through a granular medium (Carman, 1937; Jefferis, 1972). Such a medium can be regarded as a series of tortuous interconnected channels in which the flow generally is not parallel to the direction of the applied gradient. The effect of tortuosity is to increase the length of channels relative to the length of the medium. Thus, if the interconnections are ignored, a channel is represented by a straight tube inclined at an angle with respect to the direction of the applied gradient.

Prediction of Failure Gradient

Figure 3-21 shows a tube in a granular medium. The volume of the tube is $al/(\cos \theta_i)$. For a cross-sectional area A of the medium the void volume is Aln; hence

$$\sum_{i=1}^{r} \frac{a_i l}{\cos \theta_i} = Aln \tag{3-14}$$

in which r is the number of tubes in area A. If all these tubes can be represented by a single channel of area A_m inclined at an angle θ, then $A_m = An \cos \theta$. If the same bed consists of a gravel-slurry mixture, the hydraulic gra-

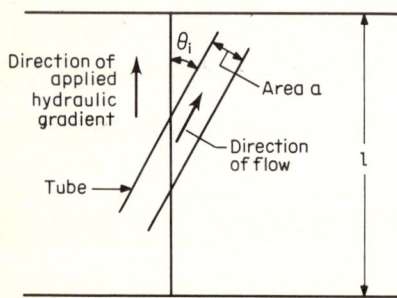

Fig. 3-21 Flow through a granular bed.

dient will exert a force tending to displace the slurry in area A_m. This is due to the pressure difference $\Delta p = \gamma_f h = \gamma_f i l$. The total force is $F = \Delta p \, A_m$, or

$$F = (\gamma_f i l)(A n \cos \theta) \tag{3-15}$$

If S is the specific surface of the gravel, the area of the channel walls is AlS. Evidently, the maximum force which the slurry can withstand is

$$F = \tau_f A l S \tag{3-16}$$

in which τ_f is the gel strength of the slurry in the pores of the gravel. From (3-15) and (3-16) it follows that

$$i = \frac{\tau_f S}{\gamma_f n \cos \theta} \tag{3-17}$$

If the flow is not horizontal, another term will appear in the right-hand side of Eq. (3-17), so that

$$i = \frac{\tau_f S}{\gamma_f n \cos \theta} + \frac{\gamma_f'}{\gamma_f} \tag{3-17a}$$

where $\gamma_f' = \gamma_f - \gamma_w$ is the "effective" weight of slurry. The term γ_f'/γ_f usually is very small and may be disregarded. If the gravel particles are idealized as spheres of diameter D, $S = \pi D^2 N$, where N is the number of particles per unit volume; but also $1 - n = N\pi D^3/6$, from which it follows that

$$S = \frac{6(1 - n)}{D} \tag{3-18}$$

Combining Eqs. (3-17) and (3-18) yields

$$i = \frac{6\tau_f}{\cos \theta} \frac{1}{D\gamma_f} \frac{1 - n}{n} \tag{3-19}$$

Typical values of the tortuosity angle θ for viscous flow through a granular medium range from 48 to 51° (Carman, 1937), so that the term $6/(\cos \theta)$ varies numerically from 9 to 9.5. A useful expression is also obtained for the theoretical penetration l by noting that $i = \Delta p/\gamma_f l$, so that Eq. (3-19) becomes

$$l = \frac{\Delta p}{\tau_f} \frac{D n \cos \theta}{6(1 - n)} \tag{3-20}$$

The foregoing relations have practical value only if it is possible to estimate τ_f and D. The value of D is sometimes taken as the mean size of the upper and lower limits of the gravel fraction, but in some instances it can be taken as the D_{20} size. Jefferis (1972) recommends taking τ_f equal to $K\tau_{10}$, where τ_{10} is the 10-min gel strength of the slurry and K is an empirical factor. From certain permeameter tests the average value of K was 2.7, but tube tests gave a value of K consistently lower and close to 1.8. On this basis, it might be reasonable to use $\tau_f = 2\tau_{10}$ as a first approximation of the blowout gradient.

Nash (1974) used the τ_{10} value for blowout tests, and this gave consistent results. When the theory was checked against laboratory "blowouts," the critical gradient was almost 3 times greater than the value given by Eq. (3-19), implying that the use of the 10-min strength was too conservative. The same investigator also conducted tests to check Eq. (3-20) using the yield stress τ_0. This again gave theoretical penetrations in which l was underestimated by a factor of about 2.5.

The author has attempted to use the foregoing data in making theoretical predictions for blowout failures in earth cutoffs. However, more difficulties arise in this case if the backfill contains clay, which makes the in situ gel strength of the slurry very difficult to predict, and if the backfill is evenly graded, which makes the choice of a representative D value mere speculation.

3-8 CHANGES IN SLURRY DENSITY

During excavation an appreciable quantity of solids is left in the trench and is mixed with the slurry. This soil aggregate imparts to the slurry a much higher density than that of the original mix. Therefore, in the process of excavation the actual thrust of the slurry changes to include the effect of the extra material in it, and in many instances this increased density partially accounts for the stability of the trench.

The amount and type of soil particles left in the trench depend on (1) the gel strength of the slurry; (2) the method of excavation and type of equipment; (3) the gradation of the surrounding soil; and (4) the method of materials handling, including the method of separating the solids from the slurry.

Figure 3-22 shows the average change in the density of the initial slurry at the end of excavation. These results are taken from actual construction and represent average conditions but different excavating techniques. Graphs of this type are useful and provide statistical data. Where the excavation is in ground which does not exhibit major variations in its profile, this information is immediately useful and is obtained by monitoring the excavation of the first panel.

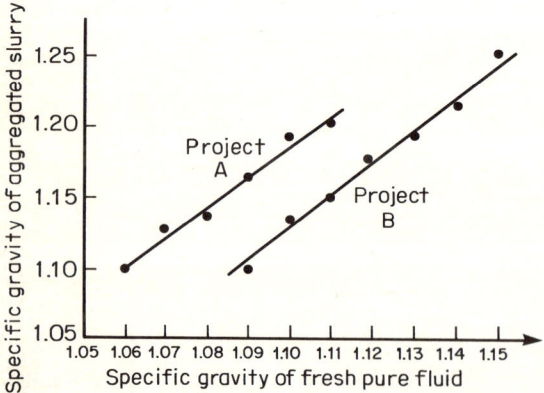

Fig. 3-22 Variation in unit weight of slurry during excavation. Data from actual construction.

3-9 FILTER CAKE

As mentioned earlier, the formation of the filter cake is time-dependent. The filter cake is assumed to be formed when there is a sudden decrease in the rate of slurry loss. For a relatively impermeable cake the flow continues as the extent of penetration tends to diminish. It can be inferred from this process that the thickness of the filter cake increases not only with depth but also with time under continuous flow. Thus, the measurements taken in a stagnation-gradient test refer to the exact time this information was recorded. For the initial phase without filter cake, a few minutes and sometimes seconds are sufficient since the cake builds up rather rapidly. The second phase may take several hours, and sometimes days, until the extent of penetration becomes nearly constant. Usually, it is a matter of judgment to decide when to conclude the test, after considering the type of soil and the practical aspects of construction.

In general, two theories have been suggested to explain the formation of filter cake. The first, and better understood, suggests that under pressure difference the slurry is filtered through the exposed face and deposits colloid aggregate and fine soil particles in the soil pores.

The second theory is based on the negative charge of montmorillonite clay particles. When the slurry is delivered to the trench, a potential difference is spontaneously created through electrochemical effects, generated from the interaction of slurry, groundwater, and soil, which stimulate bentonite particles to move vigorously toward the face of the trench to form deposits.

For continuous filtration the filter-cake process can be analyzed theoretically from a consideration of the three times given in Fig. 3-23 (Nash, 1974). This leads to an expression for the thickness u of the filter cake at depth z_1 and time t

$$u = \sqrt{\frac{2k_c(1 - n_f)(\gamma_f z_1 - \gamma_w z_2)}{(n_f - n_c)\gamma_w}} \sqrt{t}$$

where n_f, γ_f relate to slurry
k_c, n_c relate to cake
γ_w, z_2 relate to groundwater

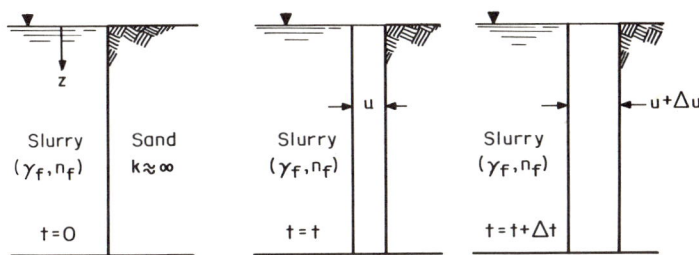

Fig. 3-23 Theoretical formation of filter cake with time.

Filter-Cake Test

Figure 3-24 shows results from a filter-cake test in fine granular soil, carried out using bentonite concentrations 6, 8, and 10 percent under pressure differences 0.5, 1.0, and 2.0 kg/cm², respectively. The data are recorded for 1, 5, 10, 20, 30, 60, and 120 min after starting the test. For all practical purposes this filter cake is seen to reach its full strength approximately 1 min after the slurry is delivered into the trench. The coefficient of permeability varies with time, but approximately 4 min after the formation of filter cake it attains a nearly constant value of 1×10^{-8} cm/s.

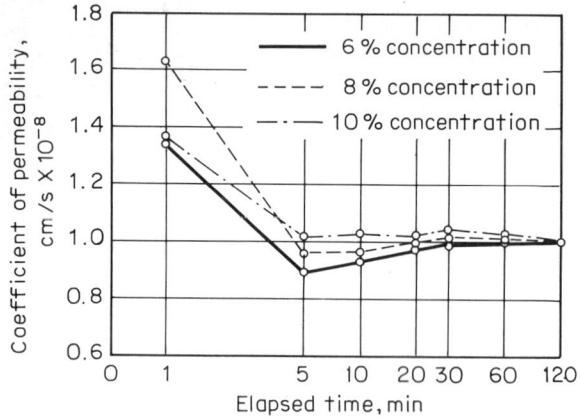

Fig. 3-24 Variation in the permeability of filter cake with time. Data obtained with Japanese bentonite.

Such a filter cake is essentially impermeable and watertight and allows the full thrust of the slurry to be exerted against the face. If it can be provided in actual field excavations, it is not necessary to consider a reduction in the theoretical thrust for stability analyses, and the actual slurry loss should be negligible.

3-10 ELECTROOSMOTIC PHENOMENA

The concentration of bentonite particles at the interface where they build a filter cake must be distinguished from the process of flocculation. The latter is essentially an electrochemical phenomenon occurring when electric charges on the surface of clay particles are reduced to a level where repulsion between them is small. Under these conditions collision between particles leads to permanent coagulation, which can be terminated only by altering the basic composition of the suspending slurry (see also Chap. 4). Unlike a flocculated system, the filter cake from a bentonite suspension usually redisperses in water to give a system exhibiting the original flow properties.

According to the second theory regarding the formation of filter cake, when a bentonite slurry is in contact with soil, electrical phenomena are spontaneously created and their action contributes to the buildup of the cake. The existence of negative electric charges on the surface of clay particles causes them to move under the influence of an electric potential. Veder (1961) has shown that a deposit similar to the filter cake can be created by placing electrodes in the mud and the soil and applying an electric potential across them. When the anode is placed in the soil, electrophoretic migration of clay particles causes them to be deposited at the interface, and vice versa. It thus is possible to modify the filter cake formed at the junction merely by creating a passage of continuous electric current. If the direction of flow is the same as the natural flow existing between the slurry and the soil, more particles are deposited on the cake. If the flow is arranged in the opposite direction, the cake is gradually reduced to the point of disappearing. Its destruction may then disturb the equilibrium of the exposed face and cause it to collapse.

There is an important field observation which supports the theory of electric charges: some trenches supported with bentonite slurry caved in when the steel reinforcing cages were welded while inserted in the trench, creating a short circuit with the surrounding soil. This experience is supplemented by small-scale tests showing that if an electrode is placed in a slurry and another in adjacent saturated sand, a high-impedance voltmeter will record a potential difference (Veder, 1969).

Tests on some cylindrical sand samples submerged in thin bentonite suspensions show that initially the sand is stable because of the formation of a filter cake around the specimen. However, in the course of time the cylinders may bulge and eventually collapse (Veder, 1969). This observation raises the question of rheological deformations in the soil. Using this as a starting point, it may be possible to explain the effect of electroosmotic forces by assuming that they initially set to keep the soil in a stable condition, but with time a small margin of safety is compensated by rheological deformations.

It is rather difficult to understand the application of electrochemical phenomena to trench-stability problems. An example of electroosmotic flow is the application of electric potential to saturated silts causing the groundwater to flow in one direction or another (Casagrande, 1949). Nash and Jones (1963) have reported similar findings in laboratory tests and offered a probable explanation for these effects (Xanthakos, 1974). On the other hand, if charged bentonite particles could migrate, it is difficult to say whether they would do so toward the face or away from it. Model tests reported by Nash and Jones (1963) show that when the potential difference between bentonite and saturated sand was reduced to zero, decrease in stability was not encountered.

There seems to be an area of doubt about how these potentials may contribute to trench stability; however, the following comments are appropriate: (1) fluid flowing through a filter cake implies a drop in the hydraulic head, and this may lead to a rise in the electric potential; (2) the filter cake sometimes acts as a membrane which is selectively permeable to anions or cations, and this causes an electric potential difference; (3) the existence of a voltage is the result of fluid flow through filtration rather than its cause; however, an externally applied voltage can help form a filter cake or cause its destruction.

3-11 FLUCTUATIONS IN SLURRY LEVEL AND WATER TABLE

One of the advantages of the slurry-trench method is the feasibility of constructing underground walls without altering the groundwater level. In granular material the location of the water table may sometimes be critical, and, as was shown in Chap. 2, the slurry level generally should be at least 2 m (6 ft) higher than the water table after making an allowance for some fluctuations during excavation. This differential usually is provided by lowering the groundwater level somewhat in the vicinity of the excavation. As an alternative, the existing grade may be raised by building a berm along the alignment of the trench to provide a higher construction platform if this solution is feasible and more economical.

Slurry trenches on urban sites are frequently excavated from a platform below the street level in connection with the so-called *under-the-roof method* to keep streets usable during construction. Thus, the excavation of the trench begins at a sublevel some 8 m, or 25 ft, below grade in order to provide the necessary headroom for the excavating equipment. In this case the groundwater level invariably is higher than the top of the trench, rendering the operation impracticable from the viewpoint of stability. The excavation now must be preceded by dewatering to a level safely below the slurry level. If thereafter dewatering should continue simultaneously with trenching, the location of dewatering points must be carefully established so that the system does not draw slurry from the trench.

Under tidal conditions the response of the water table generally is spontaneous and must be considered together with the induced tidal flow; hence this combination may influence trench stability to a very different degree. Moving groundwater reduces the differential hydrostatic head by the equivalent head of seepage pressure, and therefore dynamic flow must be considered in the analysis. If the alignment of a trench intercepts groundwater flow, for some length the trench will function as dam and change the flow pattern accordingly. Similar changes in the flow and the level of groundwater will occur where a long trench is excavated across a valley. The same effects have also been observed at the closure position on the perimeter of a site which is completely enclosed by a slurry trench. A typical example is the incident recorded during the construction of a deep basement; in this case the water table inside the enclosed area rose more than 1 m (3.5 ft) above the design level toward the end of construction, and the face collapsed while the last panel was excavated.

Consequences of Lowering the Slurry Level

Figure 3-25 shows the effects of the combined action of slurry loss and trench infiltration by groundwater. In this example the excavation is carried out in

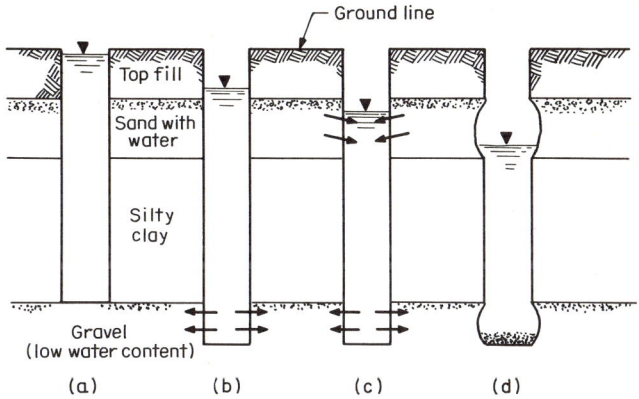

Fig. 3-25 Consequence of lowering the slurry level. Loss of ground caused by water infiltration.

multilayered soil, which is either stiff or dense, so that the excavation is stable against slip failure.

Trenching begins as shown in Fig. 3-25a and continues through the water-bearing sand and the silty clay without incident. When the excavation reaches the gravel layer, slurry begins to escape toward the ground, as shown in Fig. 3-25b by the direction of the arrows. This situation is intensified because of the high permeability of the gravel, the absence of groundwater in this layer, and the favorable pressure differential. Thus, appreciable slurry loss occurs before rheological blocking can stop the flow and causes the slurry level to drop below the water table, as shown in Fig. 3-25c. At this stage pore-water pressure exerts an inward force on the filter cake and causes its disintegration. Groundwater now flows freely into the trench, the sand caves in, as shown in Fig. 3-25d, and the excavation ends with a loss of ground. Such incidents are common and difficult to deal with, even if there is an extra supply of slurry at hand to restore the slurry level. Besides the loss of ground, which can propagate into the layers above and below the initial occurrence, mixing slurry with soil causes its contamination and may lead to the loss of colloidal stability.

A serious drop in the slurry level can also be caused regardless of the ground conditions around an excavation. An example is misjudgment in the supply of slurry in connection with reverse circulation (in this situation the slurry from a panel is continuously recirculated to transport excavated materials to the separation units). For a panel 0.6 m (2 ft) wide and 3 m (10 ft) long it is conceivable that a suction pump will recirculate the slurry at a rate of 4 m^3/min (about 140 ft^3/min). If for any reason the supply of slurry from the central plant is interrupted, the slurry level in the trench will be lowered at the rate of 2.2 m/min (or 7 ft/min). Unless this is immediately remedied, it will take only a short time for the slurry level to drop below the water table, and at this stage collapse is imminent regardless of other considerations.

Another example demonstrating the effect of construction operations is excavation with a large clamshell bucket. In many cases the total volume occupied by the grab and the upper chamber is several cubic meters. Every time the bucket is lifted to discharge the excavated soil the slurry level may drop considerably, merely to fill the volume released when the grab is out of the panel. Unless special procedures are followed (see also the sections on excavating techniques), erratic fluctuations in the slurry level can influence the stability of the trench.

REVIEW PROBLEMS

3-1 Explain thixotropy and its relevance to trench-stability problems. Discuss examples in soil engineering where thixotropy is important.

3-2 A trench is excavated in gravel. Estimate the minimum shear strength of filter cake necessary to (a) prevent grains from falling into the slurry and (b) prevent peel-off. Use a tangential angle 40°, $\gamma = 145$ lb/ft³, $\gamma_f = 75$ lb/ft³, and average gravel size 3 in.

3-3 Repeat Example 3-1 keeping all parameters the same but use $i_{01} = 20$.

3-4 Repeat Example 3-1 assuming that the thickness of filter cake increases linearly with depth as shown in Fig. 3-16a.

3-5 Two stagnation-gradient tests are carried out on a soil with a porosity ratio of 0.25. The slurry weighs 72 lb/ft³ and has a colloid concentration of 8 percent by volume. The first test gives $h_a = 50$ in and $l_a = 25$ in. Direct measurements from the second test give $h' = 10$ in, $l_b = 10$ in, and $h'' = 7$ in. Estimate the stagnation gradients and the thickness of filter cake.

3-6 The 10-min gel strength of a slurry is 50 lb per 100 ft². The slurry weighs 70 lb/ft³ and is used in soil with $D_{20} = \frac{1}{4}$ in and a porosity ratio of 0.30. Estimate the blowout gradient if $\theta = 50°$ and $K = 2$.

3-7 Find the theoretical penetration for Prob. 3-6 under a pressure head of 20 ft. State your assumptions.

3-8 Estimate the blowout gradient for gravel of average size 2 mm. State your assumptions.

3-9 Describe three significant actions that may cause the filter cake to disintegrate.

3-10 Conduct peel-off and stagnation-gradient tests. Obtain suitable samples by blending earth material in the laboratory. Develop appropriate procedures for recording data and results from the tests.

3-11 A slurry-trench excavation produces cuttings which can be approximated as 1-in-diameter uniform particles. If the excavation is interrupted from some reason, estimate the minimum gel strength of slurry required to keep the cuttings in suspension. State your assumptions.

3-12 Compare the stagnation-gradient tests with standard-permeability tests.

3-13 Discuss the requirements for the formation of filter cake.

3-14 A slurry trench is to be excavated near a building and to a depth below the existing foundation. The upper portion of the excavation is a layer of sand and gravel with water underlain by stiff clay. Explain how you would investigate the feasibility of the construction and list the main steps of the investigation.

3-15 For excavations in water-bearing formations it usually is necessary either to raise the top of the trench or to lower the water table. This problem, for example, is quite common in Brussels. Discuss in some detail the factors that may influence the decision.

3-16 List situations which may make slurry-trench excavation impracticable from the stability point of view.

3-17 Discuss the possible causes of lateral soil deformation in slurry-trench excavations, especially under varying conditions.

3-18 Give examples where you might consider the effect of the following factors on stability: (*a*) shear strength of slurry; (*b*) increase in the strength of the soil; (*c*) electroosmotic effects; (*d*) increased slurry weight due to soil retention; and (*e*) filter-cake strength.

3-19 Cite examples in which theory does not allow satisfactory predictions. Explain other procedures necessary to judge stability.

3-20 Discuss situations in which you might consider a factor of safety higher or lower than the customary overall safety margin. Be specific and give ample explanation.

REFERENCES

Bjerrum, L., 1954: Geotechnical Properties of Norwegian Marine Clay, *Geotechnique*, **4**:49.
────── and J. K. T. L. Nash, 1972: Hydraulic Fracture in Field Permeability Testing, *Geotechnique*, vol. 22, no. 2.
Carman, P. C., 1937: Fluid Flow through Granular Beds, *Trans. Inst. Chem. Eng. Lond.* p. 150.
Casagrande, L., 1949: Electro-osmosis in Soils, *Geotechnique*, **1**(3): 159-177.
DiBiagio, E., and F. Myrvoll, 1972: Full Scale Field Tests of a Slurry Trench Excavation in Clay, *Proc. 5th Eur. Conf. Soil Mech. Found. Eng., Madrid*, vol. 1, pp. 461-471.
Farkas, J., 1971: Stability of Slurry Trench Walls, *Proc. 4th Budapest Conf. Soil Mech.*, pp. 397-403.
Freundlich, H., 1935: "Thixotropy," Hermann, Paris.
Fyedorov, I. V., 1965: Slope Stability in Hydraulic Fill Structures, *Proc. 6th Int. Conf. Soil Mech. Found. Eng., Montreal*, vol. II, pp. 472-476.
Grim, R. E., 1953: "Clay Mineralogy," McGraw-Hill, New York.
Hutchinson, J. N., 1961: A Landslide on a Thin Layer of Quick Clay at Furra, Central Norway, *Geotechnique*, **11**: 69-94.
Hutchinson, M. T., et al., 1974: The Properties of Bentonite Slurries Used in Diaphragm Walling and Their Control, *Proc. Diaphragm Walls Anchorages. Inst. Civ. Eng., Lond.*
Jefferis, S. A., 1972: The Composition and Uses of Slurries in Civil Engineering Practice, Ph.D. Thesis, University of London.
Kruyt, H. R., 1952: "Colloid Science," vol. I, "Irreversible Systems," Elsevier, New York.
Lambe, T. W., 1958: The Structure of Compacted Clay, *Proc. ASCE*, vol. 64, no. SM2.
Marshall, C. E., 1949: "The Colloid Chemistry of the Silicate Minerals," vol. I, "Agronomy," Academic, New York.
Meyerhof, G. G., 1961: The Mechanism of Flow Slides in Cohesive Soils, *Geotechnique*, **7**:41-49.
Mitchell, J. K., 1956: The Importance of Structure to the Engineering Behavior of Clays, D.Sc. Dissertation, Massachusetts Institute of Technology, Cambridge, Mass.
──────, 1960: Fundamental Aspects of Thixotropy in Soils, *ASCE Soil Mech. Found. Div.*, June, **SM3**:19-52.
Müller-Kirchenbauer, H., 1972: Stability of Slurry Trenches, *Proc. 5th Eur. Conf. Soil Mech. Found. Eng., Madrid*, vol. 1, pp. 543-553.
Nash, J. K. T. L., 1972: The Design of Slurries for Use in Civil Engineering Projects, *1st Iranian Congr. Civ. Eng. Pahlavi Univ., Shirar, Iran.*
──────, 1974: The Stability of Trenches Filled with Fluids, *Symp. Underground Constr. Fluid Trenches, Univ. Ill., Chicago Circle.*
────── and G. K. Jones, 1963: "The Support of Trenches Using Fluid Mud," Butterworth's London.
Puller, M. J., 1974: Slurry Trench Stability: Theoretical and Practical Aspects, *Ground Eng.*, **7**(5):34-36.
Renard, J., 1969: The Use of Cast in Place Diaphragm Walls at Zemst Lock, *Proc. 7th Int. Conf. Soil Mech. Found. Eng., Mexico City, Spec. Sess.* 14, 15, pp. 30-36.
Tamaro, G., 1975: Chemically pretreated bentonite in slurry-trench excavations, personal communication.

Van Olphen, H., 1956: Forces between Suspended Bentonite Particles, Clays and Minerals, *Proc. 4th Natl. Conf. Clays Clay Min., Natl. Acad. Sci. Publ.* 456.

Veder, C., 1961: Discussion on Electrical Phenomena between Bentonite Mud and Cohesionless Material, *Proc. 5th Int. Conf. Soil Mech. Found. Eng., Paris,* vol. 3, pp. 146–149.

———, 1969: Testing Results on the Behavior of Bentonite Suspensions in Trenches, *Proc. 7th Int. Conf. Soil Mech. Found. Eng., Mexico City, Spec. Sess.* 14, 15, pp. 20–22.

Weiss, F., 1967: Die Standsicherheit flüssigkeitsgestützter Erdwände, "Bauingenieur-Parxis," Heft 70, Ernst, Berlin and Munich.

Whitman, R. V., 1968: Hydraulic Fills to Support Structural Loads, *Proc. Placement Improvement Soil Support Struct. Soil Mech. Found. Div., ASCE,* pp. 169–193.

Xanthakos: P. P., 1974: "Underground Construction, Fluid Trenches," Colleges of Engineering, University of Illinois, Chicago.

Chapter Four

TECHNOLOGY, PREPARATION, USES, AND CONTROL OF SLURRIES

The technology, preparation, and control of slurries have become a specialized field because of their complexity and because of new and different applications in underground works. The uses and functions of the early drilling muds are hardly the same in trenches excavated to build underground structures and foundations. Thus, among the problems associated with the present applications are (1) deeper structural excavations, in which the cost of slurries represents a serious factor; (2) interrelated effects regarding temperature, pressure, stability requirements, and displacement by concrete; (3) additional phases in the operation which did not exist in the early stages, e.g., mud treatment and disposal; and (4) the effect of slurry on the appearance and strength of the finished structure. Accordingly, slurry systems have received more attention, and specifications for the preparation and control of slurries are no longer the responsibility of colloid specialists but often are drafted by geotechnical and structural engineers.

4-1 INTRODUCTION TO SLURRY SYSTEMS

Until the early 1940s most operations were confined to water-well and oil-well drilling. For this type of work, the low-pH and phosphate muds were relatively simple and inexpensive. They were not suited to deeper drilling, however, since they could not tolerate severe contamination or higher temperatures. The development of lime fluids made deeper drilling possible in temperatures exceeding 200°F under high pressure and in the presence of contaminants. The advantage of lime muds were lower viscosity and gel, and since they were more inert to shale hydration, they resulted in considerable economy. Cooke (1963) has reported case histories where for the same conditions and depth the conversion from the low-pH phosphate to lime systems resulted in more than 50 percent reduction in the cost of slurry. However, with deeper drilling lime fluids became expensive, and again excessive gelation occurred. Probably the most important advance during the transition period from the simple phosphate to lime systems was the introduction of oil to increase drilling rates and improve the overall efficiency.

In deeper and more troublesome soils more complications resulted from the difficulty of well-bore stability, excessive hole enlargement, and expensive stuck-pipe problems requiring extensive watering back and reconditioning because of excessive clay solid uptake. Some of these problems were remedied with the introduction of cyclones and centrifuges as mechanical aids.

By the end of 1955 more developments had occurred, namely (1) the use of lost-circulation techniques including time-setting clay cements, diesel-oil–bentonite combinations, and various mechanical aids to reduce slurry loss to the ground; (2) invert-emulsion muds to protect the formation during all the phases (drilling, completion, and workovers); (3) materials for increasing the penetration rates; and (4) pressure lubricants to extend the bit-bearing life. By the late 1950s the slurry market included surfactant muds, shale-inhibited muds, and gypsum fluids. A further development was the application of materials of the ferrochrome lignosulfonate type. The year 1961 saw two important advances: the use of detergents and the introduction of chrome lignite–chrome lignosulfonate systems, also known as CL-CLS fluids. A brief summary of the main drilling fluids follows.

Detergents

Detergents are synthetic surface-active agents compatible with all water systems. Their use can be expected to influence the operation (1) by reducing the surface tension of the drilling fluid and filtrate, which in effect means an increase in the wettability of the latter and therefore a better penetration rate; (2) by reducing the tendency of cuttings to cling in place so that they can be removed by the jetting action more easily, and they pass over the shaker as larger and better-preserved samples; (3) by improving the stabilizing effect as the surfactant character of the detergent inhibits the clay hydration; (4) by improving productivity since low-surface-tension fluids are better recovered and less liable to cause water-blocking problems; and (5) by controlling viscosity and gel for a more efficient collection of entrained sand and clay cuttings.

Surfactant Muds

These were specifically developed to overcome the problem of drilling deep wells at high temperatures. Their basic constituent is a nonionic surface-active agent which is an ethylene oxide derivative of phenol used together with an electrolyte, a weighting agent, and a polyanionic fluid-loss agent. Thermal stability primarily is derived by excluding agents which contribute to cementation and through the protection afforded to the clays by the surfactant. Elimination of gelation and solidification at high temperatures generally is complete, and tolerable viscosity and fluid loss are attained economically. Applications of surfactant materials also include their use in salt water, where fluid loss, otherwise critical, is significantly improved in relation to the amount of surfactant added.

Shale-inhibited Fluids

These systems were designed to inhibit the hydration of shale more effectively than lime muds. Considerable cost improvements were achieved through the reduced uptake of formation clays which eliminated the need for water dilution and reweighting. The fluids basically are alkaline solutions consisting of lime, calcium chloride, and a thinner. The high degree of shale stability implied by these fluids allows massive shale formations to be drilled using muds of lower weight than is used with lime muds. The combined effect of these advantages is reflected in substantial cost savings (Cooke, 1963). However, since shale-inhibited muds are complex and have temperature limitations similar to those of lime muds, they lack the flexibility necessary for very deep drilling.

Gypsum Fluids

These were first used for drilling through massive anhydrite formations. They are relatively simple but exhibit high viscosities and gels which initially could not be controlled with conventional thinners. This situation was remedied with the development of thinners of the ferrochrome lignosulfonate type, which also improved the fluid-loss characteristics of the system, making lightweight gypsum muds practical for use in deep drilling. The main components are gypsum, a modified lignosulfonate, and carboxymethyl cellulose (CMC). Because of the low pH this fluid is more temperature-resistant than lime muds.

Invert-Emulsion Systems

A usual application of these muds is as workover and completion fluids to keep the formation free from damage. Other uses are (1) as stabilizing fluids in salt, gypsum, or potash formations when drilling holes; (2) as slim-hole drilling fluids to minimize differential pressure sticking; (3) as fluids for inserting casing; and (4) as a means of freeing wall-stuck pipes, collars, and casings.

Both oil-base and invert-emulsion muds are effective as soak fluids in stuck-pipe problems. Recent successes include the ability to withstand elevated temperatures of the order of 300°F and still exhibit gel and viscosity of controllable range.

CL-CLS Systems

The formulation of chrome lignite and its use with chrome lignosulfonate in the early 1960s gave the answer to simplicity and economy. This combination imparts great stability to the system at high temperatures and makes the use of gypsum, lime, quebracho, starch, and many other materials used for treating muds unnecessary. The CL-CLS mud is simple to maintain and can be prepared from fresh and brackish water or seawater. There are further advantages:

1. Improved inhibition of shale dispersion and hydration. Samples of shale tubes immersed for several days in gypsum fluid, distilled water, and CL-CLS fluids disclose that the soil remained better preserved in the CL-CLS system while the samples almost disintegrated in gypsum and distilled water (Cooke, 1963).

2. Improved control of flow properties. Figure 4-1a shows CL-CLS systems vs. gypsum and lime fluids. The highest-temperature pumpability limit is for the CL-CLS system which has 36 percent solids by volume. Figure 4-1b shows the gel strength of the CL-CLS system vs. lime and gypsum fluids after aging at 300°F for 16 h.

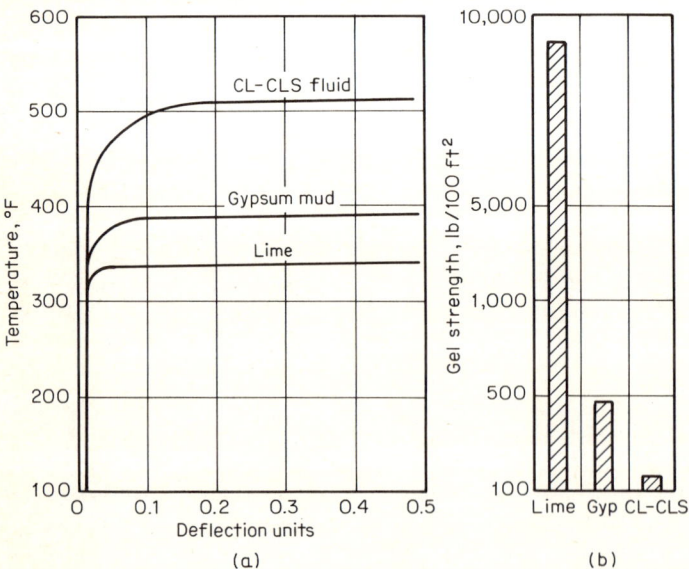

Fig. 4-1 Flow properties of CL-CLS systems vs. gypsum and lime fluids. (*From Cooke, 1963.*)

3. Improved filtration and fluid-loss characteristics. Figure 4-2 shows a CL-CLS system characterized by a downslope fluid-loss curve as the pressure increases and a calcium-treated mud for which the filtrate loss increases with pressure.

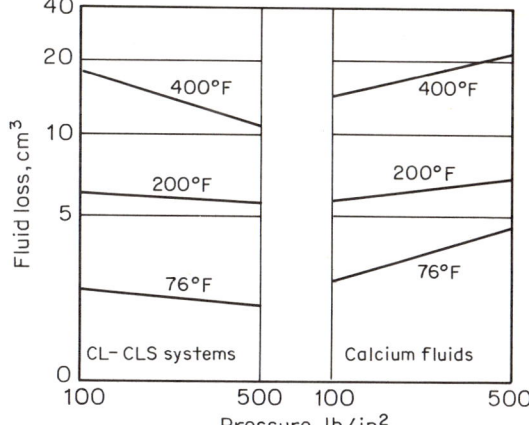

Fig. 4-2 Fluid-loss characteristics of compressible filter cakes; CL-CLS vs. calcium fluids. (*From Cooke, 1963.*)

4. Better protection of the hole afforded by the high rheological stability and simplicity of maintenance involving treatment with bentonite, caustic, and water.

The successful performance of this fluid is due to the synergistic effect of the two chrome complexes and their protective action on the mud solids, which control viscosity and fluid loss and also resist contaminants under extreme conditions.

4-2 FUNDAMENTALS OF CLAY COLLOIDAL SYSTEMS

Almost invariably slurries for trench excavations are clay colloidal systems. Such a system is obtained when a small amount of clay powder, e.g., bentonite, is stirred in a large volume of water. The powdered clay appears to dissolve in water just like common salt, but microscopic observations show that the clay actually produces homogeneous dispersion of tiny particles. If the particles do not settle in a reasonable time, the dispersion is called a *colloidal solution* or *sol*. When the dispersed particles are large and settle rather rapidly, the dispersion is called a *suspension*. The distinction is arbitrary; in principle there is little difference, and the terminology often is a matter of convenience. The particle size in natural-clay dispersions varies widely and covers a range which is typical for both sols and suspensions.

Materials which can be mixed with a liquid to form a colloidal dispersion are termed *colloids*. Bentonite is a widely used colloid material. A colloidal state is not necessarily a different state of matter, distinct from the crystalline state. Graham (1861) described gelatine as representative of the colloidal group. Although colloid is often taken to mean glue, the study of colloidal suspensions is not the study of glue types of materials. Many colloidally dispersed materials exist in the crystalline state, e.g., clays. In other instances the colloidal state is defined as a homogeneous system in which particles are dispersed kinetic units larger than simple small molecules. These kinetic units may be small soil particles, macromolecules, small droplets of liquids, or small gas bubbles.

Particle Interaction Dispersed clay particles collide frequently because of the motion of the particles in the system. Collision occurs even if the suspending system is at rest, but the particles can separate again. This situation changes completely if a small amount of salt (generally a few tenths of 1 percent) is added to the dispersion; upon collision the particles stick together and form agglomerates. However, when the salt is removed and the suspension stirred, the *flocks* disappear and the system is fully restored.

The total energy interaction between colloid particles can be found by the addition of repulsion and attraction forces. Repulsion has the features of an exponential function and remains finite for all values of the distance between particles. Attraction, however, decreases as an inverse power of the distance. For very small distances it goes to very large negative values, and consequently attraction will predominate at very small or at very large distances. At intermediate distances repulsion may govern, but whether this is actually the case will depend upon the real numerical values of attraction and repulsion. In principle there are two different types of total interaction, one with a maximum at intermediate distances and a minimum at larger distances and the other showing a monotonic decrease of energy with decreasing distance.

Flocculation and Gelation

Both these concepts have immediate practical implications in slurry-wall construction. A flocculated slurry is colloidally unstable and exhibits marked variation in the flow characteristics. Flocculation is of frequent occurrence in a slurry system and leads to particle association which is governed by three distinct modes: face-to-face (FF), edge-to-face (EF), and edge-to-edge (EE). Since the electric interaction energy is governed by each combination, a different geometry must be considered when summing the attraction of approaching plates (Vold, 1954). The net potential interaction varies with each mode of association, and the three forms may not occur simultaneously or to the same degree whenever a clay suspension is flocculated. The physical results from each mode have different effects and can change the properties of flocculated slurries in different ways.

FF association merely leads to thicker and probably larger flakes, whereas EE and EF association cause the formation of a three-dimensional voluminous structure like a house of cards, called a *cardhouse structure*. A representation of these systems is shown in Fig. 4-3. Among the three basic forms of flocculation, only the EE and EF types lead to agglomerates which actually are flocs. The thicker particles resulting from FF association are essentially aggregated forms. Aggregation and flocculation do not proceed simultaneously or to the same degree. Thus a colloidal system may well be dispersed but not deflocculated, or it may be deflocculated but not well dispersed.

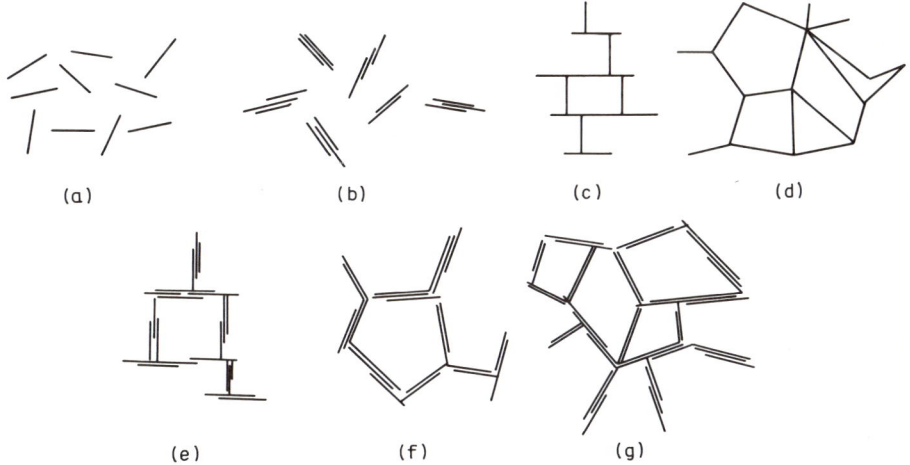

Fig. 4-3 Modes and terminology of particle association in colloidal suspensions: (*a*) dispersed and deflocculated; (*b*) aggregated and deflocculated (FF association or parallel aggregation); (*c*) EF-flocculated but dispersed; (*d*) EE-flocculated but dispersed; (*e*) EF-flocculated but aggregated; (*f*) EE-floccualted but aggregated; and (*g*) EF- and EE-flocculated but aggregated. (*From Van Olphen, 1963.*)

The viscosity of slurries generally increases when conglomerates are formed by EE and EF association but decreases when the particles become thicker by FF association. In more concentrated nonsalt systems or in less concentrated salt systems EE and EF association cause the formation of continuously linked cardhouse structures occupying the entire volume, and a gel is thus obtained. If FF association occurs simultaneously, the number of units forming the cardhouse structures is reduced and the yield stress is decreased.

Figure 4-4 shows the relative viscosity (viscosity of suspension divided by the viscosity of medium) as a function of the NaCl concentration. It is apparent that this viscosity changes considerably with increasing electrolyte concentration. An even more dramatic change in the flow properties is shown in Fig. 4-5, where the yield stress is plotted vs. the salt concentration. The addition of only a few milliequivalents of salt causes a sharp drop in the viscosity, with an accompanying decrease of the yield stress. However, as more NaCl is added, the yield stress increases, particularly as the flocculating concentration of NaCl is approached. At very high NaCl concentrations the yield stress tends to decrease again, but this is not shown.

The flow behavior for various regions of NaCl concentration can be summarized as follows (Van Olphen, 1963):

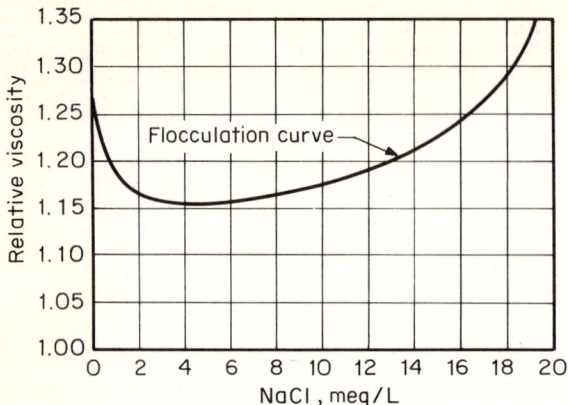

Fig. 4-4 Relative viscosity of a 3% sodium montmorillonite solution plotted vs. the amount of NaCl added.

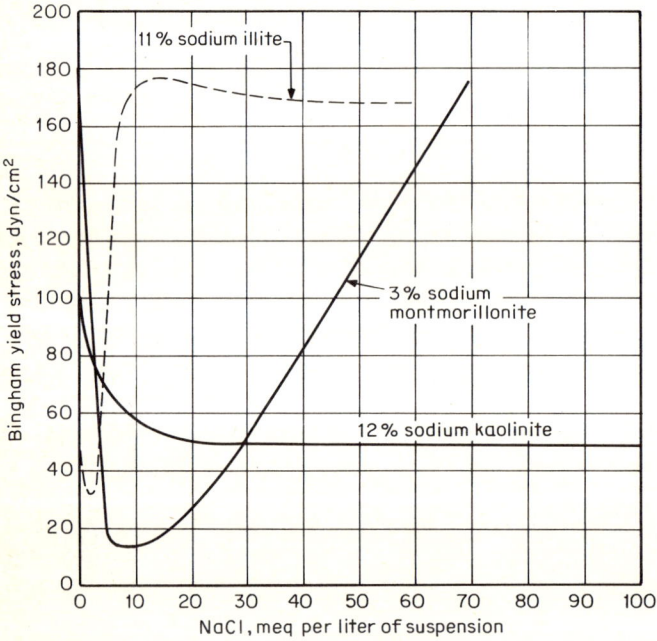

Fig. 4-5 Bingham yield stress of colloidal suspensions plotted vs. the salt concentration.

1. In electrolyte-free (nonsalt) systems the opposite EF charge (attraction between negative surfaces and positive edges) may lead to EF association and array. This is internal mutual flocculation and imparts good viscosity and gel strength to the suspension.

2. With very small quantities of salt the effective charges are reduced, hence both EF attraction and FF repulsion tend to diminish. The particles

thus become disengaged, the cardhouse structure breaks, and the yield stress is sharply reduced. In some dilute suspensions the conglomerates are dispersed and the viscosity is decreased, although the system is not necessarily aggregated.

3. As the amount of salt is increased, the attraction between edges and faces stimulates what is left of the opposite-charge attraction, and once more the balance favors the formation of the cardhouse structure.

4. At very high salt concentrations the yield stress decreases because of the simultaneous occurrence of FF association.

Reaction of Clay Systems The sodium montmorillonite system is rather unusual in the variation of flow behavior with changing salt concentration. For other clay minerals the salt effect is less pronounced and often absent because of the domination of one type of association. For example, the sodium kaolinite group hardly shows any spectacular change in viscosity and gel when treated with salt (see also Fig. 4-5), and for calcium montmorillonite the initial sharp increase does not even occur. The different reaction of various clay groups to electrolyte addition is due to differences in the initial double-layer structure and its effects upon particle association.

Minimum Gel Structure If the continuous particle linkage in a pure gel is idealized as EF associaion resulting in a cubic network of clay plates, it is possible to estimate the actual quantity of clay necessary to make a complete gel merely by knowing the size of colloid particles. The minimum concentration necessary to give the system a measurable yield stress is therefore important, and the gelating concentration is lower for thinner (smaller) colloid particles. For example, 2% sodium montmorillonite may be sufficient to just fill the volume with a continuous cubic network and provide a measurable gel strength. If converted to calcium form, the particles are 3 to 4 times thicker and the required minimum concentration for a gel may be 3 to 4 times greater (Van Olphen, 1956, 1959). The extremely thin plates of sodium bentonite make it quite suitable for technological applications.

Since the minimum gel structure may be disengaged by FF association, in practice it is advantageous to estimate the degree to which FF association may occur, as suggested by Darley (1957). In this test the suspension is instantaneously flocculated by the addition of an electrolyte. The flocculated system is then centrifuged and the volume of the mass measured. Since it can be assumed that in the rapid coagulation the particle association is EF and EE before changing to FF, the sediment volume will be a relative measure of the original number of plates in suspension. The thinner the plates the larger their number and the larger the sediment volume.

Peptization and Deflocculation
These processes are the opposite of flocculation and gelation. In actual applications a variety of agents are used to liquefy stiff, flocculated slurries. In many instances the addition of a very small quantity of these materials (probably

0.3 percent) can convert a stiff slurry into a free-flowing system. Since these agents also disperse flocs in more dilute systems, they act as deflocculants or peptizers. Added to a pure clay gel, they cause the yield stress to decrease. The yield stress is insensitive to further addition of peptizer, however, unless the latter reaches very high concentrations, in which case the yield stress may increase once again. If salt is added to a system after it is treated with peptizers, the yield stress is not affected unless the salt concentration is excessive.

These effects can be represented by a diagram in which the salt-flocculation value of a clay is plotted vs. the amount of peptizer. In Fig. 4-6a curve 1 shows the general effect. The resistance to salt is very low without peptizer, but the treated suspension exhibits a flocculation resistance which increases rapidly with increasing peptizer concentration until an optimum effect is achieved. Further addition of peptizer stimulates flocculation action and results in overtreatment. At very high peptizer concentrations the suspension

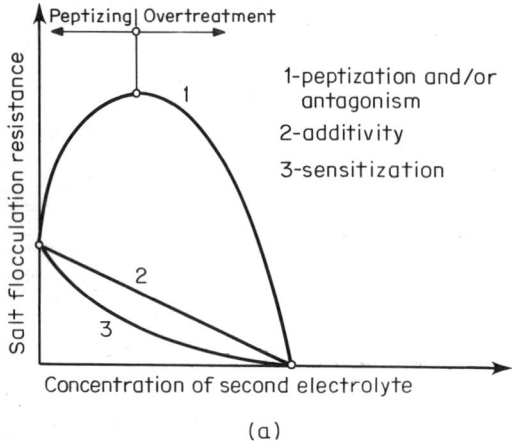

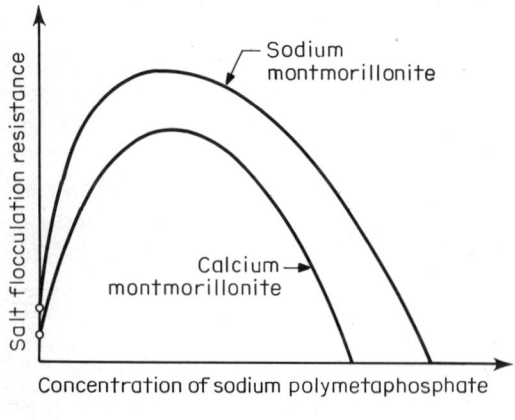

Fig. 4-6 (a) Effect of two electrolytes (salts) on the stability of a colloidal suspension; (b) peptization of sodium and calcium montmorillonite suspensions with sodium polymetaphosphate.

undergoes flocculation with the peptizer alone, but the flocculation resistance of the peptizing salt is much higher than that of salt, as shown in Fig. 4-6b.

Sensitization Possible related effects when a clay suspension is flocculated with two different electrolytes are represented by the two lower curves of Fig. 4-6a. Excluding any special interaction between the two electrolytes, the flocculation values usually are additive, as shown by curve 2. If the second electrolyte makes the system more sensitive to flocculation, the process is known as *sensitization* and is depicted by curve 3.

Peptization Mechanism It is possible that the usual peptizing agents reverse the edge double-layer charge of clay particles. All evidence seems to indicate that the edge charging by the anions of the peptizing salt forms the basic mechanism for the stabilization phenomenon. The fact that only a very small fraction of peptizer is required in a system that has small adsorption capacity for the anions of the peptizer suggests their adsorption on the clay crystal edges. Thus the effect on the flow properties is explained by the disorientation and the breakdown of the linked cardhouse structures involving the edge surface. The increase in mobility means a charging effect, and this behavior makes the peptization mechanism one of the principal factors in the stability of slurries.

Two more factors possibly contribute to the increased salt tolerance of chemically peptized suspensions (Van Olphen, 1950a, 1963), i.e., cation-activity reduction and conversion of the clay into sodium form by ion exchange with the added sodium salts.

Swelling of Montmorillonite

Dry clay generally imbibes water to form a gel and can be stirred up with more water to yield a suspension or sol. Montmorillonite powder swells spontaneously when placed in contact with water in a process which is rather spectacular. The swelling may continue as result of the double-layer repulsion between the surfaces of individual particles, which forces them apart. Some evidence indicates that this swelling is osmotic since water tends to equalize the high ion concentration between two particles with the low ion concentration away from the particle surfaces. Under suitable conditions a fluid pressure is created, which is called the *osmotic* or *swelling pressure* of the clay; it is a direct measure of the balance of the forces between particle faces.

The question often arises whether spontaneous swelling of montmorillonite in a bentonite slurry may lead to complete disintegration of a gel to a sol or may stop as soon as a certain gel strength is established. In principle, the transformation of a gel into a sol by swelling is spontaneous only if at any particle distance and configuration in space repulsion predominates. However, most clays and even certain sodium montmorillonites are not spontaneously dispersed since at a certain particle separation and configuration attractive forces cancel the repulsion.

4-3 APPLICATIONS OF STABILITY CONTROLS

Sedimentation and Separation of Solids

Sedimentation of suspended clay particles and the rate at which it occurs increase upon flocculation because of the layer size of aggregates. Sediments of flocculated suspensions usually are more voluminous than those of stable systems for the same concentration, and although this seems paradoxical, it has a valid reason. When individual particles of a stable system reach the bottom, they still slide and roll over each other since they are free to do so; hence they continuously adjust position until they finally settle to a dense packing of sedimentary material, as shown in Fig. 4-7a. In a flocculated suspension the flocs settle and pile up without breaking the linkage, keeping the voids since individual particles are unable to adjust position, as shown in Fig. 4-7b. For nearly spherical particles the flocs and the sediment are likely to have a string-of-beads arrangement, whereas flat plates and rods form a scaffolding or cardhouse sediment structure.

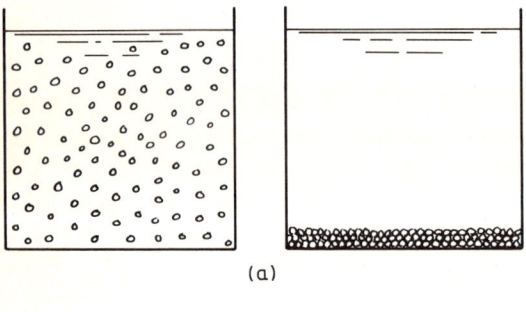

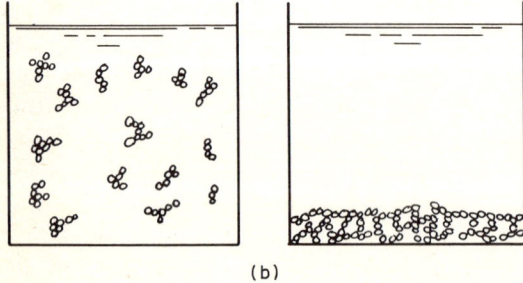

Fig. 4-7 Sedimentation of peptized and flocculated suspension: (a) peptized state, showing dense, close packing of the sediment; (b) flocculated state, showing loose and voluminous sediment.

This variation in the sedimentation will not occur if solid particles are large enough to break the link by simple gravity (see also Sec. 3-3). For example, coarse sand might be expected to give the same sediment volume in both fresh and salt water. The relationship between sedimentation and degree of stability is shown in Fig. 4-8 as a function of the salt concentration. It is evident that curves 1 and 2 are compatible; the sediment volume first decreases

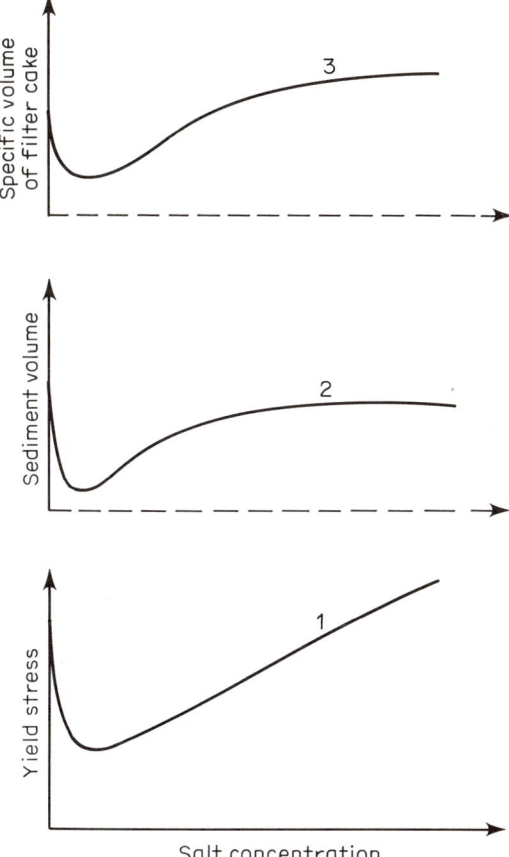

Fig. 4-8 Effect of salt concentration on yield stress, sediment volume, and filter cake for sodium montmorillonite suspensions.

sharply and then increases with increasing NaCl concentration, indicating an initial degree of deflocculation followed by progressive flocculation, precisely like the yield stress.

Separation of Solids This is quite important in all applications. Noncolloid solids (generally sand and clay cuttings accidentally mixed with the slurry during excavation) must at the end be separated and removed. This is done either by recirculating the slurry in special tanks, where these solids settle down, or by centrifuging and screening. In any case the separation is facilitated if the slurry is in a peptized condition. However, when the slurry is still in the trench, it is desirable to keep sand and clay cuttings in suspension, and this is possible if the slurry has a certain shear strength. Thus, in order to accommodate these situations a compromise is necessary.

Stability and Filtration of Cake

Likewise, when slurry is filtered to deposit a cake, the density and porosity of the latter will depend on the degree of peptization or flocculation of the system, much as sedimentation does. The filtration of a stable slurry results in a thin and compact cake, whereas a flocculated suspension deposits a relatively porous and thicker layer which has appreciable permeability. In some instances it is possible to have the pores of the cake clogged with colloid particles as more slurry is filtered, but it is quite doubtful whether this situation can be predicted or controlled. Thus, the dependence of the filter-cake characteristics on the salt concentration is usually represented by a curve similar to curve 3 of Fig. 4-8 and follows the same pattern as the sediment volume.

The effectiveness of montmorillonite clay in reducing filtration, provided salt effects are not present, is probably due to the size and shape of particles. These comprise very thin and flexible flakes which can close pore openings by building a compact impervious complex across the void space. Thicker particles, e.g., kaolinites, are less effective in forming impermeable barriers.

Pumpability of Slurry

Pumpability is quite important in reverse circulation. High recirculation rates are required, and the flow energy must be adequate to overcome the total pressure loss in the circuit. Figure 4-9c shows the pressure loss in a reverse pipe as a function of the volume rate of flow. The curves have been plotted from the τ_f vs. D and W vs. ω curves shown in Fig. 4-9b and a, respectively. In all cases curve 1 is for a flocculated slurry, and curve 2 is for the same slurry after it is reconditioned by the addition of peptizers. The rate of circulation Q_1 in Fig. 4-9c corresponds to an average rate of shear D_1 and a rate of rotation ω_1.

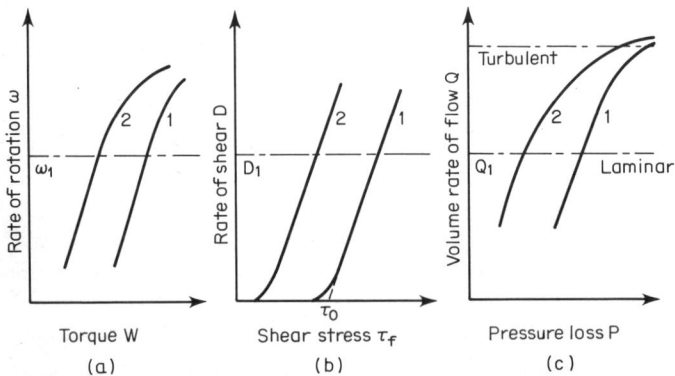

Fig. 4-9 Relationship between rheological and pipe-flow characteristics of peptized and flocculated slurry: (a) rotational-viscometer diagram; (b) shear stress rate of shear diagram; and (c) pipe-flow diagram.

It is evident that the pumping pressure required to maintain a certain rate of flow is determined from the curves and depends on the magnitude of the yield stress τ_0, that is, the state of peptization or flocculation. Hence, the pressure requirements for the flow of slurry vary according to the yield stress and are reduced as the latter decreases. At higher rates of circulation the mud flow may become turbulent, resulting in pressure loss which increases much faster with increasing flow rates, as evidenced by the relatively flat curve sections. Furthermore, the pressure loss becomes less dependent upon the flow properties of slurry, as is seen from the close approach of curves 1 and 2. In the turbulent region inertia forces become dominant, so that the density rather than the viscosity of the slurry becomes important.

No direct empirical relations of general applicability exist for predicting the turbulent-flow pressure loss, but appropriate reduction factors are available for reverse-circulation systems for use in practical situations (see also Chap. 5). These factors relate the reduction coefficient to an equivalent turbulent viscosity and the relative volume of dispersed solids in the slurry. Although this approach ignores the state of peptization of flocculation, it gives a better accuracy (Havenaar, 1954). Mud flow in other parts of the circuit (connections, tool joints, etc.) is discussed by Van Olphen (1950*b*).

If for any reason the flow in the reverse hose is interrupted, the requirements for restarting it must be correlated with the flow properties of the slurry.

4-4 PROPERTIES OF SLURRIES

The main properties of slurries are (1) viscosity, gel strength, and thixotropy; (2) rheological characteristics; and (3) density or specific gravity. The first two sets of properties constitute the so-called *flow properties*. Density and specific gravity are physical properties.

Viscosity, Gel Strength, and Thixotropy

Consider a plane section of a body of slurry, the lower boundary being fixed by line AB in Fig. 4-10, which is also taken as the reference line. If a tangential force is applied along free surface CD, it will cause this layer to move in the same direction, so that point C will move to D. By friction the next layer will move in the same direction but slower, and so on until section CE takes the

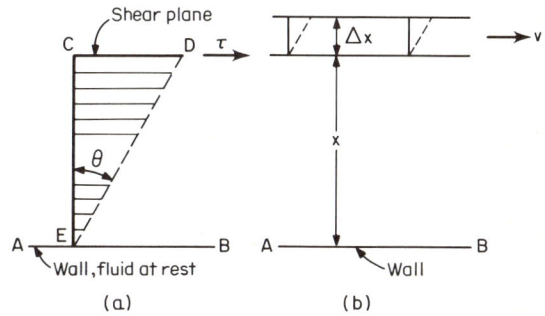

Fig. 4-10 Definition of shear stress and rate of shear; laminar flow.

position DE in a unit time. Thus, a velocity gradient is established perpendicular to the plane of movement. This is called the *rate of shear D*. As can be seen from Fig. 4-10b, it causes a slurry element Δx to shear from a rectangle at one instant to a parallelogram an instant later, as shown by the dashed line. The applied unit tangential forces is the shear stress τ, for which the notation τ_f has already been used. The rate of shear is measured in reciprocal seconds, and the shear stress may be expressed in dynes per square centimeter, pounds per square inch, or pounds per square foot.

The state of flow is expressed by the relation between the shear stress and the resulting rate of shear. The general expression is $\tau = f(D)$, and the simplest form is the linear relation

$$\tau = \eta D \qquad (4\text{-}1)$$

indicating a straight line through the origin in the τ-vs.-D diagram. Liquids obeying the flow law expressed by Eq. (4-1) are called *newtonian fluids*. The proportionality constant η is called the *coefficient of viscosity* or simply *viscosity*. If a force F acts on a slurry element of area A located at distance x from the fixed boundary AB, causing it to move with a velocity v, another expression for the viscosity is

$$\eta = \frac{F/A}{dv/dx} \qquad (4\text{-}2)$$

The unit of viscosity in the metric system is gram-centimeters per second, dyne-seconds per square centimeter, or poises (P). The viscosity of water at 20°C is about 0.01 P, or 1 centipoise (cP).

The flow of slurries is more complex than the simple pattern expressed by Eq. (4-1). Slurries behave essentially as nonnewtonian fluids. Four representative types of flow are illustrated in Fig. 4-11. At any point along these curves an *apparent viscosity* exists, corresponding to the given rate of shear, and is equal to τ/D. Alternatively, the flow can be expressed in terms of the *plastic viscosity* $d\tau/dD$, which is numerically equal to the cotangent of the angle

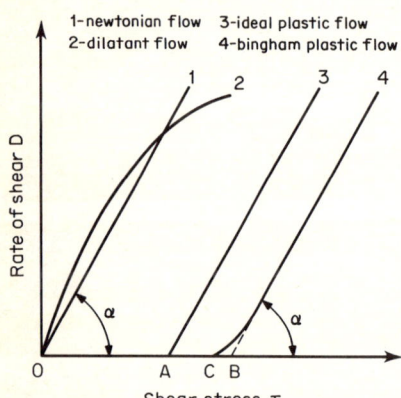

Fig. 4-11 Relations between shear stress and rate of shear.

between the tangent to the curve at the given point and the shear-stress ordinate. Curve 1 represents a newtonian fluid. The flow behavior represented by curve 2 indicates an increase in the apparent viscosity with increasing shear rates; such a system resists deformations better at higher shear rates (dilatancy). For the system represented by curve 3 flow does not begin until the applied shear stress reaches a certain magnitude OA, but beyond that the rate of shear is proportional to the shear stress. This is sometimes called *ideal plastic flow*.

A more complicated flow is exemplified by curve 4. This is of special interest since it is common in dispersed colloidal systems and therefore slurries. It is called *Bingham plastic flow*, and systems exhibiting this flow are called *Bingham bodies* or *Bingham fluids*.

Figure 4-12 shows the flow behavior of a Bingham fluid. There are several stresses of particular significance. The flow curve intersects the shear-stress ordinate at a point designated as τ_s. This is what is called *gel strength* of slurry, and it represents the minimum shear stress required to produce flow. When

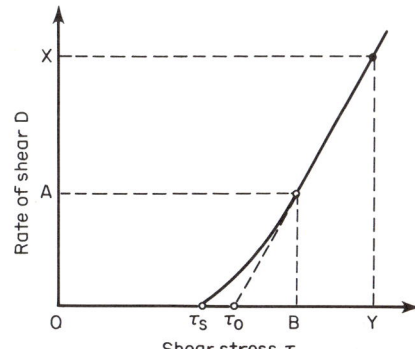

Fig. 4-12 Flow curve for a Bingham fluid.

the straight-line portion of the flow curve is extrapolated to low shear rates, it intersects the shear-stress ordinate at a point marked τ_0. This is called *Bingham yield stress* or simply yield stress and yield point. If τ_0 coincides with τ_s, the system is in ideal plastic flow. For most slurries, however, the yield stress exceeds the gel strength, or $\tau_0 > \tau_s$. The linear part of the curve obeys the relation

$$\tau = \tau_0 + \eta_p D \tag{4-1a}$$

in which η_p is the plastic viscosity. According to Eq. (4-1a), there are two components of the shear stress; one is the shear-dependent viscosity contribution $\eta_p D$ and the other the Bingham yield stress τ_0, which is independent of the rate of shear.

From Fig. 4-12 an apparent viscosity is obtained regarding the slurry as newtonian, or

$$\eta_a = \frac{OY}{OX} = \frac{\text{shear stress}}{\text{rate of shear}} \tag{4-3}$$

whereas the plastic viscosity for the straight-line portion of the curve is

$$\eta_p = \frac{BY}{AX} \qquad (4\text{-}4)$$

Gel Strength and Thixotropy As already mentioned, the gel strength of slurries is caused by a particular linkage of colloid particles. If a shear stress is applied to such a system, no flow will occur until the gel strength is just exceeded, and at this point the gel structure breaks down although not completely destroyed. The degree of breakdown depends on the rate of shear and the time over which it is applied (Jefferis, 1972). If the slurry is allowed to stand, the gel structure will reform, according to the process called thixotropy. Regardless of disturbance, colloid particles tend to flocculate as long as they are free to choose their position. However, a common complication is the dependence on the previous shear history of the system. Whereas the occurrence of thixotropy is characteristic of dispersed colloidal systems, for some slurries the restoration of the original state is accelerated by applying lower shear rates. On the other hand, the original consistency of certain slurries is not attained until after they have been liquefied by stirring, in a process called *shear breakdown*.

From Fig. 4-13 it is seen that at constant rate of shear D_c the shear stress decreases until the equilibrium value τ_D is reached so that the rate of breakdown of the gel structure is equal to the rate of formation. The curve *DCBA* is then obtained by rapidly reducing the rate of shear so that the gel structure does not have time to reform. Upon reaching point A, if the dispersion is allowed to stand for some rest time and the shear rate is then rapidly increased, so that the structure has no time to break down, the curve *FD* is obtained. If a shorter time is allowed, the curve *EBCD* results. Jefferis (1972) has shown that these curves can be obtained with a recording viscometer. In practice only the equilibrium curve *GCD* is obtained by allowing sufficient time for equilibrium at each shear rate. Points A and F are commonly measured. Point A is the initial gel strength, and point F is the 10-min gel strength. The difference between F and A gives an indication of thixotropy.

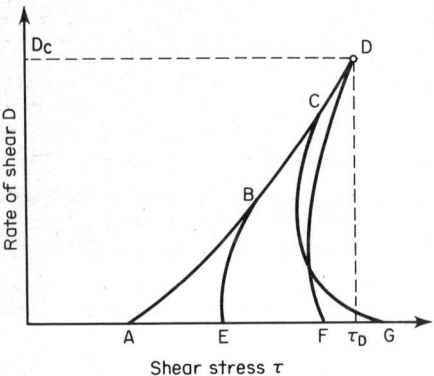

Fig. 4-13 Flow-time characteristics of a Bingham fluid. (*From Jefferis, 1972.*)

Rheological Characteristics

The deformation-time behavior of colloidal suspensions is dependent upon the state of flocculation or peptization. Therefore, the viscosity, colloid concentration, size and shape of colloid particles, and type of particle interaction are relevant to this process.

For a slurry to behave as a newtonian fluid the colloid concentration must be very low (probably less than 2 percent). For such a slurry the relative viscosity η/η_0 (viscosity of slurry/viscosity of liquid medium) is always greater than unity. Van Olphen (1963) has proposed the relation

$$2.5m = \frac{\eta}{\eta_0} - 1 = \frac{\eta - \eta_0}{\eta_0} \qquad (4\text{-}5)$$

in which m is the colloid concentration by volume. This is based on the assumption that the slurry is sufficiently dilute, thus precluding particle interaction.

Effect of Peptization or Flocculation on the Flow Curve

When particle interaction exists, its effect is quite significant and Eq. (4-5) is not valid. The yield stress now depends on the force necessary to break the links as well as on the number of such links per unit volume. The link-forming process is reversed when an originally stiff slurry is peptized and the links are in effect weakened and eventually broken. Therefore, peptization results in a lower yield stress, which may even completely disappear. This is shown in Fig. 4-14. Initially the flocculated slurry has the flow properties expressed by curve 3. The use of peptizer causes a reduction in the yield stress, and the flow curve is now shifted to position 2. More addition of peptizer can move the

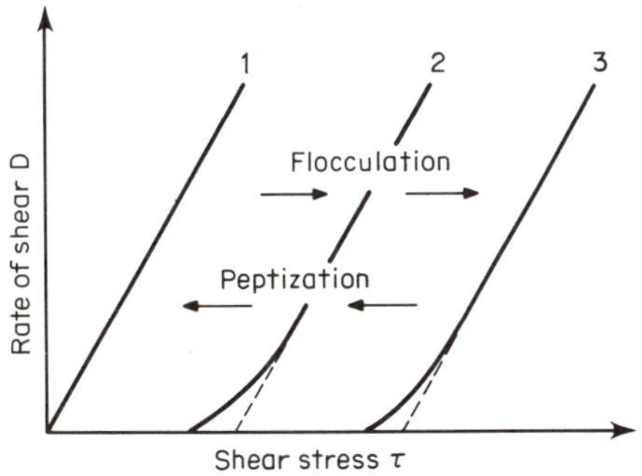

Fig. 4-14 Effect of peptization and flocculation on the flow curve of clay suspensions.

curve to position 1 so that the slurry is now newtonian. During this transformation the differential viscosity as given by the slope of the curve (which is also a measure of the plastic viscosity) remains essentially parallel; hence the thinning action of peptization is not a viscosity reduction as such but mainly a yield-stress reduction.

In a stirred system the corresponding reduction of the yield stress indicates that the application of shear causes the breakdown of particle links in the space frame of the flocculated system, and this is identical to the effect of peptization. However, after stirring and during the rest period the links are reestablished and the yield stress is restored. A stiff suspension becomes thinner upon stirring and thicker upon standing, and the thixotropic process involves changes caused mainly by temporary changes in the yield stress. In order to make these changes permanent the use of peptizers is necessary.

Density and Specific Gravity

The density or specific gravity is quite relevant because of its effects on two important phases of the construction, namely the support of the face and the displacement of slurry by fresh concrete. The density is controlled by the presence of solid materials, both colloid and noncolloid. For ordinary trench excavations some soil retention usually is sufficient to raise the density to the level required for stability. When very high densities are required at the start of the excavation, barite (barium sulfate) is sometimes used as weighting agent. Its density is 4.2 g/cm³ (about 260 lb/ft³). Barite, however, is known to have erratic effects on the flow properties and often causes a loss of colloidal stability.

The initial density of slurries generally ranges from 1.04 to 1.15 g/cm³ (65 to 72 lb/ft³) but may rise to 1.25 g/cm³ (about 80 lb/ft³) toward the end of excavation.

4-5 MEASUREMENT OF FLOW AND PHYSICAL PROPERTIES

Standard procedures for measuring the properties of slurries are given in API RP 13B (API, 1974) and by Van Wazer et al. (1963). Only certain details will be considered in this section.

Measurement of Flow Rate and Pressure Drop in Pipes

Equation (4-1a) applies to ideal plastic flow and to the straight portion of Bingham flow. If a suspension flows through a pipe of radius R and length L, for an annular streamline of radius r and velocity v the rate of shear is

$$D = \frac{dv}{dr} \tag{4-6}$$

Since the shear stress τ increases from zero at the center of the pipe to a maximum at the wall, there is a plug of radius r_0 in the central region of the pipe behaving like a solid, or $r_0 = 2\tau_0 L/\Delta p$, in which Δp is the pressure drop in length L.

Figure 4-15 shows the velocity profile for such flow. As the pressure drop increases, r_0 decreases but does not become zero. Starting with the Buckingham-Reiner equation (Reiner, 1949), expressing the volume rate of circulation or discharge Q, and noting that $D = 4Q/\pi R^3$ and $\tau_R = R\,\Delta p/2L$, we get

$$D = \frac{1}{\eta_p}\left(\tau_R - \tfrac{4}{3}\tau_0 + \frac{1}{3}\frac{\tau_0^4}{\tau_R^3}\right) \tag{4-7}$$

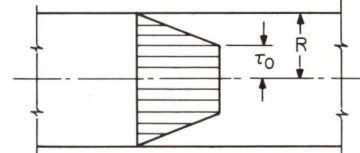

Fig. 4-15 Velocity diagram for the flow of a slurry in a pipe.

Plotting D vs. τ_R gives the curve shown in Fig. 4-16, which is similar to the Bingham flow curve. For $D = 0$ it follows that $\tau_R = \tau_0$; hence the intercept of the curve equals the yield stress. When τ_R becomes very large, the factor τ_0^4/τ_R^3 approaches zero and the curve approximates a straight line. The intercept produced by extrapolating this straight line to the shear-stress ordinate is $\frac{4}{3}\tau_0$.

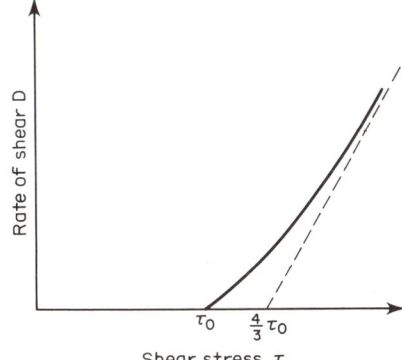

Fig. 4-16 Flow diagram of slurry in a pipe.

The curvature of the D-vs.-τ_R diagram is caused by the gradual transition from plug flow to streamline flow; hence it does not imply that the fluid is not ideal. For dispersed colloidal slurries the curvature is more pronounced, and the intercept is the gel strength τ_s. There is no simple relationship between τ_s and τ_0. Thus for a slurry flowing in a pipe τ_s is estimated by measuring the minimum pressure required to produce flow, and η_p and τ_0 are established from flow rates at various pressure gradients.

Marsh Funnel Viscometer

This simple apparatus (Fig. 4-17) consists of a funnel of standard dimensions which is filled with 1500 cm³ of slurry after it is screened to remove coarse material. The time for 946 cm³ (equivalent to 1 U.S. quart) to flow through it is then measured and quoted as the viscosity in seconds. Thus water at 21°C has a viscosity of 26 s for 946 cm³. In Japan the cone viscosity is the time for 500 cm³ to flow through a standard funnel of 500 cm³ volume capacity. If a distinction must be made, it is necessary to indicate the number of seconds followed by 946/1500 cm³ or by 500/500 cm³.

The measured viscosity is influenced by the rate of gelation and by the density, which affects the hydrostatic head in the funnel. Because of these constraints viscosity measurements obtained with the funnel viscometer cannot be directly correlated with those obtained by other procedures, but it provides a simple and practical means for routine slurry tests at the site. An

Fig. 4-17 Marsh funnel and cup. (*NL Baroid/NL Industries, Inc.*)

approximate conversion of plastic viscosity, expressed in centipoises, to funnel viscosity, expressed in 946/1500 cm³ or 500/500 cm³, can be made with the help of Fig. 4-18.

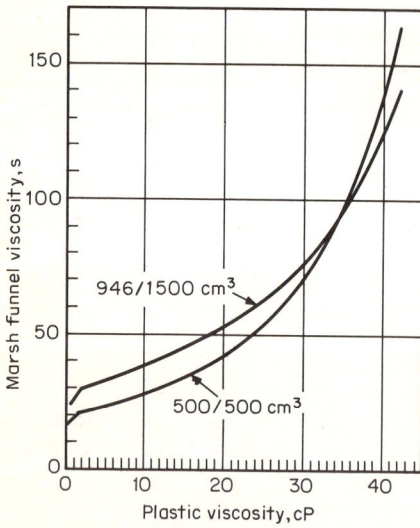

Fig. 4-18 Correlation of plastic viscosity and funnel viscosity.

Rotational Viscometers

In these devices a slurry sample is sheared between two cylinders. At a given rate of rotation the rate of shear is not necessarily constant but varies within narrow limits. The shear stress is derived from the torque on one of the cylinders, and the average rate of shear is estimated from the measured rate of rotation and the diameters of the two cylinders.

The Fann V-G meter shown in Fig. 4-19 is widely used for viscosity mea-

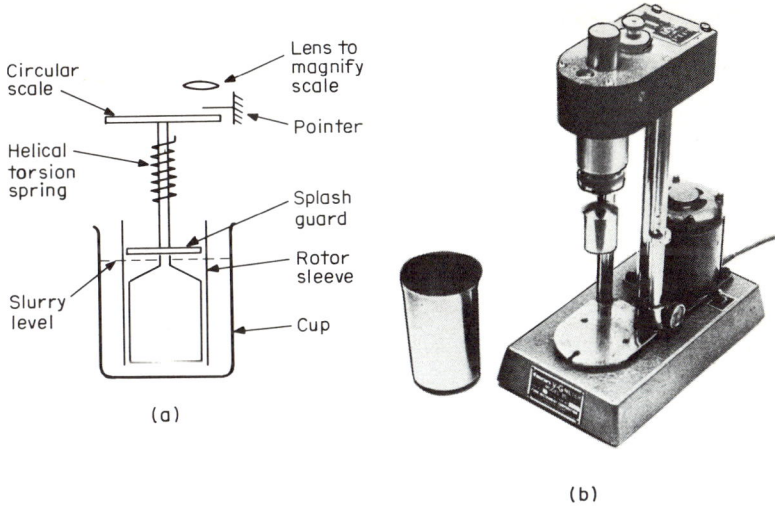

Fig. 4-19 Fann V-G meter: (*a*) schematic diagram; (*b*) meter with sleeve removed.

surements. A cup of slurry is positioned so that the slurry is brought to a certain height in the annular space between two cylinders, the outer cylinder, or rotor sleeve, and the inner cylinder, or bob. Rotation of the sleeve produces a torque on the bob which is restrained by a torsion spring, while a scale attached to the bob indicates its movement against the spring restraint. The sleeve generally can be rotated at six different speeds (600, 300, 200, 100, 6, and 3 r/min) by an electric motor and also by hand at very low speeds. Starting at 600 r/min, the scale reading is recorded for each of the six speeds when the equilibrium value is reached. The initial gel strength is measured by rotating the sleeve at 600 r/min until equilibrium is reached, then rotating very slowly by hand and recording the maximum reading before the gel breaks. The 10-min gel is likewise obtained by allowing 10 min to elapse between stirring and reading. The instrument is calibrated so that the difference in the torque readings at 600 and 300 r/min is the plastic viscosity directly in centipoises. The apparent viscosity in centipoises equals the 600 r/min reading divided by 2. The gel-strength reading generally is given in pounds per 100 ft^2.

For the theory of rotational viscometers see Jefferis (1972).

Density Measurement

Density is expressed in pounds per gallon, pounds per cubic foot, grams per cubic centimeter, specific gravity, or pressure gradient. The value in grams per cubic centimeter is also the specific gravity. Conversion tables are available in API Standard RP 13B (API, 1974). Any instrument of sufficient accuracy to permit measurement within ±0.1 lb/gal or ±0.5 lb/ft^3 may be used. For field measurements the mud balance shown in Fig. 4-20 is com-

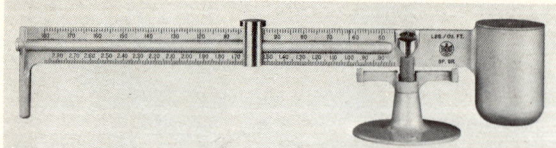

Fig. 4-20 Mud balance for measuring density. (*NL Baroid/NL Industries, Inc.*)

monly used and gives direct readings. The instrument should be frequently calibrated with fresh water. The accuracy is sufficient if fresh water gives a reading of 8.33 lb/gal or 62.3 lb/ft^3 (1.00 g/cm^3) at 70°F (21°C). If this is not obtained, adjustments may be necessary.

pH Measurement

The pH of fluids and slurries is measured according to two procedures: (1) a modified colorimetric method, using paper test strips; and (2) an electrometric method, using the glass electrode. The paper-strip method generally is not considered reliable if the salt concentration of the sample is high. The electrometric method can yield errors in suspensions containing high concentrations of sodium ions unless special electrodes are used. Furthermore, a temperature correction should be made when using the electrometric method.

For common slurries the pH value generally ranges from 7 to 10. Mixing with cement during concrete placement will raise the pH, but its value should not be allowed to exceed 12.

Filtration Measurement

The current procedures for measuring filtration are interpreted to give an indication of fluid loss and presumably relate a material to the filter-cake characteristics. Thus low filtrate loss during the test is taken to indicate a filter cake of low permeability. However, the results are quite difficult to interpret in terms of the soil-slurry interaction in situ and often are misleading. Accordingly some contractors rely on their own experience with similar slurries and soil conditions, while others prefer the tests described in Chap. 3.

A standard filter-press test is carried out using the instrument shown in Fig. 4-21. The base cap is fitted to the cell with a Whatman no. 50 or equivalent filter paper covering the screen, and the unit is filled with a slurry sample to within ¼ in of the top. Next the top cap is fitted, the entire unit is placed in the frame, and the T screw is tightened. A pressure of 100 lb/in^2 (7.03 kg/cm^2) is applied through the pressure inlet, and the volume of filtrate collected in 30 min is recorded. The filter-cake thickness is measured after the excess slurry has been drained out and the cake is gently washed. The cake consistency is also described using such notations as hard, soft, tough, firm, etc.

A loss of 5 cm^3 in 30 min is interpreted to correspond to a permeability of 10^{-7} cm/s, but the test procedure hardly allows this correlation. For the usual

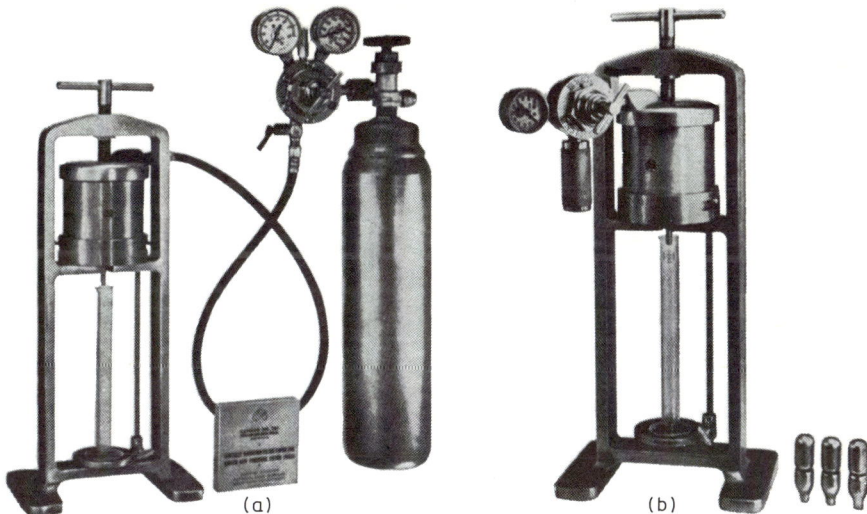

Fig. 4-21 Standard filter-press apparatus: (*a*) with nitrogen pressurization; (*b*) with carbon dioxide pressurization. (*NL Baroid/NL Industries, Inc.*)

situations the recommended allowable filtrate loss is from 15 to 25 cm^3 for a pressure of 100 lb/in^2. The author recommends an allowable loss from 12 to 15 cm^3 for a pressure of 3 kg/cm^2 (42.6 lb/in^2) and a maximum film thickness less than 2 mm.

4-6 MATERIALS USED IN SLURRIES

Among clay minerals bentonite and attapulgite are finely dispersed and therefore are suitable for slurries. Most illite clays exhibit nonswelling behavior, whereas members of the kaolinite group produce nonexpandable minerals; hence these materials tend to remain aggregated in the suspension. Furthermore, at concentrations showing satisfactory gel strength the plastic viscosity is too high, and treatment in this case is not always possible.

Bentonite as the Main Colloid Solid

Bentonite is produced from natural clays containing the montmorillonite mineral and resembles a gray powder. Different grades are available, better ones attaining the proper mixing levels and flow properties at lower concentrations. The selection of a suitable bentonite must be based on the conditions to be contended with but is also governed by local availability and cost.

As mentioned earlier, montmorillonite displays the highest surface activity and capacity to expand. This mineral constitutes the largest part of naturally occurring bentonite clays derived from the alteration of volcanic dust and ash deposits.

Because of increased uses, the demand for bentonite has led to large-scale development of material resources and resulted in the commercial availability of the material in most local and foreign markets. In the United States the chief commercial deposits are in the Black Hills of Wyoming and South Dakota. Besides the United States as major producer, commercial deposits exist in Algeria, Morocco, Italy, Yugoslavia, Argentina, South Africa, Great Britain, France, Greece, Cyprus, India, Japan, and New Zealand, of which the first four are regarded as major exporters.

The superior characteristics of bentonite made it commercially expedient to standardize its production. Thus the product is graded for the intended uses, and its quality is controlled and maintained with better limits than is generally possible with borrowed natural clays. Bentonite ores are usually mined as soft rock crushed and dried before grinding to a fine powder and then separated to produce the required fineness according to the intended use.

Effect of Salt Bentonite is suitable for all applications except when unusually high salt concentrations are encountered in the excavation. Rogers (1967) recommends 10 g of NaCl per liter of groundwater as the salt concentration of no serious consequence. However, there is no real basis for establishing absolute limits, and in practice some problems may arise in connection with the presence of seawater (which normally contains 35 g of NaCl per liter) if the slurry is not treated. These effects become more pronounced when excavating in chalk (calcium carbonate) or gypsum (calcium sulfate).

The properties of a bentonite slurry are affected by seawater dilution. Excluding very small salt concentrations, salt water generally acts as a mild flocculating agent and initially raises the apparent viscosity and yield stress. Further addition of salt water may lead to complete gelation, high fluid loss, and the subsequent disappearance of thixotropy. These effects are exemplified in Fig. 4-22a and b. Whereas bentonite swells in pure water, swelling is restrained if it is added to salt solutions, and at high salt concentrations it does not occur at all, particularly in calcium chloride or other solutions containing high-valency cations.

Flow Properties Bentonite suspensions in water containing less than 1 percent of colloid solids generally are free-flowing fluids. At higher concentrations, but depending on the bentonite activity, the systems exhibit thixotropic characteristics and the flow properties become anomalous. This behavior is exhibited in Fig. 4-23, in which both the plastic viscosity and yield stress increase exponentially with colloid concentration. Carefully prepared suspensions of pure sodium bentonite show virtually infinite swelling in water of low electrolyte content.

Bentonite possesses ion-exchange capacity; when the exchangeable ion is sodium, the mineral produces thixotropic suspensions. In a state of rest bentonite suspensions undergo thixotropic gelation, and if stirred vigorously, they flow like free liquids, gradually setting again to a gel which stiffens progressively on cessation of stirring. The strength of thixotropic gels depends

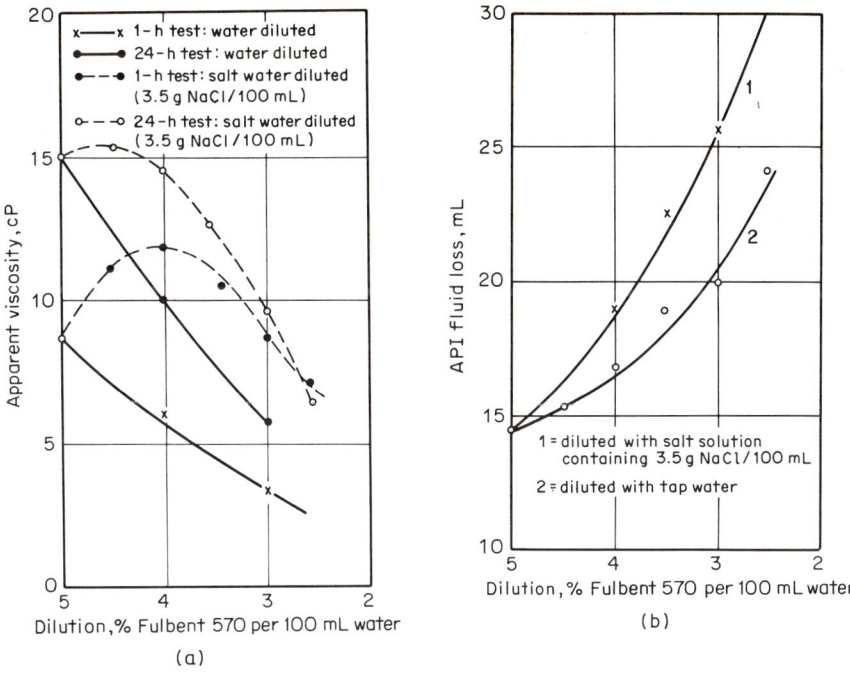

Fig. 4-22 Dilution of a 5% Fulbent suspension: (*a*) changes in apparent viscosity; (*b*) changes in fluid loss. (*From Fuchsberger, 1974.*)

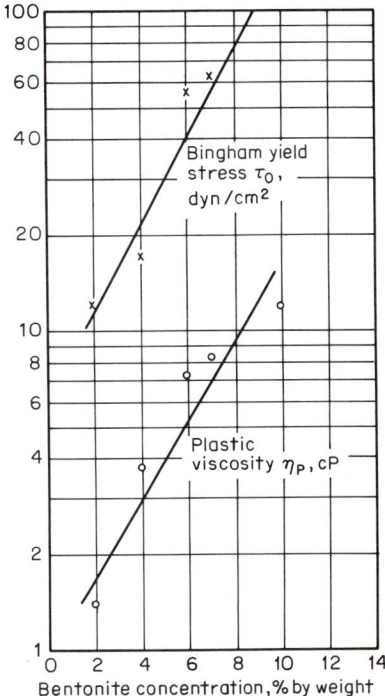

Fig. 4-23 Effect of bentonite concentration (Fulbent 570) on the flow properties of suspensions. (*From Jones, 1963.*)

119

on the setting time, the colloid concentration, and the composition of the suspending fluid. The variation in strength and viscosity with time is shown in Fig. 4-24. A distinction is again apparent between the yield stress τ_0 of the flowing suspension and the shear strength τ of the thixotropic gel, which is the time-dependent variable.

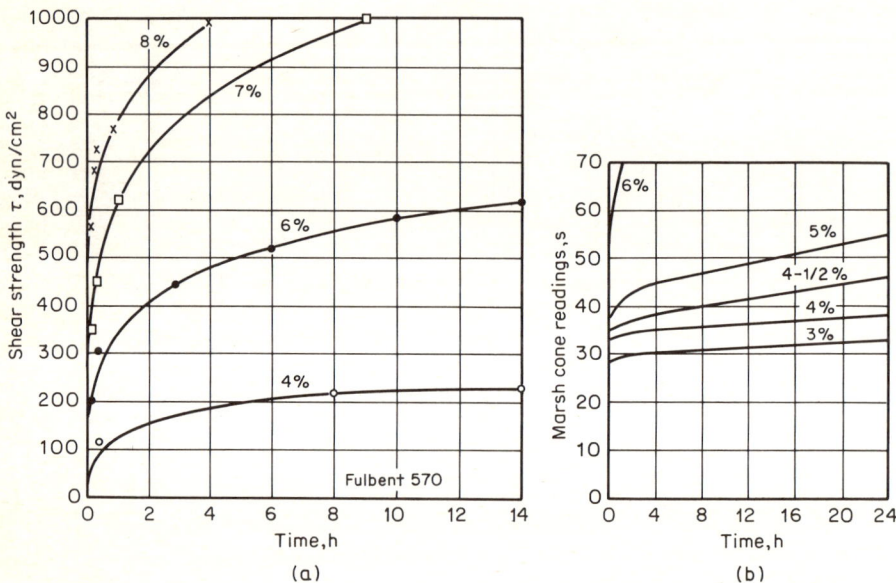

Fig. 4-24 Rheological characteristics of bentonite suspensions: (*a*) thixotropic gelation; (*b*) buildup of Marsh cone viscosity.

Bentonite suspensions can be treated with flocculants and peptizers. The plastic viscosity does not change significantly, but the yield stress is markedly influenced, as shown in Fig. 4-14. The flocculating tendency of salt is thus offset by the addition of peptizing agents. This provides a simple means for controlling the flow properties under a variety of conditions and therefore increases the resistance of slurries to deterioration.

Analysis of Bentonite and Attapulgite Whereas bentonite is found on almost every continent, attapulgite exists almost exclusively in a small region of the United States. Analysis generally involves x-ray and electron-diffraction techniques, which are laborious and expensive. Alternatively the liquid-limit test provides a simple indication of the presence of bentonite in a clay sample. If the liquid limit is over 100 percent bentonite probably is present. If it exceeds 250 percent, bentonite probably exists in sodium form, and this is confirmed if the sample gives measurable viscosity and gel when dispersed in water. The percentage of bentonite in a clay is estimated by viscometric analysis.

Jones (1964) has used a procedure based on the high cation-exchange capacity of bentonite, which is measured by the adsorption of a colored cation, e.g., methylene blue dye. In this procedure a clay suspension is treated until it no longer adsorbs dye. The quantity of methylene blue required to reach this point provides an indication of the cation-exchange capacity of the clay. Table 4-1 shows typical methylene blue adsorption capacities of common clay minerals. For a mixed clay the contribution from illite and kaolinite can be ignored, and since attapulgite is unlikely to be present, the percentage of bentonite can be calculated from the weight adsorbed by a known weight of clay.

TABLE 4-1 Adsorption Capacity of Common Clay Minerals (from Jefferis, 1972).

Clay mineral	Adsorption capacity, g dye/g clay
Bentonite	0.27
Kaolinite	<0.05
Illite	<0.05
Attapulgite	0.12

Certain standards are available to help in the procurement of materials and their use therefrom (API, 1974). These give a recommended practice for the laboratory and field testing of bentonite and attapulgite clay and establish the requirements for the flow and physical properties of colloidal suspensions. These requirements include viscometer tests, filtrate measurements, wet-screen analysis, and moisture measurements, whereas limiting numerical values of these properties are given as guidelines.

Thinners and Dispersants

The most frequently used materials for thinning or dispersing agents are sodium polyphosphates, tannins and quebracho, and lignins and calcium or chromium lignosulfonate. To these must be added sodium humate and sodium ferrochrome lignosulfonate (FCL). Basically these agents are long-chain polymers.

The use of certain sodium polyphosphates is not recommended when large quantities of calcium ions are present, as in cement, since they react to precipitate calcium polyphosphate and release sodium ions (Jefferis, 1972). The calcium polyphosphate has no action on the clay edges, but the replacement of Ca^{2+} by Na^+ can cause deflocculation as well as dispersion and breaking up of the aggregates. Thus polyphosphates may act as dispersants rather than as thinners. Sodium hexametaphosphate is recommended for such cases by the British Standards Institution (1967).

Tannins, lignins, and quebracho perform as thinning agents at high electrolyte concentration. In deep drilling a high pH is occasionally required to prevent dispersion of natural clays from the formation. Tannin-treated slurries tend to solidify at the high temperatures frequently present in deep drilling.

FCL is a derivative of the CLS systems and is available as black powder and in a variety of grades. It is useful in applications involving replacement by fresh concrete, particularly if natural untreated bentonite is used. It imparts to the slurry the ability to effectively resist the effects of cement and also neutralizes the effects of seawater. Therefore, it is used to control viscosity and prevent excessive gelation caused by the action of calcium ions and salt. Although a small quantity generally is required (from 0.1 to 0.3 percent by weight), it may add appreciably to the cost of the application.

Flocculants and Polyelectrolytes

These are used to achieve the conversion of a thin slurry into a flocculated system. This effect is greatly enhanced if a clay is first sensitized by the addition of an organic polyelectrolyte. Sensitization (discussed also in Sec. 4-2) is explained as follows. At higher concentrations polyelectrolytes envelop the colloid particles and lend them stability and high salt resistance. In the region of sensitization, where a small amount of polyions is available, the polyion becomes adsorbed on the clay particles and forms a bridge between them. Thus particle linking, or flocculation, is promoted. This process is facilitated if a certain amount of salt is initially added to the suspension before the polyelectrolyte is used.

Sodium Carboxymethyl Cellulose This material, abbreviated CMC, is commercially produced in several varieties and grades according to the particle size. Three basic types are available, regular, coarse, and fine grade. Solutions of CMC provide a wide range of viscosities, and the flow properties resemble those of Bingham fluids. The viscosity of aqueous solutions of CMC increases quite rapidly with the concentration, and a favorable factor in this case is the random orientation of molecules, which presents increased resistance to flow. Because of the wide variation in the concentration of the various CMC grades in obtaining the desired effects, it usually is necessary to carry out tests to determine the proper blending of CMC and bentonite suspensions. CMC is effective where it is necessary to thicken, stabilize, gel, or modify slurries, including the filtration characteristics. The commerical brands produce a viscosity range from 100 to 300 cP in 1 to 2% aqueous solutions at 25°C.

When added to slurries, CMC causes an obvious immediate change by increasing the viscosity of the system, but a single solution will appear to have a different viscosity when different conditions of physical force are imposed on it. Because the long-chain molecules of CMC tend to orient themselves in the direction of the applied force or flow, they impart a pseudoplastic behavior to the system. This tendency enables CMC particles to flow toward the voids and seal the cake effectively, thus improving the filtration characteristics.

On the other hand, CMC has both a sensitizing and protective action, as shown in Fig. 4-25. The salt-flocculation resistance of a sodium montmorillonite suspension is plotted vs. the CMC concentration. At low CMC ratios the

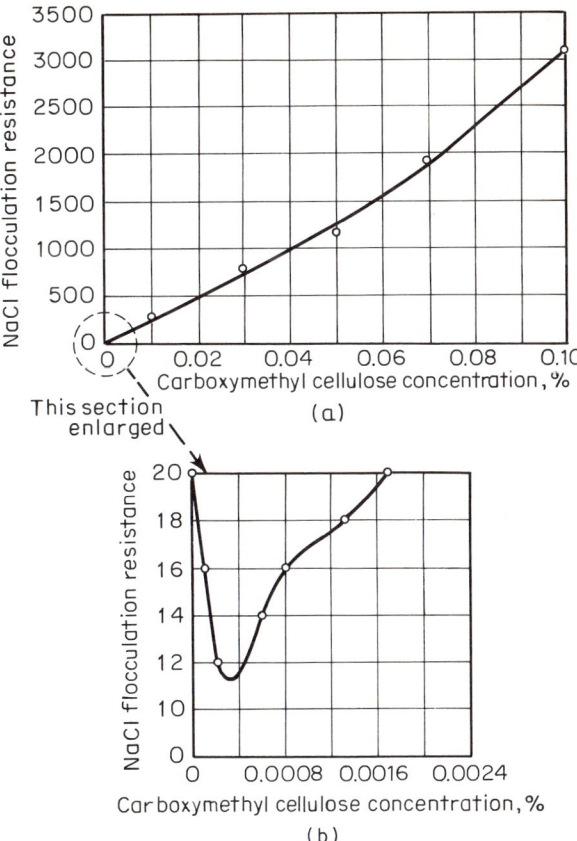

Fig. 4-25 Effect of polyelectrolyte (CMC) on a bentonite suspension: (a) protective effect; (b) sensitizing effect.

suspension becomes more sensitive toward flocculation by salt, whereas at higher concentrations CMC coats the clay particles, imparting a protective action and higher resistance to salt. This forms the basis for the use of CMC and other polyelectrolytes for applications in salt water, salt domes, and offshore drilling. The stabilizing effect under these conditions becomes quite important and far exceeds the effect of peptizers. If the polyelectrolyte is not salted out, it can be used in saturated salt conditions, and modified starches or CMC offer one of the best choices. The higher cost generally implied by the use of CMC usually is offset by the prolonged use of slurry.

Water

Almost invariably slurries for trench excavations are water-base systems. The water should be of good quality, contain a minimum amount of impurities, and have neutral consistency. As far as possible, it should be substantially clean, fresh, and free from oil, acid, alkali, organic matter, or other deleterious substances. Because water forms the major component of slurries, the required flow properties depend to a great extent on the quality of water. If other than city water is used, it is necessary to test the liquid. Hard water should be avoided where possible since it requires more bentonite and longer mixing time.

Whenever possible the use of salt water or seawater as the base liquid also should be avoided. Although the effects vary with the application, sodium chloride in excess of 500 ppm or calcium salt in excess of 100 ppm are likely to reduce the swelling ability of bentonite and produce low viscosity and gel.

Slurry-Loss-Control Agents and Lost-Circulation Materials

In gravel formations the addition of intermediate-sized particles to the initial slurry (mainly silt and sand) is very helpful. These particles are deposited in the pores and cause their blocking. Alternatively, low slurry loss is maintained by the use of starch, CMC, potassium aluminate, and aluminum chloride. A preservative is usually employed with starch to prevent fermentation. Most of these materials are organic polymers.

In very porous formations the addition of fibrous or flaky materials can improve the sealing process. A list of these materials is given in subsequent sections. Their use in a suitable slurry facilitates the effective filling of the pores and closes fissures. The process, however, involves a trial-and-error method with a wide range of selection.

When slurry loss is caused by excess hydrostatic head in relatively permeable soil, the addition of mica and cellophane is quite effective. In many instances and in average ground conditions it is possible to use sugar-cane fibers, wood shavings, and ground nutshells. The sealing operation must be carried out at the slurry-loss-producing zone and may involve a hit-and-miss operation. In extreme cases bentonite-cement mixtures or cement plugs are necessary. The portion of the panel involved is excavated, and the mixture is allowed to penetrate the loss area and then set. The remainder of the panel is then excavated under normal slurry. In badly fissured formations or in open cavities total loss of slurry may have to be accepted, and in this case it will be necessary to grout the formation before excavation.

Polymer Stabilizers

The rising cost of bentonite and the frequently high transportation costs have resulted in the development of new types of colloid materials. These consist mainly of long-chain organic polymers or inorganic silicate salts. Their use is expected to be beneficial in the following ways: (1) they are presumably easier to prepare and control; (2) they require only a fraction of the equivalent amount of natural bentonite, usually from 10 to 20 percent; (3) they yield overall economy because they facilitate recycling and thereby extend the use of slurry; and (4) they remedy some of the problems with regard to the disposal of slurries and their acceptance in public drains.

Another advantage of suspensions prepared from synthetic materials is their colloidal stability in the presence of salt and cement. This lack of sensitivity to the nature of the suspending electrolyte makes them suitable for salt-water and concrete applications.

Telmarch This is a highly successful stabilizer made in Japan, available both as organic polymer and as inorganic silicate salt. There are two types of Telmarch, one for general use and the other for salt formations or for seawater-based slurries. As shown in Fig. 4-26, satisfactory viscosity is attained at very low colloid concentrations. Because of this, the initial specific gravity of Telmarch slurries is also very low and of the order of 1.01, as opposed to bentonite slurries, which have an initial specific gravity of the order of 1.04 or higher.

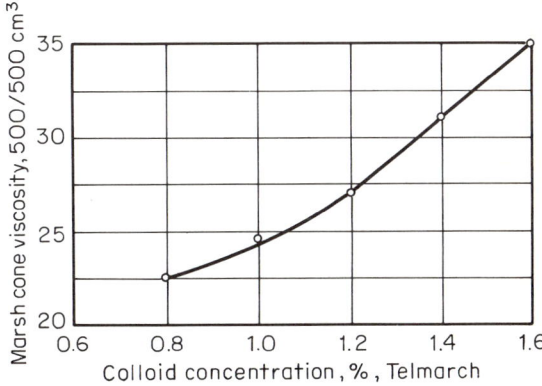

Fig. 4-26 Variation in viscosity with colloid concentration for Telmarch.

The long-chain polymers penetrate vigorously into the soil pores to form a filter cake of unusually high shear strength, and this gives the face a remarkable stability even in regions of overhanging boundaries. Because of the protective action, soil cuttings do not disperse easily in the slurry; hence the separation of solids is better facilitated, and this implies a lower rise of solid content in the slurry during excavation.

Changes in viscosity are less pronounced and often absent under conditions affecting conventional bentonite slurries. Thus, the viscosity of pure Telmarch suspensions hardly undergoes any change with time and is almost the same many days after the initial mixing. During excavation the viscosity tends to remain the same or undergo only minor changes even when the noncolloid solid concentration becomes as high as 15 to 20 percent. During concrete placement, which generally raises the pH and the viscosity of bentonite slurries and causes gelation, the viscosity of Telmarch suspensions tends to decrease. The pH, however, may rise as high as 12 when the cement content in the suspension reaches a ratio of 0.20, but this can be controlled by treatment with $NaHCO_3$.

4-7 FUNCTIONS OF SLURRIES

In the usual trench excavations slurries must perform a variety of functions and have certain characteristics, as follows:

1. Support the face of the excavation and also prevent the soil from sloughing and peeling off
2. Seal the formation and form the filter cake, preventing slurry loss to the ground
3. Suspend detritus, thereby preventing sludgy unconsolidated layers from accumulating at the bottom of the trench
4. Carry the cuttings in the slurry volume, thereby preventing sedimentation in the mud circuit

These functions must be compatible with certain associated operations. In addition, slurries must:

5. Ensure free flow of concrete from tremie pipes to allow complete displacement by fresh concrete without affecting the development of bond
6. Flow in pipes to facilitate materials handling from the excavation (the ability of slurries to lift the cuttings to the surface has a greater effect on the speed of excavation than any other single controllable factor)
7. Aid sedimentation in tanks and permit the separation of solids in shaker screens or cyclones
8. Facilitate their own disposal in dump areas or in public drains

Slurries must also reduce cavitation caused by tool disturbance and produce a filter cake that does not interfere with the in-and-out passage of tools; minimize fluid loss in excavated spoil; restart flow which is interrupted in the mud circuit; and perform more functions related to the conditions at hand. Thus the slurry requirements are typically conflicting and in some respects opposite. The first four operations generally imply relatively high viscosity and gel, but the last four functions require light and free-flowing slurries. A reasonable compromise in the flow and physical properties is therefore mandatory.

Although it is customary to indicate the slurry properties, it is hardly feasible to specify and proportion a slurry because of the wide variations in the properties of different bentonites and control agents. Although the usual range of bentonite concentration is from 3 to 10 percent, this gives no indication of the flow properties and erratic variations must be expected from different bentonites at the same concentration. This is shown in Table 4-2 for different bentonites at 6 percent concentration and for tap water. A further contrast is obvious by comparing these data with Table 4-3, which shows the properties of slurries prepared from selected Japanese bentonite for concentrations from 10 to 18 percent.

TABLE 4-2 Slurry Properties for Various Bentonite Brands at 6% Concentration

Source of bentonite clay	Plastic viscosity, cP	Gel strength, Initial	lb/100 ft² 10-min
Wyoming	16	0	0.5
Polymer treated	23	113	>200
England	12	32	87
Cyprus	3	1	1
London tap water	1	0	0

TABLE 4-3 Slurry Properties for Various Bentonite Concentrations (Japanese Bentonite)

Concentration, %	Density, g/cm³	Viscosity, cP		Yield strength, lb/100 ft²	Gel strength, lb/100 ft²	
		Plastic	Apparent		Initial	10-min
10	1.055	21	28	15	1.2	1.2
12	1.065	32	53	40	3.5	4.5
14	1.075	51	94	81	14	18
16	1.085	63	134	106	28	35
18	1.095	*	*	*	68	81

*Could not be measured.

4-8 CONTROL LIMITS

To help establish control limits for the slurry properties, Table 4-4 gives a summary of the physical and flow properties and also indicates the current test method. Since the bentonite concentration gives rise to the concept of density, it is listed as physical property. The relation between density and concentration depends on the specific gravity of bentonite and how it is expressed. Japanese bentonite has a specific gravity of 2.2, and at 6 percent concentration this would give the slurry a specific gravity of 1.035, while a change of 1 percent in concentration would change the specific gravity by 0.005. Wyoming bentonite has a specific gravity of 2.5

The colloid concentration C_c is usually expressed by weight. Thus

$$C_c = \frac{\text{lb bentonite}}{100 \text{ lb of water}} \times 100 \qquad (4\text{-}8)$$

Other convenient expressions are kilograms per 100 kg of water, pounds per cubic foot of water, or bags per 600 L of water. The expression of C_c as the ratio of the weight of bentonite to the weight of slurry (weight of water plus weight of bentonite) is seldom used.

Face Support Where stability requirements indicate the need for a heavier slurry than can normally be provided, noncolloid materials (soil or weighting

TABLE 4-4 Common Slurry Properties

Property	Definition or measurement	Current test method
Concentration	lb bentonite/100 lb water kg bentonite/100 kg water lb bentonite/ft^3 water	
Density	Mass of given volume of slurry	Mud balance
Plastic viscosity, apparent viscosity, yield stress	For a slurry behaving as a Bingham body, the flow law is $$\tau = \tau_0 + \eta_p D$$ where τ = shear stress τ_0 = yield stress η_p = plastic viscosity D = rate of shear τ/D = apparent viscosity	Fann V-G viscometer
Marsh cone viscosity	Time for 946 cm^3 (1 U.S. quart) of 1500-cm^3 volume to drain from a standard cone or time for 500 cm^3 of the 500-cm^3 volume to drain from cone (Japan)	Marsh funnel viscometer
Marsh cone gelation	Time for remainder of the 1500 cm^3 to drain from same cone	Marsh funnel viscometer
Initial gel strength	Minimum shear stress to produce flow designated as τ_s	Rotational viscometer
10-min gel strength	Shear strength obtained by allowing 10 min to elapse between stirring and reading	Rotational viscometer
pH	Logarithm of reciprocal of hydrogen-ion concentration	pH electrometer, pH papers usually not reliable
Filtration or fluid loss	Volume of fluid lost in given time from fixed volume of slurry when filtered at given pressure through standard filter	Filter-press test (but this procedure does not permit exact estimation); stagnation-gradient test more appropriate
Filter cake	Thickness and strength of filter cake for standard or actual conditions	Thickness measured in fluid-loss test, strength estimated from triaxial tests
Sand content	Percentage of sand greater than 200-mesh in suspension	API standard sand-content test using a sand-screen set

agents) are necessary. Whereas practical considerations control the maximum slurry density that cannot be exceeded, face stability usually determines the lowest possible unit weight of slurry.

Sealing Process and Slurry Loss Section 3-1 discussed the concept of cutoff concentration, which for many natural bentonites ranges from 4 to 5 percent. Since no direct procedure exists for estimation of the cutoff value, it is necessary to carry out tests. In addition to laboratory tests it often is advantageous to monitor the first panel or excavate a test panel in the same vicinity. Figure 4-27 shows the results from such a panel, and evidently the fluid loss rises sharply below a certain concentration, which in this case may be taken as 4.5 percent.

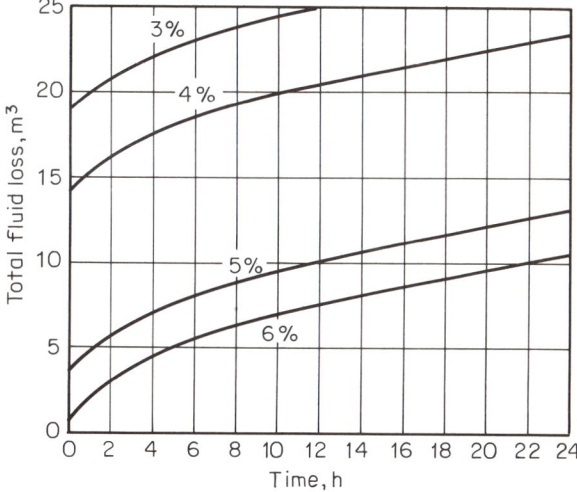

Fig. 4-27 Variation in total fluid loss with time from a slurry-trench panel 15 by 100 by 2 ft. Ground permeability 5 × 10^{-3} cm/s. (*From Hutchinson et al., 1974.*)

Suspension of Excavated Material If the earth materials from the excavation settle to the bottom, they will form a soft sludge layer which generally is not removed or displaced by the advancing concrete. In order to avoid this problem it is necessary to minimize detritus settlement by keeping in suspension any soil accidentally mixed with the slurry during excavation. According to Eq. (3-3), a shear strength of 75 dyn/cm² (about 15 lb per 100 ft²) will support sand particles 1 mm in size, which is the average particle size of coarse sand. Some doubt exists whether this should be the initial or the 10-min gel strength, but according to experience the latter gives satisfactory results. A 10-min gel strength of 15 lb per 100 ft² can be provided by most slurries at 4% bentonite concentration. At lower concentrations the suspending ability of the slurry is sharply reduced, and this allows considerable quantities of sand to settle out, as shown in Fig. 4-28.

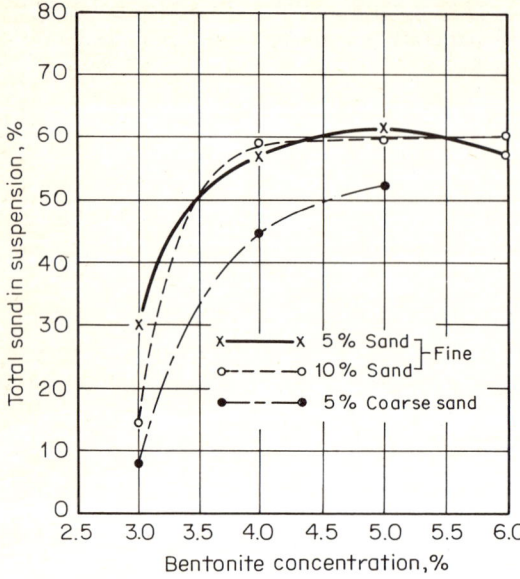

Fig. 4-28 Variation of sand in suspension with bentonite concentration. (*From Hutchinson et al., 1974.*)

Alternatively, in gravel formations higher gel strengths are necessary to prevent earth materials from settling out, but a compromise usually is necessary to avoid conflict with other operations. In this case the bottom is cleaned by passing an air lift before concrete placement. Likewise, settlement in reverse hoses, pipes, pumps, and the associated appurtenances can cause their blockage and result in material-handling difficulties. Again, the problem is avoided if the transporting slurry has a certain gel strength.

Displacement by Concrete This process is one of the most complex phases. The slurry must be displaced from the bottom, from around inserts, boxes, steel bars, and panel joints, leaving clean contact surfaces. So far it is confirmed that the density of slurry is a prime factor, but the viscosity and gel strength also affect the process.

It generally is assumed that displacement occurs in two distinct modes: initially at the bottom of the excavation, where the fresh concrete flows laterally, and then by pushing the slurry upward from the wall and from around the reinforcement bars. The density of the slurry should not rise above 78 to 80 lb/ft^3, which is equivalent to a maximum specific gravity of 1.25. This has been determined empirically by observing the occurrence of incomplete placement and is based on normal viscosity and gel-strength ranges.

The plastic viscosity should not be greater than 20 cP during concrete placement (Hutchinson et al., 1974), which in some instances means a maximum bentonite concentration of 10 to 12 percent. This was derived from a theoretical analysis of the displacement of slurry from vertical surfaces and was supported by laboratory tests. According to these tests, the thickness of slurry left on a vertical face is

$$\delta = \left[\frac{2\eta_p v}{\Delta_\partial g}\left(\frac{2z}{z_0} - 1\right)\right]^{1/2} \quad (4\text{-}9)$$

where δ = thickness of layer
η_p = plastic viscosity
v = velocity of displaying front
Δ_γ = differential density between concrete and slurry
z_0 = height of interface above initial position
z = distance above initial position

Figure 4-29 shows the ratio zv/z_0 plotted vs. the factor δ^2 from Eq. (4-9) and also from laboratory tests (Hutchinson et al., 1974). The control value of 20 cP is obtained from the foregoing relation assuming that a slurry coating along a vertical face is absorbed or swept by the fresh concrete if its thickness does not exceed 0.01 cm. This is not intended, however, as general criterion but as a first guideline; indeed, the condition of the vertical surface, the direction of flow, and concrete contamination at the interface are some of the factors that could change these control limits.

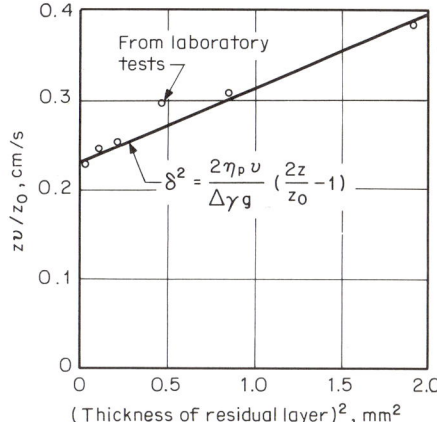

Fig. 4-29 Variation in the thickness of residual bentonite layer. (*From Hutchinson et al., 1974.*)

Separation of Noncolloid Fraction Slurries are recirculated, cleaned, and reconditioned for repeated use. This generally is possible if solid particles larger than 100-mesh are removed, but sometimes this may give rise to high levels of silt concentration. Larger cuttings and sand are separated in vibrating screens, and finer particles are intercepted in cyclones or allowed to settle in special tanks. The efficiency of vibroscreens depends on the solid content and on the viscosity and is reduced if the slurry contains more than 30 percent sand. Whereas no strict viscosity limits exist above which screening becomes slow, a more efficient collection of entrained sand is possible with a relatively thin slurry.

No controls can be established for the separation of clay cuttings since only well-preserved samples are susceptible to mechanical separation. The only cleaning process for slurries contaminated with natural clays is treatment with thinners, but this is not always possible to the degree desired.

Pumping of Slurry This is important in reverse circulation. Streamline flow is expressed by Eq. (4-7). Hydraulic pipelines and the transport of dispersed systems through mud circuits are discussed by Cheng (1970). For turbulent flow predictions are based on empirical data. For effective pumping sand particles and clay cuttings must not settle in the lines if flow is interrupted, and it must be possible to restart flow after a shutdown.

Because of thixotropy higher pumping pressures are required to start flow than to maintain it. Therefore, when the pump is restarted, the applied unit pressure p must overcome the shear strength of the slurry attained during shutdown. This requires

$$\frac{pr}{2L} > \tau_f \tag{4-10}$$

in which r is the pipe radius and L the pipe length. The shear stress τ_f in this case is taken as the 10-min gel strength, but often it is much higher. For example, given $p = 50$ lb/in^2, $r = 3$ in, and $L = 1000$ ft, the flow can be restarted if the shear strength of the slurry is less than 90 lb per 100 ft^2. If it exceeds this value, the pipeline may have to be emptied if left shut down for more than 1 h.

Summary of Control Limits

Certain control limits based on the preceding discussion are summarized in Table 4-5. These limits are quite general, but the range can be further narrowed down or refined where the scope of work is known more specifically. There is some question whether and when the apparent viscosity, Marsh cone viscosity, plastic viscosity, yield strength, and 10-min gel strength are relevant and must be used for the control of slurries.

Reference to the flow curve shows that the apparent viscosity depends on the rate of shear of the measuring system. Plastic viscosity measures resistance to flow for ideal slurries or for slurries of very low initial shear strength. For the usual conditions the plastic viscosity must be combined with the 10-min gel strength to describe the flow behavior.

Although Marsh cone tests do not provide absolute viscosity measurements, they furnish useful data for routine site work. They are therefore simple and practical in relating the slurry properties to the soil conditions on a comparative basis.

The Bingham yield stress is important when studying the flow curve and useful for a theoretical analysis of colloidal behavior. However, in practice it is seldom necessary to estimate this stress, and furthermore its correlation to the initial or the 10-min gel strength is difficult.

TABLE 4-5 Control Limits for the Properties of Slurries*

Function	Average bentonite concentration,† %	Density, lb/ft³	sp gr	Plastic viscosity, cP	Marsh cone viscosity	10-min gel strength (Fann), lb/100 ft²	pH	Sand content, %
Face support	>3–4	>64.3	>1.03	.....	Limits established by soil type	‡	.....	>1§
Sealing process	>3–4	.....	.....	.....		.....	.....	1
Suspension of detritus	>3–4	.....	.....	.....		>12–15	.....	
Displacement by concrete	<15	<78	<1.25	<20		.....	<12	<25
Separation of noncolloids	.....	.....	.....	.....		.....	.....	<30
Physical cleaning	<15	<78	<1.25	.....		.....	.....	<25
Pumping of slurry	.....	.....	.....	.....		Variable	.....	
Limits	>3–4	>64.3	>1.03	.....		>12–15	<12	>1
	<15	<78	<1.25	<20		.....	.....	<25

*Controls are not considered necessary for apparent viscosity and yield stress. Whereas fluid loss commonly is judged by standard filtration test and a maximum film thickness of 2 mm, better control limits are established by stagnation-gradient tests.
†Should be expected to vary widely because of different bentonite brands.
‡The shear strength of filter cake is more applicable to peel-off control (also the time required for its formation).
§Optional.

4-9 EFFECT OF SOIL CONDITIONS ON CONTROL LIMITS

Besides the control limits imposed by the functions of slurries, the type of soil also dictates the composition of slurry. Two considerations generally are important: the basic classification and strength of soil and the permeability of soil, together with the presence or absence of groundwater.

A slurry having bentonite concentration from 3 to 6 percent will be retained in ground of permeability 10^{-1} to 10^{-2} cm/s. In more pervious formations a denser slurry may be more effective. Where the permeability is exceptionally high, slurry-loss-control agents should be used, together with higher than normal bentonite concentrations. If the permeability exceeds about 5 cm/s, it may be difficult to obtain a seal (Sliwinski and Fleming, 1974). On the other hand, if the ground permeability is exceptionally low, filtration may not occur. Samples of clay taken from the face of diaphragm walls during excavation show little if any trace of bentonite when examined closely.

Other soil characteristics that influence the composition of slurry are (1) the sticky limit, e.g., sticky clays are troublesome and can cause shutdowns in the drilling bits and pipelines; (2) the cohesion limit, which is the lowest moisture content below which soil crumbs do not stick together; and (3) the type and percentage of salts and other aggressive material existing in a natural state or contained in the groundwater. In erratic deposits it is necessary to locate lenses and pockets of very pervious formations, gross fissures, and underground water-bearing seams.

The following are examples of potentially troublesome situations. Excavation in very pervious ground with low water table can result in high slurry loss, particularly if there has been a rapid drawdown before the operation. This situation deserves extra care, and the main excavation should be preceded by an exploratory test panel. The slurry should be light, to keep the differential pressure low, and should contain more bentonite, sealants, and plugging materials than soil.

Excavation in very soft clay with low cohesion may give rise to stability problems. More important, however, is that clay cuttings are likely to be dispersed in the slurry before reaching the separation units. This will raise the contamination level to a point where treatment is no longer practicable, and the only alternative is to replace the slurry with fresh suspension.

Excavation in natural salt formations or in the presence of salt water requires the use of special bentonite and control agents, and the best control measure is the frequent testing of slurry. Despite the broad range of effects, practical experience shows that the tendency is toward flocculation.

If the upper crust of the ground consists of cohesive materials and is underlain by a layer of granular soil, it may be sufficient initially to prepare a thin slurry. As the excavation proceeds, the slurry will retain enough natural clay and thus attain good viscosity and gel before reaching the granular layer. If the slurry fails to develop a suitable colloidal condition, it should be replaced with normal bentonite suspension.

4-10 PROPORTIONING THE SLURRY

The goal of maximum economy in the preparation and use of slurries is fulfilled by selecting the thinnest slurry possible provided it is contamination-resistant and can perform the intended functions. A conventional slurry contains the following materials, proportioned according to the effect on flow properties and control limits: (1) a liquid, usually fresh water; (2) a colloid solid, usually bentonite; (3) control agents; and (4) noncolloid solids, initially added or entering the slurry as by-product of the excavation.

The current practice of proportioning these materials varies widely. Visual inspection and judgment based on practical experience often produce slurries by the rule of thumb, and this is acceptable as one extreme. At the other extreme lies a complex specification requiring many and different laboratory tests. The first approach is oversimplified and can be deceptive and dangerous; the second approach often produces a mass of data which are difficult or impossible to apply to practical site controls. Strict control limits often have no basis and simply make the work more difficult and costly. Therefore, it is common sense to rule out these extremes and proportion slurries on the basis of simple steps and procedures.

Initially a slurry must be proportioned for the soil layer which has the highest permeability and is most vulnerable to sloughing and peel-off. This tendency of the exposed face is shown in Table 4-6 for the most common soil types. The entry "none" indicates that the face may stay stable but not indefinitely; "some" means that sloughing and peel-off do not occur for some time after the face is exposed; "appreciable" indicates that caving can occur at any time, whereas "high" and "very high" preclude excavation unless the face is protected. The tendencies exemplified in Table 4-6 are for unsupported faces.

TABLE 4-6 Relation between Soil Grains and Tendency to Collapse

Type of soil	Tendency to collapse	
	Dry soil*	Soil with water*
Clay	None	None
Silt	Usually none	Some
Silty sand	Some	Appreciable
Fine sand (moist)	Appreciable	Appreciably higher
Coarse sand	Appreciably higher	High
Sandy gravel	High	Very high
Gravel	Very high	Very high

*See text for elaboration of meaning of these terms.

The most critical soil layer can be correlated with the viscosity requirements by means of Table 4-7. This correlation is based on practical experience, and although it is given in terms of the funnel viscosity, it provides entirely satisfactory results. Thus, the proportioning of slurry involves the following steps.

TABLE 4-7 Funnel Viscosity for Common Types of Soil

Type of soil	Funnel viscosity, s/946 cm³	
	Excavation in dry soil	Excavation with groundwater
Clay	27–32	
Silty sand, sandy clay	29–35	
Sand, with silt	32–37	38–43
Fine to coarse	38–43	41–47
And gravel	42–47	55–65
Gravel	46–52	60–70

1. Determine the noncolloid fraction (from density requirements) necessary for trench stability. This may include some soil from the excavation; note, however, that excessive retention may spoil the slurry.

2. Select the funnel viscosity by reference to Table 4-7.

3. Establish the applicable control limits from Table 5-5. Refine these limits, if necessary, for the particular aspects of the operation.

4. Determine whether control agents (peptizers, polyelectrolytes, fluid-loss-control materials, etc.) are necessary and economically justified.

5. Proportion the constituent materials. This phase merely consists of a quantitative estimation of bentonite, noncolloid fraction, and control agents. The proportioning may be empirical and depend on experience if the properties of the materials selected are known, or it may have a technical basis of tests and estimations.

The relative economy of slurries depends on the actual cost at the site of the basic materials, namely bentonite or other colloid, chemicals, weighting agents, and water, hence on the availability of materials in the immediate locality. Where good-grade bentonites are not available locally, transportation costs often are the main factor. Thus, in some instances it may be better to transport high-grade, contamination-resistant bentonites, whereas in other cases it may be more economical to use local bentonites and blend them with control agents. In general, bentonites of widely varying properties may be used alone or in conjunction with other materials to produce slurries of suitable flow properties and consistency.

4-11 PREPARATION AND CONTROL OF SLURRIES

Bentonite slurries usually are prepared from fresh soft water unless it is unavailable. As we have mentioned, the degree of hardness has an influence on the flow properties. Slurries in soft water (hardness less than 50 ppm or 3 grains) attain the required viscosity and gel at lower colloid concentrations. The same slurries in hard water require 10 to 12 percent more bentonite to produce the same flow properties and longer mixing time. With fresh water the composition is relatively simple, mixing time is reduced, and the flow properties are better controlled. Additionally, the slurry may be used repeatedly.

Mixing

The time taken for full hydration, at which bentonite attains the desired flow properties, depends on the method of mixing. A well-mixed slurry must be fully hydrated. A sensitive measure of the degree of hydration is the 10-min gel strength, measured on the Fann viscometer. The minimum acceptable level of hydration corresponds to 50 percent of the fully hydrated value for a 4 to 5 percent slurry. Thus, mixing is considered satisfactory if the minimum 10-min gel strength is about 36 dyn/cm^2 or from 7 to 8 lb per 100 ft^2. Bentonite dispersions prepared with high-shear mixers have a much higher rate of hydration than those prepared with anchor stirrer and give slurries of higher final shear strength.

Bentonite is not readily dispersed in water. Because of its remarkable ability to absorb water, certain difficulties may arise from the formation of partially wetted lumps of colloid material. Bentonite must therefore be added to a stream of moving or agitated water slowly. The mixing is then continued through mechanical mixers or by passing the mix through a pump.

Standard mixing units include mechanical mixers or a combination of tank with a conical bottom which flows into a pump (Fig. 4-30). The pump assures complete mixing-tank turnover every 2 to 3 min. In this operation, about 60 percent of the pump output is recirculated into the mixing unit, and the remaining is pumped into the trench.

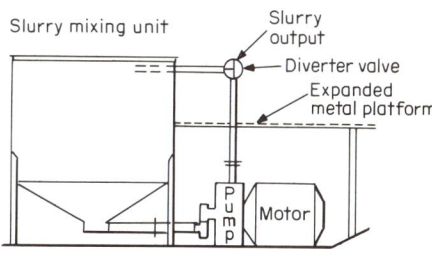

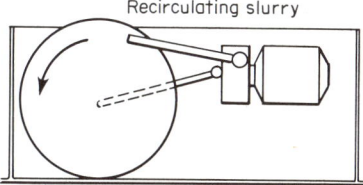

Fig. 4-30 Diagram showing facilities for mixing bentonite slurries.

Control agents and materials are mixed separately in suitable solutions and slowly streamed into the bentonite suspension while the latter remains in circulation. When all solid materials have been introduced into the suspension, it is advisable to keep the mixer in motion for additional time to produce further swelling and give uniform consistency. Mixers generally allow the swelling of bentonite to reach 92 to 95 percent expansion, but bentonite suspensions continue to expand for several hours after mixing. It is therefore advantageous to have the mixed slurry stand for some time to attain full swelling before use.

Volume of Slurry The volume of slurry required for an excavation is quite important and should be taken into consideration since it may require more storage space than usual. An ample supply of slurry must be at hand to replace slurry lost. The main causes of slurry loss are (1) initial filtration and continuous leakage to the ground; (2) waste of slurry with cuttings at the separation units; (3) disposal due to deterioration; and (4) some spilling and waste over the guide walls.

In some instances contractors prefer to rely on past experience with similar jobs for an estimation of the total volume of slurry, after also taking into consideration the number of intended uses. A semiempirical approach, on the other hand, allows an estimation of the total slurry volume in terms of all the factors affecting slurry loss. In this case the required volume of slurry is (Nakayama, 1976)

$$V_s = \frac{V_e}{n} + \frac{V_e}{n}\left(1 - \frac{K_1}{100}\right)(n - 1) + \frac{K_2}{100}V_e \qquad (4\text{-}11)$$

where V_s = required volume of slurry
V_e = total volume of excavation
n = number of panels
K_1 = rate of recovery during concrete placement, %
K_2 = slurry-loss ratio, %; slurry lost due to any of reasons listed

Field Testing

Efficient field control of slurries must be maintained by simple and manageable procedures. This evaluation of the useful and important field tests is reflected in the FPS (1973) specification, which shows a reasonable balance in the field program, but is also confirmed by contractors.

Tests must be carried out before, during, and after certain operations. The standard program includes density (specific gravity), viscosity, filter-press, sand-content, and pH measurements. The slurry sample as well as the sampling point must be consistent with the purpose of the test and relate to the phase for which an estimation of the slurry properties is desired. Thus, appropriate sampling points are rotated to include the mixer, the trench, and the mud-treatment plant. For average situations a reasonable testing program is as follows:

1. After the slurry is fully mixed, measure the specific gravity, funnel viscosity, and filtration. Before use measure the 10-min gel strength.
2. As a daily routine check the specific gravity, funnel viscosity, and filtration.

3. Before concrete placement check the specific gravity, plastic viscosity, and pH.

4. Repeat the measurements listed in step 3 at the end of a pour (slurry in storage tanks).

5. Check the specific gravity, funnel viscosity, and filtration during excavation.

6. After a slurry is reconditioned for reuse, repeat the measurements listed in step 1.

7. Check the slurry upon the occurrence of certain events (rainfall, contamination with materials from the ground, trench left open for a long time, etc.).

Contamination and Treatment

During the operation bentonite slurries change in color and overall appearance. They generally become denser and may partly lose the thixotropic characteristics; toward the end of the excavation they become difficult to pump. The cause and extent of these changes vary with the operation, but they are mainly due to the introduction of soil and cement particles into the slurry.

Effect of Soil Physically inert soil particles raise the slurry density but can also affect the flow properties. Clay from natural formations usually raises the viscosity and can even result in unreasonably high gel strength. Sand causes viscosity humps and at higher contents gives rise to low gels. Since groundwater from soil being excavated or rainwater causes dilution, it lowers the viscosity and gel strength. These changes typically occur in different ways and to different degrees.

Lower viscosity and gel resulting from mixing with sand and pore water or rainwater generally are adjusted by a balanced addition of bentonite or CMC. When the solid content becomes excessive, the entire suspension must be recirculated and treated with mechanical dispersion systems to remove the sand. If the viscosity and gel increase due to clay uptake, treatment, which may begin with the addition of fresh water, is gradually completed while the slurry is recirculated in a tank. If this fails to restore the flow properties, mud thinners in weight ratios from 0.05 to 0.30 percent may be effective. The thinning process should never be applied in the trench. Excessive clay uptake may necessitate the disposal of the slurry and its replacement with fresh solution.

Effect of Organic Material Experience with slurries in which the base water contains a small amount of sewage shows that the effect on the flow properties is only minor and may be disregarded (Xanthakos and Bailey, 1975). This conclusion is valid for slurries prepared from natural, untreated, pure bentonites which have a remarkable ability to resist contamination by sewage. However, if the sewage content in these slurries becomes excessive, it may seriously affect the flow properties.

Great caution should be exercised to avoid any sewage in slurries prepared with treated (peptized) bentonites. For these slurries mixing with sewage has an immediate adverse effect on the flow properties, and there have been instances where this resulted in complete loss of colloidal stability, causing the trench to collapse (Tamaro, 1975).

Effect of Cement Contamination by cement is the most serious and frequent problem in slurries. It occurs when calcium ions are exchanged for sodium ions. It is the most common form of contamination and invariably is observed to some degree during concreting. It results in thick, permeable filter cake, high viscosity, and high slurry loss. Calcium may come from soil deposits such as gypsum or anhydrite, but the common source is the cement in the concrete.

The effect of cement on fluid loss is shown in Fig. 4-31 for a 5 percent slurry prepared from natural bentonite. Fluid loss is seen to increase sharply at about 0.3 percent cement contamination, which is equivalent to a pH of about 11.7.

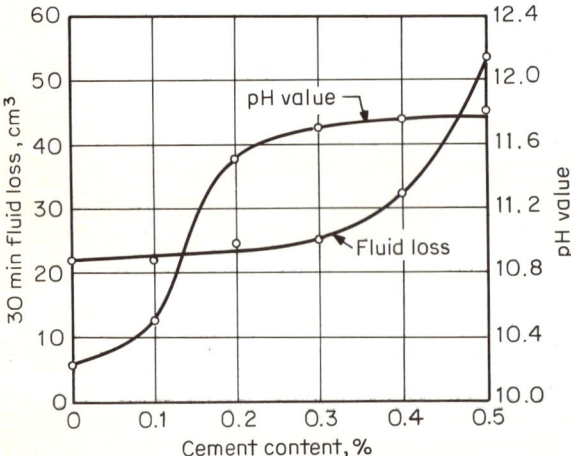

Fig. 4-31 Effect of cement contamination on fluid loss. (*From Hutchinson et al., 1974.*)

The effects of contamination by cement are avoided by treating the slurry with control agents before use. Figure 4-32 shows the results of treating with FCL and the effective control of slurry prepared with Japanese bentonite. The increase in viscosity due to mixing with cement is shown by curve a for plain slurry. Curves b_1, b_2, and b_3 illustrate the effect of FCL at various concentrations. The better bentonite grades resist contamination by cement more effectively and respond to treatment with FCL more vigorously.

Contamination effects are completely absent with certain polymer stabilizers. In Japan the problem of contamination by cement has been solved with the use of Telmarch, which remains substantially free from change when admixed with cement (see also Sec. 4-6).

Fig. 4-32 Control of cement contamination with FCL. (*Tone Boring Co., 1974.*)

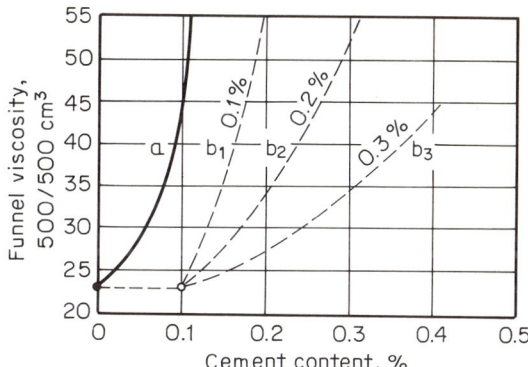

Effect of Salt Bentonite added to a salt solution does not swell readily and sometimes may even fail to swell completely. However, if an acceptable dispersion can be prepared with seawater as base fluid, the addition of polyelectrolytes benefits the colloidal stability and reduces the filtration rate. For such a slurry further mixing with salt water does not change the flow properties significantly.

If a fresh-water-based slurry encounters seawater or pore water containing NaCl, it will retain a certain amount of salt but at a ratio much lower than in the seawater. The corresponding result may be (1) a decrease in viscosity and gel; (2) a sharp increase in viscosity and gel due to flocculation effects; or (3) a succession of both in that order. Experience, however, seems to indicate that the usual salt effect is tendency toward flocculation; for relatively thin slurries the viscosity increases; and for more concentrated slurries the addition of salt leads to complete gelation. Accordingly, where salt effects are expected the common practice is to use peptizers and monitor the slurry closely as soon as the salt formation is reached.

When excavating in ground reclaimed with sea-dredged sand only limited contamination should be expected. Experience with such coastal sites shows that the basic functions of slurry still are attainable and the effects of salinity are mild and seldom detrimental to these functions.

Tables 4-8 and 4-9 present a summary of the most common problems and the most common slurry materials. The preferred treatment is also indicated. These tables are quite useful for immediate reference.

4-12 SLURRY DISPOSAL AND SALVAGE

Bentonite slurries are pumped out of the trench for disposal or for cleaning, reconditioning, and reuse. At some point it is not possible to avoid the disposal of a certain residual quantity. Even if the very fine contaminating particles can be removed and the flow properties regained, the process will eventually be possible only at considerable expense.

TABLE 4-8 Summary of Slurry Problems and Treatment

Problem	Control and treatment
To increase viscosity and gel in fresh water	Add bentonite, CMC, or both
To reduce viscosity and gel when slurry has adequate colloid material	Add water slowly or treat with thinners
To reduce viscosity and gel due to high noncolloid solid content	If solids are not completely dispersed, use mechanical separation; add water slowly and thinners
To reduce viscosity and gel when dilution is inadvisable because of inadequate colloid material or weight reduction	Add thinners; if viscosity drops appreciably and overtreatment occurs, adjust using CMC
To increase viscosity and gel due to high noncolloid solid content (sand)	Remove solids by mechanical separation; add bentonite or CMC
To decrease density	Recirculate fluid to remove solids by mechanical separation or by allowing them to settle; do not add water, but adjust flow properties if required after the density is decreased
To reduce filtration rate and thickness, i.e., reduce fluid loss	Add bentonite and CMC; if viscosity becomes too high, treat with FCL or other thinners
To handle large volumes of entrained sand and cuttings	Use mechanical dispersion; avoid adding water and chemicals
Salt flocculation from contamination by seawater	Add FCL but keep close control of slurry
Salt flocculation in offshore drilling and excavation in salt formations	Stabilize solution through the protective action of CMC or use thinners
Trench excavation in sand and gravel (sand will increase density, decrease viscosity, and aggravate tendency toward lost circulation)	Provide adequate initial gel strength to keep sand in suspension; build good filter cake and film to keep fluid loss low; use higher bentonite concentrations and add CMC
Trench excavation in clay	Keep viscosity and gel low; use thinnest suspension colloidally stable; use thinners
Trench excavation in shale	Reduce filtration rate to prevent hydrous disintegration or sloughing of formation; add bentonite and CMC; monitor slurry level to control sudden loss of fluid
Excavation in erratic formations	Base selection of slurry on most critical formation; make periodic adjustments
Lost circulation	Use lost-circulation materials; maintain minimum safe slurry weight
Contamination with cement	Add FCL or other thinning agents; if restoration is not achieved, reject slurry; use pretreated bentonites
Contamination with organic matter and sewage	Avoid peptized brands; use natural bentonite and monitor slurry closely

TABLE 4-9 Common Slurry Materials and Additives

Weight materials	Barite (barium sulfate) or soil (sand)
Colloid materials	Bentonite (Wyoming, Fulbent, Aquagel, Algerian, Japanese, etc.), basic freshwater slurry constituent
	Attapulgite, for saltwater slurries
	Organic polymers and pretreated brands
Thinners and dispersing agents	Quebracho, organic dispersant mixture (tannin)
	Lignite, mineral lignin
	Sodium tetraphosphate
	Sodium humate (sodium humic acid)
	Ferrochrome lignosulfonate (FCL)
	Nitrophemin acid chloride
	Calcium lignosulfonate
	Reacted caustic, tannin (dry)
	Reacted caustic, lignite (dry)
	Sodium acid pyrophosphate
	Sodium hexametaphosphate
Intermediate-sized particles	Clay, silt, and sand
Flocculants and polyelectrolytes	Sodium carboxymethyl cellulose (CMC)
	Salts
	Starches
	Potassium aluminate
	Aluminum chloride
	Calcium
Fluid-loss-control agents	CMC or other flocculants
	Pregelatinized starch
	Sand in small proportions
Lost-circulation materials	Graded fibrous or flake materials; shredded cellophane flakes, shredded tree bark, plant fibers, glass, rayon, graded mica, ground walnut shells, rubber tires, perlite, time-setting cement, and many others

There are guidelines for the reconditioning and reuse of bentonite slurries, but often the principles of recycling have little practical value. Some contractors prefer a single use followed by disposal, but this preference is rather academic since the same standards and guidelines cannot be applied to every situation. In urban areas mud disposal has become an important and costly operation because of the difficulty of finding acceptable places for dumping or disposing of the slurry. In such cases it is expedient to extend the use of slurries, and this is even more desirable if the availability of mixing facilities is limited. Thus, a final decision often cannot be made until all the factors have been considered and the cost of reconditioning has been compared with the cost of single application.

Used slurry cannot be discharged in public drains and sewers at the same concentration and with its contaminant loads. The contention is that it may gel and cause environmental problems. A simple and inexpensive way is to dilute the slurry to 1 to 2 percent since at this concentration it will generally behave like free-flowing fluid. Because of loss of its suspending power most of the colloid solids will pass out of suspension and deposit to the bottom of settling tanks.

In Japan the problem of mud disposal was a prime factor in the development of new colloid systems. An example is the development of Telmarch, which, because of its ability to resist changes, including density rise, allows the recycling of slurry and eventually its conversion into disposable forms. Elsewhere mud-disposal techniques have not changed appreciably and include (1) direct dumping at the site if space allows and the owner approves; (2) mechanical dispersion and free precipitation, or both; and (3) disposal in dump trucks if acceptance of diluted slurries in public disposal systems is categorically prohibited by local ordinances. These limited options in some instances leave no choice but to use slurries of the lowest possible colloid concentration.

There is a trend toward salvaging slurries, from which the noncolloid fraction is removed, and mud salvage may become a new industry. The basic principle is to salvage used slurries to the extent possible and recondition and convert them into usable slurries in a central plant. Road tankers with capacity up to 100 barrels are used to deliver reconditioned slurries, and this operation is successful particularly with invert-emulsion muds.

4-13 BLEEDING OF SLURRIES

The permanence of the permeability reduction produced by a colloidal slurry in the voids of gravel depends to a great extent on the base liquid. Freshwater slurries do not show much change with time, but slurries prepared from Dead Sea water have shown considerable bleeding when left at rest in a beaker or tube; i.e., the solids settle, leaving a free water layer at the surface. If bleeding occurs in the slurry-gravel mixture, e.g., in rheological blocking, pockets of free water can develop in the gravel voids. If adjacent pockets can be connected, the formation will no longer be sealed in the horizontal direction. This has immediate practical significance in earth cutoff walls consisting of gravel and slurry only; bleeding in this case may render the cutoff inoperative with time. The vertical permeability of a bleeding system should not be affected unless settlement occurs, leading to the vertical linking of water pockets.

Jefferis (1972) used attapulgite clay slurries prepared from Dead Sea water to test the permeability of $\frac{3}{8}$- to $\frac{1}{4}$-in Thames gravel. Blocking of the gravel voids was achieved at 4 and 5 percent attapulgite concentration. The results are shown in Fig. 4-33, and evidently there is only slight difference between the vertical and horizontal permeability. This difference shows a slight tendency to decrease with increasing gradient, especially for the 4 percent slurry, and might cause some bleeding, explained as the preferred orientation of attapulgite particles under the action of gravity. This phenomenon is known to occur in dilute clay suspensions and is utilized to prepare oriented samples for x-ray diffraction (Grim, 1962).

Technology, Preparation, Uses, and Control of Slurries

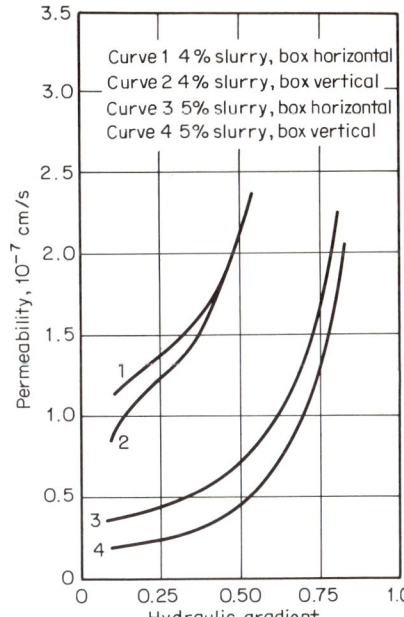

Fig. 4-33 Permeability of gravel mixed with Dead Sea attapulgite slurry. (*From Jefferis, 1972.*)

Bleeding of Slurries in Glass Tubes Jefferis (1972) has tested dry-clay slurries in six glass tubes (diameter from 0.2 to 1.5 cm) to study the bleeding phenomenon. The tubes were filled to a depth of 25 cm, placed vertically, and capped with wax. Settlement was observed for more than 6 months.

In the two smallest tubes (0.2 and 0.3 cm) bleeding did not occur. The largest sample (1.5 cm) continued to bleed approximately for 3 months, but the total settlement for this time was only 0.8 cm. The three intermediate tubes (0.4, 0.6, and 0.8 cm) settled about 0.4 cm over a period of 2 weeks but the settlement stopped thereafter. These observations indicate that the tube diameter influences bleeding, and the slurry is in this case regarded as free body supported along the wall of the tube by its gel (shear) strength.

Figure 4-34 shows a column of slurry of uniform density and gel strength. The slurry undergoes a certain bleeding which produces pore fluid of depth h. For a section at depth z below the slurry surface the total vertical stress is

$$\sigma_v = \gamma_f z + \gamma_e h - \frac{S}{A} \qquad (4\text{-}12)$$

where γ_f = slurry density
  γ_e = pore-liquid density
  A = column cross-sectional area
  S = total shear force at wall for depth z

Rewriting Eq. (4-12) in terms of the effective stress $\sigma'_v = \sigma_v - \gamma_e(h + z)$ gives

$$\sigma'_v = (\gamma_f - \gamma_e)z - \frac{S}{A} \qquad (4\text{-}13)$$

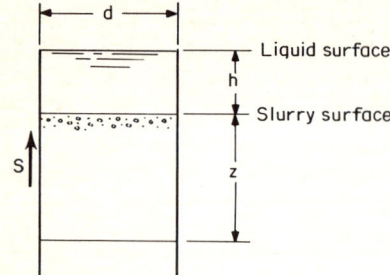

Fig. 4-34 Bleeding of slurry in a column.

If the gel strength of the slurry is τ_f, then

$$S = \pi dz \tau_f \quad \text{and} \quad A = \frac{\pi d^2}{4}$$

which, combined with Eq. (4-13), yield

$$d = \frac{4\tau_f}{(\gamma_f - \gamma_e) - \sigma'_v/z} \tag{4-14}$$

If the slurry develops a gel strength which allows its support at the wall by shear forces, bleeding should be prevented. If during bleeding the slurry attains a certain gel strength (because of thixotropy or increased concentration), bleeding should be stopped. When the slurry is supported by wall shear, the effective stress is zero; hence

$$d = \frac{4\tau_f}{\gamma_f - \gamma_e} \tag{4-15}$$

However, experience shows that Eq. (4-15) expresses bleeding only approximately, and does not take into account all the variables involved, including temperature equilibrium, interparticle interaction, and interparticle forces. It is even more difficult to express bleeding theoretically in a gravel medium because of the difficulty of selecting a representative pore size and of predicting the shear strength of slurry.

REVIEW PROBLEMS

4-1 A colloid is added to a liquid at 2 percent concentration by volume. If the suspension is assumed to be dilute, estimate the probable viscosity.

4-2 Discuss the disadvantages of the early fluid systems. Describe the significant developments in fluid technology.

4-3 Distinguish the main functions of slurries in drilling and trenching.

4-4 Explain the colloidal condition of matter and give examples of colloidal systems.

4-5 Describe flocculation, peptization, and deflocculation. Distinguish between the effect of peptizers and the effect of salts and describe all the factors contributing to peptization.

4-6 Two suspensions are prepared from fresh water and 3 percent Wyoming bentonite. One is natural montmorillonite, the other polymer-treated. Samples are placed in glass containers and allowed to settle for some time. Which sample should have a more voluminous settlement and why? What will happen if a very small amount of salt is added?

4-7 Give an example in which the impermeability of filter cake is independent of the degree of stability or flocculation.

4-8 How would you relate the hydraulic principles of turbulent flow in pipe to the properties of the suspension? Explain what factors are important in turbulent flow.

4-9 Prepare a 5% bentonite suspension and carry out density, viscosity, filtration, and pH measurements. Repeat 2 days later. Test again if the suspension is contaminated with (*a*) 4 percent of sand; (*b*) 0.3 percent of cement; and (*c*) 0.4 percent of salt.

4-10 Repeat the tests in Prob. 4-9 after first treating the suspension with 0.4 percent of FCL.

4-11 What are the difficulties in adapting rotational-viscometer measurements to the prediction of flow in a mud circuit?

4-12 Describe the main factors you would consider before selecting the colloid material for the construction of a diaphragm wall in slurry trenches.

4-13 For excavation through formation domes containing calcium, thinners are necessary to control the flow properties. State which thinners should not be used and explain why.

4-14 Describe situations in which the use of polyelectrolytes is indicated and explain the effects on the slurry.

4-15 Through tests determine the increase in bentonite concentration required to produce the desired flow properties if mixing is carried out at water temperature from 35 to 50°F. From these results formulate a statement on the effects of temperature on mixing.

4-16 The flow of slurry in a mud circuit is temporarily stopped. A pump delivers a pressure of 40 lb/in². Estimate the maximum gel strength that should not interfere with restarting the flow if the pipe diameter is 4 in and the pipe length 300 ft. State your assumptions.

4-17 Discuss the basis for expressing the solid proportions (concentration) in a slurry.

4-18 Overtreatment must be carefully avoided in the preparation and control of slurries. Discuss examples in which overtreatment has adverse effects.

4-19 A fresh-water-based bentonite suspension weighing 68 lb/ft³ is placed in a 1-in-diameter tube. Estimate the probable gel strength required to prevent bleeding.

4-20 A fresh-water-based bentonite suspension weighs 66.4 lb/ft³ and has a gel strength of 8 lb per 100 ft². A sample is placed in a $\frac{3}{4}$-in-diameter glass tube, filling it up to 24 in. Estimate the probable amount of bleeding.

4-21 A trench is excavated to elevation -40 from the ground level. The soil is dense sand (porosity ratio 0.3), and the groundwater level is at elevation -28. Tests show that groundwater contains 4 percent salt. What might be the probable effects on the flow properties of bentonite toward the end of excavation? State your assumptions.

4-22 After excavation a panel is infiltrated with rainwater. Describe the control tests and the adjustments that might be necessary before placing concrete.

4-23 Discuss the requirements for used-slurry disposal in your locality and the effects on the preparation and control of slurries.

4-24 When would you use attapulgite in lieu of bentonite?

4-25 What is the difference between specifying the slurry properties and proportioning the slurry? If the latter is not your responsibility, how would you ascertain compliance with the slurry-control limits?

4-26 If 5 lb of bentonite is mixed with 100 lb of fresh water, estimate the specific gravity of the suspension if the specific gravity of bentonite is 2.4.

4-27 Discuss the effect of EE, EF, and FF particle association on flow properties.

4-28 Discuss in some detail the relevance of the initial gel strength, the 10-min gel strength, the apparent viscosity, the plastic viscosity, and the yield stress to the functions and applications of slurries.

4-29 Equation (4-1a) gives the shear stress for the linear portion of the flow curve. Investigate the two components of this stress and determine their physical meaning. *Hint*: Begin with Goodeve (1939).

4-30 Discuss the similarities and differences in the effects of peptization and stirring.

4-31 Proportion a slurry for the construction of a reinforced-concrete diaphragm wall. The soil is medium clay. The cost of bentonite in the area is the main cost factor.

4-32 A plain-concrete diaphragm wall is built in a slurry trench 85 ft deep. The ground consists of dense alluvial deposits, premeability 10^{-4} cm/s, and the water table is about 7 ft from the ground level. Rail transportation costs favor the use of high-grade pretreated bentonite without control agents. Proportion the slurry.

4-33 Explain why CMC, which generally raises the viscosity of bentonite slurries, also improves the filtration characteristics and therefore is used as a slurry-loss-control agent.

REFERENCES

API, 1974: Standard Procedure for Testing Drilling Fluids, API Standard RP 13B, American Petroleum Institute, Dallas.

British Standards Institution, 1967: Methods of Testing Soils for Civil Engineering Purposes, no. 1377, London, p. 67.

Cheng, D. C. H., 1970: A Design Procedure for Pipeline Flow of Non-newtonian Dispersed Systems, *1st Int. Conf. Hydraul. Transp. Solids Pipes, Coventry, G. B., Hydromech. Res. Assoc., Cranfield.*

CIRIA, 1967: The Effect of Bentonite on the Bond between Steel Reinforcement and Concrete, *Constr. Ind. Res. Inf. Asso., Lond. Interim Res. Rep.* 9.

Cooke, P. W., 1963: Up-to-date Techniques with Drilling Muds, in "Grouts and Drilling Muds in Engineering Practice," Butterworths London, pp. 211–217.

Darley, H. C. H., 1957: A Test for Degree of Dispersion in Drilling Muds, *J. Pet. Technol. Trans. AIME*, **210:** 93-96.

FPS, 1973: Specifications for Cast in Place Diaphragm Walling, Federation of Piling Specialists, London.

Fuchsberger, M., 1974: Some Practical Aspects in Diaphragm Wall Construction, *Proc. Diaphragm Walls Anchorages, Inst. Civ. Eng., Lond.*

Goodeve, C. F., 1939: A General Theory of Thixotropy and Viscosity, *Trans. Faraday Soc.*, **35:** 342-358.

Graham, T., 1861: Liquid Diffusion Applied to Analysis, *Phil. Trans. Soc., Lond.*, **151:** 183-224.

Grim, R. E., 1962: "Applied Clay Mineralogy," McGraw-Hill, New York.

Havenaar, I., 1954: The Pumpability of Clay-Water Drilling Fluids, *J. Pet. Technol.*, **6:** 49-55.

Hetherington, H. A., 1963: Drilling Muds for Mineral Drilling and Water-Well Construction, in "Grouts and Drilling Muds in Engineering Practice," Butterworths, London, pp. 206–210.

Hutchinson, M. T., et al., 1974: The Properties of Bentonite Slurries Used in Diaphragm Walling and Their Control, *Proc. Diaphragm Walls Anchorages, Inst. Civ. Eng. Lond.*

Jefferis, S. A., 1972: The Composition and Uses of Slurries in Civil Engineering Practice, PhD. Thesis, University of London.

Jones, F. O., 1964: New Fast Accurate Test Measures of Bentonite in Drilling Muds, *Oil Gas J.*, June, pp. 76-78.

Jones, G. K., 1963: Chemistry and Flow Properties of Bentonite Grouts, in "Grouts and Drilling Muds in Engineering Practice," Butterworths, London, pp. 22-28.

Nakayama, J., 1976: BW System Diaphragm Walling, *Tone Boring Co., Spec. Bull.*, Tokyo.

Reiner, M., 1949: "Deformation Strain and Flow," Lewis, London, p. 17.

Rogers, W. F., 1967: "Composition and Properties of Oil Well Drilling Fluids," 3d ed., Gulf Publishing Co., Houston.

Sliwinski, Z., and W. G. K. Fleming, 1974: Practical Considerations Affecting the Construction of Diaphragm Walls, *Proc. Diaphragm Walls Anchorages, Inst. Civ. Eng., Lond.*

Tamaro, G., 1975: Pretreated bentonites in trench excavations, personal communication.

Tone Boring Co., 1974: Stabilizing Solutions for Diaphragm Walling, *Bull.* BW107, Tokyo.

Van Olphen, H., 1950a: Stabilization of Montmorillonite Sols by Chemical Treatment, *Rec. Trav. Chim.*, **69**(1): 1308-1312, (2): 1313-1322.

——,1950b: Pumpability, Rheological Properties, and Viscometry of Drilling Muds, *J. Inst. Pet.*, **36**: 223-234.

——,1956: Forces between Suspended Bentonite Particles, *Clays Clay Minerals, Proc. Natl. Conf., 4th Conf.*, pp. 204-224.

——,1959: Forces between Suspended Bentonite Particles, II, *Clays Clay Minerals, Proc. Natl. Conf., 6th Conf.*, pp. 196-206.

——,1963: "Clay Colloid Chemistry," Wiley, New York.

Van Wazer, J. R., et al., 1963: "Viscosity and Flow Measurement," Wiley, New York.

Vold, M. J., 1954: Van der Waals Attraction between Anisometric Particles, *Colloid Sci.*, **9:** 451-459.

Xanthakos, P. P., and B. Bailey, 1975: Report of Geotechnical Information, Slurry Trench Cutoff, Southport AWT Facilities, ATEC Associates, Indianapolis.

Chapter Five

EXCAVATION AND EQUIPMENT

5-1 INTRODUCTION TO EXCAVATING SYSTEMS

Probably no other type of construction is influenced so much by the process of excavation. Thus the production planning and success of a job depend largely on the efficiency of the excavating system, which governs the operational output and therefore the construction rate. The excavating system in this case is not only the machine and equipment but also the method of materials handling, separation of excavated soil from the slurry, incidental plant facilities, and control of face stability and verticality.

Little is to be gained by identifying the method of excavation or the type of equipment when a job is in the design stage, since this can only restrict the trade and limit the number of bidders. Leaving this matter to the contractor presumably allows more competitive bidding and indicates a more diversified range of choice. However, these favorable results are not always achieved, and often the actual construction cost is not known until a job is completed. On the other hand, in certain sites and soil conditions certain types of equipment may be more efficient than others, and in this case the engineer should plan the construction details accordingly.

The ideal endeavor for any contractor is to use a process and equipment that will permit the excavation to be completed in the shortest possible time and at the least possible cost. An important factor in planning the excavation is therefore the direct cost of equipment to the contractor. This includes not only the original investment or the rental charge but also operating and maintenance costs, which are appreciable for this type of work. If contractors own some construction equipment, they may be tempted to try it in slurry-trench excavations after making certain changes and modifications to suit the job requirements. In this manner the initial equipment-ownership cost is eliminated, but this saving as a rule is offset by incidental costs and unforeseen complications since the conversion of equipment is seldom, if ever, successful. Thus, the first advice to engineers and contractors is to use only excavating systems especially developed for this type of work. Furthermore, contractors should always use equipment with which they are familiar and only with competent technical advice.

Classification of Excavating Systems

Classification of slurry-trench equipment is primarily a matter of convenience, but for practical purposes equipment is identified according to the functional and operational details of the excavation, e.g., whether the excavation is intended for a cutoff or for a structural retaining wall, how the materials are handled, etc.

Conventional Trench Excavators In this category are backhoes and dragline buckets, used primarily for the construction of certain types of slurry-trench cutoffs (see also Sec. 6-3). Equipment of this type is essentially simple, and the excavation is a single-process operation.

Machines of the Bucket and Grab Type This equipment has a high weight-to-volume ratio to overcome drag and flotation effects of the slurry gel during excavation. The grabs are usually round or egg-shaped to provide half-rounded joints between panels. The excavation is carried out directly, and all excavated materials are discharged by the grab in a cyclic process. Bentonite slurry in the trench is used only to improve stability.

Examples of these machines are mechanical diggers, either back- or forward-acting; bucket excavators of special design; shovels and auxiliary tools such as augers; and special trenching grabs and clamshells.

The last type is in wide use, and economical and efficient machines have recently appeared in the market using grabs. Cable-supported (or rope-operated) grabs generally are lighter and more maneuverable, and the digging action can be controlled by experienced operators. These machines give fast excavation speeds, but unless the ground is relatively soft, they require bouncing at the trench bottom to give sufficient bite. To remedy this problem power-closing grabs were developed. The power is either hydraulic and supplied by ram-operated mechanisms or electric, with motors attached to the main body. This offers the advantage of higher operating speeds in relatively dense ground and results in increased efficiency. Furthermore, the full weight of the grab presses the jaw into the ground during closing, whereas in cable-suspended machines only part of the weight is utilized to close the grab.

Further improvements include the use of a kelly, to which the main equipment is fixed. The kelly is guided above the ground and functions to position and control the equipment above the ground; guide the machine in a true vertical line during lowering; and provide additional weight. Kelly-bar and boom-operated grabs are considerably heavier and suitable in conditions where restricted headroom, difficult perimeter plan, and weight restrictions are not a problem. They can deal with much tougher soil conditions because of their stronger bite.

Percussive Tools These are extra heavy and rigid, often of special design: they are used to break rock or loosen hard ground and in situations where other types of equipment are not effective. Excavation with percussive tools is slow and therefore expensive. In this respect it is important to determine carefully whether a diaphragm wall should penetrate into rock or other hard formation for either bearing or for sealing an excavation. This requirement can cause serious excavation problems and raise the construction cost accordingly, since it generally means conversion of the excavating system to percussive tools as soon as the hard formation is reached.

Rotary Drilling Equipment Also referred to as hollow-stem large-diameter bit-drilling rigs and machines, this equipment can perform either slot or circular excavations. Several rotary drilling bits loosen the soil through simultaneous action, helped by side cutters, which move vertically to cut through soil not reached by the bits. The soil is consolidated into individual grains or cuttings, as opposed to the bulk excavation possible with grabs and clamshells. These cuttings are then mixed with slurry at the cutting face, held in suspension, and circulated up through the drill stem and reverse-circulation hose to be separated out in screens and cyclones.

This type of equipment is used primarily in the construction of diaphragm walls and load-bearing elements, and seldom, if ever, in slurry-trench-cutoff walls. It is a relatively recent development made possible by the expansion of the market, and promising features are the control mechanisms and devices in the main machine to monitor the various operations.

Reverse-Circulation Machines Almost all rotary drilling tools and most percussive tools are equipped with reverse circulation. This process consists of using the drill stem and hydraulic pipeline for the direct removal of sand and clay cuttings from the excavation. Reverse circulation is thus a built-in feature whereby the slurry sustains all excavated materials in its volume and through its continuous recirculation conveys them to the surface. This requires the use of direct suction or air lift to force the slurry up through a reverse hose in order to carry all excavated soil to the mechanical separation units. There the solid materials are intercepted and collected for disposal while the slurry is returned to the trench through the supply line.

In this process the slurry functions in two ways, initially as the supporting medium to keep the face stable and then as the transporting agent for the excavated materials. Accordingly compatibility is necessary between drilling and material-handling rates and between the supply and suction pumps. Representative types of equipment are shown in Fig. 5-1.

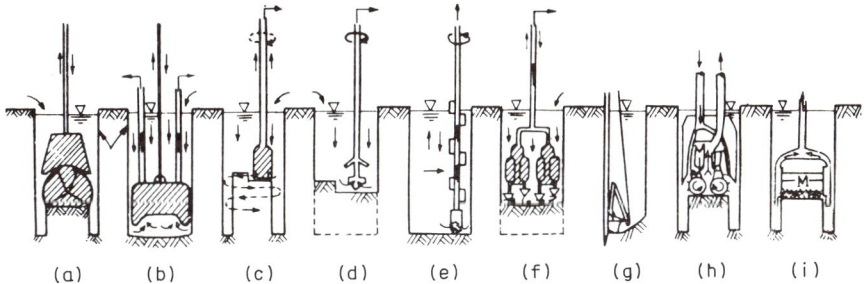

Fig. 5-1 Equipment for slurry-trench excavation: (*a*) clamshell bucket attached to a kelly; (*b*) vertical percussive bit with reverse circulation; (*c*) percussive benching bit; (*d*) rotary benching bit; (*e*) rotary bit with vertical cutter; (*f*) rotary drilling machine with reverse circulation; (*g*) bucket scraper; (*h*) bell-mouth suction rotary cutter with direct circulation; and (*i*) horizontal auger machine.

5-2 FACTORS AFFECTING THE SELECTION OF EQUIPMENT

Satisfactory excavation rates, accuracy in the horizontal and vertical alignment, and ability to deal with the special site and soil conditions are the main criteria for the selection of equipment. To the contractor the selection must be justified in terms of cost and maintainability; i.e., equipment is chosen because it can do the job at the least total cost. Contractors, however, must be cautioned that the cost of different types of equipment varies widely, but so does the total cost of a job under different types of equipment.

Other factors, described below, can influence the choice of an excavating system.

Purpose of Construction Certain excavations, e.g., cutoffs with earth backfills, require trenches wider than usual but excavated in a fairly simple process. In this case the usual implements include backhoes, draglines, and clamshells. Diaphragm walls require relatively narrow trenches, better accuracy in the excavation, and ample coordination in materials handling, hence more elaborate equipment.

Type of Soil The soil at the site may range from very soft to very hard and often encompasses the entire range between these extremes. Most clay and sand formations can be worked with any excavator, but difficulties will be encountered if the equipment must work in hard cohesive material, dense gravel, bouldery ground, and bedrock. Under these conditions a serious drop in efficiency should be expected for the usual equipment and even complete inability to perform. Percussive and rotary tools are often the only equipment that can work in rock.

The problem of penetrating different strata is critical from the viewpoint of construction efficiency. Figure 5-2 shows the effect of difficult ground layers on construction time. The ratio D of the difficult layers hard cohesive soil, very dense granular soil, or harder materials) to the total depth of excavation (expressed in percent) is plotted vs. the ratio T of excavation time to overall time (until the concrete pour is completed). These data are for several bucket- and cutter-type excavators and show that when the difficult layers exceed about 15 to 20 percent of the total excavation volume, the time of excavation increases exponentially (Ikuta, 1974).

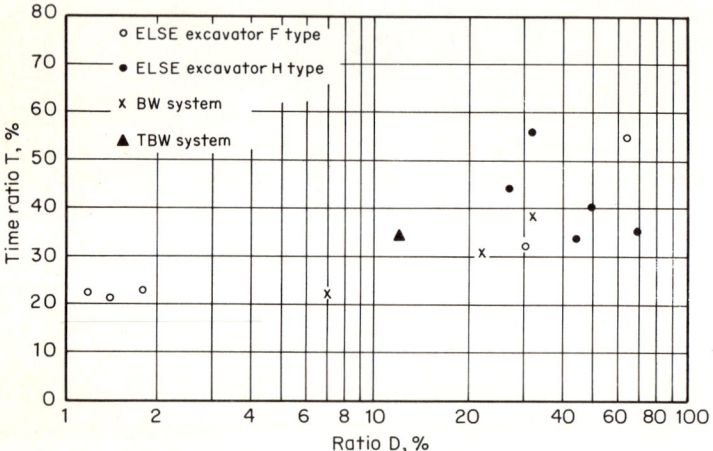

Fig. 5-2 Effect of difficult ground on excavation time. (*From Ikuta, 1974.*)

Excavation Depth As shown in subsequent sections, almost all excavating systems have depth limitations or at some point efficiency decreases to the extent of rendering the equipment unsuitable.

Physical Layout and Special Conditions These are space availability, proximity of excavation to existing structures, headroom, size and weight limitations, mobility of equipment, accuracy of excavation, time restrictions, and environmental constraints such as noise and vibrations.

Specification requirements ideally should only dictate or describe the results to be achieved. In some instances, however, it is necessary to indicate certain intermediate steps in the construction in order to ensure the final details, and to some extent this can restrict the choice of equipment. Finally, the commercial availability of equipment should be taken in consideration.

Operating Skills Mobile construction equipment for slurry-trench excavations is not easy to operate, and exceptional skill is required of the operator. Although many machines have automatic control devices, some operations still rely on hand controls, levers, and pedals. For example, maintaining the accuracy of the vertical alignment requires utmost coordination of man and equipment, and thus contractors can gain many advantages by using operators experienced with the equipment at hand.

5-3 EXCAVATION FOR SLURRY-TRENCH CUTOFFS

The general procedure for the construction of earth cutoffs is discussed in Sec. 6-3. Two factors generally influence the operational characteristics of equipment used for this type of work; the width of trench, which usually ranges from 1 to 3 m (3 to 10 ft); and the relative simplicity of the excavation process and the greater allowable tolerance in the accuracy of construction.

The basic types of equipment are backhoes, dragline buckets, and clamshells. The choice is governed mainly by the depth and width of excavation. Guide walls are not considered necessary, and structural bracing generally is omitted; the face of the trench therefore has a greater tendency to collapse or cave in, especially close to the ground surface and under surcharge loading close to the edge. In order to prevent these problems equipment and operations must be planned accordingly, and it is generally necessary to mount the excavator on crawlers or crane tractors to distribute the load and thus reduce the surcharge per unit area.

Backhoes

Backhoes are special attachments or pull shovels carried on cranes or tractors. A unit mounted on standard crane is shown in Fig. 5-3. Backhoes are advantageous and economical usually to a depth of 10 m (about 33 ft). The reach, depth, and dump range vary with the model and usually are included in the manufacturer's catalog. Digging stability and load distribution are achieved if the main implement is carried on crawler tracks or wheels. The digging assembly has a boom, a dipper stick with the bucket mounted on its outer end, and cables or hydraulic cylinders to control motions. The bite width of most standard sizes ranges up to about 1.7 m (5.6 ft), but adjustments can be made to accommodate wider trenches.

The arrangement of the digging mechanism allows the backhoe to reach out horizontally or down into the trench with the boom, dipper stick, and bucket

Fig. 5-3 A backhoe used for trenching. (*Bucyrys-Erie Company.*)

extended to start excavation. Once the bucket is full, the boom is raised to clear the sides of the trench and then moved horizontally to discharge the load. The cycle is repeated from the same position until the pass is completed and the excavation reaches the intended depth.

The following factors should be considered in selecting a backhoe for slurry-trench excavations: (1) although the dipper-bucket width can be adjusted to handle wider trenches, there is a limit imposed by the density and toughness of the soil and the power available; (2) likewise the capacity of cable-controlled or hydraulically operated units must be compatible with the operating load of the equipment at the site; (3) mobility of the equipment must be sufficient to reach the intended depth and discharge excavated materials at preselected places; and (4) it may be possible to carry out incidental operations with the same equipment, e.g., mixing earth materials and bentonite and backfilling the trench.

Draglines

In most instances slurry-trench-cutoff walls are deeper than can be excavated with backhoes and usually are within the effective range of draglines. Contractors regard this equipment as versatile machines with great reach and width capability. The maximum depth that can be excavated is about 25 m, or 80 ft, but most contractors prefer to operate in the 60- to 70-ft range. A clamshell has an advantage over a dragline in that it has a four-sided container which is better for excavating soil under slurry since it minimizes waste and spillage. However, clamshells cost more than dragline buckets.

Operational Characteristics A dragline excavator is made by hanging a drag bucket from the hoist line of a crane and providing a drag cable from the bucket to an operating drum, as shown in Fig. 5-4. The conversion from the basic crane to a dragline can be made from standard-power excavator units. The front-end attachment in the basic crane structure is the boom and the dragline bucket. The bucket is made to handle the loads and forces for this type of work, and commonly it is made 50 to 100 percent heavier than the standard heavy-duty sizes. Furthermore, it is much longer and has a taper at the cutting end. These features make the penetration of dense alluvium and even hard strata easier. Side cutters with cutting teeth can be mounted on the main body to extend the width range and excavate the trench in one pass. Both perforated and nonperforated buckets are used, but spilling from nonperforated buckets often results in more slurry returning to the trench.

Excavation proceeds as shown in Fig. 5-5. The crane moves backward, and as the trench is excavated at one end, backfill is continuously placed at the other (see also Sec. 6-3). The direction of excavation usually is established with reference to the rock profile; thus in sloping rock bed the dragline operates more efficiently if it moves opposite the downslope of the rock profile. Since the crane moves ahead of, rather than alongside of, the trench, its weight poses no serious problems to trench stability, and live-load surcharge diminishes before the trench boundary is reached.

Fig. 5-4 View of dragline excavation in operation. (*Bucyrys-Erie Co.*)

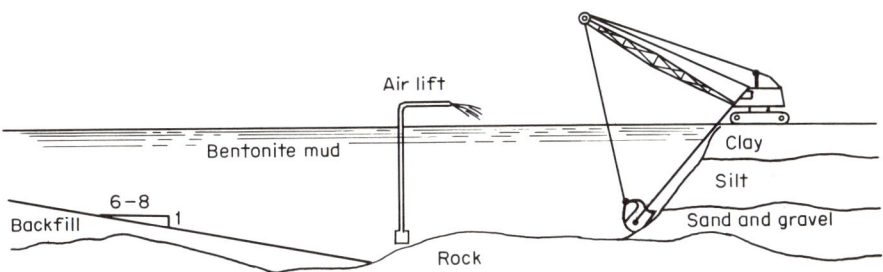

Fig. 5-5 Slurry-trench excavation by dragline.

The operation requires coordination of all moving parts. The bucket is loaded as it is pulled toward the crane with the drag cable. When filled, the bucket is lifted while the drag cable is let out. A holding combination of the drag cable and the chain keeps the bucket from dumping until reaching the proper place. The empty bucket is then swung back and dropped at the next suitable position to repeat the cycle. The excavator is usually operated with its boom at 40° with the horizontal.

Dragline and Crane Stability The free motion and swinging of the bucket can influence the stability of the crane and hence the maximum load that can be handled. According to PCSA (1968a), the recommended safe load should not exceed two-thirds, or about 65 percent, of the tipping load. Stability should be first ascertained in the direction in which the digging force is applied. The tension in the drag cable must be resisted by the crawler tracks at the ground line, and this combination will redistribute and confine the total load acting upon the ground to a smaller area. This is shown in Fig. 5-6. During peak digging the maximum tension in the drag cable is T_d, and this must be resisted by some horizontal reaction F_h as friction at the ground line. The force F_h together with the vertical load yields a resultant R eccentrically applied and causing a bearing pressure as shown.

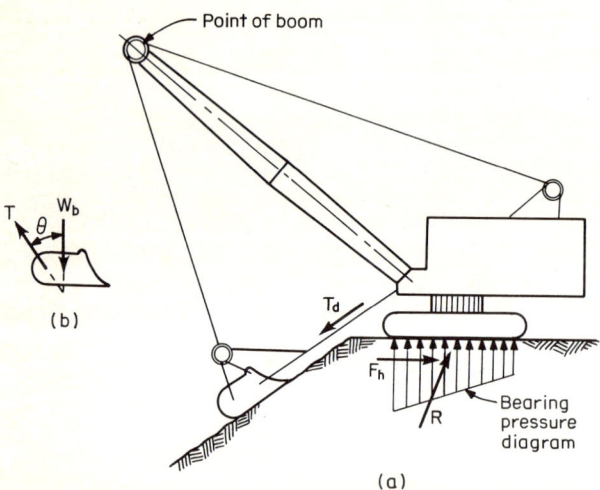

Fig. 5-6 Forces mobilized during dragline operations: (*a*) dragline performing excavation, forces at peak digging; (*b*) bucket in raised position.

The second dragline effect that must be checked is the maximum pull in the hoist line with the bucket in the raised position, as shown in Fig. 5-6*b*. If the total weight of the loaded bucket is W_b, the tension in the hoist line is

$$T = \frac{W_b}{\cos \theta} \qquad (5\text{-}1)$$

in which θ is the angle which the hoist makes with the vertical. This angle usually varies from 15 to 45°, and for the 45° angle the tension in the hoist is $1.42 W_b$ or 42 percent greater than the weight of the loaded bucket.

Another extreme position of the boom is when the assembly is swung 90° from the normal direction of excavation and the bucket is lowered to discharge the load. In this position the crane must be checked against tipping over. This can occur when the overturning moment approaches the stabilizing moment of the equipment for a crane firmly supported on a level surface.

In loose or soft ground serious problems will develop if the crane undergoes excessive settlement, particularly if the settlement is uneven and occurs at peak digging, in which case the chances for tipping over increase suddenly. Because of these combined consequences it is necessary to establish the maximum probable weight of the crane loaded, determine the eccentricity in various critical positions, and analyze the effect on the crawler-track area.

This should be checked against the allowable bearing at ground level to give an indication of probable settlement.

Efficiency   In firm ground the digging force is improved by moving the drag-chain hitch higher on the bucket, so that the bucket teeth can cut deeper and apply greater digging force. In loose material a lower hitch produces a shallower but longer cut. As already mentioned, the bucket is made much heavier than standard heavy duty and has reinforced metal plates which enable the implement to handle broken rock, boulders, and other abrasive materials. Bulk materials are likely to have greater void space in the loaded bucket and thereby reduce productivity. Ideal excavation is in fine-grained or granular soil and even common earth. When such soil is excavated under slurry, the particles are likely to remain suspended in the slurry volume taken into the bucket. Wet soil will not heap or swell much, and therefore the actual amount in the bucket will be very close to the volume of soil in the natural state, whereas its weight will be that of soil in the saturated condition.

Other factors influencing the efficiency of a dragline are the depth of excavation and the uniformity of the bottom profile; the angle at which the boom must swing to discharge the load; the loading cycle time of the excavator; the clearance and sequence of movements required to carry out the operation; and the degree of dependence on other phases of the job.

Storage Space   Since a dragline bucket hangs free and operates free with a large horizontal swing it cannot dump the load with much precision. If it is used for loading haul units, production will be slower and will result in considerable spillage. In slurry-trench cutoffs most of the material is retained at the site, where it is mixed with clay and bentonite to produce blended backfill; therefore it is only necessary to make the target area in the dump places larger than required for shovel units. Slurry draining from piled materials is reclaimed by small channels that conduct the slurry back to the trench.

Draglines and Clamshells

Although with appropriate modifications draglines can excavate deeper than 80 ft (25 m), most contractors prefer to use clamshells in this depth range, the contention being that dragline efficiency drops seriously in deeper trenches. In many instances it is possible to combine draglines and clamshells, and a usual conversion depth is 70 ft (20 m).

Standard clamshells can operate from a hoist line of mobile cranes. When they are in the closed position, there is little opening for earth material to wash out and the slurry can drain through special perforations. Clamshells from hoist lines exert a dynamic load as the bucket drops vertically into the trench. When the clamshell has surrounded enough material, the operator applies closing power to complete the job and the implement is then lifted to discharge the load.

Slurry-trench excavation by dragline and clamshell is shown in Fig. 5-7.

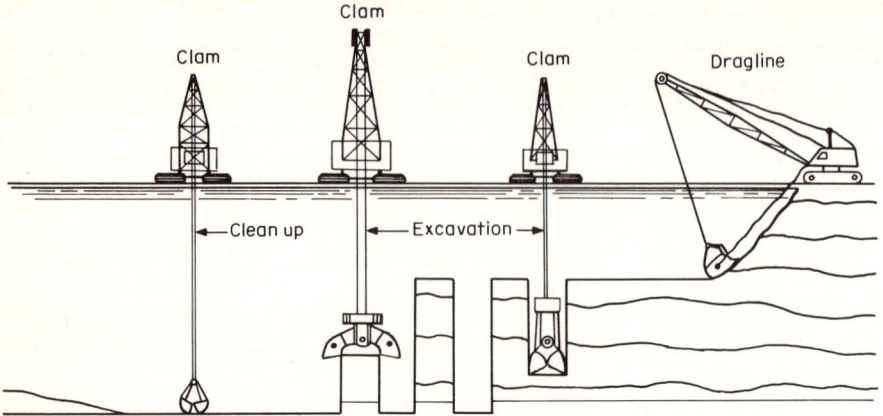

Fig. 5-7 Slurry-trench excavation by dragline and clamshell.

The upper 65 to 70 ft of soil are removed by dragline excavation, and clamshells complete the operation below that depth. For this example two different types of clamshells and rigs are used; the first follows the dragline and opens primary slots or holes, and the main machine grabs the soil and excavates the tongues thus produced.

Slurry Plant

The large quantities of slurry that must be produced continuously require large facilities and special bentonite mixing plants. These usually consist of a concrete bin feeding a vertical silo which drops bentonite powder into a series of mixers. The mixers must often produce slurry at more than 70 gal/min on a continuous basis. The slurry is then discharged into open pits, where it undergoes further hydration, and from there it is pumped into the excavation. As mentioned in Sec. 6-3, slurry-disposal facilities are seldom needed since most slurry is either used up or mixed with backfill materials.

5-4 EXCAVATION WITH BUCKET SCRAPER

Figure 5-8 shows excavation by scooping action at the end of a trench which is accomplished with the aid of a bucket scraper attached to a vertical member of the kelly bar. The machine shown (ELSE) consists essentially of a trenching face shovel or bucket scraper traveling on a mobile vertical mast, running in turn on a fixed mast at the forward end of a structural frame. The base frame moves on rails laid on guidewalls and carries electrically operated winches which transmit the power to the working parts by wire cables.

Excavation is arranged with the bucket starting its downward movement from against the mobile mast and then traveling forward with a curved path. The bucket scrapes away the soil within the range of movement and then folds back to be drawn to the surface and discharge the spoil into a conveyor. This

Excavation and Equipment   161

Fig. 5-8 Slurry-trench excavation with bucket scraper. (*ELSE Machine.*)

action is shown schematically in Fig. 5-9. In phase 1 the bucket position is at the foot of the mast, ready to begin the scraping operation. Phase 2 shows the bucket in movement, scraping the soil. In phase 3 the bucket completes its full reach. Phase 4 shows the bucket lifted to the surface, and in phase 5 the bucket discharges the load.

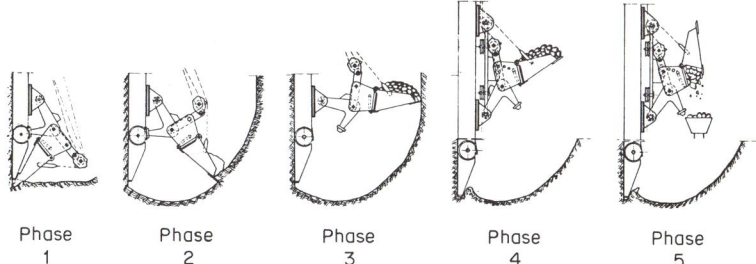

Fig. 5-9 Action of a scraping-bucket excavator.

The cutting power is supplied by the winches and the strength of the wire rope, and therefore the machine can excavate in material that can be worked with a surface excavator of a similar size. Verticality is maintained by plumbing the vertical mast, which controls the line of travel for the bucket, and the overall alignment is maintained by the guide walls on which the system moves. The conveyor may consist of a mobile unit like that shown in Fig. 5-8, on which a hopper receives the load and then travels to discharge it into a mud vehicle for disposal.

The machine is worked with the shovel pointing opposite the direction of excavation, presumably to allow the fresh concrete of the nearest panel to set. In the final phase the bucket can remove the intervening earth material and come into contact with concrete that has reached a safe curing point.

In hard ground or where the excavation is very deep the machine works in the same direction as the general excavation to prevent the thrust of the bucket from overloading and distorting the mobile mast. In this case the end of the previously concreted panel is cleaned with the help of the chisel point of the mast. Upon completing the excavation the bucket gives the base a final sweep to remove soft materials and by-products that may have reached the bottom.

In normal conditions this machine can reach depths of 25 m (about 80 ft) with trench width usually in the range from 380 to 840 mm (15 to 33 in). The trenching speed depends more on the type of soil and is independent of the trench width. Obstructions not wider than the trench can be removed in place, but larger ones must be broken by the mast point or by chiseling. Close to existing buildings the machine can be worked in a fore-and-aft direction, and if necessary the mast and bucket unit can be turned at right angles.

5-5 BASIC TYPES OF CLAMSHELLS

Cable-suspended Clamshells

These are fairly simple in their structural details. They were among the first tools to be used in slurry-trench excavation and are still popular. Grabbing types of equipment operated on a cable are easy to handle, and in most cases they can alleviate the problems and difficulties commonly encountered in trenching operations. Their popularity is also attributed to their efficiency in bulk excavation for average soil conditions and to the smooth shearing operation when cutting the soil.

Cable-suspended clamshells are quite maneuverable and therefore suited to sites having physical constraints. They can work under limited headroom, and when crane-operated they do not require a fixed relationship between the excavating clam and the position of the crane. The crane can be placed in any suitable location in the vicinity of excavation as long as the boom can reach the trench. Increased mobility is quite important in congested sites. This advantage also allows the crane to handle the excavating implement for trenching directly adjacent and close to existing structures while positioned away from the trench, so that live-load surcharge has little effect on trench stability.

In cable-suspended clamshells the verticality of the tool and therefore the accuracy of excavation are primarily controlled by gravity. In this context a heavy tool performs better than a light one, and the suspending mechanism from the winching rig is most effective if it utilizes the continuous influence of gravity during the excavation process. The repeated lifting and lowering of the tool under gravity in the cyclic operation has in fact a rectifying effect on deviations from verticality (Fuchsberger, 1974), and a check on verticality with increasing depth may be required more frequently for a rigid than for a freely suspended tool.

Since conventional clamshells are seldom if ever suitable for this type of work, they are modified or completely redesigned to handle the soil and depth conditions usually encountered in slurry-trench excavations. The penetration of the grab depends on its design and construction features, more specifically on the closing force and weight, as shown in Fig. 5-10.

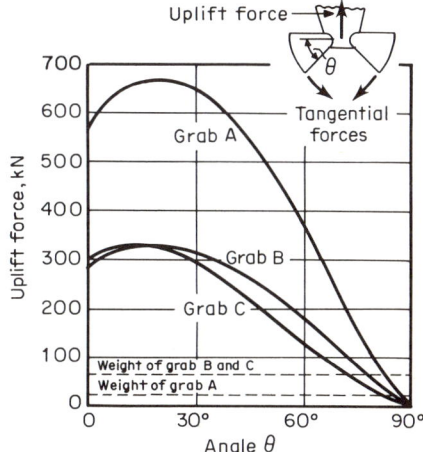

Fig. 5-10 Uplift force on grab during closing cycle. (*From Sliwinski and Fleming, 1974.*)

Mechanical (or rope-operated) clamshells depend on their own weight to close the grab, and in fact only a fraction of this weight can be mobilized when attempting to close the grab since the actual force available is the weight minus the lifting force. To assist closing, pulleys are often used to bring the two grabs together. The first mechanical clamshells have been improved to excavate in a wide range of site and soil conditions; they are much heavier and usually are carried on cranes. Such a clamshell is shown in Fig. 5-11. A usual feature is the outer skirt, or skin, which provides a relatively smooth and vertical face.

Hydraulic or power-operated clamshells, with the advantage of greater closing power, provide better operating characteristics and improved efficiency. Both hydraulic and electric motors are self-contained in the main body. In addition to the suspending cables another cable goes down to the clamshell to carry electric power and controls. The choice between the two types depends on several factors. However, hydraulic grabs are favored by most contractors because they are more flexible, their operation does not conflict with union and local regulations, and the operating power is readily available. Power-

Fig. 5-11 Cable-operated and self-guided heavy-duty 17-t clamshell excavating a slurry trench at the Manhattan Center, Brussels. (*Bachy.*)

operated clamshells, on the other land, are better controlled because of the unlimited power, thus providing more efficiency, but their use often is prevented by other considerations. Figure 5-12 shows a cable-suspended hydraulic grab.

In ground marked with obstructions, boulders, and difficult formations cable-suspended grabs of the extra-heavy type offer certain advantages. On the other hand, in soil with boulders mechanical clamshells are more flexible, and many contractors consider them more effective than hydraulic grabs (Ressi, 1974). Furthermore, rope-operated mechanical clamshells have been successful in excavating trenches more than 75 m deep (about 250 ft), whereas hydraulic and power grabs have a maximum depth range of about 55 m (about 180 ft). The depth limitations are not related to the excavating capability of the tool but are imposed by the desired tolerance in the verticality of the excavation; this becomes a serious factor in executing satisfactory construction joints and therefore affects the watertightness of the final structure.

Kelly Bars

Kelly bars are used either to guide and control the vertical line of the excavation or to provide additional weight during the closing of the grabs. The clamshells usually are hydraulic or power-operated. Figure 5-13 shows a hy-

Excavation and Equipment 165

Fig. 5-12 Hydraulic cable-suspended clamshell: (*a*) the clamshell is moved to the trench area; (*b*) the clamshell is shown entering the mud as slurry is simultaneously supplied to the excavation; (*c*) the clamshell discharges the earth load. Note the long guide skirt. (*Franki.*)

draulic clamshell attached to a kelly excavating slurry-trench panels for load-bearing elements.

The choice between kelly bars and cable-suspended machines usually is a matter of preference and convenience on the part of the contractor but often is dictated by factors related to productivity. Obstructions and boulders are better handled by freely suspended clamshells since this gives them flexibility to adjust position if necessary in order to surround and grab the obstruction, whereas homogeneous soil of average density and consistency is more efficiently excavated by clamshells fixed to kelly bars.

The depth of excavation can also be a limiting factor. Single-piece kelly-bar

166 Slurry Walls

Fig. 5-13 Hydraulic clamshell attached to a kelly bar excavating slurry-trench panels for load-bearing elements. (*Franki.*)

rigs cannot perform economically in trenches exceeding 40 m (130 ft) in depth. Telescoping kelly bars have a usual depth range of about 60 m (200 ft).

Structural Details Most contractors order the structural details of kellys and kelly extensions to suit their own requirements and excavating machines. Kelly bars usually carry hoisting and down-thrust blocks at the top and bottom. Single-piece kellys can be square shafts or H sections of solid steel, although hollow kellys are often used to increase resistance to buckling. Single-piece kellys are made for average depth ranges, and for deeper excavations telescoping kellys are preferred. They have an inner shaft sliding in a larger hollow section, like the one shown in Fig. 5-14. If it is necessary to excavate battered trenches (Fig. 5-15), a kelly bar provides the only possible method.

The combined weight of the grabbing tool and the kelly generally provides sufficient downward force to close the grab and yield high productivity. The hydraulic rams operating the jaws of the grab can be backed up by the thrust of the winch on the kelly bar to give a force at the jaw teeth sometimes in excess of 30 tons. Other features that speed up excavation is the short time required to raise and dump the grab and the all-around slewing motion and adjustable dumping height.

Horizontal deviations can sometimes occur in slurry-trench excavations by

Excavation and Equipment 167

Fig. 5-14 Telescoping kelly guiding a hydraulic clamshell. This kelly has a depth range of 32 m. (*Franki.*)

Fig. 5-15 Slurry-trench excavation at 15° batter using a kelly bar to guide the alignment. (*Franki.*)

168 Slurry Walls

rotating the excavating tool about its axis of suspension despite the guide walls, and this tendency is more pronounced with depth. A large tool rigidly controlled by a kelly thus has advantages and can prevent irregular or wavy trench alignment.

5-6 TOOLS AND EQUIPMENT FOR HARD GROUND AND BOULDERS

Rock, boulders, cobbles, and other hard consolidated materials are difficult to excavate and remove with standard equipment. Where the size of obstructions and boulders does not exceed the trench width, they can be picked up with an extractor type of tool like a cable-suspended grab. If the obstruction is too large or too tightly held to be removed in this manner, it must be broken up or loosened first, often by dropping on it a kelly equipped with a chisel. Loosened or broken pieces are then picked up and extracted by the grab. A standard chisel (Fig. 5-16) consists of several hard metal plates beveled to form cutting teeth and welded to form a compact solid frame embracing almost the entire width of trench. The frame usually is attached to a steel H beam, as shown.

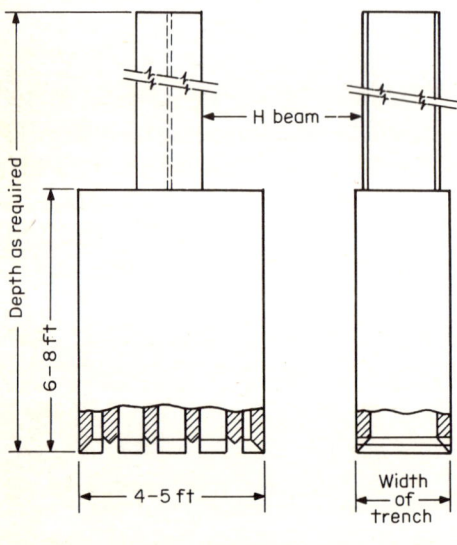

Fig. 5-16 Typical chisel details for breaking embedded boulders in slot excavations.

Percussive Tools Layered rock or cemented soil of moderate thickness can be consolidated and broken up by chiseling and then removed with a clamshell. When penetration cannot be affected by this technique, it may be necessary to use percussion either to assist the clamshell excavation or as an independent method.

Percussive tools derive their name from the violent striking of hard bodies and the shock produced as they fall upon such bodies. They may be cable-suspended, in which case they are freely dropped to break and disintegrate rock and hard obstacles by the action of their own weight, or they may be attached to a drill rod. Percussive tools can also be driven by special impact or vibratory hammers.

It is advantageous to provide reverse circulation with the tools for the removal of disintegrated rock or other soil material and thus leave a relatively free surface for the action of the chisel. The main operational details are worked out from experience with similar ground conditions. A free-falling chisel can work effectively in flat rock, but in steep bedrock profile the same tool is likely to keep sliding along the surface and do limited useful work. This situation is handled better by a steel guide used to keep the tool firm as it moves up and down. Tools for holes can be made by assembling several thick hard plates, as shown in Fig. 5-17, and beveling them to provide sharp cutting edges. Figure 5-18 shows percussive tools used for hole excavation in the cutoff of the Manicouagan 3 dam.

Fig. 5-17 A percussive tool for hole excavation. (*ICOS.*)

Control of Verticality In deep excavations deviations from the allowable tolerance (often 6 in, or 15 cm) can always occur and must be rectified. Sometimes this means backfilling with lean concrete to the level where deviation occurred and redrilling. Alternatively, control of verticality is achieved by using a plumb bob introduced into the hole after removing the drilling rod, but this can be time-consuming.

A new percussive tool was used in the Manicouagan 3 dam (Pigeon, 1974) to solve the problem of verticality control. Although the basic drilling rig was

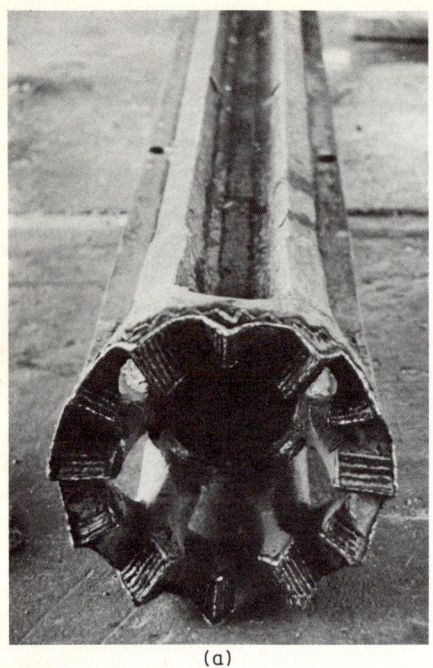
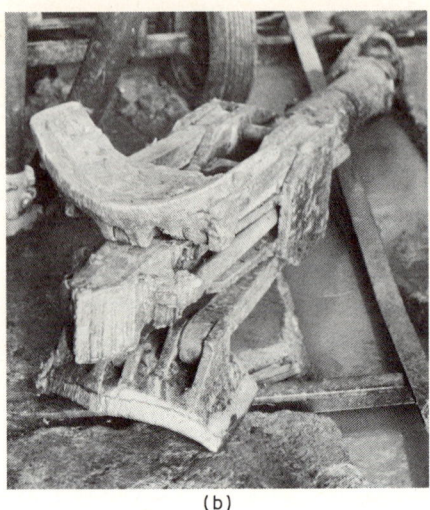

Fig. 5-18 Percussive tools used in the Manicouagan 3 dam: (*a*) chisel used in the primary holes; (*b*) hydraulically expandable chisel used in the secondary elements. (*ICOS.*)

not changed, two concentric chisels were used. In this arrangement a pilot chisel is attached to a flush-jointed casing, which serves to guide the main chisel. Inside the casing is a knob to which a cable is hooked (see Fig. 5-19). After tensioning the cable is aligned vertically and gives direct indication of

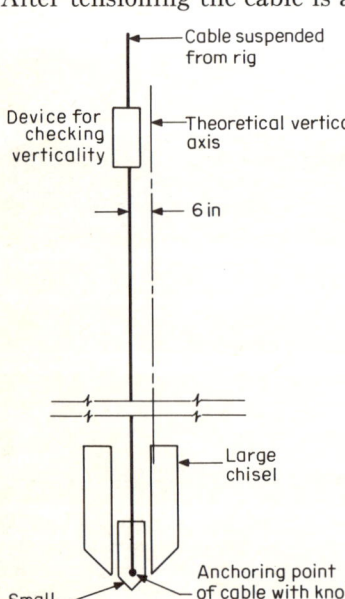

Fig. 5-19 Double chisel for checking vertical alignment of primary piles. (*From Pigeon, 1974.*)

any deviation between the top of the hole and the chisel. Since measurements can be taken without removing the drill rods, verticality is readily checked and corrected before deviations become excessive.

5-7 EXCAVATION WITH CLAMSHELLS AND PERCUSSIVE TOOLS

Single-Stage Panel Excavation

The excavation is in this case carried out in a series of panels and requires clamshells only. It is feasible in soil which is soft to medium hard, either granular or cohesive, but without boulders or other obstructions. The minimum linear element that can be excavated is the minimum grab bite (usually 2 m, or 7 ft), and the maximum panel that can be excavated in one pass is the maximum grab bite (about 6 m or 20 ft).

Average panels usually are excavated in three passes, as shown in Fig. 5-20. The middle tongue is made 1 to 2 ft less than the grab opening to allow the bucket to embrace it. The first pass generally begins away from the last concreted panel to allow the concrete to harden. Round-end clamshells are used in conjunction with round-end tubes at the construction joints, and square-end clamshells are used for composite panels where a steel beam forms the joint. The usual width range is from 2 to 5 ft (60 cm to 1.5 m) but can be exceeded with suitable modifications.

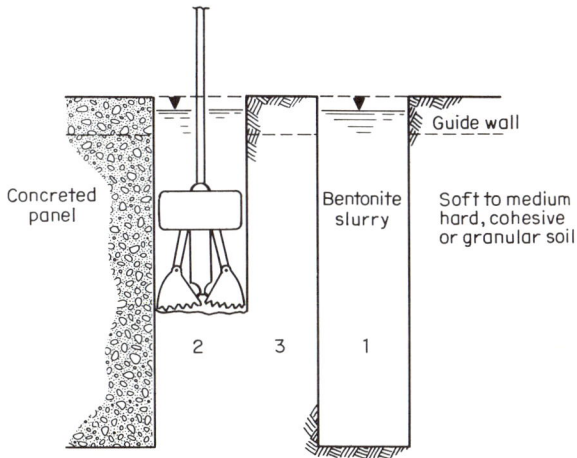

Fig. 5-20 Single-stage panel excavation with clamshell bucket: passes 1 and 2: spread of clamshell bucket; pass 3: spread of clamshell bucket minus clearance for grab to embrace soil.

It is not practical or economical to use clamshells to excavate trenches deeper than 180 ft (about 55 m), although the excavation capability of cable-suspended grabs exceeds 200 ft (60 m). Thus, when all factors are taken into consideration, excavations deeper than 200 ft are carried out solely with percussive tools.

Desanding Operations

Desanding is indicated at the conclusion of the excavation and before concrete is poured or after the concrete is placed and the slurry is withdrawn to storage tanks for reuse. If desanding is necessary, it is advantageous to use an air lift with the suction end near the bottom of the trench, where most of the loose materials commonly accumulate. An air lift requires the use of compressed air to draw slurry to the preselected location. Once the slurry arrives at the surface, it goes through mechanical separation units. Sand separators include a vibrating screen to collect gravel and coarse sand particles and a cyclone to intercept the finer fraction; the slurry is returned to the trench through the supply line. Vibrating screens and cyclones are discussed in subsequent sections.

If specific-gravity measurements on slurry samples collected near the bottom of the trench show no unusual amounts of soil in the slurry, it is sufficient to give a final sweeping pass to clean the bottom of loose materials.

Panel Excavation with Pilot Holes

If the ground contains some boulders and obstructions, it still is possible to carry out most of the excavation using clamshell buckets and grabs but it is advantageous to do the work in two stages. Initially holes are predrilled at suitable intervals, using either percussive tools or rotary drills, as shown in Fig. 5-21. This preboring is quite useful in overcoming the difficulty of grabbing soil when rock and boulders are present. The hole spacing is selected so that the intermediate tongue including tolerance is about 1 to 2 ft less than the grab in the open position, and this allows the implement to slide along the soil column to secure full bite. Two-stage panel excavation is feasible in predominantly soft ground with boulders and cobbles as long as these obstruc-

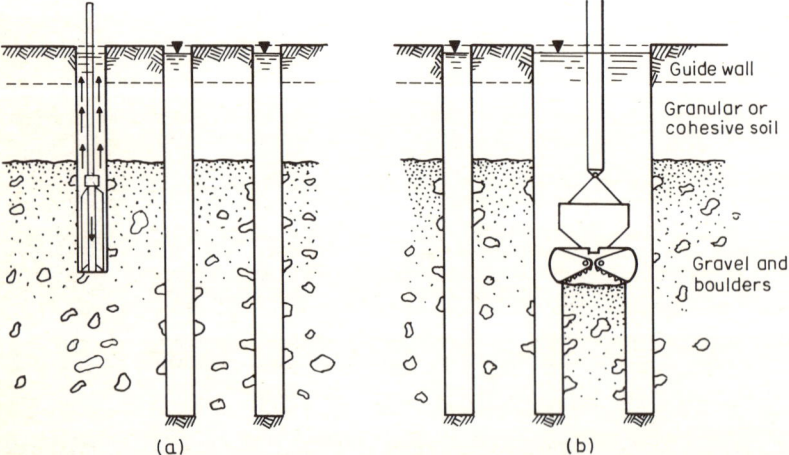

Fig. 5-21 Two-stage panel excavation with percussive tools and clamshell grab: (*a*) excavation of pilot holes with percussive tools; (*b*) panel excavation with clamshell bucket.

tions can be removed with rotary boring machines or percussive tools and the pilot holes can be bored.

Interlocking Elements

If the ground is very hard or strewn with boulders and the excavation must reach considerable depth, the only feasible method is the use of percussive tools. The construction sequence is shown in Fig. 5-22. Primary holes are first drilled with the aid of percussive tools or rotary drilling equipment using direct circulation of slurry to remove consolidated materials. The holes are then reinforced and concreted by means of tremie pipes. A special hydraulically expandable chisel shaped to fit the primary holes is inserted between them. Its construction allows the chisel to embrace and scrape the concreted holes and thus remove all hard and soft material as it is driven. When the excavation is completed, the secondary elements are reinforced and concreted.

The interlocking-element method is inevitably slow, particularly in boulder layers, and far more expensive than excavation with clamshell buckets only. The difficulties involved should not be underestimated, particularly in view of the great depths usually involved. The most serious problems are deviations from the true vertical line of the primary holes, slurry-mud communication between adjacent pile holes, difficulties in installing steel casings where they must be used, actual watertightness of the finished structure, and slow progress in the overall construction.

When several drilling rigs work simultaneously to excavate pile holes at relatively close spacing (in the range of 10 ft or 3 m), slurry-mud communication is likely to occur between adjacent holes because of the penetration of slurry into the ground before rheological blocking occurs. This situation is more pronounced when drilling is carried out in coarse alluvial deposits and mainly in gravel, cobble, and boulder formations that cannot retain slurry. The result is slowdown in production rates, as the work must often be interrupted in one of the communicating piles until the other is completed. Experience shows that this problem can be avoided if the spacing between two simultaneously bored holes is greater than 15 ft (about 5 m) and if a relatively thick slurry is used in the excavation (Pigeon, 1974). On the other hand, slurry loss occurs preferentially under a pressure differential, and in this respect it is advantageous to use chisels that provide larger open areas for the mud flow. Falling chisels act as pistons, forcing slurry out of the hole. Larger openings in the chisel frame reduce this effect and also allow a more efficient collection of cuttings and excavated materials.

The primary holes frequently must be cased in the upper part in order to withstand unusual loading conditions and stresses and to support caving and sloughing soil. When casing is practicable, the installation is carried out by dropping in the casing freely once the hole is advanced to a suitable depth. In this case it is sufficient to drill the hole with a chisel of a diameter 2 in larger than the outside diameter of the casing. Sometimes this cannot be done because of deviations or protrusions into the actual hole, and it may be necessary

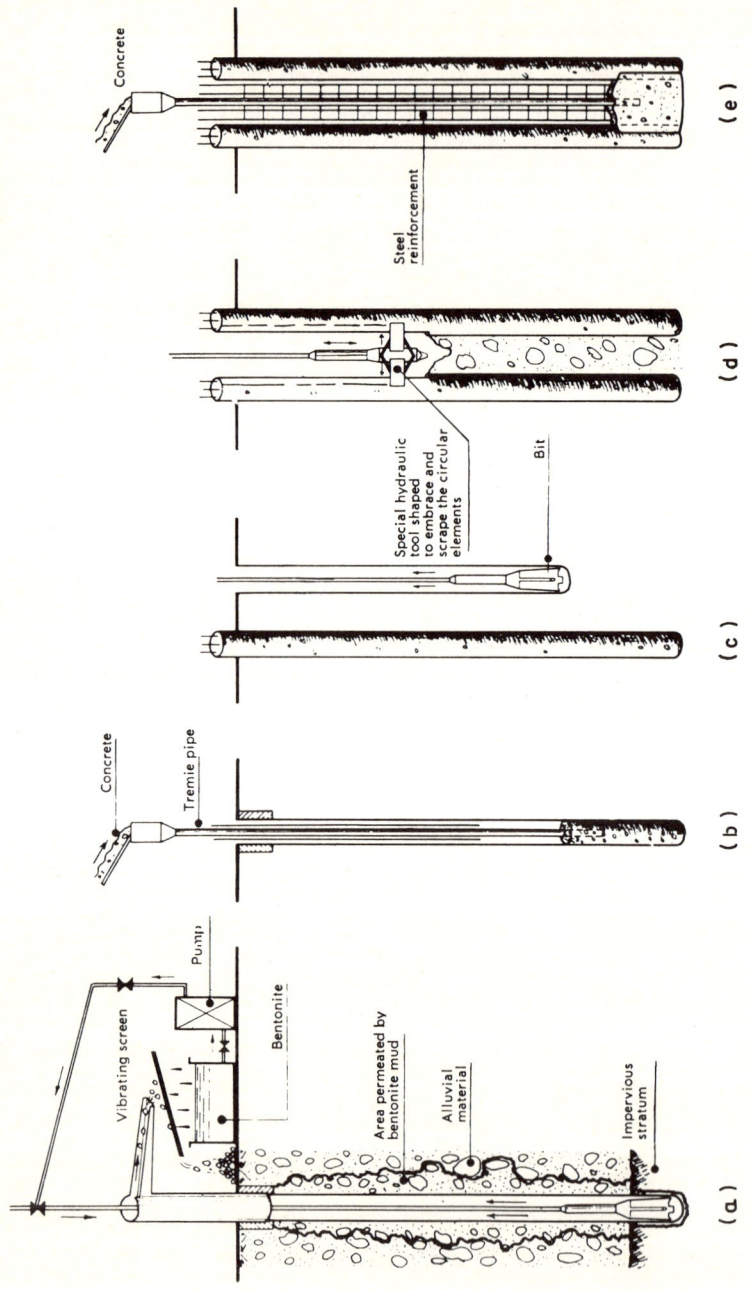

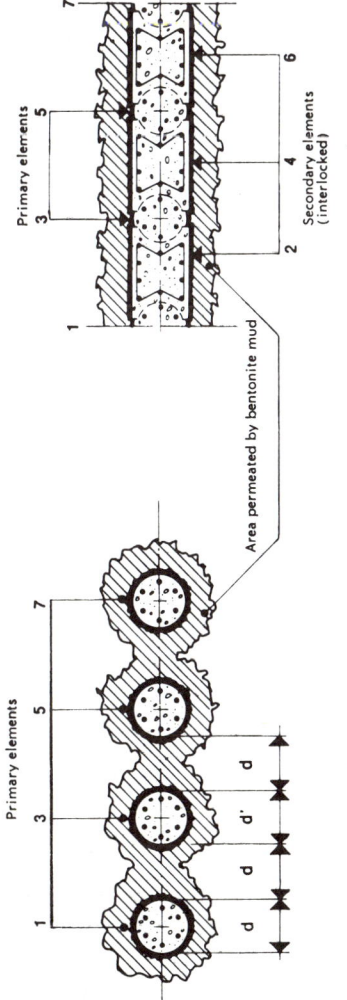

Fig. 5-22 The interlocking-element method: (*a*) drilling of primary holes 1, 3, 5, . . . ; (*b*) pouring of primary elements; (*c*) upon completion of element 1, construction proceeds for element 3; (*d*) drilling of secondary elements; (*e*) pouring a secondary element after lowering the reinforcement; (*f*) sectional plan showing primary holes (dimension *d* should be selected to avoid slurry-mud communication between adjacent primary holes); (*g*) sectional plan showing the interlocking of primary and secondary elements.

175

to drive the casing or use a larger chisel. In either case production is seriously slowed down.

Excavation Rates

It is rather difficult to establish standard rates for the productivity of clamshell buckets and percussive tools other than referring to the average record from jobs already done. In general maximum efficiency is attained with single-stage panel excavation where soil conditions are favorable. For example, in soil which is neither dense nor very stiff a 40- by 20-ft panel can be excavated in four 8-h shifts excluding shutdowns due to machine breakdown or other reasons. This rate is only an average figure and should be used only for rule-of-thumb estimates. In most instances it is advantageous to use several rigs in order to improve efficiency, even on relatively small jobs.

Table 5-1 shows excavation rates for the Manicouagan 3 dam. The excavation was carried out in deposits of sand and gravel with cobbles and boulders. Up to a depth of 170 ft (52 m) the excavation was done with clamshell buckets helped by percussive tools. Column A of Table 5-1 shows the excavation record of panels using percussive tools for the pilot holes and a 6-ton mechanical bucket. Column B shows the rate of progress for panels excavated with two hydraulic buckets guided by telescoping kellys, assisted by two percussive rigs which were continuously available to crush boulders and loosen very dense materials. The panels were 24 in (60 cm) wide.

TABLE 5-1 Excavation Record for Manicouagan 3 Dam* (from Pigeon, 1974)

Depth, ft	Rate of progress, ft²/h	
	A†	B†
0–100	7.2	14.3
100–rock	3.0	4.4
Rock key	0.45	0.69

*Panels up to 170 ft deep.
†See text for explanation of differences.

Table 5-2 shows average drilling rates for primary and enveloping piles of the interlocking-element method applied to the same project. This method was used when the depth exceeded 170 ft. The rates reflect only actual drilling, excluding machine breakdown, power failure, and other causes of shutdown. For this condition the rate of progress should be expected to change substantially with depth only with respect to changes in the type of soil. Thus, it is evident that soil materials become coarser and denser with depth. It is also evident that the penetration rates are not very different between primary and enveloping piles, although the panel area of the latter was much greater and the excavation included the use of expandable chisels.

TABLE 5-2 **Excavation Record for Manicouagan 3 Dam*** (from Pigeon, 1974)

Depth, ft	Drilling rate, ft/h	
	Primary piles	Enveloping piles
0–100	1.80	2.41
100–200	1.47	1.19
200–300	0.86	0.74
300–400	0.63	0.68
400–430	0.31	0.37
Rock key	0.44	0.32

*Interlocking-element method. Panel depth > 170 ft.

Rock Sockets

Tamaro (1975) describes two methods for forming a socket in rock. The choice depends on the extent of penetration and the hardness of rock material. Shallow sockets in soft to medium-hard rock can be formed with lightweight chisels dropped from a clamshell or from a percussive rig. If deep sockets are required, or if the material is medium hard to hard, the most suitable tool usually is a rotary drill or a churn drill.

Such keys into bedrock are often difficult to achieve, and even if the actual penetration is 2 ft, it still may not solve all the problems of potential seepage at this location. Hence, the goal of attaining a completely watertight rock socket should also be judged from practical limitations.

5-8 EXCAVATION WITH ROTARY DRILLING EQUIPMENT

Figure 5-1d to f and h shows several commercially produced rotary drilling machines for slot excavation available in a variety of mountings and driving arrangements. The cutting elements are rotary drilling bits provided with teeth of hard metal. The drilling is done either by cutting the soil or by crushing soft rock. The cuttings thus produced are suspended in the slurry and are immediately removed or flushed out of the trench. A sufficient number of cutters or bits are extended over the face of the machine to cover the entire area as the equipment is operated. This type of operation usually is accompanied by reverse circulation, e.g., the BW system, or it may have suction and direct circulation, e.g., the TBW system. Several rotary drilling systems are described in the following sections.

The BW System

The BW long drill, shown in Fig. 5-23, is a power-operated excavator with reverse circulation as standard feature. It was developed in Japan and has operational characteristics suited to urban construction conditions and requirements. Thus the system is designed to provide manageable size and

178 Slurry Walls

Fig. 5-23 Rotary drilling equipment with reverse circulation; the BW long drill. (*Tone Boring Co.*)

overall height to work under limited overhead room and in confined space; nominal weight to avoid live-load surcharge problems in very soft or in very loose ground; adjustable components and varying motor drill sizes to accommodate variations in panel width and length; and freedom from noise and vibrations.

The motor drill usually is operated from the standard drilling rig shown in Fig. 5-24a and b, but for increased mobility around corners and irregular excavations the drill can be mounted on crawler tractors. On the standard derrick the overall height is about 7 m (23 ft), and the width in the direction of excavation is slightly in excess of 4 m (13 ft). The submersible motor drill is available in three models, BWN-4055, BWN-5580, and BWN-80120. The first two and the last two (or three) numerals in the model number indicate the lower and upper width range respectively in centimeters. Each model has adjustable drill bits, the diameter of which increases in 5-cm (2½-in) increments. Figure 5-24c and d shows mechanical drawings for model BWN-5580. Specifications and dimensions are shown in Table 5-3.

Equipment and Instrumentation The assembly consists of the following parts:

1. Long drill, equipped with a prime motor. Main parts are the drill bits, side cutters and guide plates, and reverse-circulation hose.

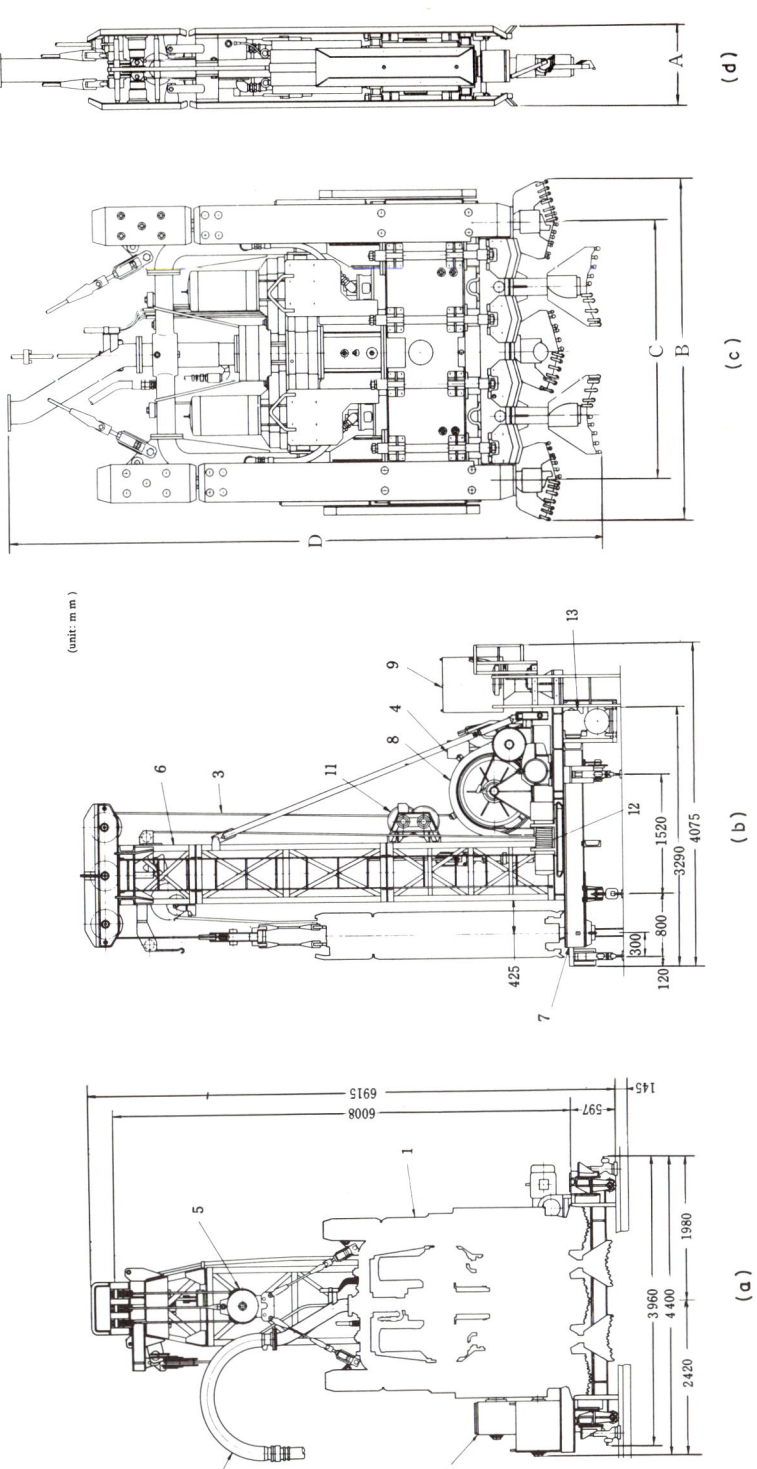

Fig. 5-24 The BW system: (a) and (b) drilling rig and assembly parts; (c) and (d) submersible motor drill; 1 = submersible motor drill, 2 = reverse hose, 3 = hoisting wire rope, 4 = feed indicator, 5 = running block, 6 = derrick, 7 = frame, 8 = hoist, 9 = switchboard, 10 = cable reel, 11 = rope guide, 12 = derrick hoist, 13 = baby compressor for adjustable guide. (*Tone Boring Co.*)

TABLE 5-3 Specifications and Dimensions for BW Long Drill

Model	Bit diameter A (wall width) mm	Bit diameter A (wall width) in	Single excavation length B mm	Single excavation length B in	Effective length C mm	Effective length C in	Height of motor drill D mm	Height of motor drill D in
BWN-4055	400	15.75	2500	98.43	2100	82.68	4300–4320	169.29–170.08
	450	17.72	2550	100.39				
	500	19.69	2600	102.36				
	550	21.65	2650	104.33				
BWN-5580	550	21.65	2470	97.24	1920	75.59	4525–4555	178.15–179.33
	600	23.62	2520	99.21				
	650	25.59	2570	101.18				
	700	27.56	2620	103.15				
	750	29.53	2670	105.12				
	800	31.50	2720	107.09				
BWN-80120	800	31.50	3600	141.73	2800	110.24	5505–5555	216.73–218.70
	900	35.43	3700	145.67				
	1000	39.37	3800	149.61				
	1100	43.31	3900	153.54				
	1200	47.24	4000	157.48				

Model	Excavation depth m	Excavation depth ft	No. of drill bits	Bit rotation at 50 Hz, r/min	ID of reverse pipe mm	ID of reverse pipe in	Power required,* kW	Weight of motor drill kg	Weight of motor drill lb
BWN-4055	50	164	7	50	150	6	15	7,500	16,500
BWN-5580	50	164	5	35	150	6	15	10,000	22,000
BWN-80120	50	164	5	20	200	8	18.5	18,000	39,700

*× 2 sets (6 poles).

Excavation and Equipment   181

2. Control system, usually mounted on a derrick. This includes the hoist, cable reel, and miscellaneous operating instruments such as the switchboard, feed indicator, and deflection indicator. The last is a sensitive mechanism detecting deviations from verticality of a magnitude of $\frac{1}{500}$. Another important control device is the slurry-level indicator, agitated when the slurry level suddenly is lowered.

3. Suction pump, used to maintain reverse circulation. This suction is used to eject soil and cuttings mixed with slurry through the reverse hose and convey them to the mechanical separation units. Alternatively, an air lift may be used to remove excavated materials.

4. Mud-circulation units, usually including a vibrating screen to intercept coarse granular soil and dry-clay cuttings and a cyclone to remove finer soil particles.

5. Slurry mixers and storage tanks.

Excavation This is shown in the flow diagram of Fig. 5-25. The fresh slurry is supplied to the trench usually through two hoses; one delivers slurry continuously while the other is an emergency line to control the slurry level. Drilling is performed by the bits at the selected speed while the bits are arranged in two levels to cover the entire panel and provide adequate overlapping. Side cutters oscillate up and down to trim off soil not reached by the bits, producing a reasonably smooth face. Since only a portion of the weight of the motor drill is utilized for feed pressure on the bits, the remaining downward force tends to keep the equipment in a true vertical line by gravity, which is best utilized by the cable suspension of the motor drill.

Inside the watertight chamber two electric motors provide the power for drilling. Gear trains drive half the drills clockwise and the other half counterclockwise, so that the torque on the unit as a whole is minimized.

Cuttings and excavated soil particles are sucked through the center drill stem and forced up to the vibrating screen, where oversized grains and cuttings are separated and discharged into a conveyor for disposal. The slurry continues its circulation through the cyclone, which intercepts finer particles, and then it is discharged into storage tanks, from which it is returned to the trench. If the supply of slurry is shut off, the emergency line must take over; otherwise the slurry level can drop quickly.

The continuous recirculation of slurry is maintained through turbulent flow in the pipe circuit, and this agitates the slurry in the trench continuously. Since the colloidal slurry seldom has a chance to gel in the trench during excavation, the slurry control limits should be based on the initial rather than on the 10-min gel strength.

Reverse Circulation In general reverse circulation is more efficient with pump suction than air lift. High efficiency is attained when soil materials are handled at the same rate at which they are excavated. In most practical situations there is no limit on the pump capacity required to accommodate the

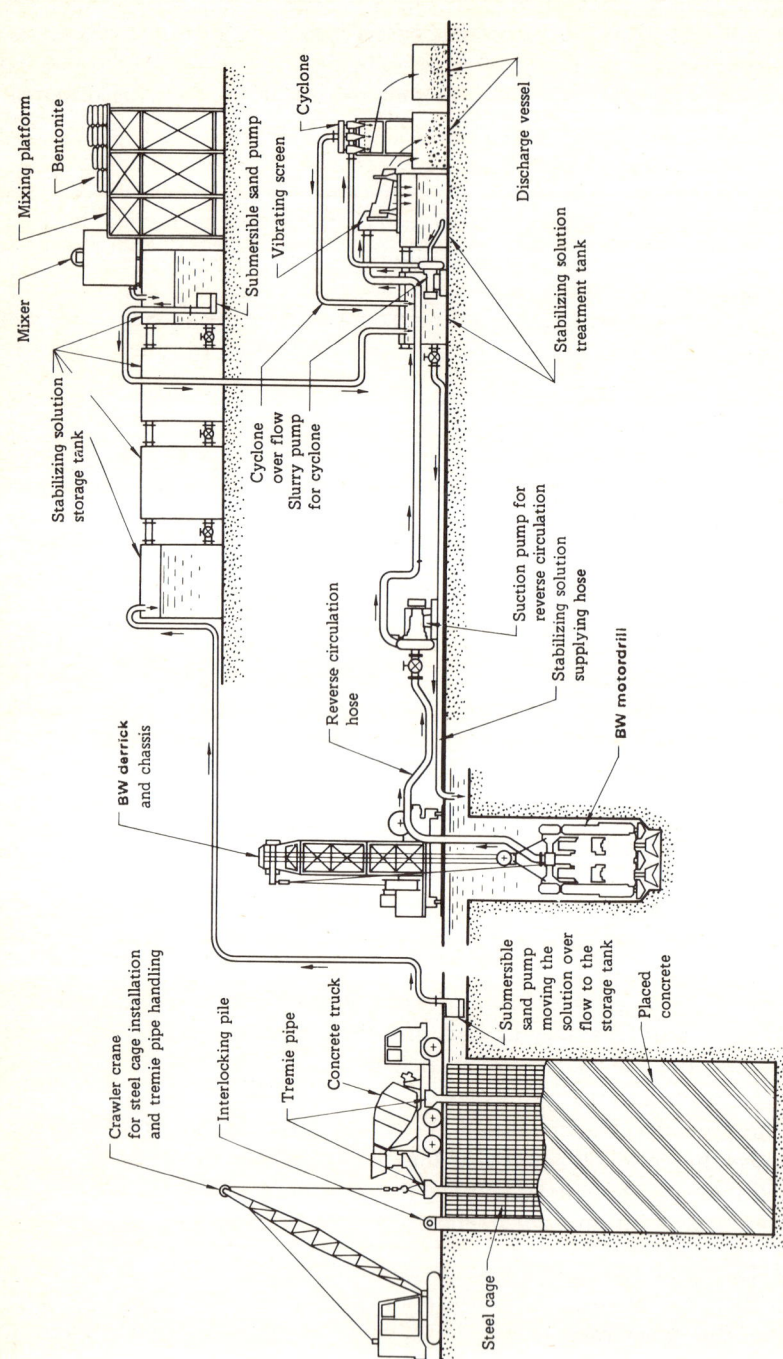

Fig. 5-25 Flow diagram of BW system showing excavation and recirculation of slurry.

expected maximum excavation rate, and thus it is possible to provide a suitable set of pumps, power motors, and appurtenances.

However, when a pump operates at speeds much higher than normal, cavitation may occur and cause the suction to decrease. Thus the suction is restrained at some specific speed designated as the critical speed for the permissible height of suction lift up to which cavitation is avoided. If v_c designates the critical speed,

$$v_c = \left(\frac{k}{h}\right)^n \qquad (5\text{-}2)$$

where h is the total suction head and k and n are appropriate constants. From Eq. (5-2) it follows that the initial speed of the flow will at some point become critical velocity v_c once the critical height has been reached in the course of excavation. When this occurs, there will be a gradual but continuous reduction in the speed of flow, resulting in a loss of efficiency. Accordingly, there is a maximum excavation depth up to which the suction lift is effective.

Since the flow in the reverse pipe is turbulent, it is difficult to express it theoretically; hence determination of flow rates and critical suction-lift heights is provided experimentally. Pumps with a freshwater capacity of 5 m³/min have been used and found effective in providing suction lift for excavation depth up to 45 m (about 150 ft). The basic capacity Q in fresh water is reduced, however, when pumps handle slurry mixed with soil because of the greater friction loss in the pipe circuit and the much heavier fluid; hence it is necessary to multiply Q by an empirical reduction coefficient k_1, which ranges from 0.40 to 0.55 (Ueda, 1974). On the other hand the slurry can sustain only a certain volume of earth materials as it travels through the pipe system. The maximum ratio of solid content that can be sustained in the flow, designated as k_2, is approximately three-tenths of the slurry volume for suction pumps but much less for air lift (Ueda, 1974). Since soil usually expands when excavated to about 130 percent of its volume in the undisturbed state, it follows that the practical rate at which it is removed is

$$Q_p = \frac{Q k_1 k_2}{1.3} \qquad (5\text{-}3)$$

where Q is the initial freshwater capacity of the pump and k_1 and k_2 are given empirically. For example, for a pump with initial capacity $Q = 5$ m³/min, $k_1 = 0.40$, and $k_2 = 0.30$, the rate of materials handling is

$$Q_p = \frac{5(0.40)(0.30)}{1.3} = 0.46 \text{ m}^3/\text{min} = 27.6 \text{ m}^3/\text{h}$$

The practical rate is compared to the maximum expected drilling rate, and if the two rates are not compatible, a pump of greater capacity is selected. The recommended diameter for the reverse-circulation pipe is 6 in and preferably 8 in.

Comparison of Suction Pump and Air Lift When the panel depth becomes 40 m (about 130 ft) or greater, the removal of excavated materials is done with an air lift. Compressed air applied to the center bit forces mud and cuttings to enter the reverse hose. In very cohesive soil water jetting usually is combined with air lift to prevent cuttings from sticking to the drill bits and the slurry is delivered to the trench with the help of a mud slush pump.

Air lift is not efficient for shallow depths, particularly in the first 20 ft of excavation. In this depth range it is better to slow the drilling rate down or to allow most of the excavated materials to remain in the trench until the efficiency of the air lift is improved. If the upper formations consist of gravel or coarse sand, the excavation is very difficult with air lift and the system may even become inoperative. However, the efficiency of air lift is considerably improved with depth, as shown in Fig. 5-26. The rates indicated are for raw slurry. These diagrams suggest the following: (1) for excavation depths less than 30 m (100 ft) suction pumps are more efficient; (2) for excavation depths exceeding 50 m (165 ft) the efficiency of suction pumps is greatly reduced; and (3) if conversion from one system to another is necessary, it should be made at depth of 30 m, where the two systems have almost the same efficiency.

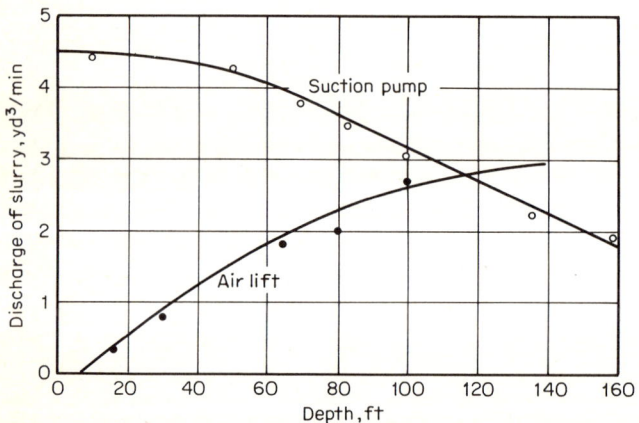

Fig. 5-26 Efficiency of suction pump vs. air lift in reverse circulation; suction pump: 6-in-diameter suction, 6-in-diameter delivery, capacity 6.5 yd³/min; air lift 100 lb/in²; air-compressor capacity 13 yd³/min. (*From Ueda, 1974.*)

In the range of suction pumps the operation has further advantages; the immediate removal of cuttings gives them little chance to dissolve and mix with slurry, and this controls the contamination level better; since cuttings absorb less water, they arrive at the vibrating screen well preserved and in solid compact state and hence better separated; the system can transport slurry up to 500 ft horizontally from the excavation point; and both cohesive and granular soil can be handled by suction pump, whereas in an air lift water jetting is necessary with sticky clays.

Drilling Rates Table 5-4 shows excavation rates from tests for the BWN-5580 model in a panel 2.52 m long and 0.6 m wide (total cross-sectional area 1.44 m^2). The drilling rate is the rate of vertical advance, which multiplied by the panel area gives the excavation rate. The data represent the actual operation time excluding incidental breakdowns and mandatory shutdowns to service and move the equipment. Variations can be expected for other models, panel size, and configuration. It is evident that the best rates are attained in soft or loose soil, but this does not necessarily mean that construction as a whole is more economical in this type of ground.

TABLE 5-4 Drilling Rates of BW Long Drill*

Type of soil	Drilling rate, m/h	Excavation rate, m^3/h
Stiff clay	15	21.6
Very stiff clay	12	17.3
Hard clay	8	11.5
Medium sand	12	17.3
Dense sand and gravel	5.5	8.0

*Data taken from actual field tests.

The time required for drilling becomes the most important single factor influencing total cost when excavation is in solid rock. Although rotary drilling equipment usually is employed in urban sites, where the ground is primarily of sedimentary formation, it is sometimes used to penetrate and disintegrate rock or cemented materials. In this case the side cutters are removed since they can act as brakes and slow down drilling, but the penetration rate still is very slow and averages 0.3 to 2.0 m/h (about 1.0 to 6.5 ft/h). Drilling usually is supplemented by percussive tools or chisels to break boulders and consolidate rock layers.

Separation Units These include mechanical separators or simple tanks where suspended soil particles are separated by settling down to the bottom. As already mentioned, the settling process is effective only with peptized slurries and a relatively low gel strength.

The first mechanical separator is a vibrating screen which intercepts coarse sand and gravel or clay cuttings, as shown in Fig. 5-27a and b. This structure has a screen, frame, separator, switchboard, and piping. The average capacity of mud screens is from 1 to 2 m^3/min, which is sufficient to accommodate the maximum attainable drilling rates.

From the screen the slurry continues its circulation to the cyclone. This unit is used in two ways, either to clean the overflow of slurry by removing sand particles not intercepted at the screen or to reduce the solid content by condensing the underflow of slurry. Either process is feasible if the specific

186　Slurry Walls

Fig. 5-27 Mechanical separation of excavated soil in a vibrating screen: (*a*) sand and gravel; (*b*) clay cuttings.

gravity of slurry is less than 1.15 and its viscosity does not exceed 60 s; hence cyclones generally must be used together with a mud screen for maximum efficiency. Suitable models of mud screens and cyclones and methods for checking their efficiency are described by Tone Boring Co. (1974). A mud plant consisting of mud screen, cyclone, and slurry tank for the BW machine is shown in Fig. 5-28.

Panel Sequence and Arrangement　The panel length is influenced by the factors discussed in Sec. 9-2. A panel can be processed as shown in Fig. 5-29, but the three- and five-pass excavation is most common. The clearance between alternate passes usually is less than 3 ft (about 1 m), so that the resulting thin earth column will allow the intermediate or short pass to overlap with the full passes (Ueda, 1974). If adequate overlapping is not provided, it may be necessary to reset the machine over the alternate panels and work the full depth once more. For the arrangement shown in Fig. 5-29 the average panel length is about 5 m (16 ft) for three passes, 7.5 m (25 ft) for five passes, and 10.5 m (35 ft) for seven passes.

Excavation and Equipment

Fig. 5-28 Assembly of mud plant and separation units for the BW machine; shown are a vibrating screen, a cyclone, and a slurry tank.

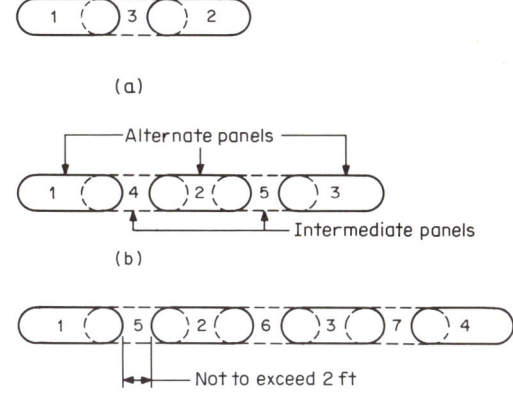

Fig. 5-29 Panel arrangement and sequence of passes: (*a*) three-pass excavation; (*b*) five-pass excavation; and (*c*) seven-pass excavation.

The time for completing a full pass depends on the type of soil and can be judged from Table 5-4. The time for a 3-ft short (intermediate) pass is approximately 30 percent of the time for a full pass. The time allowed for drilling should be less than 70 percent of each shift time. For example, for an 8-h shift the drilling time should not exceed 5½ h. It usually is advantageous to complete a pass, either full or short, within the shift time.

The TBW System

This excavator is operated from a kelly bar set on special derrick as shown in Fig. 5-30. It excavates rectangular panels with square ends for a width of 0.6 m (about 24 in) and a basic length of 1.500 or 1.940 m (4.9 or 6.4 ft). The excavation is feasible for depths up to 30 m (about 100 ft).

188 Slurry Walls

(a)

(b)

Fig. 5-30 The TBW system: (*a*) excavator and rig; (*b*) mechanical separation units and hose reels, (*Takenaka Komuten Co.*)

Equipment and Instrumentation The excavator has the following main units:

  1. Main drill with a pair of rotary cutters facing each other and rotating in opposite directions. The speed of revolution and penetration is decided accord-

ing to the soil conditions. Sliding side cutters with cutting teeth complete the excavation and give reasonably smooth face and verticality. The cutters move up and down in reciprocating movements, and besides trimming the face they also serve as a shield to protect the walls from the jetting action of water fed to the rotary cutters.

2. Hose reels, which control the accuracy and efficiency of excavation. These are mounted at the rear of a frame tower and are powered by hydraulic motor. The upper reel, shown in Fig. 5-30b, is for the blow or supply hose and the lower for the suction hose.

3. Dual pumping system, a combination of suction pump and blow pump to remove and supply slurry.

4. Slurry treatment plant, which includes separation units and a high-speed coil mixer.

5. Miscellaneous controls including ultrasonic devices to check the accuracy of excavation and deviations from the true vertical face.

Excavation The excavator is advanced into the ground by penetrating cylinders attached to the tower through a supporting frame. Two supporting pillars for the rotary cutters are kept in position by guide rollers on the tower. Water jetting is forced through the supply line to prevent clogging of the rotary drills with earth materials. The mixture of slurry and cuttings travels through the suction hose and pump to the slurry treatment plant for the two-stage separation in the vibrating screen and cyclone, and the slurry is recycled to the rotary cutters. Since there is little agitation of soil in the suction line, almost all the cohesive soil is recovered in the vibrating screen in the form of well-preserved cuttings.

The condensed soil dropping from the outlet at the bottom of the separation units absorbs some bentonite and is not solidified for some time. It usually requires mixing with cement or other hardening agents in the high-speed coil mixer. One day after mixing the soil may attain a strength of 2 to 4 kg/cm^2, which makes its disposal in ordinary dump tracks possible.

The two hoses for blow and suction are drawn in and out at constant speeds. For this purpose a traverse apparatus is provided and synchronized with the reel drum. A hose guide is further provided along the mast of the tower to allow automatic insertion of the hose in the supporting pillar of the excavator. The hose reels are operated automatically from the control panel.

Efficiency and Drilling Rates Each of the two models available has maximum penetration capacity of 25 tons, and the excavation is free from vibrations and shock. The quoted efficiency indicates excavation speeds of 15 to 25 m^2/h corresponding to penetration rates of 10 to 15 ft/h. The excavator can be used in a variety of soil conditions including gravel and hard clay. The rotary cutters can consolidate gravel up to 8 cm (about 3 in) in size. Boulders and other obstructions must be broken and consolidated with a chisel and then removed with a grab. These tools must therefore be available at the site as stand-by equipment.

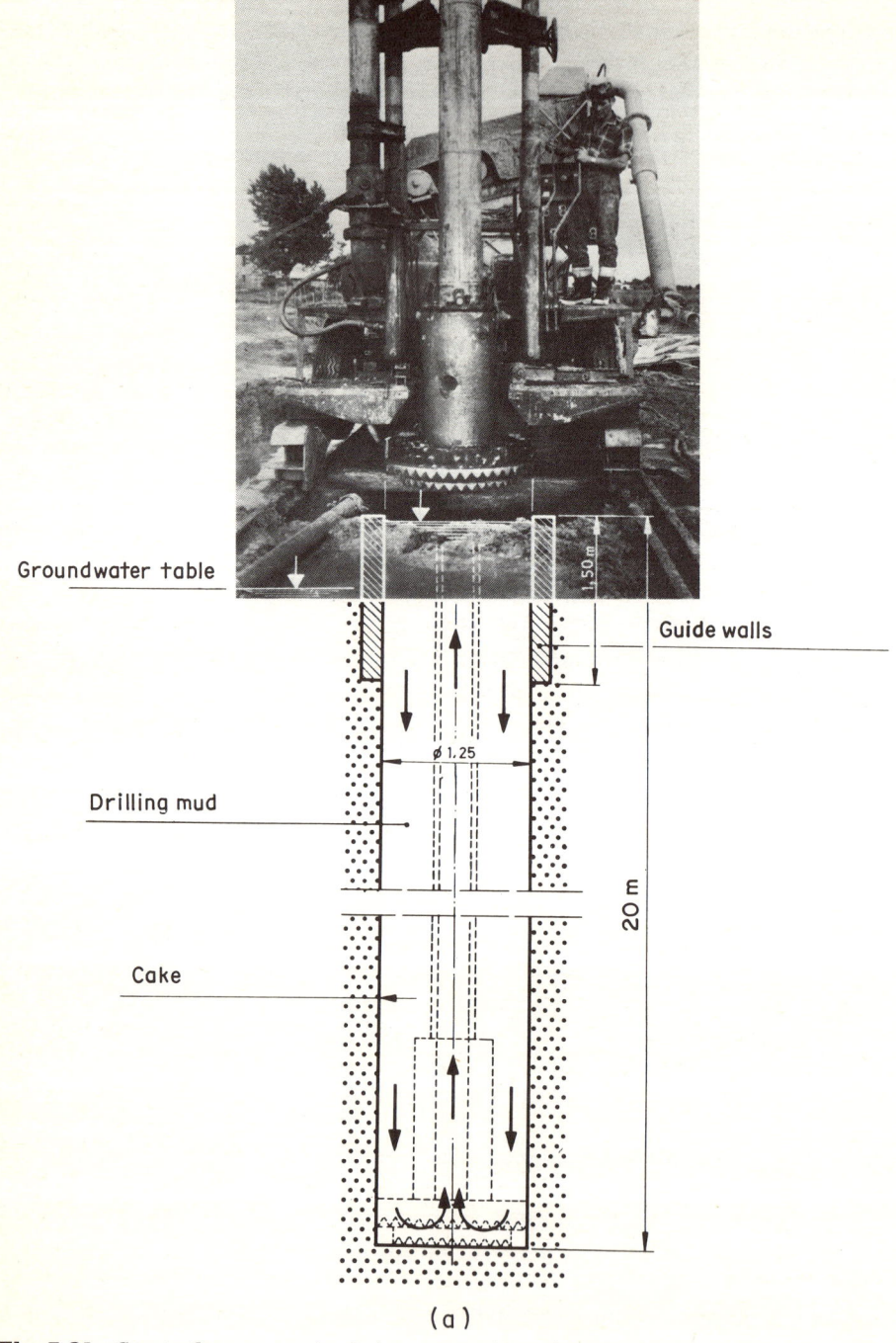

Fig. 5-31 Counterflow reverse-circulation system: (*a*) equipment and section through trench; (*b*) diagramatic illustration of the excavation process. (*Radio-Marconi.*)

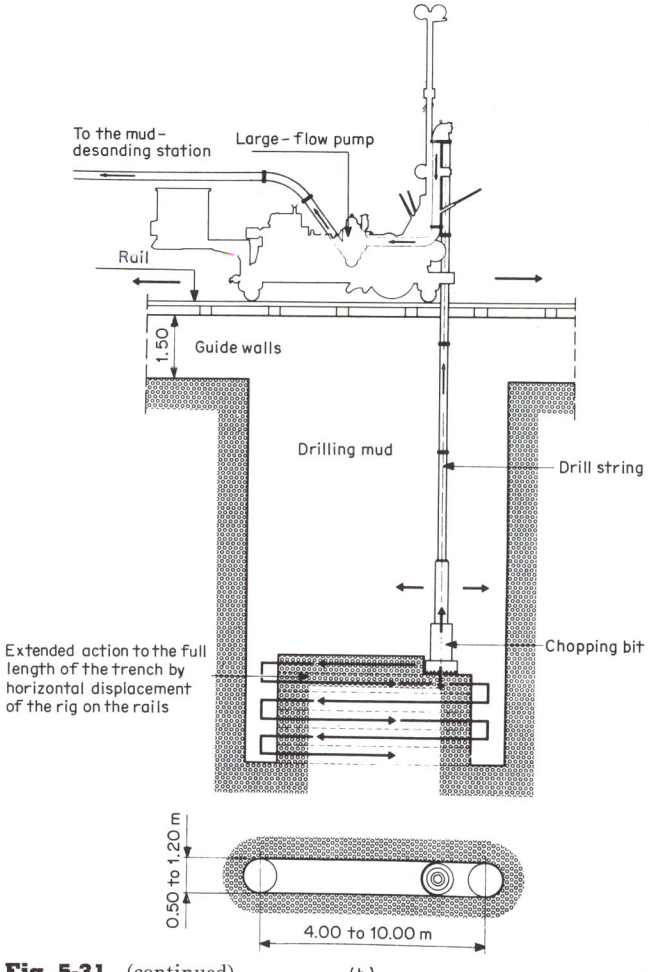

Fig. 5-31 (continued)   (b)

Counterflow Reverse-Circulation Systems

Figure 5-31 shows excavation with a counterflow reverse-circulation system, usually employed in hard compact soils or in thick stratified formations. The slurry again serves a dual purpose, for face support and as transporting agent for excavated materials. The cutter may be a drilling bit of diameter equal to the width of trench or a percussive tool penetrating the soil by up-and-down movement. Linear elements are excavated by extended action of the cutter when the rig is horizontally displaced, as shown in Fig. 5-31b, while the bit works vertically.

Circular Drills

RRC Drill The rodless reverse-circulation (RRC) drill shown in Fig. 5-32 is used for large-diameter foundation elements excavated with slurry. The drill

192 Slurry Walls

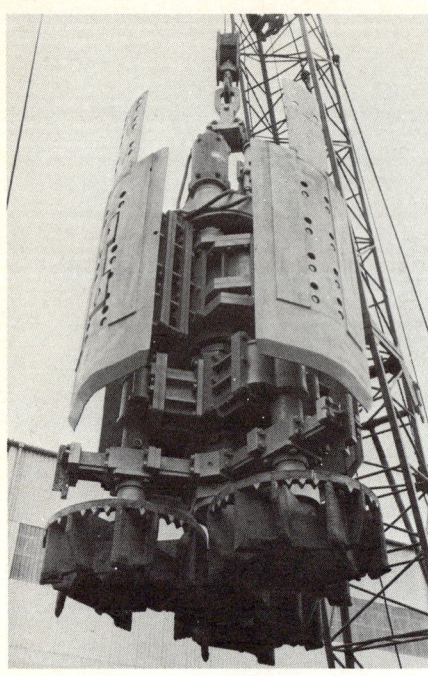

Fig. 5-32 The RRC drill for large-diameter foundation elements excavated with slurry. (*Tone Boring Co.*)

is available in three different models that can excavate holes of diameter from 1 to 3 m (3.3 to 10 ft) and to a maximum depth of 80 m (260 ft). Because the torque on the drilling bits is offset by the motor drill, the machine does not require drill rods but can be cable-suspended from ordinary service cranes. As the bits are rotated to cut the soil, fresh slurry is delivered to the excavation as the side-cutter assembly moves up and down to form a circular shape. Cuttings are continuously ejected mixed with slurry through the center bit by air lift. Because of the special motion of the cutting blades of the bits, gravel or rock fragments are forced toward the reverse gate and large boulders are crushed against the walls of the hole before the smaller pieces enter the reverse circulation.

The drill can excavate the bottom of the hole flat and level and finish it according to the bearing requirements. Cleanout is done with the application of air lift. Since it is not feasible to make a visual inspection of the bottom or lower a workman into the hole with the appropriate cleanout tools, the excavation must rely on the skills of the operator and the efficiency of reverse circulation to produce a good bearing surface.

Other Circular Drills In the United States in addition to the auger-type drilling rigs there are several models of rotary rigs using a circulating drilling slurry to carry the cuttings out of the hole. These are used for drilling pier foundations. Examples are the Acker Model WA and the Wabco 1500 HD.

Direct Circulation

In certain conditions and for relatively shallow trenches it may be possible to remove excavated materials by direct circulation. In this case slurry is pumped into the trench under pressure through a supply line which is extended almost to the bottom. As materials are excavated, they are forced by the flow of slurry to travel up near the surface and stay in suspension by the shear strength of the system. Once they arrive at the surface, they are overflown into sand pits, where they are collected while the slurry is returned to the trench. Direct circulation is used in conjunction with rotary drilling equipment, and its main disadvantage is that it results in unusually high slurry densities due to the high percentage of soil retained in the slurry volume even after separation.

REFERENCES

Fuchsberger, M., 1974: Some Practical Aspects of Diaphragm Wall Construction, *Proc. Diaphragm Walls Anchorages, Inst. Civ. Eng., Lond.*
Ikuta, Y., 1974: Diaphragm Walling in Japan, *Ground Eng.,* **7**(5): 39–44.
PCSA, 1968a: Cable-controlled Power Cranes, Draglines, Hoes, Shovels, Clamshells, *Tech. Bull.* 4, Power Crane and Shovel Association, Milwaukee, Wis.
———,1968b: Crane and Excavator Standards No. 1, 2, and 3, Power Crane and Shovel Association, Milwaukee, Wis.
Pigeon, Y., 1974: Manicouagan 3, The Main Dam Cutoff, *Proc. Eng. Found. Conf., Found. Dams, Calif., March.*
Ressi, A., 1974: Excavating Systems, *Conf. Underground Constr. Fluid Trenches, Univ. Ill., Chicago Circles,* April.
Sliwinski, Z., and W. G. K. Fleming, 1974: Practical Considerations Affecting the Construction of Diaphragm Walls, *Proc. Diaphragm Walls Anchorages, Inst. Civ. Eng., Lond.*
Tamaro, G., 1975: Slurry Walls, *Proc. Sem. Open Cut Constr. Metrop. Sec. ASCE, New York, February.*
Tone Boring Co., 1974: Manual for Stabilizing Fluids, *Bull.* BW 121, Tokyo.
Ueda, S., 1974: Excavation with the BW system, personal communication.

Chapter Six

SLURRY-TRENCH-CUTOFF WALLS

6-1 INTRODUCTION TO CUTOFF SYSTEMS

In this context slurry-trench cutoffs encompass the entire range of continuous-earth, semirigid, and rigid structures built below grade for the control of groundwater in a continuous process impeding lateral flow. The process generally consists of excavating a trench, adding bentonite slurry as needed to keep the face stable, and then backfilling with selected materials to produce an impervious barrier. The operation is carried out with special equipment but usually is within the capability of dragline buckets, backhoes, clamshells, or conventional trenchers. Alternatively, a watertight screen can be built from the ground surface by injecting clay-cement grout under pressure into a preformed narrow slot and then letting the mixture harden.

Control of groundwater movement is essential in maintaining a balance in the water supply, where it is necessary to prevent pollution of clean water and to protect deep excavations. Examples from these applications are cutoff walls built as seepage barriers beneath the main body of canal embankments and earth dams; impervious barriers for pollution control in alluvial terrains and industrial sites; the prevention of landslips, which are invariably facilitated and often caused by the percolation of groundwater; the prevention of saline intrusion into water supply; groundwater and aquifer recharge schemes; and special cutoff walls built for the control of underground erosion.

In these applications the main advantage of the technique is the ability to provide a positive permanent control measure without altering the groundwater level during operations. In addition the slurry method offers the following benefits: (1) the cutoff wall can be inserted below ground even if the ground is very mobile; (2) the cutoff can be guaranteed to be continuous; i.e., there are no seams and joints to cause potential leakage; (3) because the final cutoff structure can be made sufficiently flexible, it can adjust to local ground movement without cracking; and (4) the construction is rapid and relatively inexpensive. In certain applications and underground works slurry-trench-cutoff walls can be supplemented by other alternatives, namely impervious blankets, chemical grouting, relief wells and surface drains, sheet-pile walls, conventional dewatering and pumping, cofferdams, and freezing applications.

Cutoff walls are classified according to the degree of stiffness, the type of backfill materials, and the construction methods. In this context they can be grouped as follows:

1. Earth cutoffs, in which the backfill consists of selected earth materials blended with bentonite; these are suitable where a completely impervious barrier is neither practical nor economically justified, when the cutoff must be flexible, and where space at the site does not present construction problems.

2. Cement-bentonite cutoffs, sometimes called solidified walls; these are a good choice where construction space is limited, e.g., a city site, and the installation must be completed with a minimum sequence of construction events.

3. Special systems such as injected-grout curtains; these are used when horizontal differential ground movement is not anticipated and the cutoff is preferentially of moderate depth. These systems, more than others, require specialist contractors.

4. Plastic concrete cutoffs, also called flexible walls; these are generally adopted where both strength and deformability are desired features.

5. Concrete diaphragm walls, either plain or reinforced, including the range of interlocking elements; used when rigidity is the prime factor for the conditions under which the structure will be in service or where the cutoff is exceptionally deep. The last four types are discussed in some detail in Chap. 7.

Initially a cutoff is selected for the intended purpose and function. The design and construction features are determined after a study of the site conditions and an analysis of ground movement under the expected consolidation, transient flow, and gradient. On the other hand, an engineering solution must be compatible with the probable construction imperfections inherent in each type and project. For instance, a concrete cutoff can seemingly have a lower permeability than an earth backfill, but in a concrete structure water may leak through cracks, defective construction joints, or areas of segregated and porous concrete, sometimes causing more leakage in the actual service. Many cutoffs bear the evidence of action which as a problem does not relate to theoretical or design aspects but shows the need of securing a construction conforming to the conditions of service.

If a cutoff is placed in compressible soil that is to be overloaded or subjected to a hydraulic gradient, it must be sufficiently deformable to adapt itself to ground movement without failing, sufficiently strong to withstand stress concentrations where they occur, and sufficiently impermeable to control the flow of groundwater. Occasionally, other conditions must be satisfied; e.g., when a slurry trench is excavated at the toe of an embankment, the overall stability must be ascertained, as explained in other sections. Thus, for the general case it is necessary to study various types of cutoffs and substantiate the conclusions through the laboratory testing of suitable backfill materials to determine that (1) it is possible to select materials that will conform to the elastic properties of the soil if necessary; (2) it is practicable to insert a cutoff that can withstand deformations and also exhibit ductile behavior; (3) it is possible to adapt the construction to the site conditions; and (4) it is possible to provide the desired reduction in permeability.

This program does not exhaust the possibilities or the scope of engineering services. For instance, even at sites regarded as reasonably safe chemical attack on unprotected cutoffs sometimes can cause considerable damage; cutoffs adjacent to organic soil or exposed to thermal pollution, acids, and other toxic waste can be damaged by surface and deep erosion that will reduce efficiency and may also render the structure unsuitable with time.

6-2 EARTH BACKFILL

Cutoffs built with earth backfill were probably the first application of the slurry-trench method (see also Chap. 1). Since the early uses their popularity has increased because they are fairly inexpensive, effective, and relatively easy to construct, particularly if the materials are available at the site. The process consists of excavating a continuous trench using bentonite slurry and then backfilling it with a selected blend of earth materials to form an impervious barrier. The same slurry used in the excavation is also added to the backfill, mainly to facilitate its mixing and placement and keep it homogeneous. In the final position the slurry gives the earth materials better watertightness and resistance to displacement. However, the backfill must be properly graded since mixing with slurry alone cannot ensure the desired reduction in permeability.

Clay Mixes

The impermeability of an earth backfill is substantially improved if the constituent materials are supplemented with fines, usually silt or clay, of a suitable type and source. Despite the importance of a well-graded backfill, the design of earth cutoffs sometimes specifies only the degree of watertightness to be achieved, leaving the choice of the fine material to the specialist contractor, the idea being that this flexibility is consistent with the scope of the work and the economics of the project.

The important functions of clay in a mix of earth materials justify a precise analysis of the physical and flow properties of the clay in the design stage. In the first place the selected clay must form a successful blend with the bulk of the backfill, and for this it must be comparatively inert, i.e., chemically inactive. Second, the physical and rheological properties of the clay will affect the properties of the backfill, which must display characteristics during construction and at rest consistent with the conditions it will experience during mixing, storage, placement, and compaction. Finally, when it is in place, the backfill must seal the ground and resist displacement.

The first function of the fine material is to block the voids of the aggregate fraction without causing segregation or displacement of soil particles. If this material can expand upon hydration and form a gel while at rest, it will help the operation by reducing the bulk quantity while giving the backfill the ability to resist displacement without the addition of rigidifying agents. This fine material is generally available among the naturally occurring clays, particularly the prepared clays of sodium montmorillonite.

Sources of Clays Clays may be used as they occur naturally or treated and conditioned. Natural clays are generally preferred in view of the economy implied, particularly when they are locally available. Maximum economy is attained when the fine fraction is provided by clayey and silty soil from the excavation, blended with sand and some gravel and then mixed with bentonite slurry. A frequent treatment of natural clays is the removal of coarse particles and organic matter. Wet plastic clay is very difficult to mix with the backfill and should be avoided. When the source of natural clays is not reliable, the more expensive prepared clays may offer a more economical construction considering the overall efficiency. Among these sodium bentonite clays offer the best choice.

In the search for natural clays it is expedient to concentrate first on the locally known deposits. Marine or alluvial clays are more suitable than glacial clays because the latter sometimes contain a high proportion of sand and silt and also are erratic in distribution. Once the better grades are identified, the selection is based on the type of clay minerals and mainly on the content of colloid-sized particles. In most clays these particles constitute the three basic groups of montmorillonite, illite, and kaolinite. As we have mentioned, the most active group is the montmorillonite clay because of its ability to swell by taking water molecules directly into the space lattice.

Natural clays may be selected for either the backfill or the slurry. When extended use is contemplated, a standard-quality laboratory analysis is indicated to determine the suitability of the material.

Physical Properties The particle size distribution is important, especially where the variation of the content of the coarser fraction is considerable. Next, the Atterberg liquid limit is useful. A clay with a liquid limit less than 60 should not be considered unless it is practical and economical to remove the coarser fraction forming the skeleton of the soil and thereby improve its fineness. Clays containing mostly bentonite have a liquid limit between 300 and

400, but this should not be the sole criterion since it does not describe the rheological behavior.

Chemical Properties Relevant chemical properties are the surface activity, base-exchange capacity, and pH. In natural clays the base-exchange capacity can vary considerably, and similar variations might be expected in the backfill or the slurry. Such variations are difficult to eliminate or regulate economically, and this often means the acceptance of wider limits for the properties of the backfill, e.g., a relatively minor change in composition determined by chemical analysis may result in a major change in the actual properties of the backfill.

Clay suspensions usually tend to be slightly alkaline and thus flocculate upon the addition of acidic materials. The pH should be in excess of 7, which is the neutral point on the pH scale, but for many jobs a minimum pH value of 8 is quoted.

Rheological Properties These sometimes are identified with the flow properties discussed in Sec. 4-4 and include the yield stress, apparent and plastic viscosity, gel strength, filtration loss, and thixotropy. Initially the slurry must protect the face of the excavation and then form a gel in the backfill to withstand the applied gradient. If the clay does not possess thixotropic characteristics, it may not provide sufficient shear strength, in which case rigidifying agents will be necessary.

Certain clays display a property whereby the gelling action is speeded up by agitation, a situation which can cause serious trouble in the mixing and placement process. Adequate use of dispersants will remedy mild occurrences, but clays exhibiting this property to a high degree should be abandoned at the outset (Leonard and Dempsey, 1963).

Permeability and Composition of Backfill

The in situ permeability of earth cutoffs varies considerably with the project and the application but depends mainly on the gradation of the backfill, the mixing procedures and the conditions of placement. On the basis of permeability tests on samples of the backfill material, the range of permeability is between 2×10^{-6} cm/s (Katowicz, 1967) and 1×10^{-6} cm/s (Xanthakos and Bailey, 1975), although values as low as 5×10^{-9} cm/s have been quoted (LaRusso, 1963). For a permeability of the order of 10^{-6} cm/s the backfill must contain a significant clay fraction; at best the grading must be such that there is sufficient fine material to fill the pores created by the coarser particles. Besides the low permeability, this grading gives also the backfill low compressibility characteristics. Extra fines sometimes are provided to prevent segregation, but if the portion of fines becomes excessive, it may cause consolidation of the backfill.

Backfills for cutoffs constructed in Europe and Australia have the gradation limits shown in Table 6-1. The gradation requirements for cutoffs built in the

TABLE 6-1 Typical Gradation Limits for Backfills in Europe and Australia

Screen size, British standard sieve	Percentage passing by weight
3 in	80–100
3/4 in	40–100
1/4 in	30–70
No. 25	20–50
No. 200	10–25

United States are exemplified in Table 6-2. Comparison of the data from these tables shows fairly good agreement.

Addition of Fines From Tables 6-1 and 6-2 it is evident that at least 10 percent of the backfill material should pass the no. 200 sieve, but quite often the minimum percentage of fines is much higher and close to 20 percent by weight including the entrapped bentonite. When the fraction of fines becomes excessive, it can cause consolidation, although in the actual trench the degree of consolidation is limited by arching. In this respect it is advantageous to carry out consolidation tests under low overburden pressures when selecting backfill materials, except for well-graded gravel-sand-clay backfills, which show less variation in permeability with overburden pressure.

TABLE 6-2 Typical Gradation Limits for Backfills in the United States

	Percentage passing by weight		
U.S. standard sieve	Wanapum Development	Camanche Dike	Southport AWT Facilities, Indianapolis
3 in	80–100	80–100	100
3/4 in	40–100	60–100	55–100
No. 4	30–70	40–80	40–75
No. 30	20–50	20–60	25–50
No. 200	10–25	10–30	20–30

When suitable silt or clay deposits are not available at the site or from borrow areas, it may be necessary to resort to more plastic or cohesive materials for the fines. In some instances, although the fine materials are plastic and lumpy, they will break up fairly well by blading and disking by dozer and grader, and the desired backfill will be produced without difficulty. In other examples the plastic clay may be too wet to blend well with the excavated sand and gravel, and the presence of capillary moisture and organic matter will prevent its use. Thus, if the suitability of fines is doubtful, field tests with construction equipment should be carried out to determine whether they will blend.

6-3 CONSTRUCTION OF EARTH CUTOFFS

Excavation

The excavation is usually done with dragline buckets, but clamshells, backhoes, and conventional trenchers are also used. The construction requirements mean that the trench for an earth cutoff wall generally is wider than can be excavated with standard slurry-wall equipment. The final structure requires less accuracy both in verticality and in plan, and the actual smoothness of the face is not critical. These factors have resulted in the development of new techniques and equipment, less complicated and more suitable for this type of work, described in some detail in Chap. 5.

The basic purpose of an economical construction is achieved by excavating a continuous trench to the specified depth, backfilling with suitable blend, and carrying out a supplementary field-control program. Thus contractors may be tempted to utilize the fastest excavation method available in order to minimize cost. On the other hand, the actual depth of trench will dictate the choice of equipment, since most machines can perform only within a certain depth range. It is therefore important to establish the bottom profile of the cutoff accurately so that a suitable machine and sequence of trenching can be selected before work is begun.

The excavation should be carried down to final depth immediately where it is started, and the entire depth of trench should then be carried along the trench line. Where this operation is not practicable, the slurry trench can be excavated and constructed in sections, leaving temporary plugs of undisturbed earth along its base line. The joints between consecutive sections can be made like butt joints by adequate overlapping to ensure the continuity of the backfill. If the cutoff is constructed in sections, the dragline must straddle the previously backfilled portions, which must be reexcavated for a certain length. Experience shows that the face of the reexcavated section will not slough as long as it is supported by slurry and if it has been left in place for some time to consolidate.

Although the control limits for the slurry density in the trench are not so critical as those in the construction of diaphragm walls, the slurry in the trench must occasionally be recirculated to remove the coarser fraction and prevent unreasonably high density level. If the latter becomes too high, the dragline bucket will tend to float in the slurry, causing delays and other problems in the excavation.

Special Features of Dragline Bucket The standard dragline bucket with vertical side walls is effective in providing vertical trench sides without wedging. Contractors report that it is not necessary to use buckets fitted with an eye or swivel to excavate deep trenches. It is advantageous to use buckets slightly narrower than the specified trench width since the side cutters tend to cause some raveling on the sides of the trench.

Some preference has been expressed for nonperforated buckets because they remove more sand than is possible with perforated buckets. The latter are sometimes selected on the basis that they remove less slurry, but experience again shows that little slurry is wasted by the use of nonperforated buckets provided that the dragline operator keeps the bucket at the correct angle when hoisting and sloshes the slurry within the bucket so that it can spill back into the trench and that the slurry in the stockpiles is drained back into the trench.

Bottom Cleaning In order to ensure full contact at the bottom with bedrock or other impervious formations, the base of the excavation generally is probed for unconsolidated material, cracks, and potholes using an air lift. Pervious material in natural depressions and sand or sediment that settles out of the slurry are removed by an air-lift pump, which has proved to be quite effective in cleaning the bottom and even removing cobbles. When the sand-slurry mix is blown out from the air-lift pump and onto the bank, the sand settles out and the slurry is drained back into the trench.

Occasionally, the contact of the backfill and the underlying material is checked by drilling holes through the backfill and performing pressure tests. If the water loss is excessive, the problem must be remedied by grouting.

Mixing and Backfilling

The optimum gradation of the backfill cannot always be achieved unless the unsuitable materials are separated and removed from the main spoil and suitable borrowed materials are added to form a balanced mix. On the other hand, it is advantageous to utilize as much material from the excavation as possible and remove only the unsuitable portion. In some cases the method of excavation has a certain influence upon the composition of the backfill since it is relatively easier to separate earth materials with some machines than it is with others.

This type of work requires large volumes of slurry to be available at the site continuously. Contractors report that this quantity may be eight times the slurry required for diaphragm walls of the same surface area. This requires unusually large facilities for mixing, storing, and recirculating the bentonite slurry. The much greater consumption of slurry is due to wider than usual trenches; considerably higher fluid loss as slurry is filtered toward the ground; mixing of slurry with the backfill material, and a certain waste in the spoil dikes. As a consequence, it is seldom necessary to pump any mud out of the trench for disposal until the completion of the job, when any slurry left is merely wasted.

The backfill is mixed by windrowing, dozing, or blading to remove lumps of clay and silt and pockets of sand and gravel. The earth mix is then sluiced with slurry in a process which must be thorough without wasting either soil or slurry. Some contractors achieve this by dumping the spoil into diked areas from which any excess slurry is reclaimed by draining back to the trench through channels. When seepage control is most critical, mechanical mixing under adequately controlled conditions may be specified to ensure the uniformity of the backfill.

The backfill mix should not be allowed to become segregated, which means that it should not be too wet. On the other hand, if it is too dry, the slurry will not be completely mixed with the earth materials. If it is too stiff, it should be made more workable and flowable by adding more slurry and then mixing again; otherwise it will leave earth lumps and pockets of slurry in the trench. Mixing with water should not be permitted. Optimum consistency for placement and self-compaction is attained when the mix has a slump of 6 in measured with a standard concrete cone according to ASTM C143, although there is some concern that such a mix may be too wet and cause the larger gravels to segregate.

All particles and soil grains should be coated with slurry to provide a mix which is viscous so that it will slide slowly into the trench after bulldozing but without rolling down the slope. Such a mix displays high plasticity. Stockpiled earth materials composed predominantly of granular soil and having enough friction to produce an angle of repose of about 1.3 horizontal to 1 vertical come to rest at a slope of 6 horizontal to 1 vertical or flatter when mixed with slurry. This is attributed to a reduction in the friction angle caused by the lubrication of individual soil grains, as discussed earlier.

Placement Backfilling should follow the excavation but not too closely. When the excavated portion is about twice the depth, backfilling may start, but in stable ground much longer unfilled lengths are allowed provided the slurry facilities can accommodate the excavation. If the backfill becomes too fluid upon placement, either the unfilled sections must be made longer or a barrier must be inserted across the trench to prevent excavating materials already placed. Some specifications require the end of the excavation slope to precede the toe of the backfill by no less than 50 ft and no more than 150 ft, although there is no clear basis for these limitations. In general, enough space should be provided behind the excavation for cleaning the bottom.

The backfill should not be dropped into the trench since this will cause segregation, forcing the coarse fraction to the bottom and leaving the fines in suspension. Falling material will also trap pockets of slurry. The backfill is initially lowered to the bottom of the trench using clamshell buckets or similar equipment from one location only and is deposited according to an angle of repose about 6 to 8 horizontal to 1 vertical. This continues until the backfill emerges from below the slurry surface, reaching the top at one end and sloping down toward the other. The placing of more material then continues by placing an earth pile at the starting point and pushing it into the trench with a bulldozer. This material slides progressively down the slope of the previously placed backfill without trapping pockets of slurry. If the backfill is too cohesive and fails to slide, a clamshell bucket must be used throughout.

The earth materials must form the cutoff as placed and without further compaction, since this can have a stirring action and disperse the fines in the mix. Surface rolling may be introduced at the top of the trench but is likely to have only a localized effect due to arching of the backfill. At the extreme case the material may be placed dry if it is friable enough.

If it is not possible to control the distance between the toe of excavation and backfill, e.g., where the placed fill has a very flat slope, the latter may override the coarser materials that settle through the slurry and force them to be piped into the foundation material, especially if it consists of open gravel. Piping may in turn cause sink holes to develop at the surface of the backfill.

Temperature during Construction According to experience from the construction of earth cutoffs in cold climates, backfill can be mixed and placed when the minimum daily temperature is as low as 15°F and when the maximum daily temperature is only 25°F. These are absolute limits and can be accepted only in extreme cases, whereas for the usual cutoffs the mixing and placing of backfill should be limited to days when the average temperature is not less than 25°F. The construction of an earth cutoff is shown in Figs. 6-1 to 6-4.

Fig. 6-1 Excavation of a slurry-trench cutoff with dragline bucket. (*INQUIP Corp.*)

Fig. 6-2 Air-lift cleaning of trench bottom. (*INQUIP Corp.*)

Fig. 6-3 Placement of backfill for a slurry-trench-cutoff wall. (*INQUIP Corp.*)

Fig. 6-4 Entire operation from the construction of a slurry cutoff wall. (*INQUIP Corp.*)

Special Treatment

Protection of Base The method of removing unconsolidated materials from the bottom by passing an air lift is effective in cleaning the bedrock surface but also can be used to excavate uncemented sands and silts from depths slightly below the capability of draglines. Besides removing all overburden material from irregular bedrock surfaces, it is sometimes advantageous to place a concrete pad 2 to 3 ft thick at the bottom to provide a solid base and as protection against the action of potential seepage at this location. The fresh concrete is lowered with open bucket and receives a layer of prepared impervious backfill on its top, also lowered by bucket.

Protection of Top Protection of the completed cutoff at the top is provided by covering with a 2- to 3-ft-thick layer of impervious clay-soil backfill core material. This is placed as soon as the backfill reaches the intended level and before it is allowed to dry out. The material should have a moisture content slightly above the optimum value, and it is compacted by the gentle application of rolling equipment; care must be taken not to disturb the bulk of backfill.

Trenches wider than usual are gradually widened at the top by flaring the uppermost part at a suitable slope. This is very practical when a cutoff is placed beneath an earth embankment and allows the backfill to fit the base of the main impervious core. The smooth transition provides a good connection, allowing the core to subside together with any settlement of the backfill in the trench.

Impervious blankets over the top may be used for protection. Their placement and compaction, however, deserve special attention due to the difficult passage of construction equipment (Anton and Dayton, 1972). Besides the creation of potential seepage paths, settlement of the backfill may cause it to separate from the relatively rigid fill placed over the trench. A good connection can be obtained if at least 50 to 70 percent of the blanket depth is installed and compacted first and the trench cutoff is installed through the blanket.

Protection below the Base If the depth of ground to be sealed off exceeds the depth range of conventional equipment, it may be practicable to supplement the construction with a grouted cutoff injected below the base of the slurry trench. This has been done in many instances and found effective. The grout is injected using a special sleeve grout pipe known as the *tube à manchette*, which is inserted into a cased hole through the finished cutoff and the alluvial underneath. Examples of this application are the Wanapum development (LaRusso, 1963) and the Balderhead dam (Little, 1974).

Settlement The recorded settlement of earth backfills caused by consolidation of the earth material generally varies from 1 to 6 in for a depth of 50 to 90 ft. This settlement appears to occur in trenches wider than 8 ft. Relatively narrow backfills settle much less, due to a more pronounced arching effect. The settlement usually is completed 6 months after placement; it may sometimes continue thereafter but at a negligible rate. If a cutoff is placed beneath an earth embankment, a normal amount of subsidence will be observed at the crest of the completed structure.

Connections When it is necessary to provide a junction between an earth cutoff and a different structure, e.g., a gravity dam, the terminal portion of the backfill must be reexcavated and replaced by a rigid cutoff. Reexcavation is carried out under conventional dewatering and bracing techniques. The same treatment has been incorporated for connecting an earth cutoff to the outlet works of a dam project.

Construction Period In extremely adverse climates construction should be scheduled to take advantage of seasonal benefits, such as low groundwater table and favorable temperature. On the other hand, ample time should be provided to allow the completion of the cutoff in one construction season.

6-4 DESIGN OF EARTH CUTOFFS

During excavation the bentonite slurry must keep the excavation stable without arching since cutoffs usually are constructed in continuous trenches. Since some soil will generally be left in the trench and mixed with the slurry, a density somewhat greater than that of fresh slurry is admissible in stability analyses.

The design of a cutoff must ensure that the slurry-impregnated earth backfill will provide the desired reduction in permeability and that slurry or fine particles will not be forced out of the backfill under the maximum hydraulic head. For the average field conditions, a coefficient of permeability of the order of 2×10^{-6} cm/s (about 2 ft/year), often taken as a design criterion, is justified for most applications. Such a cutoff is not completely impervious but represents a soil with very poor drainage. Alternatively, when complete seepage and pollution control are required, it will be necessary to provide cutoffs of lower permeability.

For a first analysis of blowout failure reference can be made to Sec. 3-7; the critical hydraulic gradient which controls this type of failure is approximated by Eq. (3-19). The accuracy of this prediction depends on (1) the actual gel strength of slurry in its final position in the backfill, which is often taken as 2 or 3 times the 10-min gel strength but is much higher if the backfill contains clay or other cohesive material; and (2) the representative size of the gravel fraction most relevant to blowout, which is even more uncertain if the backfill consists of soil taken from erratic granular deposits and the final material is not graded. More often, therefore, the result of these theoretical predictions is a gross overestimation; less often it is an underestimation of the actual factor of safety against blowout.

When the coefficient of permeability and the critical gradient are tentatively known, it is possible to check the total flow of water through the cutoff and compare it with the desired reduction in the percolation volume by reference to the relationship

$$v = ki \tag{6-1}$$

known as *Darcy's law*. In this expression v is the discharge velocity, defined as the quantity of water that percolates in a unit time across a unit area. If the discharge velocity is too high for the conditions involved, it is possible to adjust the factors k and i merely by redesigning the mix and changing the thickness of the cutoff.

Ordinarily the flow of seepage through the cutoff is based on the backfill material only. Current theories tend to dispute this approach, suggesting that the cutoff derives its watertightness from three contributing elements, namely the zone outside the trench which is impregnated with slurry, the filter cake at the interface, and the backfill itself. This is true in many instances, but it is equally true that neither the extent of penetration nor the long-term condition of the filter cake is known or controlled. On the other hand, the rheological characteristics of bentonite gels can diminish or even disappear if the water is allowed to evaporate, and reversal from swelling to shrinking can occur under progressive dehydration.

The bentonite concentration is decided mainly from three considerations: filter-cake and stability requirements during excavation; the intended reduction in permeability; and protection of the fines against blowout. If the first requirement is satisfied, this is usually sufficient to satisfy the other two conditions as well.

Special precautions must be taken to keep the slurry loss to a minimum since the trench usually is so long that the replacement of lost slurry can become a serious problem. In some cases it will be advisable to anticipate cave-ins or partial collapse of the trench, particularly in the upper part of the excavation. This situation can conceivably arise from a combination of careless construction procedures, the passage and application of heavy surcharge loads near the face, and the usually high groundwater table at the site of cutoff walls. In most instances the consequences are not serious except for the use of more material and therefore higher cost to the contractor; however, since cutoffs wider than specified can affect the configurations of other improvements at the site, the problem warrants ample stability investigations.

Design of Backfill The proportioning of the slurry-clay-gravel mix to provide the stipulated control of groundwater and resistance against blowout can be simplified with the special tests described in the following section. In addition, the mix must be flowable and lack excessive cohesion to allow its placement. Although a higher gel strength means that the slurry is more resistant to blowout, such a slurry is hardly workable and is difficult to mix with earth materials. Besides the minimum slump requirements, the placement of the backfill is facilitated if the slurry density is less than 85 lb/ft^3. Contractors complain that this is difficult to maintain without recirculating the slurry; hence it imposes undue costs to the construction. In spite of that, this author suggests a maximum slurry density not exceeding 90 lb/ft^3 under all circumstances.

Configuration of Cutoff Whereas the location, layout, profile, and depth of cutoff are determined mainly by the site conditions and the project characteristics, the width must be established to provide adequate safety against blowout failure for the exit gradient under the maximum differential head. Because of the many uncertainties involved in the determination of the blowout gradient the recommended factor of safety against blowout should not be less than 4 to compensate for construction imperfections and other limitations. Alternatively, where the differential head is not critical, the minimum trench width is based on practical considerations, including the availability of excavating equipment.

6-5 BLOWOUT TESTS

Blowout tests, also called *pressure tests*, are very useful if the procedures and the materials used in the conduct of the test represent the in situ conditions. They are performed on representative samples of backfill material obtained by mixing bulk soil samples from the site with a suitable bentonite slurry. The purpose of the tests is to establish the probable hydraulic gradient at which failure will occur and also to provide an estimation of the permeability of the backfill material.

Test Apparatus This consists of a compressed-air cylindrical tank of a suitable size, usually 28 to 30 in high and 15 to 18 in OD. The tank is provided with a drain, discharge pipe, and collector vessel and is connected to a water source to which air pressure can be applied (Figs. 6-5 and 6-6).

Slurry-Trench-Cutoff Walls   209

Fig. 6-5 Steel tank and miscellaneous equipment for blowout test.

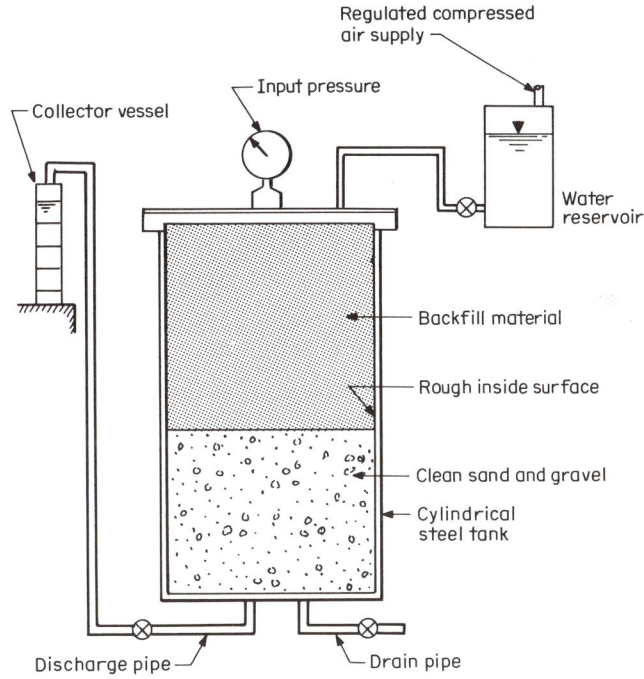

Fig. 6-6 Schematic section of blowout apparatus.

The test involves placing a clean, saturated sand-and-gravel filter at the base of the tank within 9 to 10 in of the bottom and filling the tank with bentonite slurry, which is allowed to stand until a seal is obtained in the gravel bed. The backfill is then placed to fill the tank, simultaneously displacing the slurry, and the lid is securely fastened and sealed to the top of the tank.

Pressure increments are applied to the system, and the resulting discharge of water from the base of the tank is measured at suitable intervals. The critical hydraulic gradient is assumed to have been reached when a sudden or substantial increase in the flow rate occurs or when malfunctioning of the test becomes apparent. Besides monitoring the applied pressure and the flow from the tank it is also useful to observe the presence and type of solids in the discharge water and examine the condition of backfill and gravel bed at the conclusion of the test.

The permeability of backfill in the tank is estimated directly from the relation

$$k = \frac{Q}{tiA} \tag{6-2}$$

where k = coefficient of permeability, cm/s
Q = discharge volume of water, cm^3
t = time over which discharge is measured, s
i = hydraulic gradient
A = cross-sectional area of backfill normal to flow, cm^2

During the test, for a pressure increment the water outflow through the discharge line first increases rapidly and then decreases gradually with time to almost a steady-state flow (Xanthakos and Bailey, 1975). Thus, an apparent permeability is obtained as the slope of the chord between two points in the discharge-time curve divided by the product iA. This apparent permeability approaches the true permeability of the backfill after a consolidation equilibrium is reached for all hydraulic gradients less than the critical.

Test Limitations Major discrepancies will result during the conduct of the test if (1) the backfill is loosely placed, contains entrapped pockets of slurry, and has poor contact with the lining of the tank, all causing potential seepage paths; (2) the backfill absorbs too much slurry as it is placed in the tank; or (3) it is not possible to apply uniform pressure increments and maintain constant pressure. On the other hand, some previous experience with the test is quite helpful, particularly in deciding when the consolidation equilibrium has been reached.

Localized piping along the backfill–tank-wall interface can be avoided if suitable materials are used at this joint. A good connection is obtained if a mortar lining is applied to the wall of the tank and the perimeter is sealed at the top with a poured-in-place polyvinyl chloride rubber seal. Application of a silicone caulking material to roughen the wall has been tried but is not always successful, and epoxy filler cement mixture has been found more effective.

The backfill should be thoroughly mixed before placement in the tank, and this process is facilitated if it has a slump of 4 to 5 in. As the material is placed in the tank, it is likely to absorb some slurry, which will raise the slump to about 6 in. Loose or too fluid backfill having entrapped pockets of slurry can be avoided if the slurry is removed from the tank before placement.

Test Results Variations in the boundary and consolidation conditions between the testing apparatus and the actual cutoff give rise to difficulties in the interpretation of results from blowout tests, but the following conclusions can be used on a general basis:

1. The presence of fines can reduce the permeability of the backfill to less than 1×10^{-6} cm/s.
2. In the lower range of gradients the apparent permeability decreases with increased gradient, whereas in the upper range the permeability increases with increased gradient. The division between lower and upper range is not always the same.
3. Consolidation effects on the backfill in the tank influence its true permeability for pressure increments of short duration; however, for long-duration pressure increments the measured permeability agrees fairly well with permeability measurements from standard tests.
4. For cutoffs with gradation limits similar to those of Tables 6-1 and 6-2 blowout failure appears to occur at gradients from 30 to 40. For preliminary estimates a gradient of 32 may be used, and thus for a factor of safety of 4 an approximate cutoff thickness is found by dividing the differential head by 8.

6-6 A CASE STUDY OF EARTH BACKFILL

From the practical standpoint some concern is always raised regarding the permeability reduction that can be achieved by bulldozing earth material into a slurry trench and how it changes with time and consolidation effects and whether the method of placement can cause pockets of slurry to be trapped in the backfill and what their significance would be.

These points have been investigated in prototype tests carried out by Jefferis (1972) for materials similar to those used in the construction of a slurry-trench cutoff in the Dead Sea dikes. These materials consisted of two basic soil types; dry clay, which is a light grey-green very silty clay, natural water content 13 percent, LL = 35, and PL = 17; and asphaltic clay, which is a grey-brown friable material containing occasional hard lumps, water content 14 percent, LL = 50, and PL = 26. Tests on the dry clay for both the dry and wet condition showed that it broke down considerably when dispersed in water. Tests on the asphaltic clay showed that it was difficult to wet and that when it was placed in water, pockets of air remained entrapped.

Slurry and Cutoff Combinations Dead Sea water is quite unsuitable for bentonite slurries because of the excessive amount of salt (120 g of sodium chloride and 140 g of magnesium chloride per liter of water). Dry bentonite is not dispersed in such a system, whereas bentonite suspensions prepared from fresh water would immediately flocculate upon contact with salt. Hence, attapulgite was chosen as the colloid matter because it is not affected by the electrolyte content and requires no initial hydration in fresh water. The following slurries were prepared:

1. Dry-clay and asphaltic-clay slurries having the following composition by weight: attapulgite, 39.9 g; Dead Sea water, 1070.4 g; dry or asphaltic clay, 169.7 g; and fine to medium sand, 79.7 g, added to simulate trench conditions toward the end of the excavation.

2. Plastic-clay slurry containing attapulgite, 7.8 g; Dead Sea water, 341.4 g; lime, 9.7 g; and raw slurry 1157.5 g; the last was produced by mixing Dead Sea water with plastic clay until the density was raised from 1.23 to 1.62 g/cm³ for the raw slurry.

The properties of these slurries are shown in Table 6-3.

TABLE 6-3 Properties of Slurries, Slurry-Trench-Cutoff Wall, Dead Sea Dikes (from Jefferis, 1972)

Slurry type	Density, g/cm³	Plastic viscosity, cP	Gel strength, lb/100 ft²	
			Initial	10-min
Dry clay	1.36	16	14	18
Asphaltic clay	1.36	17	13	13
Plastic clay	1.52	15	20	28

For the cutoff three different soil-slurry combinations were used, dry clay in dry-clay slurry; dry clay in plastic-clay slurry; and asphaltic clay in asphaltic-clay slurry. Additionally dry and asphaltic clays were treated in Dea Sea water without slurry, thus giving five different soil-pore fluid combinations, which were investigated for changes in permeability and homogeneity with time.

The tests were conducted in a special apparatus (Jefferis, 1972) in which the samples could be subjected to a known constant overburden pressure and from which they could be removed intact at the conclusion of the test to allow consistency and water-content measurements. The tests were continued for 1 year to investigate time effects.

Permeability and Consolidation Tests Consolidation tests were carried out under an overburden pressure of 6 lb/in². The coefficients of consolidation and compressibility were calculated from direct measurements. Permeability tests were run at the same pressure, and the samples were further consolidated under an overburden pressure of 10 lb/in², which was maintained throughout the aging period. Besides the measured permeability, a calculated permeability was also obtained from the Terzaghi one-dimensional consolidation theory

$$k = c_V m_V \gamma_w \tag{6-3}$$

where k = coefficient of permeability
c_V = coefficient of consolidation
m_V = coefficient of compressibility
γ_w = density of pore fluid

For each of the five soil-pore fluid combinations the samples were aged for 1 week, 1 month, 3 months, and 1 year.

Consistency and Water-Content Measurements At the end of each aging period consistency measurements were made using the Swedish fall-cone apparatus. The penetration after 30 s is read on the scale in millimeters and recorded as the fall-cone penetration. For a given cone weight W_c in grams, the undrained shear strength s_u in tons per square meter is inversely proportional to the square of the penetration h expressed in millimeters; hence

$$s_u = \frac{W_c}{h^2} \qquad (6\text{-}4)$$

The water content was measured after removing the sample from the fall cone. The pressure of salt pore water complicated this measurement due to a salt residue left after drying. Hence, the water content was apparent water content ignoring the presence of salt.

Test Results

Table 6-4 shows typical measured and calculated permeabilities for the five combinations under three different effective stresses. The measured and calculated values show reasonable agreement. The permeability is reduced considerably with increasing effective stress, and this will have practical implications where arching in the trench can limit the consolidation of the backfill material.

TABLE 6-4 Measured and Calculated Permeabilities, Slurry-Trench-Cutoff Wall, Dead Sea Dikes (from Jefferis, 1972)

Sample type	Effective stress, lb/in²	Permeability, 10^{-7} cm/s	
		Measured	Calculated
Dry clay, in Dead Sea water	2	6.8	6.3
	4	4.3	
	8	2.9	2.7
In dry-clay slurry	2	5.1	4.8
	4	2.5	
	8	2.0	2.3
In plastic-clay slurry	2	5.0	4.4
	4	3.3	
	8	1.8	3.9
Asphaltic clay, in Dead Sea water	2	17.5	6.7
	4	11.6	
	8	6.0	9.4
In asphaltic-clay slurry	2	3.1	1.2
	4	2.4	
	8	1.4	2.0

Figure 6-7 shows the variation of permeability with time for an effective stress of 8 lb/in². For all the dry-clay samples and the asphaltic-clay-in-slurry sample the permeability decreased by about 2 in the first 90 days, but there was very little change thereafter. The effect of slurry on pure clays is to lower their permeability, but this is not pronounced for the aslphaltic clay because it

214 Slurry Walls

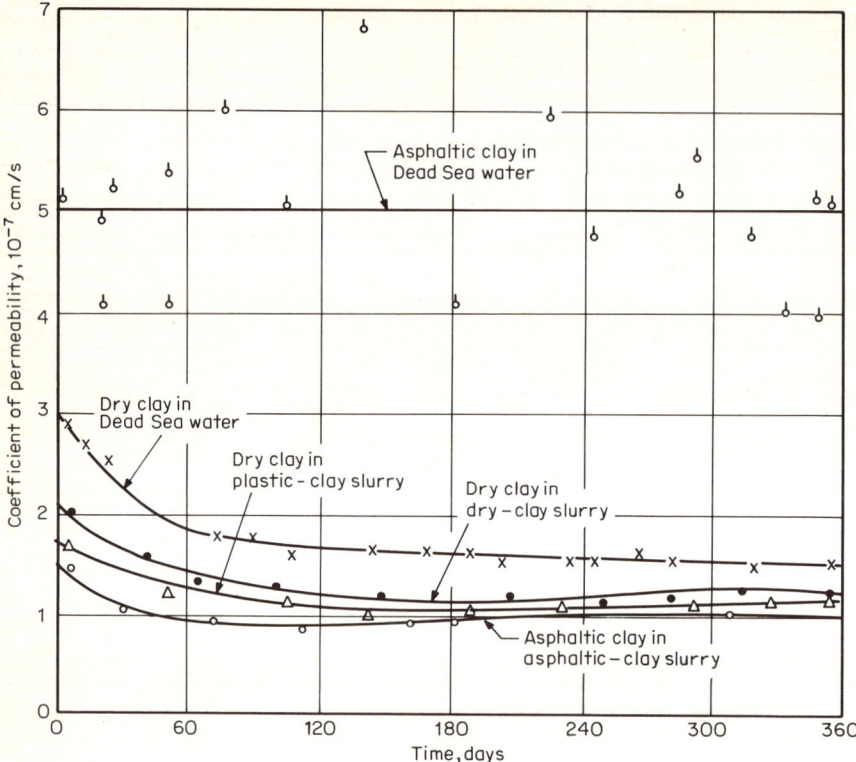

Fig. 6-7 Variation in permeability with time for Dead Sea soils. (*From Jefferis, 1972.*)

did not readily break down in water. The apparent variation in the permeability of this material is attributed to air trapped in the voids and changing in volume with pressure changes.

Table 6-5 shows results from consistency and water-content tests. According to these tests, the undrained shear strength of all samples increased by about 2 for the 1-year period. There was a corresponding decrease of about 2 percent in the water content of the dry-clay samples, but the asphaltic clay showed little change, probably because the decrease in water content due to consolidation was offset by the slow penetration of water in the harder clay lumps.

The coefficient of variation for the undrained shear is given as an index of variability, defined according to the relation

$$\text{Coefficient of variation} = \frac{\left[\sum_{i=1}^{n} \frac{(x_i - x)^2}{n}\right]^{1/2}}{x} \times 100$$

$$= \frac{\text{standard deviation}}{\text{mean}} \times 100 \quad (6\text{-}5)$$

TABLE 6-5 Changes in Consistency of Soil in Slurry-Trench-Cutoff Wall, Dead Sea Dikes, with Time (from Jefferis, 1972)

Sample type	Age, months	Water content, %	Bulk density, g/cm³	Undrained shear strength, t/m²	Coefficient of variation, %
Dry clay, in	$\frac{1}{4}$	23.64	1.85	3.2	33
Dead Sea	1	21.77	2.01	4.5	31
water	3	22.33	1.98	3.2	50
	12	21.44	2.06	6.7	42
In plastic-	$\frac{1}{4}$	24.35	1.94	2.2	70
clay slurry	1	23.85	1.82	2.7	55
	3	23.38	1.97	3.5	37
	12	22.98	2.00	4.6	48
In dry-clay	$\frac{1}{4}$	26.37	1.88	1.8	62
slurry	1	26.13	1.99	2.5	60
	3	24.80	2.01	2.8	23
	12	24.53	1.98	4.4	54
Asphaltic	$\frac{1}{4}$	27.07	1.59	3.4	69
clay, in Dead	1	28.66	1.68	2.8	23
Sea water	3	26.37	1.72	7.1	42
	12	27.58	1.73	7.9	47
	$\frac{1}{4}$	31.00	1.55	2.0	28
In asphal-	1	28.61	1.64	2.9	27
tic-clay	3	29.06	1.62	4.0	20
slurry	12	30.52	1.83	4.3	25

For the quoted values of undrained shear strength this coefficent shows no specific trend with aging. Thus the samples do not become more homogeneous with time; instead the weak and strong points gain strength in similar proportions.

6-7 EFFICIENCY OF TRENCH CUTOFFS

Seepage loss through a cutoff can occur due to imperfections inherent with any construction. The extent of these imperfections expresses the percentage area of the cutoff left unsealed. In concrete diaphragms this may be caused by faulty construction joints, and in earth cutoffs it may be due to nonuniform backfills improperly placed. Seepage loss can be estimated assuming that the openings are evenly distributed as a group of parallel slits per unit length of cutoff. The extent of imperfections is then taken as the ratio of the total area of the slits A_s to the total area of the cutoff D per unit length of the structure.

For earth cutoffs extending down to impervious strata, the efficiency E is defined as

$$E = \frac{100(Q_0 - Q)}{Q_0} \qquad (6\text{-}6)$$

in which Q is the total discharge per unit length of cutoff and Q_0 is the total discharge per unit length if the cutoff did not exist. The configuration of the problem is shown in Fig. 6-8. The cutoff has a thickness l and is located at distance $m_0 + d$ from the upstream edge of the dam whose width is B. The depth of the pervious stratum underneath the dam is D, and the soil permeability coefficient is k. The pressure head causing flow is ΔH_0.

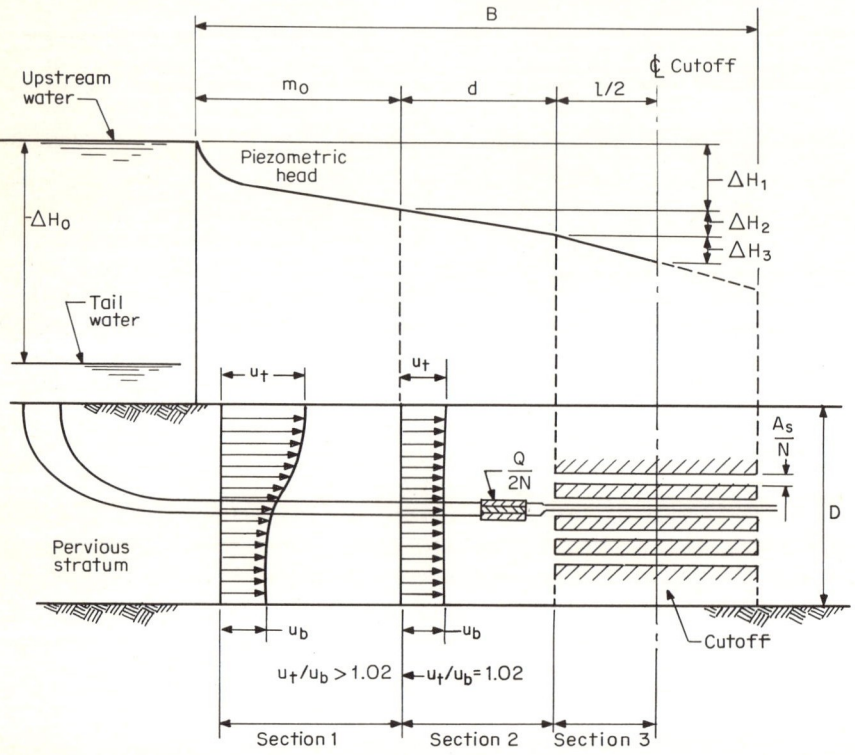

Fig. 6-8 Configuration of cutoff underneath earth dam; various zones of flow.

Knowing the values of ΔH_0, B, D, k, and l, we can derive a solution in which the efficiency is related to the extent of imperfections. Ambraseys (1963) has considered this problem by summing the known solutions of individual parts of the flow. The assumption is made that the flow is continuous and within each individual part it is parallel and uniform. In each flow part the loss associated with the discharge Q through this part can be estimated, and since the total drop of head is known, the total efficiency can be determined.

The three main parts constituting the flow are shown as sections 1, 2, and 3 in Fig. 6-8. Section 1 covers the flow part from the upstream face to a point where the velocity of flow becomes essentially constant with depth. For practical purposes this is satisfied if the top and bottom velocities, shown as u_t and u_b, have a ratio of 1.02. The distance m_0 from the upstream face which satisfies the velocity ratio of 1.02 depends on the ratio B/D. Table 6-6 gives the minimum m_0 values as a function of D for various B/D ratios.

TABLE 6-6

B/D	m_0
1.4	D/2.0
1.5	D/2.5
2.0	D/3.3
4.0	D/5.0
8.0	D/10.0

The loss of head through section 1 is ΔH_1, given by

$$\Delta H_1 = \frac{Q(m_0/D + 0.44)}{k} \tag{6-7}$$

For section 2 the flow is parallel and uniform, and away from the slits it is essentially constant with depth. If the flow is divided in $2N$ horizontal layers, each layer will carry a discharge equal to $Q/2N$ and each pair of adjacent layers must pass through one of the slits in the cutoff. The width of each slit is A_s/N, and on the basis of the assumption of constant velocity the orientation of the slits is irrelevant. The loss of head ΔH_2 in section 2 is (Nelson and Sknonr'kov, 1949)

$$\Delta H_2 = \frac{Q\{d/D - 2.3 \log [\sin (\pi A_s/2D)]/\pi N\}}{k} \tag{6-8}$$

which includes the loss in the parallel-flow zone as well as entrance loss into the slits.

Section 3 begins at the upstream edge of the slits and is terminated at the centerline of the cutoff. The head loss in this area can be computed from the parallel-flow formula

$$\Delta H_3 = \frac{Q(l/2A_s)}{k} \tag{6-9}$$

The remaining loss of head from the centerline of the cutoff to the downstream toe of the dam is likewise computed from Eqs. (6-7) to (6-9). The minimum m_0 value which satisfies constant velocity is $D/2$ for B/D greater than 1.4. This can be used for most practical purposes, and as long as the cutoff is placed outside the distance m_0, it can be assumed to be at the center of the dam.

The efficiency of the cutoff can be obtained by combining the foregoing relations and by noting that $\Delta H_1 + \Delta H_2 + \Delta H_3 + \cdots = \Delta H_0$. If N is allowed to approach infinity and the ratio A_s/D is replaced by the ratio k_1/k, where k_1 is the permeability of the backfill material in the cutoff, the efficiency is given by Ambraseys (1963) as

$$E = \frac{100 \, [l(k/k_1 - 1)/D]}{B/D + l(k/k_1 - 1)/D + 0.88} \, \% \tag{6-10}$$

which shows that the efficiency decreases with increasing B/D ratios if all the other factors are kept the same. Since for most dams this ratio is fairly great, it follows that a cutoff might have to be fairly wide to be reasonably effective.

An apparent limitation in the preceding analysis is that the efficiency is expressed in terms of the unknown factor Q_0. When the performance of the system is checked after the construction is completed, this measure of efficiency cannot be used directly since there is no way of knowing what the leakage would have been without the cutoff. Bishop (1963) suggests expressing the cutoff efficiency in terms of other direct measurements, such as the uplift pressure in the upstream and downstream portions. On this basis the loss of head is estimated by considering the following five zones, shown in Fig. 6-9:

1. Entry zone m_0, corresponding to head loss ΔH_e as given by Eq. (6-7)
2. Upstream zone of essentially parallel flow with head loss ΔH_1 and length $l_1 - m_0$.
3. Cutoff zone of length l_2, parallel flow with head loss ΔH_2
4. Downstream zone of essentially parallel flow with head loss ΔH_3 and length $l_3 - m_0$.
5. Exit zone m_0 and head loss again ΔH_e

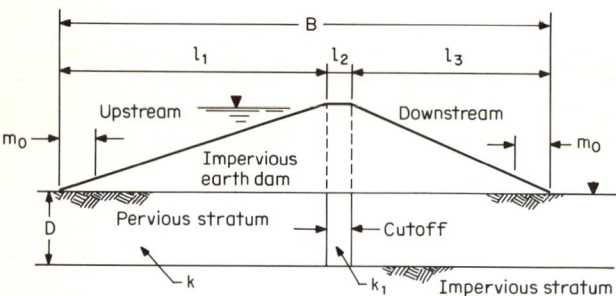

Fig. 6-9 Configuration of cutoff underneath earth dam; cross section.

With respect to the loss of head the efficiency is

$$E_h = \frac{\Delta H_2}{\Delta H_e + \Delta H_1 + \Delta H_2 + \Delta H_3 + \Delta H_e} \qquad (6\text{-}11)$$

If k and k_1 are the coefficients of permeability of the pervious layer and the soil, respectively, Eq. (6-11) becomes

$$E_h = \frac{1}{1 + (k_1/k)[(l_1 + l_3 + 0.88D)/l_2]} \;\% \qquad (6\text{-}12a)$$

or

$$E_h = \frac{1}{1 + (k_1/k)(B/l_2 - 1 + 0.88D/l_2)} \;\% \qquad (6\text{-}12b)$$

According to Eq. (6-12b) for fairly large dams where the ratio B/D is likely to be great, the denominator of the second term is governed by the ratio B/l_2.

Bishop (1963) also made an interesting observation regarding the relationship of cutoff width to effectiveness. For example, consider a cutoff such that $k_1/k = 0.02$ (which actually corresponds to a very small reduction in permeability considering the design values of k_1) and $B/D = 10$. Assume further that the cutoff is relatively thin and $l_2/D = 0.2$. According to Eq. (6-12b), the efficiency E_h is 48 percent. Now if the l_2/D ratio is decreased to 0.1 (which means a 10-ft-wide cutoff in a pervious stratum 100 ft deep), the efficiency also is decreased to 32 percent. However, consider the same situation but with the k_1/k ratio taken as 1×10^{-4}, which is much closer to the actual applications. It now is seen that the difference in efficiency between the two different thicknesses is only a small fraction of 1 percent and in fact the thinner cutoff has an efficiency of 99 percent. It would hardly be economically justified to double this thickness in order to increase the efficiency by less than 1 percent.

Comparison of Efficiency with Sheet-Pile Walls

Bishop (1963) has compared the relative efficiency of slurry-trench cutoffs with the efficiency of sheet-pile walls based on data quoted by Lane and Wohlt (1961). These data show that on three dams, steel sheet piling installed in alluvium of permeability of the order of 0.1 cm/s gave an efficiency of 8 to 18 percent in the early stages, rising later to 20 to 38 percent due to rusting and the possible movement of fines. The longest record was 17 years, but in longer service rusting might lead to deterioration. If we assume a k_1/k ratio of 0.02 for a clay backfill cutoff, which is only a very modest reduction in permeability, and an average cutoff width, Eq. (6-12b) yields an efficiency of 30 to 40 percent, making a cutoff of very modest performance equivalent to steel sheet piling driven to the substratum. On the other hand, the usual k_1/k ratios are of the order of 1×10^{-4} or smaller for earth cutoffs, making these structures 2 to 3 times more efficient than steel sheet-pile walls. It is difficult, however, to predict the long-term behavior in terms of the discharge alone, since this might also depend on such factors as the sedimentation in the upstream portion, probable changes in the permeability of the pervious stratum, etc.

The foregoing procedure for estimating cutoff efficiency is simplified by assuming that the average permeabilities of the backfill and the natural soil materials show minor variations and the substratum below the cutoff is fairly impervious. A lower vertical permeability in the alluvium will increase the importance of the entry and exit terms. If the pervious stratum is underlain by rock, the permeability of the latter may range from 10^{-3} to 10^{-5} cm/s. This will change the total discharge but is unlikely to have appreciable effect on the distribution of head across the base of the dam.

REVIEW PROBLEMS

6-1 A cutoff is built to waterproof an excavation as shown in Fig. P6-1. How would the presence of the cutoff influence the stability for the assumed failure if it consists of

(a) earth backfill, (b) cement-bentonite mix, (c) solid concrete diaphragm wall, and (d) sheet-pile wall?

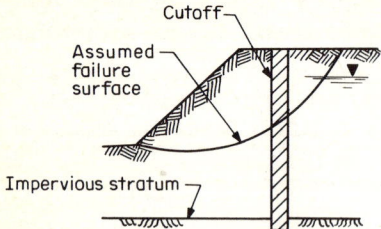

Fig. P6-1

6-2 The changes in consistency of an earth backfill with time are shown in Table P6-2. Calculate the efficiency of variation for the water content, bulk density, and undrained shear strength. What is the homogeneity of the mix with time?

TABLE P6-2 Consistency of Earth Backfill with Time

Age, months	Water content, %	Sp gr	Undrained shear strength, lb/ft²
$\frac{1}{4}$	25.4	1.92	600
1	24.8	1.85	750
3	24.1	1.95	700
6	23.5	2.00	805
12	22.9	2.05	1050

6-3 With the 400-g Swedish fall-cone apparatus the penetration of a sample is 10 mm. Calculate the probably undrained shear strength of the sample.

6-4 For an earth cutoff both natural and prepared clays are available. Discuss the factors that will influence the choice.

6-5 Explain the importance of adequate mixing and storage facilities when excavating earth-backfill trenches. How would you determine the daily capacity of the slurry plant?

6-6 The maximum head difference for a cutoff is 40 ft. If the cutoff consists of earth backfill conforming to the gradation limits of Table 6-2 and the in situ strength of the slurry is 100 lb per 100 ft², estimate the width of the trench for a factor of safety of 4. State your assumptions.

6-7 Establish general limit controls for a slurry to be used in conjunction with earth backfill. Prepare a tentative specification.

6-8 The coefficient of permeability for the earth cutoff of Prob. 6-6 is 1.8×10^{-6} cm/s. Calculate the discharge velocity and the total discharge volume if the cutoff is 40 ft deep and 300 ft long.

6-9 A slurry-trench cutoff is assumed to fit the conditions of the Dead Sea slurries. What changes in permeability and consistency would be expected with time? Describe the factors affecting these changes.

6-10 On what main factors does the efficiency of earth cutoffs depend? Would you arrive at the same conclusions from a consideration of Darcy's law?

6-11 Discuss the limitations of blowout tests. Explain the correlation of blowout tests to permeability tests.

6-12 Present a method for aggregate gradation blending of earth backfill.

REFERENCES

Ambraseys, N. N., 1963: Cutoff Efficiency of Grout Curtains and Slurry Trenches, in "Grouts and Drilling Muds in Engineering Practice," Butterworths, London.
Anton, W. F., and D. J. Dayton, 1972: Camanche Dike 2 Slurry Trench Cutoff, *Proc. Performance Earth Earth-Supported Struct., ASCE, Purdue Univ.*, vol. 1, pp. 735-749.
Bishop, A. W., 1963: Discussion of Cutoff Efficiency, in "Grouts and Drilling Muds in Engineering Practice," Butterworths, London.
Jefferis, S. A., 1972: The Composition and Uses of Slurries in Civil Engineering Practice, Ph.D. Thesis, University of London.
Katowicz, M. S., 1967: The Design and Construction of the Bentonite Trench Cutoff in Khancoban Dam, *Proc. 5th Aust.-N.Z. Conf. Soil Mech. Found. Eng.*, pp. 153-159.
Lane K. S., and P. E. Wohlt, 1961: Performance of Sheet Piling and Blankets in Missouri River Reservoirs, *Proc. 7th Congr. Large Dams, Rome.*
LaRusso, R., 1963: The Wanapum Development, in "Grouts and Drilling Muds in Engineering Practice," Butterworths, London.
Leonard, M. W., and J. A. Dempsey, 1963: Clays for Clay Grouting, in "Grouts and Drilling Muds in Engineering Practice," Butterworths, London.
Little, A. L., 1974: In Situ Diaphragm Walls for Embankment Dams, *Proc. Diaphragm Walls Anchorages, Inst. Civ. Eng., Lond.*
Nelson, F., and P. Sknonr'kov, 1949: "Seepage through Saturated Media," Sovetska Nakne, Moskow.
Xanthakos, P. P., and B. Bailey, 1975: Report of Geotechnical Information, Slurry Trench Cutoff, Southport AWT Facilities, ATEC Associates, Indianapolis.

Chapter Seven
SEMIRIGID AND RIGID CUTOFF WALLS

7-1 CLAY-CEMENT MIXES

Fundamentals of Clay-Cement Mixes

Clay and cement can be mixed in various proportions for use as base material in relatively flexible backfills or in special continuous curtains. The context in which these mixes are studied here relates to this type of construction and is not intended to apply to conventional grouting work.

The attained strength, low permeability, durability, and deformability are to considerable extent controlled by the relative proportions of the essential compounds. However, the rate of hardening (development of strength) is likely to be slower, and there may be no well-defined setting time. The thixotropic characteristics of clay serve to keep the cement in suspension and thus prevent the mix from bleeding. This is possible since the relatively small size of cement particles is easily supported by the gel strength of the colloid material. During the hardening process the clay performs no other function, and the final set is influenced mostly by the ambient conditions.

The final strength of clay-cement mixes is considerably lower than in the all-cement structure, yet it is sufficiently high to satisfy the usual strength requirements; actual strengths range from as low as 10 lb/in² to as high as 1000 lb/in² (70 kg/cm²), according to the cement content (Greenwood and Raffle, 1963). This range allows the material to resist normal working pressure gradients acting on water-barrier structures as well as stresses caused by consolidation pressures. The upper range of set strength is also required in fissured rocks and open ground, where both hardening and deformability are important in the same order.

For higher final strength the clay content must be kept to a minimum, usually a few percent of cement weight, requiring the use of very active clays such as bentonite. Mixes of this type are classified predominantly as cement-bentonite mixes. In this case the colloid is regarded as an agent giving a slight rigidity to the base liquid of the cement mix, whereas the cement still is the reaction product serving as the binder. With this combination a relatively fluid cement suspension can be obtained and freely placed and yet set to an adequate strength with negligible bleeding.

Mixes containing relatively higher proportions of clay are referred to as clay-cement mixes. They are used for sealing coarse soils having permeability in the province of 0.1 cm/s or higher; injected-grout screens are examples of this application. In this case the strength requirements are less pronounced, and the clay acts as a filler to increase the volume yield per unit cost of material while producing a stable mix. Since these mixes do not require highly active clays, they can be prepared from locally available clays modified when necessary with such additives as bentonite and sodium silicate.

Regardless of the source of the colloid material, most clay-cement suspensions are anomalous systems. The gel which develops upon mixing is useful in resisting displacement but also means a corresponding higher yield stress; hence when the mix is injected, it requires higher injection pressure. The demand for extra pressure is greater in fine soil, and if the gel is too high and the pressure limited, the mix may not penetrate the pores to the extent desired. This behavior forms the basis for designing clay-cement mixes in conjunction with injected screens.

The flow properties of clay-cement suspensions deviate considerably from the colloidal behavior discussed in Chap. 4. These suspensions are more viscous than clay-water or cement-water systems and also have an increased initial gel strength. If the addition of clay to cement can increase the penetrability of the mix, this must be attributed to the suspending action of clay rather than to an increase in fluidity. As a consequence, this action prevents cement from settling near the injection point and thus makes it possible to transfer cement particles farther than is feasible without the presence of clay.

Table 7-1 shows the characteristics of a clay-cement mix. This mix has a 7-day compressive strength of about 5 lb/in². Although in terms of strength this is a weak material, its shear strength is sufficient to resist external gradients as high as 8 in soil with permeability 1 cm/s, which is many times higher than gravitational gradients. A clay-cement mix of this class is suitable for treating relatively coarse alluvial soils and sometimes rock fissures

TABLE 7-1 Properties of a Clay-Cement Mix in Water Suspension

Type	Cement content, % of total weight	Clay content, % of weight of cement	Total solid content, % of total weight of suspension	Sp gr	Plastic viscosity, cP	Yield strength, lb/100 ft²	Initial shear, lb/100 ft²
Clay-cement	10	50	15	1.10	10	85	15
Cement-bentonite	75	5	80	1.50	90	270	125

(Greenwood and Raffle, 1963). Fluidity is in this case relatively unimportant since the material is used only in soils sufficiently open to accept the coarser cement particles without filtration. However, if the permeability of the soil to be treated is less than 0.1 cm/s, clay-cement mixes cannot be used to penetrate the pores because the cement particles are filtered out during injection. On the other hand, the small size of clay particles allows them to penetrate voids in soils with permeability 0.01 cm/s and even lower.

Cement-Bentonite Mixes

Table 7-1 also shows the characteristics of a cement-bentonite mix with a 7-day compressive strength 650 lb/in² (about 46 kg/cm²). This mix can be used to fill cavities, rock fissures, and very open granular soil for both permeability reduction and consolidation. It is conceivable that such a material might be injected into soil voids with permeability of the order of 10 cm/s under a pressure gradient of about 50. In similar ground the same material can withstand external gradients of about 25 immediately after injection.

Basically, bentonite is added in small amounts to water-cement suspensions to extend the range over which the system is free from segregation by settlement. The dry cement and bentonite powder can be mixed together and then added to water, or a pregel of bentonite can be made and added to a cement slurry. Conversion of bentonite into the calcium-exchanged form generally occurs through reaction with free lime released from the cement, and the calcium bentonite is then flocculated by the excess of calcium cations. The flocs so formed constitute a state of gelation which prevents sedimentation of the relatively coarser cement particles. Reference to previous sections will show that the gravitational forces acting on these particles are, in fact, offset by a very small shear strength. Depending on the fineness of the cement material, it is estimated that from 1.5 to 2 percent bentonite usually is sufficient to reverse the bleeding characteristics of water-cement suspensions at water-cement ratios of about 0.6.

Typically the set strength of water-cement-bentonite suspensions is lower than if the bentonite fraction is omitted, since in the presence of bentonite the cement particles are held apart and thus prevented from consolidating. On the other hand, the 7- or 28-day strength of a water-cement mix containing a small fraction of bentonite (1 or 2 percent) is primarily a function of the water-cement ratio, as in ordinary water-cement suspensions. This dependence continues to hold in such weak suspensions as those obtained with water-cement ratios as high as 20. The presence of bentonite causes the formation of a rather weak cement gel, yet much stronger than any gel produced by bentonite alone.

Limitations on the quantity of bentonite in water-cement suspensions are imposed by practical considerations:

1. Workability of mix. Increasing the bentonite concentration means increased stiffness, eventually making the mix unworkable.
2. Set strength. Overaddition of bentonite sharply reduces the compressive strength of set cement.
3. Penetrability. Material with low specific gravity may show a reduced tendency to migrate through the soil and thus limit penetration; increasing the bentonite concentration can result in such a low gravity.
4. Stability toward sedimentation. At lower cement content more bentonite is needed to produce a system resistant to bleeding.

A useful reference for the properties of cement-bentonite-water mixes is provided by Fig. 7-1, but it should be used only for a preliminary evaluation of the mix.

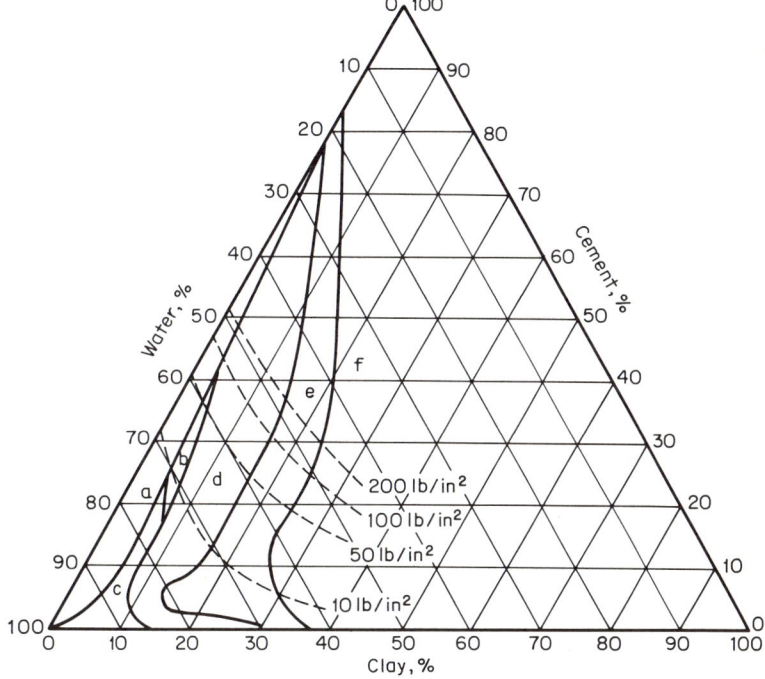

Fig. 7-1 Bentonite (Fulbent 570)-cement compositions; a = unstable suspension: settles; b = temporarily stable suspensions: settle before setting; c = clay-cement gels of low compressive strength; d = free-flowing, stable, and pumpable suspensions; e = stable puttylike suspensions; f = solid unworkable mixes, normally powders; compressive strength on 2- by 2-in cylinders. (*After Jones, 1963.*)

Deflocculated Clays and Gels

In fine-grained soils (permeability less than 0.1 cm/s) mixes containing cement cannot be used as penetrating materials since the voids are too tiny for the cement to pass through and the particles are thus filtered out. Clay mixes are suitable provided they have a low yield strength so that they can be injected. Conversely, resistance to displacement in small voids also requires relatively low shear strength. The ability of very active clay slurries to form thixotropic gels at low concentrations makes them ideal for this purpose. However, if a clay suspension stiffens excessively on standing, it can be a real nuisance if used as penetrating material since there may be no control over the rate of gelation. An extended halt in pumping can cause the complete gelation of the system in the pipe circuit and thus require unprecedented pressures to restart the flow.

Whereas bentonite clays are noted for the fineness of their particles, they often fail to produce sufficient strength in a plain and rather dilute suspension in the final position. The additives mentioned in Chap. 4 can be used to provide a treatment which, while causing the solidification of the clay, prevents its flocculation. A flocculated state means loss of the advantages associated with the fineness of particles. Deflocculated clays treated in this fashion give good results in reducing the permeability of fine-grained soils, and they are almost as penetrating as the standard gels but less expensive.

Gels   These consist of sodium silicate and a reagent. The consistency of the gel (soft or hard) depends to some extent on the reagent type. Soft gels have long been known, whereas hard gels have appeared more recently and have proved to be very successful for strengthening alluvia. Their setting time can be adjusted by simply dosing the adapted reagent. They require a single injection, and this is a great improvement over the formerly used hard gels (known as the *Joosten process*), which required two successive injections.

Other Soil-Cement Mixes

In certain areas the scarcity of natural impervious materials can yield costs unreasonably higher than justified by the purpose of the project. For such cases the use of granular-soil–cement mixes is suggested, and successful applications have been reported in earth embankments and dams (Holtz and Walker, 1962). The barrier is provided in the form of a key-cutoff trench, slope paving, or core wall. Other uses reported by Zaffle (1970) are for impervious lining at the bottom of reservoirs.

A wide range of soils can be blended with cement to make a mix fairly impermeable for all practical purposes. Shrinkage cracks at the final position presumably increase seepage moderately compared with the uncracked condition.

Results of laboratory permeability tests on predominantly granular soil-cement materials reported by Zaffle (1970) show that substantial reduction in permeability is achieved by the addition of 7 to 10 percent of cement by volume. The conclusion is that the amount of cement required to stabilize the mix in order to be sufficiently resistant to freeze-thaw and wet-dry action may normally reduce the permeability to a value as low as 1 ft/year (1×10^{-6} cm/s). This often is the permeability specified for impervious cutoffs. The soils involved in these tests included fine and coarse sand, sandy loam, and loamy sand.

7-2 CEMENT-BENTONITE CUTOFFS (SOLIDIFIED WALLS)

Cement-bentonite cutoffs, also called solidified bentonite slurry walls, basically consist of cement-bentonite mixes containing no aggregate other than some soil mixed with the slurry during excavation. Because they are sufficiently flexible to accommodate differential settlement and horizontal ground movement, they offer a practical solution in situations which fit these conditions. Furthermore, the choice is economically attractive because the construction involves limited use of materials and is completed with a minimum sequence of operations.

The method consists of reusing the bentonite slurry in the excavated trench and transforming it into a construction material by adding cement in certain ratios. When a cement-bentonite-water slurry is placed in the ground and allowed to set, it attains sufficient strength and yet remains essentially elastic to deform without cracking. The quoted strength range is between 10 and 30 kg/cm², or approximately 150 and 400 lb/in² (ICOS, 1973), and the modulus of elasticity is from 200 to 500 kg/cm², or 2800 to 7000 lb/in². Evidently no set relationship exists between strength and modulus of elasticity, and when such variables are introduced as the bentonite content and the incidental soil fraction, it is reasonable to expect broad variations in the elastic behavior. The quoted permeability is of the order of 10^{-7} cm/s or even lower. For all practical purposes the material remains impermeable and free of cracks as long as it is buried in the ground.

Construction and Panel Sequence The excavation is carried out with grab buckets, clamshells, or conventional trenchers, depending on the required depth and width. It is always desirable to use the minimum width that can be excavated with conventional equipment, and very often this width is made 0.5 m (about 20 in).

Certain disadvantages associated with the disposal of used slurries and the separation of solids are eliminated in this case. Although the construction usually is processed in alternating panels, the absence of joints between them allows full continuity, and no stop-end tubes or guide trenches are required. The installation proceeds as shown schematically in Fig. 7-2. Element *A*, having the spread of the excavator as its length, is opened in one pass and simultaneously filled with cement-bentonite slurry. The equipment is then moved to position *B* and after this panel is installed, the bucket is returned to

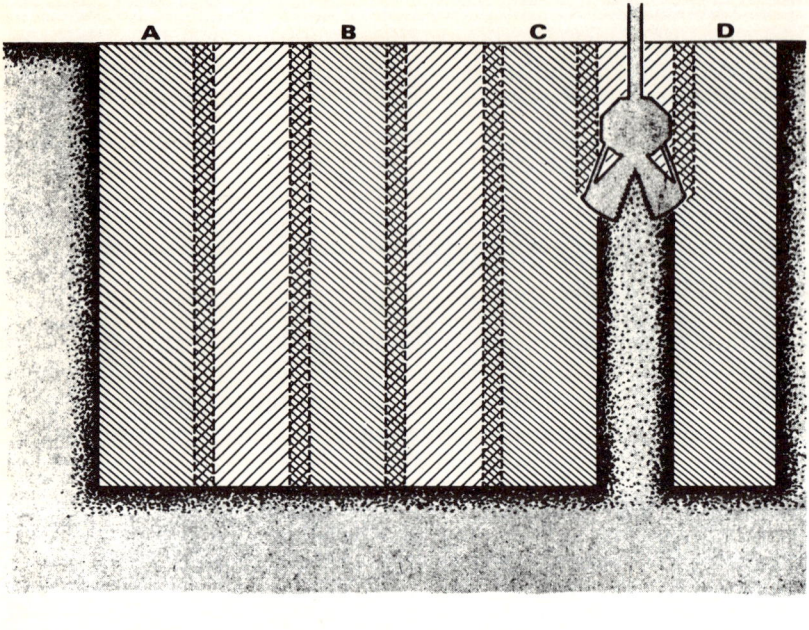

Fig. 7-2 Construction sequence of a continuous flexible wall.

remove and complete the tongue between A and B. As long as the mix in A has not hardened excessively and the mix in B has become stiff enough to be self-supported, the two elements blend with the intermediate tongue and form a jointless curtain. The sequence is repeated with elements C, D, etc.

If the depth is relatively shallow, it is possible to construct the cutoff in one continuous operation (Xanthakos, 1976a). The advantage in this case is more flexibility in scheduling the construction.

Proportioning the Mix The materials are mixed in the following proportions: bentonite, 2 to 4 percent; cement, 15 to 20 percent; and aggregate entering the mix as a by-product of the excavation, 5 to 10 percent. A retarder of the lignosulphite group is also added to the system in small ratios (0.1 percent) to control the curing process, mainly to delay the initial set. Premature hardening of the mix in the panel must be avoided since the setting time must be coordinated with the excavation time.

Although the cutoff is essentially impermeable, tests with various additives show that a mix containing fly ash maintains better watertightness and resists disintegration more effectively. This improvement is even more significant if the soil fraction from the excavation can be limited to less than 15 percent.

It is apparent from the foregoing composition that the weight of the mix is considerably lower than that of earth or concrete cutoff structures. For a water content of 65 to 70 percent the specific gravity of the mix is between 1.25 and 1.30, corresponding to a density of 80 lb/ft^3 or slightly higher.

Geotechnical Aspects The relatively lower density of the material in the trench makes it necessary to check the stability of the face under the long-term condition but taking into account the presence of considerable shear strength in the set slurry, as outlined in Chap. 2.

The colloidal stability of the slurry is controlled mainly by the presence of cement. The flow properties are rather uncertain owing to the wide and erratic effects of cement on the bentonite; however, the slurry enters the trench in a flocculated state. Thus, filtration of bentonite particles along the interface is influenced by their association and usually produces noncompact, porous, and permeable cake. Diluted slurry continues to flow toward the ground until the flow is restrained and finally stopped by the shear strength of the mix in the pores. Accordingly, the analysis of stability must be based on a reduced thrust due to a relatively permeable filter cake (see also Chap. 3).

Evidently the slurry loss toward the ground is higher than normal, and in some instances it is reported as 100 percent of the trench volume. This means that the actual quantity of slurry that must be available is twice the theoretical. Excessive slurry loss can also change the initial composition of the mix appreciably and cause variations in the expected properties.

Expected Performance The use of the initial slurry as final construction material implies attractive cost savings and thus favors applications. The low permeability and the high tenacity with which the mix is held in place allow considerable disturbance to the soil without affecting the performance of the cutoff. If necessary, the bentonite can be replaced by a bituminous emulsion to provide higher deformability for the same strength provided the higher cost is justified. When in place the material may be of the expanding type, performing satisfactorily in the confinement of the trench and in the moist environment below ground level. In the original solidified state the mix is strong and free from textural irregularities.

From the present record cutoffs of this type should be expected to respond elastically to excessive stress levels without developing the cracks typical of conventional concrete. The mix requires only a high moisture content to retain its strength and elastic properties, and certain puzzling aspects in this behavior will be better understood with further research in the cement-bentonite technology.

Microstructure Analysis Figure 7-3 is a photograph of a cement-bentonite sample showing the surface detail of the vacuum-dry microstructure at a magnification of 8000. This should not be interpreted to mean material of low density, although it is true in this case.

The investigation of three-dimensional assemblies of mineral particles

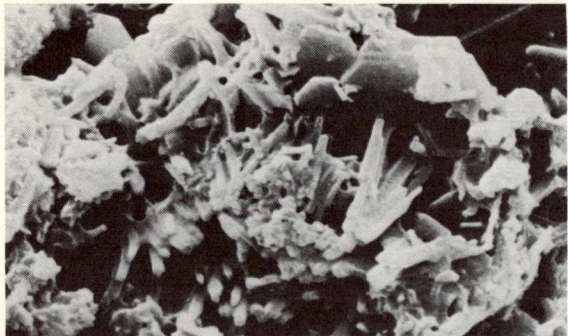

Fig. 7-3 Photograph showing microstructure analysis of a cement-bentonite mix at a magnification of 8000. (*ICOS.*)

within the 0.01- to 10-μm range in bentonite and cement-bentonite slurries has shown that particles of this size dominate the properties of the material to an extent of practical significance. The photograph of Fig. 7-3 supports the view that the cement-bentonite slurry provides an ideal environment for the growth of stable clusters of fibrous calcium silicate hydrate. These spiky crystals apparently prevent the stacks of flat, platelike clay particles with wet surfaces from sliding apart.

It also appears from the stereoscan picture that transformation of cement particles into clusters of fully developed spikes takes place under favorable conditions, as the structural skeleton in the thixotropic gel is arranged with minimum restraint. Once the skeleton of hydration products is completed, it can be assumed that the gel is effectively trapped (ICOS, 1973). The strength of the composite system is thus much greater than the shear strength of bentonite gel and the strength of the spikes alone.

The supporting action of bentonite ceases, however, if the water is allowed to evaporate. The appearance of large surface cracks confirms the reversal from swelling to shrinking under the influence of progressive dehydration.

Although present knowledge indicates that cement-bentonite mixes in cutoff walls have rather erratic variations in their properties, the following comments are appropriate for work of this type: (1) higher water-cement ratios produce more plastic but weaker mixes; (2) the presence of more soil and bentonite generally reduces strength; (3) the plasticity is decreased while strength is improved with time; and (4) neither strength nor plasticity appears to correlate with fly-ash content, and they do not seem to provide good correlation with bentonite content.

This application was originally conceived by ICOS of France as a practical solution to the slurry-disposal problem. At present it is standard practice for groundwater control, and among recent applications are solidified walls for embankment dams (Little, 1974).

7-3 INJECTED SCREENS (IMPERVIOUS WALLS) AND VIBRATED MEMBRANES

A thin impervious screen can be inserted from the ground surface by making a continuous slot in the ground and then filling the space thus created with clay-cement grout. Since slurry is neither required nor used for this purpose, this application is hardly a derivative of the slurry-trench method.

In principle the operation consists of driving a group of steel H piles with the flanges back to back through the ground which is to be sealed off until their tips reach the underlying impermeable layers, as shown in Fig. 7-4. The piles are subsequently extracted one at a time, and the void so formed is filled with clay-cement grout injected under pressure. In this fashion a screen is produced, consisting of a continuous core of impervious material occupying the space left by the extracted piles and overlapped by a cemented zone of soil penetrated by the injected grout. The thin fillet of earth usually trapped between the flanges of adjacent piles as they are driven is disrupted and removed not only as the piles are extracted but also under the pressure of the material as it is injected.

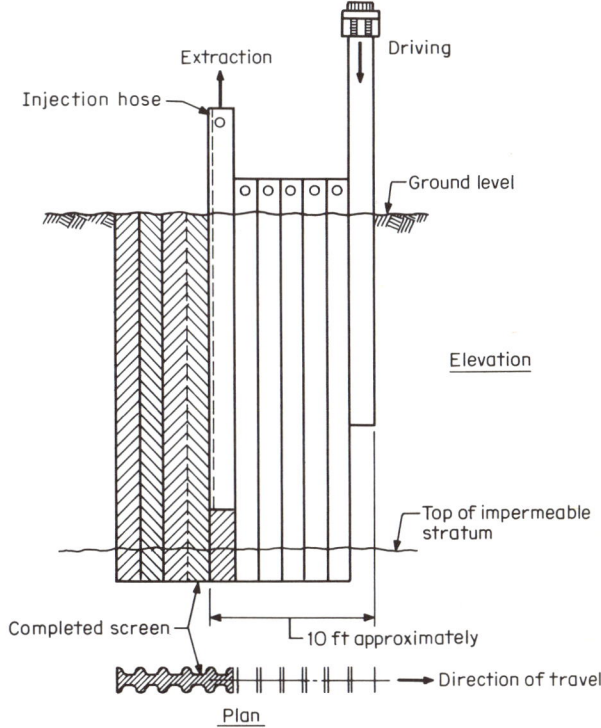

Fig. 7-4 Method of installation of an injected-grout screen.

Construction Procedure

The driving and extraction unit includes a piling rig equipped with hydraulic jack extractors mounted on a self-propelled platform. A diesel hammer commonly is used to drive the H piles, whereas the extractors can apply a pull of about 150 tons. A substantial improvement in the method of installation has been possible with the use of a powerful vibratory hammer, which is permanently attached to the mandrel for easier extraction as well as driving (White, 1975). The process is greatly facilitated and speeded up, and because of the induced vibration the grout can achieve better penetration into the ground. A further improvement is that the mandrel is driven to overlap the previous insertion and extraction, and this assures excellent continuity of the curtain.

Normally seven piles are used with each unit, and the installation proceeds by leapfrogging the piles. The unit moves along a length of rail track to provide the required accuracy in the alignment of the screen. When the end of each travel is reached, the unit is raised; it lifts the track, moves it sideways into position for the next run, and returns to the track to continue the installation.

The piles are heavy-steel sections of adequate flange and web thickness, provided with a grout pipe attached to the web. The selection of appropriate steel sections is based on the requirements of the grout curtain in service as well as on the driving conditions to be contended with. In relatively soft soils it is possible to drive light sections generally 9 in deep. In dense sands and gravels the steel sections are at least 15 in deep, usually reinforced with cover plates for the full length, and have the tops protected to withstand impact from repeated use. Practical experience shows that the installation is governed by depth limitations, and normally the maximum depth range that can be attained is between 33 and 35 ft. Under appropriate modifications of the equipment this depth can be increased to 50 ft or even higher.

Installation Upon starting the work, the entire group of piles (usually seven) is installed by driving. The work continues with the simultaneous driving and pulling of piles so that for every pile extracted one pile is driven. As the pile is driven, the lower extremity of the grout tube normally is kept closed with a rivet to prevent soil from entering the tube, but in recent modifications reported by White (1975) the rivet plug is omitted.

Upon beginning the extraction process the pile is first raised by about 6 in to create a void in the soil. The rivet is driven out, the grout hose is connected, and grout is injected under pressure until a surge of material emerges at the top of the adjacent previously filled hole. At this stage the lifting jacks and the injection pump are coordinated to work together, and by successive lifts the pile is slowly withdrawn from the ground. It is essential to coordinate the extraction of the pile and the quantity of grout pumped at a given time, which is achieved manually and is controlled by visual observation. In order to ensure that grout already placed and being placed is not disturbed by driving at the far end, the driving process must be carried out at a safe distance;

according to experience, this is about 10 ft (or 3 m), which corresponds to an assembly of seven sections for the usual applications.

As already mentioned, the rivet plug in the grout pipe can be omitted so that grouting can be done during the downward penetration as well as during pulling. The injection of material as the pile is driven speeds up the operation by exerting a lubricating effect.

The method should provide satisfactory verticality in normal alluvium, with some exceptions in ground containing cobbles. The continuity of the curtain is better assured by using sections which are stiff in both directions in order to accommodate straight driving. Deviations caused by obstructions are unavoidable and become obvious while driving; if they are excessive, they will require supplementary injections. The maximum deviation from the exact alignment that can be tolerated is apparently the width of the flange. When a pile is extracted, if it is found to be bent, the area involved should be reinjected with new grout.

Although the profile of the impermeable formation that must be penetrated by the impervious curtain is established fairly accurately from borings and subsurface explorations, examination of the toe of each pile extracted provides a visual check. The pile generally brings a trace of the lowest soil penetrated, which is trapped between the grout pipe and the pile tip.

Flow Properties of Grout

The flow characteristics of colloidal suspensions discussed in previous sections may to some extent be applied to injections in capillaries and pervious soil over a wide range of pressure gradients. Although the flow curves of cement-bentonite systems resemble those of Bingham bodies, considerable deviations must be expected in the curved portion. The initial shear strength and viscosity increase erratically but rapidly with increasing concentration of cement and clay. The penetrability of pervious soil by the mix is inversely proportional to its apparent viscosity. Thus the choice of mix for a vibrated screen involves consideration of the following factors:

1. The feasibility of pumping grout into the void created by the withdrawal of the pile and through a suitable pipe and the convenience of restarting the flow if pumping is temporarily interrupted.
2. The feasibility of injecting grout that can penetrate radially to a certain distance beyond the artificial void and into the natural voids of the soil.
3. The resistance of the screen to displacement under the expected gradient.
4. The reduction in permeability during service.

These requirements are satisfied if the clay-cement-water mix develops good penetrating characteristics; sets to a relatively plastic curtain for flexibility but is resistant to softening by immersion in water; attains a certain minimum set strength; and remains fairly watertight.

Clay-cement mixes of the type discussed in Sec. 7-1 can be proportioned to form grouts with the foregoing properties. Natural clays are suitable provided they are free of coarse particles and have an Atterberg liquid limit more than 60. The mix usually consists of 2 parts of clay to 1 part of cement, whereas the amount of water is balanced to keep the mix flowable. The density of the material is slightly in excess of 100 lb/ft³ (1.60 g/cm³), and the 28-day strength generally ranges from 100 to 250 lb/in² (7 to 17.5 kg/cm²).

Flow Behavior For a rather thin suspension of clay and cement, the most relevant property to flow is the initial gel strength τ_s. If the curved portion of the flow curve is disregarded, the initial gel strength can be taken as the yield stress τ_0; that is, the flow is regarded as ideal plastic flow. For this situation the flow in a capillary of radius R is expressed by Eq. (3-1), which in terms of the pressure gradient i_p becomes

$$\tau_s = \frac{R}{2} i_p \tag{7-1}$$

Actually, the clay pastes and cement mixes used for this soil treatment hardly obey Eq. (7-1) and typically fail to follow the idealized flow pattern expressed by Eq. (4-7). The flow is in this case better described by Fig. 7-5 (Marsland and Loudon, 1963). When the shear stress τ reaches the value τ_s, shear failure occurs near the wall of the capillary and the suspension moves forward as a plug (stage II). As the pressure gradient continues to increase, the diameter of the solid plug becomes progressively smaller (stage III) until the entire material in the capillary flows in a streamline manner like a viscous fluid. The rate of flow increases thereafter linearly with the pressure gradient (stage IV).

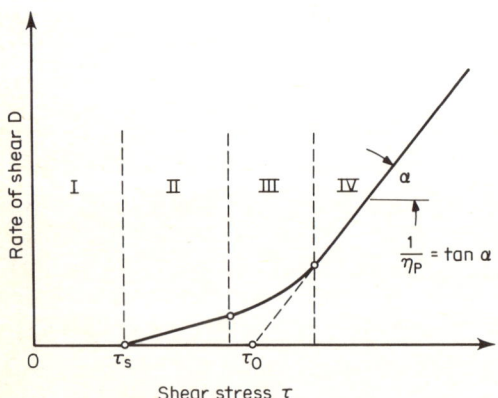

Fig. 7-5 Flow curve for clay pastes and viscous grouts.

The injection can be complicated by thixotropic thickening and changes in the gel structure occurring if the grout is allowed to rest. Tests of bentonite-cement suspensions show distinct anomalies in the pressure–rate-of-flow relationship during injection and with the measurements of the pressure drop taken at constant flow volume. Although there exists a minimum pressure which must be exceeded for flow to occur, there also exist distinct areas where an inverse relationship governs the pressure drop and volume flow rate. The same conclusion is also reached if the flow test is carried out at constant pressure and the volume flow rate is allowed to reach a steady value.

It is important to note that for impervious screens installed by this process the flow behavior is significant primarily in the low rates of shear. The injection through the grout pipe must accommodate the rate at which the pile is withdrawn without damaging the slot, and usually it must be accomplished under reduced shear rates. On the other hand, only at low concentrations do clay-cement mixes behave like Bingham bodies, so that the rate of shear can be correlated to the apparent viscosity and yield stress. However, for the mixes used in preformed slot screens the clay-cement concentration usually is high enough to preclude Bingham body behavior and rapidly increases the apparent viscosity and yield stress. This situation, together with the low rate of shear during injection, complicates the prediction of flow in the system, and thus it is better to conduct capillary viscometer tests to obtain this information.

Penetrability and Strength of Grout

A certain lateral penetration of the soil must always be expected, but the relatively high viscosity of the material may restrict the amount of fingering occurring during this penetration. Conversely, the existence of a finite initial shear strength may lead to the preferential filling of high-porosity areas in the more erratic formations. The filling of voids in the soil is further limited if instead of purified bentonites natural clays having larger particles and traces of other minerals are used.

If the grout contains even a small proportion of coarse particles, it may form a tight filter cake on the soil face near the injection source, and this will cause the penetration rate to drop. The well-known Terzaghi treatment for drainage filters applies to this problem (Taylor, 1952). When the soil surrounding the slotted voids is steeply graded, particles in the grout larger than about one-tenth of the soil particles are likely to be trapped and thus make penetration beyond the face of the slot impracticable. Since both permeability and critical filtering size depend on the average diameter of pore channels, the latter can be correlated with the size of the largest particles in the grout. Thus clay-cement suspensions containing many particles as large as $100~\mu m$ begin to form filter cake in soils with permeability as high as $1/\sqrt{10}$ cm/s (Scott, 1963). These filter cakes can clog the pores, thereby preventing penetration beyond the interface.

The flow of grout in the soil pores is thus rather complex. Studies of nonnewtonian fluids with fixed shear strength or for strength increasing with time have been made by Greenwood and Raffle (1961), Scott (1963), Marsland and Loudon (1963), and others. The results from these investigations should be used with caution since the clay-cement grouts used in these screens usually are structured to retain a shear strength typical for a weak solid. In general, however, the penetrability of the grout depends on the pressure gradient, soil permeability, grout viscosity and shear strength, and grout particle size.

The strength and resistance to displacement developed in the set grout actually favor this type of cutoff. Nonetheless, the screen can tolerate relatively small differential ground movement. The tendency of the mix to extrude under a pressure head is resisted by the shear strength of the material over the internal surface of each void passage. A shear strength of about 1 to 2 lb/in² (0.07 to 0.14 kg/cm²) can resist hydraulic gradients of 100 in soil with average grain size of 1 in (2.54 cm). Resistance to displacement is further aided by some cohesion developed in granular soil penetrated by grout. The soil in this case forms hard agglomerates offering considerable resistance to deformation by shear.

Applications

More than 15 million square feet of injected screens have been built in Europe (White, 1974), ranging in depth from 16 to 30 ft. Based on actual data, about 1.3 ft³ of void space produces 1 yd² of finished screen (9 ft²) or 0.15 ft³ per square foot of screen. From actual grout take the grout volume is 3.8 ft³ per square yard of screen, or 0.42 ft³ per square foot of screen, including the average penetration into the soil. About one-tenth of that is used to fill the gap between flanges of adjacent piles. This consumption indicates that for the average application the soil is penetrated by several inches.

Figure 7-6 shows a finished screen. Much of the soil had to be removed by pneumatic hammer in order to expose the shape of the screen as built. The loose soil fillet between flanges was disintegrated and replaced by grout. The screen has a depth of 28 ft, and was built by driving H piles and then injecting grout as the piles were withdrawn. The grout consisted of a mix of clay, silt, and cement and required a pressure of 20 lb/in² through a 1-in diameter injection pipe.

7-4 SIMILARITIES BETWEEN GROUTS AND SLURRIES

Certain important details of modern alluvial grouting can be applied to slurry-trench cutoffs. Although this technique is said to have originated independently of the development of grouting, the similarities between these areas of technology are now so evident that a certain relationship between slurries and grouts can be established beyond doubt.

Nash (1974), referring to the Joosten process, showed how it can be related to the slurry process of thixotropic clays. The Joosten method consists of sealing alluvial deposits by successively injecting alternate doses of sodium silicate and a solution of calcium chloride and by forcing sufficient liquid at each stage to saturate the soil. This produces a stonelike substance, causing the formation of masses of solidified soil. The zone involved is built up from

Fig. 7-6 Exposed impervious curtain built by injecting grout in preformed slots. Note the extent of penetration into the surrounding soil.

such masses, and since the injected material must travel only a short distance from the injection point, problems associated with the loss of grout to more permeable lenses are avoided.

The beginning of the slurry-trench process has never been confirmed historically, but drilling muds must have been used as early as 1901, or long before the Joosten process. Yet the treatment was basically the same: a simple material like clay was placed in the ground and was filtered to deposit an impermeable cakelike layer, which was solidified according to a thixotropic process. Thus a seal was obtained much as in the Joosten process. Subsequently, the technique was developed, refined, and explained. Better understanding of the concept of thixotropy and of the colloidal stability inherent in clays marked the starting point for the applications which followed. Veder, Lorenz, and others were successful in following the step-by-step procedure and thus gave the first confirmation of the thixotropic properties of clays.

7-5 PLASTIC-CONCRETE CUTOFFS

When a new project is contemplated at sites which are not ideal from the viewpoint of soil engineering, certain potential construction problems can be avoided merely by choosing a different location or by improving the initial site through soil stabilization. If this is not possible, the usual approach is to adapt the design to the actual conditions. Thus, a cutoff often must be flexible to withstand settlement and ground movement in loose and compressible soil and simultaneously attain sufficient strength to resist unusual overburden and consolidation stresses. Plastic concrete is a suitable material, and besides strength and deformability it provides adequate seepage control.

Characteristics of Plastic Concrete Plastic concrete means an ultimate strength of 1400 lb/in^2 (about 100 kg/cm^2) or less. Alternatively, it can be defined according to the proportions of the constituent materials, the desired results, or the plasticity of the set mix. Plastic concrete is considerably less stiff at the final set than conventional concrete, although it may be several times stiffer than earth backfill. This gives the cutoff two important characteristics; a degree of deformability which is not available with rigid-concrete diaphragms and a strength which earth cutoffs lack. This successful combination is achieved without reducing watertightness.

Situations Favoring Plastic Concrete Under additional overburden load, e.g., the weight of a dam, or when extra hydraulic pressure is created, relatively loose natural materials around a cutoff are squeezed and continue to deform until a new equilibrium is established. For a cutoff to withstand the corresponding distortion it must display characteristics roughly similar to those of surrounding soil; hence the soil modulus is pertinent to the problem and provides an indication of the desired flexibility. Additionally, the cutoff must not be displaced or eroded under stress or hydraulic gradient, and therefore it is expedient and advantageous to add cement to improve the stability of natural materials.

However, the choice is not always simple since it is not always possible to obtain a reliable measure of in situ conditions and the changes that may occur with time. An example is a loose formation under an earth dam. As the height of the dam is increased during construction, the formation is compacted and the stress on a soil element increases. After the dam is put into service, the saturation of the compacted layer also is likely to increase due to permeating water and the higher hydraulic gradient. Therefore, the total deformation and settlement will depend not only on the initial soil conditions but also on the subsequent changes. This example becomes more complex if a cutoff is inserted beneath the dam, thus changing the drainage and saturation from the upstream to the downstream site.

Selection of Modulus

A plastic cutoff will be most functional if the modulus of elasticity of the plastic concrete approximates the soil modulus at the selected state and the stress-strain curves for the two materials are substantially similar. In practice it is seldom if ever possible to duplicate the properties of a soil at any state or loading conditions because of variations in its elasticity, strength, and compressibility under loading, compaction, flow, saturation, and drainage. However, the investigation is aided by noting the following: (1) since in normally consolidated sedimentary soil the void ratio and water content decrease with depth, the strength and soil modulus of most alluvial deposits are likely to increase with depth; (2) compaction normally increases soil strength and reduces permeability but also increases the ratio of horizontal effective stress to vertical effective stress; (3) the soil deformation depends on many factors, but in general the total stress-strain curve of granular soil is nonlinear; and (4) there is not reliable correlation between modulus and blow count.

Soil modulus, also quoted as Young's modulus, is obtained from standard triaxial tests and is influenced by the confining stress, cycle of loading, void ratio, sample disturbance, stress history, and other factors (Lambe and Whitman, 1969). Even if a reliable measurement is available, it still has little practical value since the most relevant condition is the postconstruction period. Thus, the soil modulus can be approximated by

$$E = \sqrt{\sigma_v \frac{1 + 2K_0}{3}} \qquad (7\text{-}2)$$

where K_0 is the coefficient of lateral stress at rest. This relationship is valid if K_0 is between 0.5 and 2.0 and the factor of safety against failure exceeds 2. Values of E for granular soil are summarized in Table 7-2.

TABLE 7-2 Young's Modulus for Repeated Loadings (from Chen, 1948)

	Young's modulus, lb/in²	
Soil (1 atm confining pressure)	Loose	Dense
Screened crushed quartz, fine angular	17,000	30,000
Screened Ottawa sand, fine rounded	26,000	45,000
Ottawa standard sand, medium rounded	30,000	52,000
Screened sand, medium subangular	20,000	35,000
Screened crushed quartz, medium angular	18,000	27,000
Well-graded sand, coarse subangular	15,000	28,000

From the preceding brief discussion it is apparent that it is impracticable (and often unnecessary) to make an exact estimation of soil modulus. It is more advantageous to concentrate on the investigation of suitable plastic mixes emphasizing economy and overall performance and to match the elasticity of soil to the extent possible.

Elasticity and Strength of Plastic Concrete

Plastic-concrete mixes usually consist of gravel, sand, cement, clay, and bentonite mixed with water to produce a workable mass. In general, the stress-strain diagram follows the curve shown in Fig. 7-7. The curvature may be

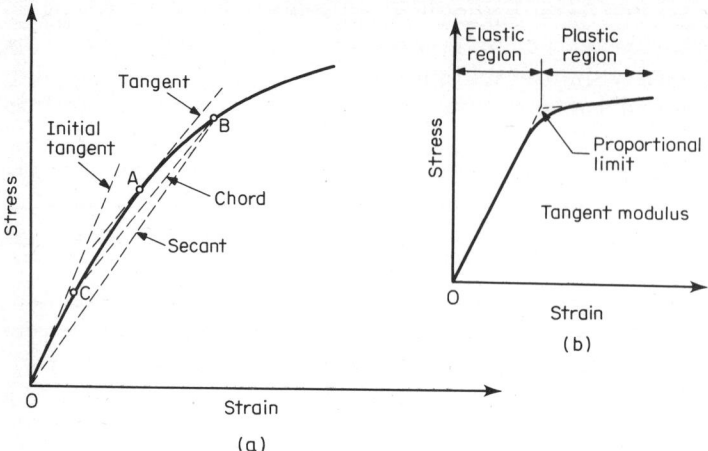

Fig. 7-7 Stress-strain diagrams and moduli of elasticity for plastic concrete.

pronounced throughout the range as depicted in Fig. 7-7a or follow the more linear pattern in Fig. 7-7b. The curve in Fig. 7-7a has the following moduli of elasticity: the *initial tangent modulus*, represented by the slope of a tangent to the curve passing through the origin; the *secant modulus*, represented by the slope of a line drawn from the origin to any point B on the curve; the *tangent modulus*, represented by the slope of a line which is tangent to the curve at any point A; and the *chord modulus*, represented by the slope of a line between any two points B and C. The secant modulus is the most practical and should be accepted for general use because it represents the actual deformation at the selected point.

The behavior represented in Fig. 7-7b is typical of elastoplastic material: the concrete is essentially elastic until the stress reaches a certain value which is the proportional limit. The strain thereafter continues to increase faster than the stress increments and increases even at constant stress. This concrete has a tangent modulus, the proportional limit, and the two regions of elastic and plastic behavior. Concrete placed in the ground can be allowed to reach the plastic region.

The flexural modulus of elasticity as determined from the deflection of a cantilever beam has a practical interest. This deformation can be approximated by the lateral translation of a cutoff built in soil becoming progressively stiffer with depth so that the deflection is zero at the bottom and maximum at the top. If the maximum deflection can be estimated or predicted, the modulus of elasticity can be determined by flexural theory.

Although in the lower stress range plastic concrete is essentially elastic, it shows a tendency to creep under permanent load; but the distinction between the two stages is not always clear. In the ground the concrete is acted upon by forces which gradually build up a steady loading condition and change little thereafter.

Semirigid and Rigid Cutoff Walls 241

Factors Affecting Strength and Modulus The same factors which influence strength also influence modulus but to a lesser degree, among these being the clay-bentonite content, the water-cement ratio, and the type and gradation of aggregate.

The effect of age is an increase in modulus, especially for richer mixes. The effect of increase in the water-cement ratio is a distinct reduction in the secant modulus for the usual water-cement range of plastic concrete. This range is from 1.5 to 3.0 but may be even higher.

The type of aggregate has an effect on modulus. Since the deformation produced in concrete is partly the elastic deformation of the aggregate fraction, a higher modulus is obtained with stiffer aggregate. Thus, limestone is more suitable than sandstone for decreased stiffness and increased plasticity. The grading of aggregate influences modulus in the same way it influences strength: if the strength is increased, so is the modulus. Owing to the high water-cement ratio, plastic concrete is never harsh or unworkable; hence its elasticity responds to the elasticity of the fines.

The effect of free moisture on mixes of the same consistency and age is a higher modulus for wet than for dry concrete. Both the modulus of elasticity and strength are improved with longer curing periods, which typically occur under slurry-trench conditions. The effect of bentonite on the concrete is lower strength and higher plasticity.

No set relationship exists between modulus of elasticity and strength, which is also true for cement-clay-bentonite mixes. In fact when all the variables are introduced, e.g., component materials and proportions, it is possible to retain the same modulus and vary strength by as much as 100 percent. This variation is manifested largely through the water-cement ratio and the bentonite content.

Examples of Plastic-Concrete Mixes

Table 7-3 provides data from plastic concrete used in cutoffs for dams (Little, 1974). This concrete has a modulus of elasticity 1.5×10^4 kg/cm^2 (210,000 lb/in^2), which is many times the modulus of any soil except rock. The data show, however, a marked difference in the permeabilities of the two cutoffs despite the similarities in the materials and the construction methods. The concrete was tremied into the panels, which were processed using round-tube end joints. A probable explanation for the variation in permeability is leakage through the joints, differences in the quality of constituent materials, and the influence of subsequent treatment of concrete.

Table 7-4 shows various data from plastic mixes in which sand is the only aggregate and constitutes the main bulk of the material. The ratio of fine fraction is varied, whereas the cement ratio is the same for all mixes. The quantity of water is enough to produce a slump of about 15 cm (6 in). The clay-bentonite ratio is 9.

Figure 7-8 shows the 7-day stress-strain curve for mix 1 of Table 7-4. The

TABLE 7-3 Materials and Properties of Plastic Concrete Cutoffs (from Little, 1974)

Material	Balderhead dam		Lluest dam	
	kg/m³	%	kg/m³	%
Weight per unit volume	2039	100	1956	100
Bentonite	44	2.2	24	1.2
Water	400	19.6	405	20.7
Cement	195	9.6	227	11.6
Aggregate	1400	68.6	1300	66.5
Water-cement ratio	2.05		1.78	
Water-solids ratio	0.24		0.26	
Permeability k, cm/s	$0.6 \times 10^{-7} - 2 \times 10^{-7}$		$10^{-3} - 10^{-4}$	

TABLE 7-4 Materials and Proportions of Plastic Mixes (from Dupeuple and Habib, 1969)

Mix no.	Sand S, %	Clay+bentonite C+B, %	Cement, % of S + C+B	Water, % of S + C+B
1	80	20	6	
2	83.3	16.7	6	10-12
3	85	15	6	
4	90	10	6	

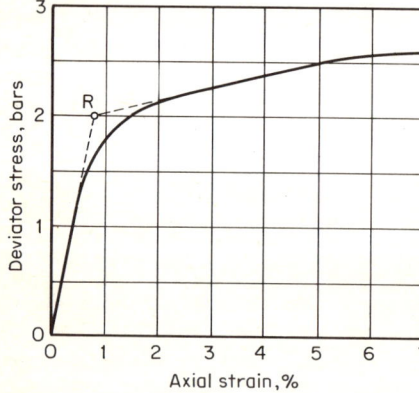

Fig. 7-8 Stress-strain data from a triaxial test. Mix 1 of Table 7-4. (*From Dupeuple and Habib, 1969.*)

set concrete appears to be almost perfectly elastoplastic and has a well-defined proportional limit, which is the intersection point of the two tangents (point R on the graph). The material can withstand 7 to 8 percent strain without failing, although in this region a permanent deformation is likely. From the graph the 7-day initial tangent modulus is 250 bars (about 3600 lb/in²).

Figure 7-9 shows the peak deviator stress at the rupture point for a confining stress of zero or 5 bars (72.5 lb/in²) at 7 and 28 days as a function of the clay-bentonite content for the mixes of Table 7-4. The relationship between modulus of elasticity and clay-bentonite fraction for the same mixes and under

Semirigid and Rigid Cutoff Walls   243

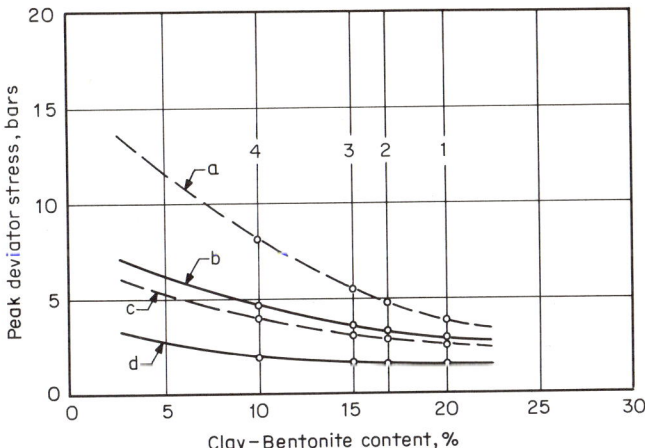

Fig. 7-9 Relationship of strength to clay-bentonite content. Mixes of Table 7-4; $a = 28$ days, confining stress of 5 bars; $b = 7$ days, confining stress of 5 bars; $c = 28$ days, 0 confining stress; $d = 7$ days, 0 confining stress. (*From Dupeuple and Habib, 1969.*)

the same conditions of confining stress and age is shown in Fig. 7-10. The curves of Figs. 7-9 and 7-10 show distinctly the effect of age, confining stress, and clay-bentonite content; all these factors apparently influence strength and modulus, although the effect on the latter is less pronounced.

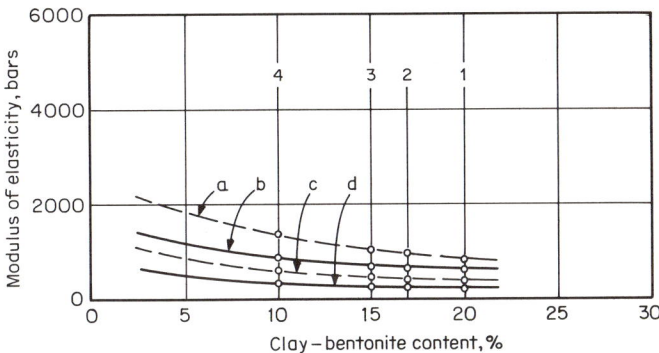

Fig. 7-10 Relationship of modulus to clay-bentonite content obtained after first cycle of loading. Mixes of Table 7-4; $a = 28$ days, confining stress of 5 bars; $b = 7$ days, confining stress of 5 bars; $c = 28$ days, 0 confining stress; $d = 7$ days, 0 confining stress. (*From Dupeuple and Habib, 1969.*)

The plastic mixes discussed in the preceding examples would be quite suitable for alluvial deposits of the type exemplified in Fig. 7-11. This soil consists of medium subangular sand, porosity 0.39. The stress-strain curve is for a confining stress of 14.3 lb/in² (about 1 bar). Although the curve is essentially nonlinear almost from the beginning of loading, up to the 0.8 percent strain it can be represented by a secant modulus equal to 7000 lb/in².

244 Slurry Walls

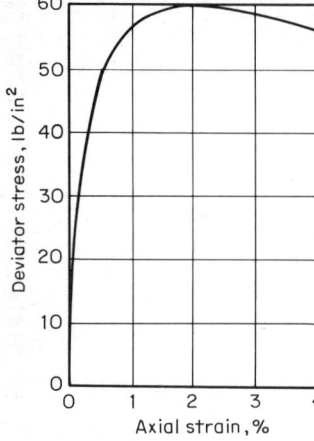

Fig. 7-11 Stress-strain data from a triaxial test for granular soil. (*From Chen, 1948.*)

In general, the strength of plastic concrete depends on the constituent materials and mainly on the water-cement ratio. Bentonite, however, influences this strength in the same way it influences the set strength of cement-bentonite grouts and suspensions. This is shown in Table 7-5, which gives miscellaneous data for three plastic-concrete mixes. The water-cement ratio is held constant and equal to 2.0, whereas the bentonite content is varied from 0 to 3.5 percent. Both the 7- and the 28-day strength are considerably reduced due to the presence of bentonite (Xanthakos, 1976b).

TABLE 7-5 Effect of Bentonite on Plastic Concrete (from Xanthakos, 1976b)

	Mix 1	Mix 2	Mix 3
Materials:			
Cement, lb	324	232	176
Sand, lb	972	1390	1790
Stone (Romeo, ASTM no. 57, lb)	1620	1620	1600
Water, lb	648	464	352
Bentonite (Volclay), lb	123	75	
%	3.5	2.0	0
Physical properties:			
Slump, in	8	8	8
Air, %	0.5	0.7	1.6
Unit weight, lb/ft^3	132	140	145.2
Water-cement ratio	2	2	2
7-day strength, lb/in^2	214	280	313
28-day strength, lb/in^2	375	407	539

7-6 RIGID CUTOFF WALLS

A rigid cutoff wall, usually plain or reinforced concrete, is used primarily for strength and durability. If high-quality concrete is available at the site and the construction details are planned and executed carefully, the structure can provide a cutoff of durable quality and satisfactory watertightness for the full period of the expected service. Alternatively, a cutoff consisting of conventional concrete may be necessary (1) where the depth of the structure is unusually great and thus excludes other types (concrete cutoffs have been built to depths exceeding 400 ft or 120 m); (2) where the site and subsoil conditions restrict the construction techniques and particularly the methods of excavation; e.g., in erratic formations with cobbles and boulders the excavation often is restricted to the so-called interlocking elements; and (3) where the wall must be connected to other structures to form a rigid system.

Basic Requirements Full-scale field tests before undertaking the main project are quite useful and often necessary in order to establish the feasibility of the construction as a whole; the effectiveness of the selected method over the entire depth range; the anticipated rate of progress and therefore the probable construction time; and the relative cost of the various depth ranges. For very large projects pullout tests are also indicated, to give an in situ evaluation of friction between the wall and the surrounding soil.

It is essential to predict the action of a rigid wall inserted in the ground, e.g., in compressible soil beneath the central impervious core of a dam. The wall can in this case punch the core, causing fissures and cracks to develop at the base of the dam as the layers underneath begin to settle, and this can become critical if sufficient friction does not develop at the interface. On the other hand, downward forces due to negative friction at the interface can induce excessive compressive loads and cause failure in the deepest sections. It is conceivable, however, that as soon as a crack is developed, the accompanying small movement along it may be sufficient to relieve the vertical load in the sheared zone and stop further movement. The resulting redistribution of stresses will propagate to another zone and cause more shear cracks to develop, and the process may continue until most of the settlement is completed. The wall is as watertight along these shear cracks as it is along the vertical construction joints.

The preceding example demonstrates the importance of the connection details at the top of the wall. These details should be developed and executed for the actual conditions at the site and be compatible with the deformation of the main core and the underlying foundation materials.

Types of Rigid-Concrete Cutoffs A rigid-concrete wall usually is processed in a series of panels. It may consist of plain concrete, or it may be partly or fully reinforced. If it is excavated in single panels to be filled with concrete to form a continuous curtain, it is basically a diaphragm wall. The construction of such a wall is discussed in other sections, together with the method of excavation, concrete placement, and structural details.

A rigid cutoff can also consist of two parallel walls made with interlocking piles or with interlocking primary and secondary elements for the deepest portions of the foundation. Two parallel walls provide better control of ground movement and allow any seepage occurring through the cutoff to be sealed off. The latter is possible by grouting the area in the intermediate confined zone, which is accomplished from a foundation gallery located on top of the structure. Since the interlocking-element process is characterized mainly by the special excavation techniques and equipment, it is discussed in Chap. 5.

Design Principles Figure 7-12 shows a rigid-concrete cutoff built in pervious ground, e.g., clean sand. The cutoff is extended to impervious formations and is assumed to intercept the flow of groundwater completely. Thus, the groundwater level on the upstream side is at the surface, whereas for the downstream face it is approximately at the bottom of the wall.

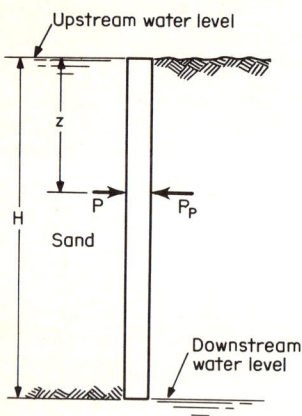

FIG. 7-12 Impervious rigid diaphragm in cohesionless clean sand.

Under these conditions the upstream face is acted upon by the active earth stress due to the weight of clean sand and by the full water pressure, which in this case is taken as static pore pressure. The downstream displacement must be resisted by the passive earth resistance of the soil.

The deformation of the wall due to the unbalanced water pressure depends on the flexural rigidity of the concrete and also on the coefficient of horizontal subgrade reaction K_h of the downstream soil. Terzaghi (1955) gives a crude conception of the factors that determine the subgrade reaction and presents an approximate solution assuming that the diaphragm is perfectly rigid. On this basis the coefficient of subgrade reaction is determined from the expression

$$K_h = l_h \frac{z}{H} \tag{7-3}$$

in which l_h is a coefficient depending only on the relative density of the sand in contact with the wall and the factors z and H correspond to Fig. 7-12. The contact pressure p_p on the downstream face at depth z is given by Terzaghi as

$$p_p = K_0' + y l_d \frac{z}{H} \tag{7-4}$$

where K_0' = earth-pressure coefficient corresponding to lateral displacement of wall of height H over distance $y_0 = 0.0002H$
y = displacement at depth z
l_d = coefficient appropriate to cutoff, usually = l_h

Terzaghi (1955) gives the probable values of K_0' and l_h for vertical walls such as anchored bulkheads and concrete diaphragms.

Equation (7-4) must be combined with the flexural theory of beams for the computation of moments and shears, and this generally leads to a differential equation of the fourth order.

7-7 HIGH-RESISTANCE NONCORROSIVE CUTOFFS

Pollution of alluvial terrains occurs not only in situations where there is continuous groundwater surface but also with a series of local perched water tables developed within existing gravel lenses. Although the usual pollution scheme within a site involves groundwater in direct connection with an aquifer or the application of substantial hydraulic heads, movement of pollutants within an alluvial deposit may occur as a result of capillary and diffusive transport through what is essentially a vadose zone. Polluting matter can be disseminated even in fine soil such as silty-clay matrix.

The effects on the permeability and durability of plastic or rigid concrete can vary but generally are caused by the disintegrating action traced to the following factors.

Aggressive Water in Alkali Regions Sodium, potassium, and magnesium sulfates in alkali soil and water are usual causes of concrete deterioration. The sulfates presumably react chemically with the hydrated lime in the cement paste to form calcium sulfate, and this reaction is accompanied by considerable expansion and disruption of the concrete. Alternatively, alkali water entering concrete may deposit salts in the larger pores. The growing crystals resulting from this deposition can eventually fill the pores and develop pressures sufficient to disrupt the concrete.

Disintegration of the first type is usually prevented by the use of sulfate-resisting cement (ASTM type V). Resistance to disintegration caused by crystal growth is improved if the concrete mix is dense and impervious, has a relatively low water-cement ratio, and preferably contains entrained air. Crystal growth is therefore more of a problem in plastic concrete than in normal concrete.

It may be argued that attack of aggressive water on fresh concrete mixes is moderated by the presence of impermeable filter cakes, which separate the trench from its soil environment. On the other hand, exposure of bentonite slurry to ground pollutants can affect the colloidal stability and therefore the stability and permeability of the filter cake.

Some people contend that if concrete is to be exposed to seawater or to underground salt water, it is better to use normal cement with sound, non-reactive aggregates to obtain uniform, impermeable, and dense concrete. Such concrete presumably can resist disintegration due to seawater attack, but it usually is necessary to specify a moderately sulfate-resistant cement for cutoffs exposed to salt.

Leaching in Hydraulic Structures Water can pass through large pores, construction joints, and cracks in improperly constructed cutoffs. This water can dissolve some of the readily soluble calcium hydroxide and other solids and cause appreciable errosion of concrete in the course of time. This action is evidenced by white deposits on the surface of exposed concrete. Problems relating to the dissolving and leaching effect of percolating water also cause permeability problems. This situation can be avoided if it is possible to provide and maintain a watertight concrete.

Chemical Attack Considerable damage can be caused by surface corrosion if concrete is directly exposed to organic acids, farm silage, polluting wastes, and other forms of chemical attack. Chemically active materials and substances can be harmful to concrete. A good discussion for the chemical effects on unprotected concrete and the associated protective treatment is given by Stanton (1948).

Fly-Ash Mixes

These provide very satisfactory results where it is necessary to improve the resistance of concrete to sulfate attack. An example of such a combination is the mix shown in Table 7-6, used for the cutoff of Withens Clough dam. The water retained in this reservoir has a pH of 3.8, necessitating the use of a plastic concrete that can resist aggressive liquids. Tests on fly-ash concrete

TABLE 7-6 Fly-Ash Plastic-Concrete Cutoff (from Little, 1974)

Materials	Withens Clough dam	
	kg/m^3	%
Weight per unit volume	1845	100
Bentonite (B)	25	1.3
Water (W)	409	22.3
Cement (C)	61	3.3
Fly ash (Fa)	300	16.3
Aggregate (A)	1050	56.9
W/C	6.7	
W/solids	0.28	
W/(C + Fa)	1.13	
B/(C + Fa)	0.07	

show that this combination can provide the required reduction in permeability and simultaneously resist disintegration (Little, 1974).

Fly ash is a pozzolan material consisting of fine solid particles of noncombustible ash carried out of a bed of solid fuel by the draft. Pozzolan is a silicious or aluminous substance which reacts chemically with slaked lime under moisture to form a cementlike material. Several commercial portland-pozzolan cements show considerable resistance to sulfate attack. Justification for the use of such cements depends on the resulting long-term economy and improvement of the properties of mass concrete. This may include increased impermeability, reduced alkali-aggregate expansion, and improved workability. Inherent disadvantages in the use of pozzolan for structural concrete are slower strength development and lower resistance to deterioration caused by freezing and thawing, unless longer than usual moist curing is available, which generally is the case in slurry trenches.

Where mass concrete is used, e.g., dams, sulfate attack on concrete is not a major problem, whereas it may be serious in concrete cutoffs of the usual thickness. On the other hand, the potentially detrimental effects of pozzolan on concrete do not constitute adverse considerations for structures which, being buried underground, are not subjected to the effects of weather changes.

The sulfate-resistant properties of fly-ash concrete have been studied by many investigators (National Ash Association, 1971; Dikeou, 1970). Generally, test results show that fly ash consistently produces significant improvement in the sulfate resistance of the mix compared to non-fly-ash concrete. Maximum sulfate resistance is desired in structures built in soils where the soluble sulfate concentrations are 5 percent and higher. Whereas distinct benefits are in this case derived from the use of fly ash, results of tests on concrete containing other pozzolans show a wide variation in the effects on sulfate resistance, ranging from marked improvement to marked deterioration (Dikeou, 1970).

Tests on Fly-Ash Concrete Figure 7-13 shows the effect of fly ash on concrete expressed as a percentage increase in sulfate resistance. Figure 7-14 shows the reduced expansion of concrete containing fly ash. These data were derived from tests on concrete mixes from portland cements, portland fly-ash cements, and fly ashes. Concretes containing cements without fly ash were used as control mixes. Figure 7-15 shows the effect of fly ash on sulfate resistance of concrete containing type I cement.

The following conclusions can be drawn:

1. Fly ash generally improves the resistance of concrete to sulfate attack regardless of the cement type in the sample. The improvement, however, varies for various cement types in the following order:
   a. Type I cement (by far the greatest improvement)
   b. Type V cement
   c. Type II cement (Fig. 7-13)

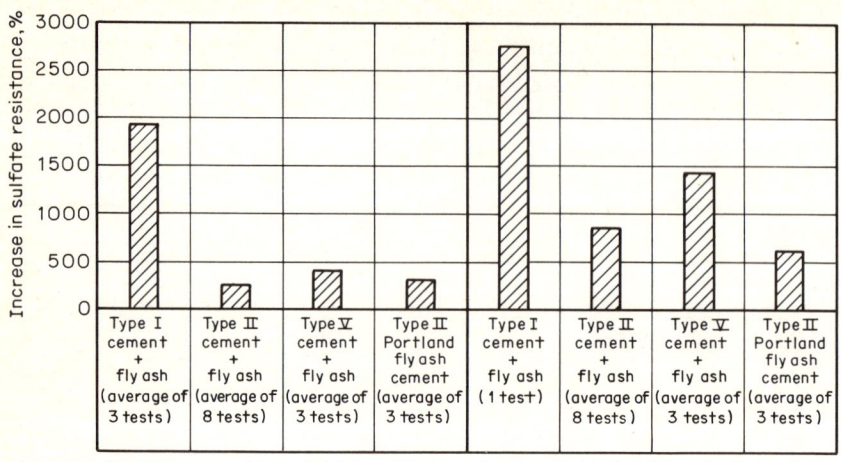

Fig. 7-13 Effect of fly ash on concrete resistance to sulfate attack. (*From Dikeou, 1970.*)

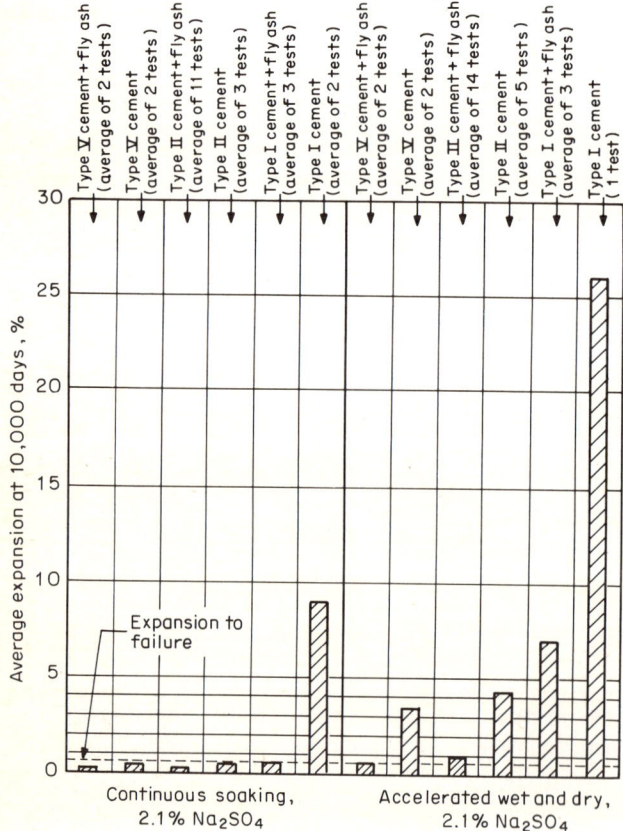

Fig. 7-14 Effect of fly ash on expansion reduction of concrete. (*From Dikeou, 1970.*)

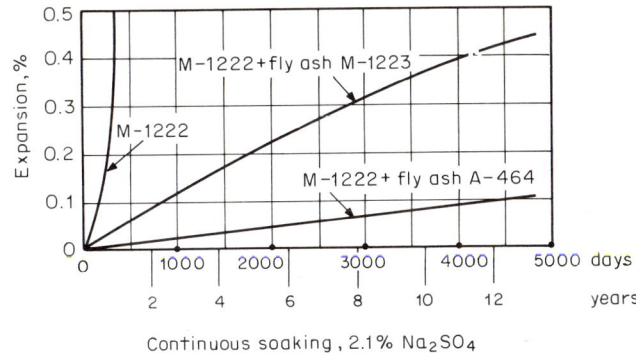

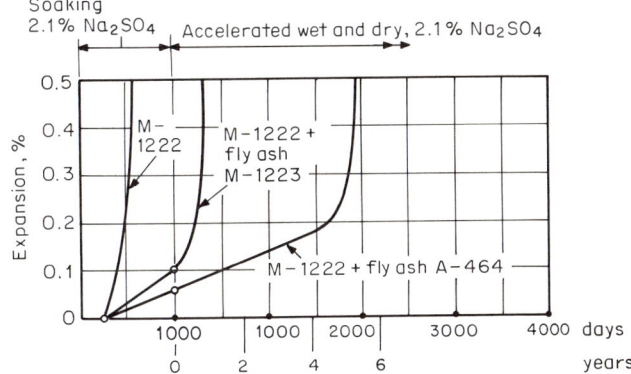

Fig. 7-15 Effect of fly ash on sulfate resistance of concrete containing type I cement; M-1222 = concrete sample; fly ash M-1223 obtained from Blend Corn Products, Argo, Illinois; fly ash A-464 obtained from Chicago Northwest Station. (*From Dikeou, 1970.*)

2. The degree of resistance of various concretes to sulfate attack is according to the following order of the cementitious materials used (Fig. 7-14):
   a. Type V cement and fly ash (by far the greatest resistance)
   b. Type II cement and fly ash
   c. Type V cement
   d. Type II cement
   e. Type I cement and fly ash
   f. Type I cement
3. Sulfate resistance varies primarily with the type of cement and depends to a lesser degree on the cement content, air content, and the variation of properties for each composition. On the other hand, the effectiveness of fly ash in improving sulfate resistance increases with the severity of sulfate exposure.

Bituminous Mixes

Among the various bituminous materials mastics are generally preferred because they melt easier, have high plasticity and complete impermeability, and are highly resistant to sulfate attack. The many problems that arise in connection with the stability and control of slurries and with the placement of bituminous mastic in continuous deep and narrow trenches under water or bentonite warrant ample investigation of the situation at hand. The work carried out thus far confirms the feasibility of inserting bituminous cutoff walls underground, but the associated problems should not be underestimated.

The bituminous mastic backfill is in many instances placed hot by means of a tremie pipe whose tip is lowered to within a few inches of the bottom of the trench, and this allows to raise its free level from the bottom up, also facilitating the continuous displacement of slurry. Whereas the bituminous material is sufficently flowable and workable to fill the excavation by its own gravity, the main difficulty relates to the flow of the material in the tremie pipe.

Bituminous mixes are not affected by the corrosive action of underground water containing sodium chloride and magnesium sulfate. The mixes generally are prepared using 70 percent aggregate, 10 percent lime filler, and 20 percent bitumen. The binder must not require excessively high temperature for placing, and once it is cooled, it should not be too fluid. A suitable temperature of the mix at placement is 160 to 180°C.

Other Types of Cutoffs for Pollution Control

Cement-bentonite solidified walls of the type discussed in Sec. 7-2 have been used successfully for pollution control of highly polluted aquifers in alluvial deposits. An example of this application, reported by Prentice (1974), involved a cement-bentonite cutoff inserted into the ground to stop pollution caused by an old millrace.

Since no previous experience was available regarding the performance of cement-bentonite mixes exposed to polluted groundwater, some doubts were expressed about the possible effects on the setting process and durability of the structure. Tests with various cement-bentonite mixes, using polluted water for mixing and allowing the setting process to be completed under polluted conditions, showed that except for an acceleration of setting time due to the presence of high electrolyte concentration, the polluted water had no significant effects upon the development of strength and impermeability. However, in view of the high salt content of the polluting water, sulfate-resisting cement was used.

A regular analysis of groundwater samples has shown an immediate reduction in pollution following the insertion of the cutoff wall. A period of heavy rains causes a slight rise in the pollution level, followed by a drop to the low level. It is anticipated that with time the natural flushing action of rainfall will finally remove all the pollution.

REVIEW PROBLEMS

7-1 Cite the factors to be considered in the injection of a grout curtain. Discuss a field-control program necessary to assess the construction.

7-2 Using illustrative diagrams, formulate a statement about fluid behavior. First consider plain water, then thin colloidal suspensions, more concentrated systems, and finally grouts.

7-3 A cutoff 40 ft deep and 2 ft wide is expected to have a deformation according to a parabolic function. The predicted deformation is zero at the bottom and 6 in at the top. Explain which concrete should be used; $f'_c = 3000$, 2000, or 1000 lb/in^2. State your assumptions.

7-4 Discuss in some detail the role of clay in earth cutoffs, solidified walls, and injected screens.

7-5 Describe the requirements for the top connection of the various cutoff walls. Discuss the significance in the method of placement and execution of construction details.

7-6 Outline the requirements and the properties of clay-cement suspensions for solidified walls and injected screens. Discuss the difference in the flow properties.

7-7 Does the fineness of a soil influence the selection of the sealing method? Explain.

7-8 Discuss the effect of the following factors upon the strength and modulus of plastic concrete: age, water-cement ratio, aggregate, free moisture, and bentonite.

7-9 Discuss the factors affecting the permeability and durability of concrete cutoffs. When would you choose high-resistant cement and concrete containing fly ash?

7-10 Describe the pollution effects on various types of cutoff walls.

REFERENCES

Chen, L. S., 1948: An Investigation of Stress-Strain and Strength Characteristics of Cohesionless Soils by Triaxial Compression Tests, *Proc. 2d Int. Conf. Soil Mech. Found. Eng.*, vol. 5, p. 35.

Dikeou, J. T., 1970: Fly Ash Increases Resistance of Concrete to Sulfate Attack *U.S. Bur. Recla. Res. Rep.* 23.

Dupeuple, P., and P. Habib, 1969: A Plastic Concrete Cutoff, *Proc. 7th Int. Conf. Soil Mech. Found. Eng., Spec. Sess.* 14, 15, *Mexico City,* pp. 71–76.

Greenwood, D. A., and J. F. Raffle, 1961: Non-newtonian Fluids, *Proc. 5th Int. Conf. Soil Mech. Found. Eng.*, vol. 1, p. 789.

———and———, 1963: Formulation and Applications of Grouts Containing Clay, in "Grouts and Drilling Muds in Engineering Practice," Butterworths, London.

Holtz, W. G., and F. C. Walker, 1962: Soil-Cement as Slope Protection for Earth Dams, *J. Soil Mech. Found. Div. ASCE, December,* pp. 122–132.

ICOS, 1973: Economics of Cutoff Walls by New Slurry Method, International Construction, Milan.

Jones, G. K., 1963: Chemistry and Flow Properties of Bentonite Grouts, in "Grouts and Drilling Muds in Engineering Practice," Butterworths, London.

Lambe, T. W., and R. V. Whitman, 1969: "Soil Mechanics," Wiley, New York.

Lane, K. S., and P. E. Wohlt, 1961: Performance of Sheet Piling and Blankets in Missouri River Reservoirs, *Proc. 7th Cong. Large Dams, Rome.*

Little, A. L., 1974: In Situ Diaphragm Walls for Embankment Dams, *Proc. Diaphragm Walls Anchorages, Inst. Civ. Eng., Lond.*

Maillard, R., and S. Serota, 1963: Screen Grouting of Alluvium by the E.T.F. Process, in "Grouts and Drilling Muds in Engineering Practice," Butterworths, London.

Marsland, A., and A. G. Loudon, 1963: The Flow Properties and Yield Gradients of Bentonite Grouts in Sands and Capillaries, in "Grouts and Drilling Muds in Engineering Practice," Butterworths, London.

Nash, J. K. T. L., 1974: The Stability of Trenches Filled with Fluids, *Semin, Underground Constr. Fluid Trenches, Univ. Ill. Chicago, April.*

National Ash Association, 1971: The VKR Lightweight Aggregate Plant and Quality Control Program for Fly Ash Utilization, *NAA Rep.* 3-71, Washington, D.C.

Nelson, F., and Sknonr'kov, 1949: "Seepage through Saturated Media," Sovetska Nakne, Moskow.

Prentice, J. E., 1974: Pollution Control in Alluvial Terrain, *Ground Eng.,* September.

Scott, R. A., 1963: Fundamental Considerations Governing the Penetrability of Grouts and Their Ultimate Resistance to Displacement, in "Grouts and Drilling Muds in Engineering Practice," Butterworths, London.

Stanton, T. E., 1948: Durability of Concrete Exposed to Sea Water and Alkali Soils, California Experience, *Proc. ACI,* **44:**821-847.

Taylor, D. W., 1952: "Soil Mechanics," Wiley, New York.

Terzaghi, K., 1955: Evaluation of Coefficients of Sugrade Reaction, *Geotechnique,* December, pp. 295-326.

White, R. E., 1974, 1975: Construction procedures for injected screens, personal communication.

Xanthakos, P. P., 1976a: Specifications for a Solidified Wall in Westmont, Ill.

———, 1976b: Tests on the Effect of Bentonite on Plastic Concrete, unpublished report.

Zaffle, J. A., 1970: Soil-Cement for Water and Sewage Works, Soil-Cement Slope Protection, *Portland Cem. Assoc. Publ.* C-24.

Chapter Eight

CONCRETE TECHNOLOGY AND CONSTRUCTION

PART A: Technology

8-1 PRACTICAL REQUIREMENTS AFFECTING MIX DESIGN

The requirements for plain or reinforced concrete placed in slurry trenches generally are different from conventional construction. The fresh concrete is exposed to certain conditions during placement and curing which can materially affect its properties. The final product, however, must satisfy the specified strength and durability, and since the walls are built below grade, the impermeability of set concrete is quite essential.

The method of concrete placement in slurry trenches differs from ordinary structural work, in which the fresh mix is vibrated and mechanically compacted in well-constructed forms. The concrete in diaphragm-wall panels is poured through tremie pipes and displaces the slurry by gravity alone. No vibrations or other mechanical processes are or can be used, and the only means available to cause the complete displacement of slurry by the concrete is the difference in density between the two materials.

The placement is more vigorous and complete if it occurs with an upward motion of the fluid concrete. This motion benefits the operation further by providing a gentle sweeping action on the face of the trench and on the reinforcement bars and incidental inserts. This action presumably cleans and removes the bentonite coating from these surfaces and allows better development of bond by full contact. As is shown in subsequent sections, the effect of bentonite on bond strength varies considerably with the flow motion of concrete, the bar type, and the properties of slurry (see also Sec. 4-8).

For reasons of practical convenience many contractors prefer the use of one tremie pipe even for panels unusually long. This method should not always be adopted, however, because the lateral flow of fresh concrete from a single tremie position generally is limited and cannot ensure the complete displacement of bentonite. Thus for relatively long panels two and sometimes three tremie pipes are used simultaneously, although this arrangement introduces certain difficulties in the continuous supply of concrete and can also impose a potential risk of inclusions in the overlapping zones.

It is evident from these brief remarks that the method of placement of fresh concrete is an important and difficult phase, owing to the wide effects which any bentonite not displaced can have on the set concrete. The process is greatly facilitated if the fresh concrete is flowable and workable. In this respect it is not sufficient to specify the initial concrete mix by strength alone, and in addition to strength workability is the governing factor in designing a mix. Experience shows that faults, reversed "hanging up," and "whirls" in which contaminated concrete is trapped can be avoided by providing a mix which is workable and yet not subject to segregation. Additionally, carefully selected aggregate, a suitable aggregate-cement ratio, and a retarder to delay initial set are desirable features in a mix.

Lack of attention to these details and failure to understand the difficulties associated with the placement of concrete often produce walls with distinct structural deficiencies. However, even if a direct basis is available for providing quality control and securing suitable construction procedures, the condition of the final structure or portions thereof usually cannot be confirmed until after the wall is exposed and visually inspected.

For a successful placement the concrete must satisfy the following conditions:

1. The mix must be flowable and display a plastic consistency. If the initial shear is too high for the conditions involved, the flow will be restrained and bentonite is likely to be trapped in areas not reached by the mix.

2. On the other hand, the mix must be cohesive enough to prevent segregation and bleeding. Concrete which bleeds or disintegrates under the pressure of its own weight is likely to block the tremie pipe or accept bentonite.

3. The mix must not set or stiffen too quickly but should remain workable until the pour is completed. The setting time must be extended to avoid adverse effects on quantities already delivered but not immediately placed or on portions already placed but not completed because of delays in delivery.

The consistency of concrete which satisfies the foregoing requirements depends largely on the quantity of water and its relation to the fines, including cement. For a mix to be workable the water must be trapped in the fines fraction, and since the particles opposing water flow are generally below the 0.5-mm size, they must be present in sufficient quantity.

Whereas the consistency of concrete is primarily a matter of design, the field control has immediate practical significance. Besides routine slump tests, fresh concrete should be examined before placement by experienced persons since segregation and bleeding are usually noticeable to the trained eye.

8-2 PROPORTIONING CONCRETE MIXES

For a complete mix design and associated details reference is made to current ACI manuals of concrete practice. Among the many excellent reviews available on the composition and properties of fresh and hardened concrete, Troxell et al. (1968) and Neville (1973) discuss concretes and describe the effects of component materials such as air-entraining agents and other admixtures. They also review curing procedures, the manufacture of concrete, and special types of concrete.

The review presented here will cover only those aspects which relate to the special requirements discussed in the preceding section, merely to show how simple adjustments in proportioning the basic mix can produce concrete which is workable without sacrificing strength and durability.

Workability

Workability is a relative property; concrete that is flowable and workable under some conditions is not necessarily workable under other conditions. The main factors influencing workability are the shear resistance or force required to start flow; the mass mobility after flow has started; cohesiveness or resistance to segregation; and the sticky limit of the material. In practice, workability is affected by the quantity and characteristics of cement, gradation and shape of aggregates, quantity of water, quantity of entrained air, and the type of admixtures.

Almost invariably for field concrete in slurry trenches the slump test is accepted as the measure of workability, but in reality this is a consistency test. Other means of determining the workability of concrete include the flow and the ball-penetration tests, but none of these procedures take account of all the inherent factors. A crude indication of workability can be obtained by tapping the side of the slumped pile with a tamping rod; concrete which is cohesive will not break apart or crumble. If slump is used to determine the mass mobility of fresh concrete, it should never be less than $17\frac{1}{2}$ cm (7 in) and preferably close to 20 cm (8 in). As shown in Fig. 8-1, the measured slump is considerably affected by the temperature of concrete, and the higher this temperature the less the slump. If conditions warrant, this variation should be taken into consideration.

An alternate procedure is the compacting-factor test, which is related to the reciprocal of workability. The degree of compaction achieved by a certain work is determined by the density ratio, which is the ratio of the density

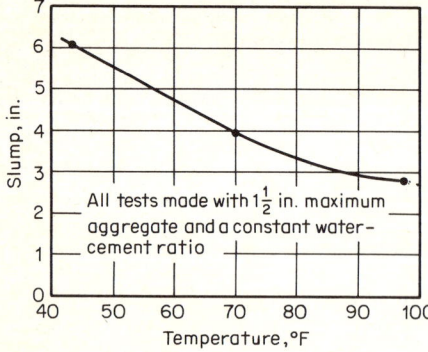

Fig. 8-1 Variation in the slump of concrete with temperature. (*U.S. Bureau of Reclamation, 1963.*)

actually achieved in the test to the density of the same concrete fully compacted (Neville, 1973). This test differs from the slump test in that variations in the workability of concrete are reflected by a substantial change in the compacting factor; however, the test is more sensitive at low workability levels than at high workability. The recommended compacting factor for the required workability of tremied concrete is 0.95 to 0.96.

Water Quantity

Neither strength nor durability should determine the water content of tremied concrete. Although an actual loss in the strength of set concrete will occur if the water-cement ratio is increased, the quantity of water should be enough to produce a mix which is workable and flowable. For a $7\frac{1}{2}$-in slump and a maximum aggregate size of $\frac{3}{4}$ in, the necessary quantity of water may be from 42 to 44 gal per cubic yard of mixture for non-air-entrained concrete and from 37 to 39 gal per cubic yard of mixture for air-entrained concrete.

These water requirements can be further reduced through the use of plasticizers of a recognized type. Finely divided workability agents are added to the fresh mix, allowing a reduction in the water content by more than 10 percent, and this increases the concrete strength providing also a pronounced antibleed action. Furthermore, it raises the specific gravity of the mix somewhat, which improves its displacing ability during the pour.

Grading of Aggregate

A flowable consistency is obtained if the water actually is trapped within the aggregate. In this respect smaller aggregate will oppose movement of water within a mix more effectively than larger aggregate. Before considering the use of plasticizers to offset deficiencies in grading that tend to produce harshness, a simple redesign of the mix focusing on the ratio of fine to coarse aggregate and also on the grading of the fine aggregate can accomplish good results.

The tendency for the mix to segregate is reduced by limiting the maximum size of aggregate to $\frac{3}{4}$ in (2 cm). The shape of the grading curve should show evenly graded aggregate since gap-graded concrete is prone to segregation.

Cement Quantity

Because of the higher than normal quantity of water and fines the cement content is also higher to satisfy the strength requirements. For the average conditions, maintaining a water-cement ratio between 0.5 and 0.6 by weight yields a 3000 to 4500 lb/in^2 concrete, or 210 to 315 kg/cm^2 (28-day standard moist-cured compressive strength). In addition to strength, the cement particles in rich mixes combine with the fines of the aggregate to produce cohesion and flowability. Higher than average cement quantity is also useful in reducing some tendency for partial segregation when concrete is poured through tremie pipes, which can occur despite adequate supervision.

Retarders

These are used to prevent the premature stiffening of some cements or to delay the stiffening under difficult placing conditions. The setting time must be checked against the time necessary to complete the pour. Concrete that sets too quickly is quite difficult to tremie, especially at high temperatures. In such cases retarders are advisable to offset the accelerating and damaging effects of quick setting and keep concrete workable for the entire pour. However, some of the chemicals and admixtures alleged to retard the setting time are likely to have variable and uncertain action. A reduction in strength usually accompanies the use of organic retarders, and creep in concrete is also influenced by the setting time.

Despite the incidental effects, retarders selected carefully and in certain proportions offset the effects of disruption during placement and may even have a beneficial influence on the concrete strength. The admixtures most commonly used are lignosulfonic acids, hydroxylated carboxylic acids and their salts, and derivatives of these chemicals. In general, they should be used with competent technical advice and after adequate testing under conditions similar to those in the trench.

Air Entraining

Concrete made from air-entraining admixtures has lower strength than non-air-entrained concrete but shows greater plasticity even at lower water content, relative freedom from bleeding and segregation, and improved durability against freezing and thawing. Air entrainment contributes to the fluidity of the plastic mass, much as added water does. Without loss of workability it is possible to reduce the quantity of mixing water and use a higher ratio of coarse to fine aggregate. Each 1 percent of entrained air generally permits a reduction of water of about 3 percent and a reduction in the sand content by an amount approximately equal to the volume of entrained air.

The tendency for the strength to decrease is partially offset and compensated for by the use of additional cement. The decision to use air entrainment or how much air to entrain therefore depends on the degree to which strength can be sacrificed in the interest of improved workability.

For the usual applications the maximum theoretical strength of concrete is not used because the wall thickness is based on practical requirements rather than on theoretical considerations and the design is seldom balanced. Therefore, a slight reduction in strength is acceptable in view of the improved workability. A slight excess of water is less harmful than a quantity which produces stiff and nonflowable concrete. Likewise excess of fines is less harmful than a mix behaving like a stony mass.

These considerations make it advantageous to specify entrained air, which should be 4 percent minimum and 5 percent desired average by volume for a maximum aggregate size $\frac{3}{4}$ in (2 cm). Given the compressive strength, the probable water-cement ratio can be estimated from Table 8-1 for both non-air-entrained and for air-entrained concrete. For constant water-cement ratio the concrete strength is reduced as the air content is increased from the usually entrapped to the usually entrained amount.

TABLE 8-1 Compressive Strength of Concrete for Various Water-Cement Ratios (from ACI, 1954)

Water/cement ratio		Probable average compressive strength at 28 days, lb/in^2	
Gal/sack of cement	By weight	Non-air-entrained concrete	Air-entrained concrete
4	0.35	6000	4800
5	0.44	5000	4000
6	0.53	4000	3200
7	0.62	3200	2600
8	0.71	2500	2000
9	0.80	2000	1600

8-3 CONCRETE PLACEMENT

In general it is advantageous to carry out the concreting operation soon after the reinforcing cage is inserted into the panel. The pour should be completed in the shortest possible time and without discontinuities or interruptions in order to avoid embedment of stop-end tubes, blockages in the pipes, and cage flotation caused by the upward drag of rising stiff concrete. For these reasons ready-mixed concrete is used on a general basis since this gives better quality control and can accommodate unusually high pouring rates, often on the order of 35 yd^3/h (27 m^3/h) or more.

Tremie Pipes

The most common method of placement is by means of tremie pipes, which are withdrawn in stages as the concrete level rises. The fresh mix is directly conveyed into the hopper of the pipe, as shown in Fig. 8-2. The operation is basically simple but requires adherence to certain rules. Tremie pipes suitable for concreting are from 6- to 10-in-diameter (15- to 25-cm) steel pipes, but

Fig. 8-2 Placement of concrete in a slurry-trench panel by means of a 33-cm-diameter tremie pipe. (*Franki.*)

pipes of larger diameter are often used. A rule of thumb is to select a diameter at least 8 times the maximum aggregate size, since this will help prevent blocking of the pipe. Since several instances of serious weakening of concrete by contamination from contact with aluminum pipe have been reported, only steel pipes should be used to convey fresh concrete.

The required length is obtained by using suitable splices. There is no standard pipe length between splices, but for simple adjustments 6-ft (2-m) lengths are practical. The splices should be watertight, easily disconnected, and have no projecting flanges that could interfere with the steel cage. Threaded pipe joints are used by many contractors and are considered effective. In relatively narrow panels elliptical pipes can be used to allow easier handling between closely spaced reinforcement bars.

The process of tremie placement is based on certain simple principles, and the most serious disadvantage is that it cannot be checked visually. In a dry trench care must be taken that no water is entering, which might cause dilution or segregation of concrete. In a slurry trench any leakage will move water out rather into the panel regardless of the location of the natural water table. When concrete is tremied in water or slurry, a common source of dilution or segregation is water or slurry in the pipe before starting the pour. At the start of the pour the pipe should be empty except for air. To satisfy this requirement the tremie pipe is lowered through the slurry until its tip rests firmly on the base of the trench. After the funnel hopper at the upper end has been assembled, a plug is placed inside which floats in the slurry and serves to separate the initial batch of concrete from the slurry that fills the pipe. As concrete enters the tremie, the plug travels down under the weight of the fresh mass and reaches the bottom. At this stage the tremie pipe is slowly lifted, allowing the concrete to push the plug out. Once this occurs concrete begins to discharge and fill the panel while the pipe is kept submerged and completely filled. The last requirement must be checked against the rate at which fresh concrete can be conveyed from the ready-mix truck.

Once the placement begins, the process requires routine checking of the concrete level around the pipe and at the ends of the panel. The initial plug is eventually recovered because it floats out. Blocking the upper end of the pipe with rags or with anything likely to lodge in the concrete and produce a weak zone should be avoided.

Recommended Procedures to Avoid Defective or Incomplete Placement

The usual defects of diaphragm walls are cold joints, zones of segregated or contaminated concrete, trappings of bentonite mud, and complete cavities appearing at the ends of panels not reached by the concrete.

The first two types of defects usually result from interruptions during concreting or premature withdrawing of the tremie pipe either partially or completely from below the concrete-bentonite interface. Mud trappings are caused by impediment to the flow of concrete because of closely spaced bars, poorly designed and placed boxing out, and concrete of low workability. At the extreme these factors can prevent the mix from filling the panel vigorously and tightly at the joints, in which case complete cavities will appear when the wall is exposed. The occurrence of faults and cavities can be explained with respect to the energy available from the flowing concrete; they are caused when this energy is less than the energy required to displace the bentonite slurry.

The fluid concrete must first flow laterally from the discharge point and then upward. The energy available for this movement decreases with increasing distance from the outlet. As the concrete level rises, the energy gradient is reduced because of an actual reduction in the differential pressure head between the concrete level in the hopper and the trench. Trappings and faults of bentonite mud are therefore most commonly found at the panel joints and in the upper portion (usually the upper third) of the panel. Shorter panels are less prone to these defects than longer panels for the same number of tremie pipes.

The nature of mud trappings and inclusions varies considerably. Usually they appear as a mix of soil, bentonite, and concrete in incidental proportions. Although they are eventually squeezed and compressed under the weight of overlying concrete, they normally fail to develop significant strength and they are not acceptable structurally. They are, however, fairly impermeable. If they occur locally, they can be left in place provided they do not appear at locations which can impair the strength of the wall. If they must be removed and replaced with grout, the area behind the wall should be sealed first; otherwise groundwater will enter the excavation.

Tremie-Pipe Spacing Despite the practical advantages associated with the use of a single tremie pipe, as opposed to two or more, it generally is essential to specify the maximum tremie pipe spacing and therefore the maximum panel length that can be poured with one tremie pipe. The pipe spacing should not exceed 14 ft, or 4 m, and the end distance should not be more than 7 ft, or about 2 m (Xanthakos, 1974). Another recommendation is to limit the lateral flow on either side of any tremie position to about 2.5 to 3.0 m (approximately 8 to 10 ft). From this it follows that the average panel length that should be poured with one tremie pipe is about 15 ft, although at the extreme it may reach 20 ft if the mix is sufficiently workable and the structural details do not impede lateral flow. Where two or more tremie pipes are used, the pouring should be simultaneous if practicable or in alternating sequence to maintain the level of rising concrete as uniform as possible (see also Sec. 8-4).

Contaminated Concrete In general, the upward flow of fresh mix combined with the difference in density between concrete and slurry prevents intermixing except for a zone 1 to 2 ft thick at the interface. This intermixing can cause weakening of the concrete in the uppermost portion of the wall. The usual practice is therefore to remove the upper 2 ft of wall and replace it with new concrete after the wall is exposed, or overflow the initial pour until the contaminated portion is eventually disposed of. The latter solution is favored by contractors.

Face Stability during Concreting

The introduction of fresh concrete in the trench changes the lateral pressure exerted against the face. The new pressure is considerably higher than the simple hydrostatic thrust exerted by the bentonite slurry and usually exceeds the initial stress at rest.

Field measurements reported by DiBiagio and Roti (1972) for a trench 18 m (59 ft) deep show that upon completion of the pour the pressure exerted by the fresh concrete was the hydrostatic pressure of liquid concrete only in the upper portion of the panel and specifically for a depth of 5 m. When the column of overlying concrete exceeded 5 m, the lateral pressure on the face gradually became less than the total overburden pressure. At a depth of 10 m (33 ft) or more the pressure was between 0.6 and 0.8 times the overburden pressure.

It is conceivable that for the example just cited the flow motion was such that the concrete in the lower part remained undisturbed after placement and began to set as the upper part was still poured. This situation is analogous to the effects caused when conventional wall forms are filled with fresh concrete (see also ACI, 1963). Although the pressure of fresh concrete also depends on the rate of pour, the temperature of the concrete, and the height of the column, it is mainly influenced by the flow motion of the liquid mix. It is quite difficult to delineate these relationships for slurry trenches; however, a certain stiffening of some parts of the placed concrete before the pour is complete is accompanied by a corresponding shrinkage, which relieves the initial pressure in these parts of the trench.

An approximate analysis of stability of the face under the pressure of fresh concrete can be carried out using limit theory. For cohesive soil with $\phi = 0$ face stability requires

$$\gamma z + N s_u = K \gamma_c z$$

or
$$s_u = \frac{z(K\gamma_c - \gamma)}{N} \tag{8-1}$$

where s_u = undrained shear strength of soil
  γ_c = unit weight or bulk density of concrete
  γ = unit weight or bulk density of soil
  N = stability factor (may be taken as 4 for shallow trenches)
  K = pressure coefficient of concrete, usually 1 for liquid mix
  z = depth of trench

For the usual conditions ($\gamma_c = 150$ lb/ft^3, $\gamma = 110$ lb/ft^3, $N = 6$, $K = 1$) and moderate depth (50 to 60 ft) the undrained shear strength must be at least 400 lb/ft^2 for the face to sustain the pressure of fresh concrete. In reality, however, it is seldom (if ever) necessary to consider this type of failure, since in very soft or loose soil the trench is likely to fail before this stage is reached.

More significant in this case is to predict the actual enlargement or widening of the trench as the face is shifted toward compressible soil under the action of the fresh mix until equilibrium in the soil mass stops this movement. The usual control is to monitor the concrete level during pouring, and if this rises unreasonably slower than expected, it probably does so because the face moves toward the soil. The author has encountered situations where face movement in loose silt resulted in a 1-ft increase in the actual wall thickness, or 50 percent increase in the theoretical concrete volume.

8-4 FLOW MOTION OF TREMIED CONCRETE

Although in conventional underwater construction the concrete usually is placed either by pumping or by the hydrovalve method (ACI, 1971; CUR, 1972), in slurry-trench work the preference, as mentioned, is for tremie placement. In many instances cost factors favor this method, but invariably the tremie process allows faster placement and better quality control. For properly tremied concrete the set product is of high compressive strength and amply sufficient for structural work. Furthermore, it is fairly homogeneous and has only few isolated loose pockets.

The importance of the flow motion of tremied concrete has long been recognized and studied in a variety of conditions. Although the effects of incidental factors can be significant, tests on prototype panels indicate that the flow motion is related mainly to the panel dimensions, number and position of tremie pipes, and the workability and flowability of the fresh mix. These points have been investigated by Ikuta et al. (1971), using activable tracers in bentonite. The procedure and results of these investigations are briefly described in this section.

Procedure for Activable-Tracer Analysis Table 8-2 provides pertinent information regarding the prototype structures. The activable tracers selected for the tests had to be readily dissolved in water and eventually absorbed by the bentonite slurry in order to trace the infiltration of concrete by bentonite. Among the highly activable elements gold (Au), samarium (Sm), scandium (Sc), and antimony (Sb) were considered suitable.

Samarium was the activable tracer for bentonite-slurry medium, i.e., for concrete tremied in bentonite slurry instead of plain water. The reason is that although this element cannot be seen in the gamma-ray spectrum of water, it can be seen in the bentonite-slurry spectrum. This means that samarium is absorbed by bentonite and is not elutriated by water. A prequalification test was in this case necessary to determine the transfer of tracer from the bentonite in the slurry to fresh concrete under static conditions.

The activable tracers consisted of nuclides which can be measured separately by nondestructive procedures under the actual irradiation conditions. Within the range of gamma-ray energy the photoelectric peaks of the various nuclides were established far apart in the gamma-ray spectrum for measurement. The four elements were added to concrete and activated, and the spectra were then measured. In the final process it was possible to measure each photoelectric peak separately.

Whereas samarium was the only element selected as activable tracer for checking the introduction of bentonite into concrete, all four elements were used for observing the flow motion and mixing of fresh concrete during tremie placement.

Sampling of Analysis of Tracers Table 8-3 shows the quantity and type of tracer for each specimen. These quantities are several times greater than the minimum detection limits and are more than adequate for the minimum amount of bentonite that must be detected, which for this case was set at 0.5 percent.

The samples were obtained from the prototypes by coring, were activated, and then were tested in compression. Smaller pieces were also taken from near the wall surface and from the interior of the wall and were ground to fine material. The gravel and coarse particles were removed by screening, and suitable samples were obtained for activation from the materials passing. These were broken down into the components shown in Table 8-4, showing a nearly constant cement ratio, which made the comparison possible.

The samples were subjected to irradiation and a 5-day cooling period, after which gamma-ray spectometry was performed. The quantity of samarium was estimated from the ratio of area of the photoelectric peak, and this gave a measurement of bentonite present in the concrete. For the conditions of flow

TABLE 8-2 Specimens and Materials for Activable Tracer Tests (from Ikuta et al., 1971)

		Specimen			
	III	I	IV	II	V
Dimensions, m	2.1 × 6.0 × 0.6	2.1 × 6.0 × 0.6	4.1 × 6.0 × 0.6	2.1 × 6.0 × 0.6	2.1 × 6.0 × 0.6
Activable tracer	Au, Sc, Sm, Sb	Sm	Sm	Sm	Sm
Tracer medium	Concrete	Slurry	Slurry	Slurry	Slurry
Slurry composition:					
Bentonite concentration, %	0	5.0	5.0	10.0	10.0
CMC, %	0	0.05	0.05	0.05	0.05
Sodium humic acid, %	0	0.10	0.10	0.10	0.10
Slurry properties:					
Specific gravity	.....	1.03	1.03	1.05	1.05
Funnel viscosity, 500/500 cm³	.....	26.0	26.0	32.5	32.5
Apparent viscosity, cP	.....	9.0	9.0	15.5	15.5
Filtration film thickness, mm	.....	1.0	1.0	1.5	1.5
pH	.....	8	8	8	8
Concrete quantity, m³	7.2	7.2	14.4	7.2	7.2
Placing time, min/m³	10	9	8	14	4

Mix ratio,* kg/m³	Cement	330
	Water	185
	Sand	843
	Gravel	938
Mix properties:*	Design strength, kg/cm³ (28-day)	225
	Water-cement ratio	0.56
	Slump, cm	20
	in	8

*All specimens.

Concrete Technology and Construction

TABLE 8-3 Quantity and Type of Activable Tracer (from Ikuta et al. 1971)

Specimen	Concrete volume, m³	Nuclide-activable tracer	Quantity added, g
I	7.2	Samarium	80
II	7.2	Samarium	160
III	7.2	Gold	15
		Scandium	25
		Samarium	65
		Antimony	25
IV	14.4	Samarium	160
V	7.2	Samarium	160

TABLE 8-4 Analysis of Samples (from Ikuta et al., 1971)

Sample no.	600°C loss, %	Insolubles, %	CaO, %	Cement, %
1	21.4	29.7	28.6	43.8
2	28.6	18.5	29.6	45.3
3	22.3	30.3	30.7	47.0
4	18.4	34.9	27.4	42.0
5	18.9	30.6	30.8	47.2
6	20.5	29.4	29.4	45.0
7	19.0	28.6	29.4	45.0
8	17.4	29.8	30.0	46.0
9	20.0	31.4	29.8	45.6
10	23.2	26.0	31.4	48.1
Average				45.5

motion and mixing of concrete during placement different energies of photoelectric peaks were used for the four different elements.

Test Results Figure 8-3a gives the content of each nuclide in various locations of specimen III of Table 8-3. Assuming that the 15 g of gold added to the initial mix were uniformly mixed with the mass of concrete, the weight of gold in the cement is 13.8 μg per sample. In Fig. 8-3a detection quantities one-tenth or less this amount are omitted. It is evident that the introduction of two or more nuclides occurs at all locations of sample collection. Figure 8-3b shows the distribution of the various concrete batches.

Irradiation results of samples obtained from speciments I to V made it possible to estimate the bentonite-cement ratio. Additionally, 28-day compressive strengths were obtained from standard tests. The intermixing of bentonite with concrete is evident in Fig. 8-4a to d, but the bentonite-cement ratio is greater at or near the wall surface, where the infiltration of concrete by bentonite is more pronounced, and almost negligible away from the face.

Method of Placement With reference to the foregoing results it is interesting to compare the exact process of placement. Panel V was concreted by smooth flow from a pump truck through a tremie pipe at uniform speed.

268 Slurry Walls

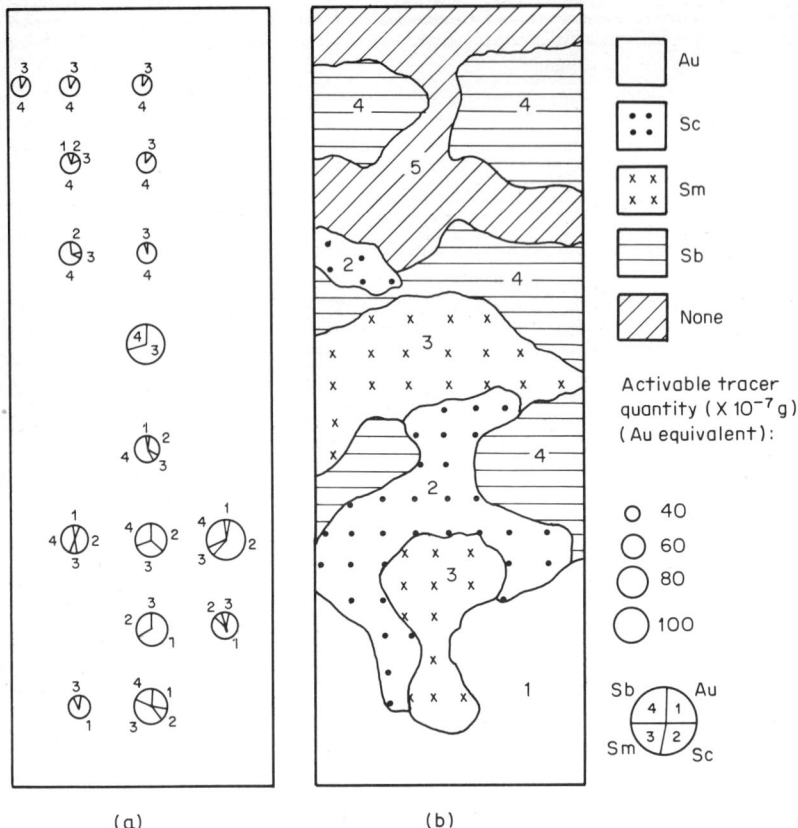

(a)   (b)

Fig. 8-3 Estimated diagram of set concrete (specimen III of activable-tracer test). Numbers in (b) indicate the sequence of concrete batches. (*Data from Ikuta et al., 1971.*)

Whereas intermixing is considerably lower than in other specimens, the mixing ratio is essentially the same near the face and in the interior of the wall. For specimens I, II, and IV the concrete was conveyed to the tremie pipe from buckets; hence the flow was considerably less uniform.

Observations on the Flow Motion of Fresh Concrete

In this type of construction a practical question is often raised: How are the conditions of placement manifested in the strength characteristics of set concrete? These conditions can sometimes be traced from the flow motion of the fresh mix and by the amount of bentonite intermixed with it.

When concrete is tremied, it usually is assumed that flow occurs in two ways, first laterally at the bottom of the panel and then upward. In this manner the initial batch is forced up by the batches which follow so that it

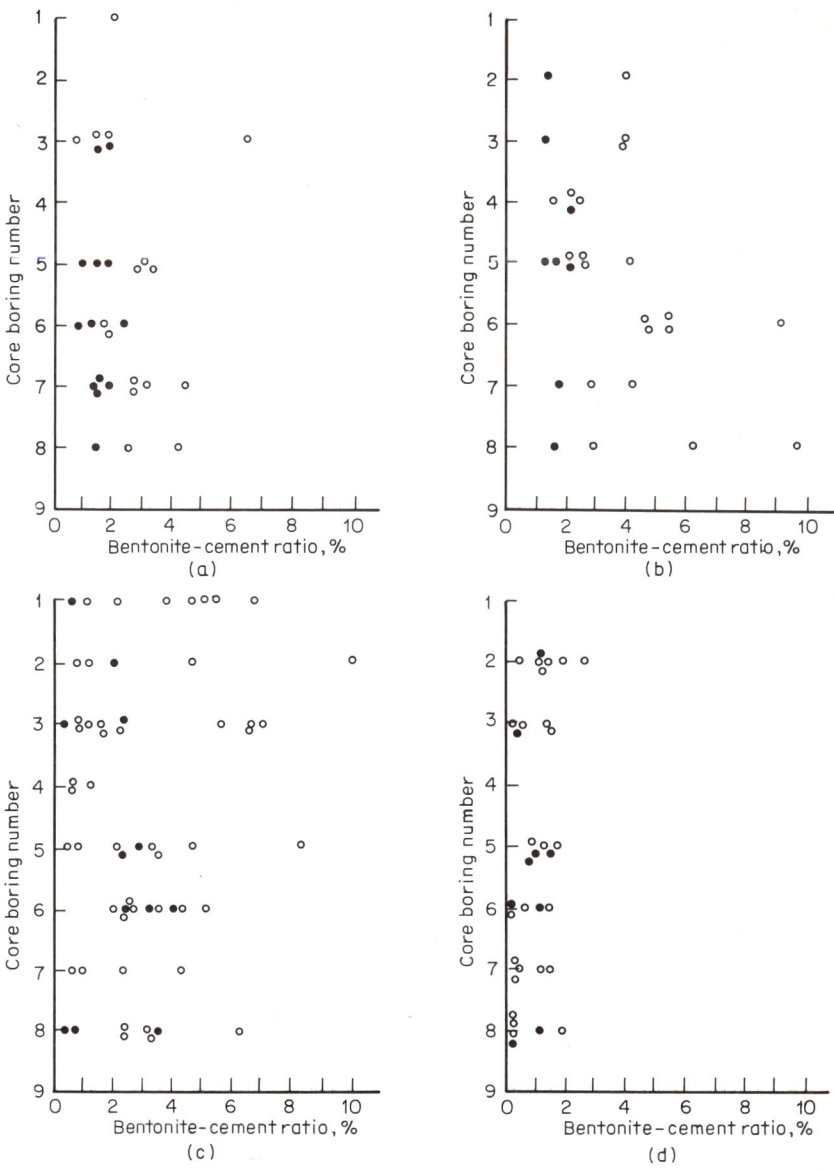

Fig. 8-4 Mixing and distribution of bentonite in concrete samples of the activable tracer test: (a) specimen I, (b) specimen II, (c) specimen IV, (d) specimen V; ● = sample from interior of wall; ○ = sample from near face. (*From Ikuta et al., 1971.*)

always stays on top of the pour. This particular flow motion is assumed to continue until the pour is completed. Accordingly, bentonite and slime from the slurry are intermixed with fresh concrete at greater proportions near the top of the pour, where contact occurs, whereas the rest of the concrete is only slightly affected. However, the random arrangement of nuclides and the flow pattern shown in Fig. 8-3a and b, respectively, suggest a different flow motion. The position of set concrete was in this case inferred from the detection of activable tracers, their relative proportions in the samples, and observations during placement. The sequence of concrete batches was gold, scandium, samarium, antimony, and no nuclide. The discharge tip of the pipe was held 40 cm (16 in) from the bottom until the block containing antimony was poured.

It is seen from Fig. 8-3b that as the gold block was placed, the fresh mix spread laterally until the bottom part was filled. The scandium block was tremied next and rose like a well above the gold block, which remained at the bottom. Likewise the samarium and antimony blocks rose alongside and around the tremie pipe, although some previous batches were pushed up, probably because at this stage the pipe was sufficiently submerged into the fresh concrete. The tremie was then withdrawn upward, so that the last block once again welled up and rose to the surface of the pour.

This example shows that tremied concrete can rise according to different patterns: like a well around the tremie pipe, in a lateral motion, by pushing from the bottom up like a plug (in which case the first batch is always on top of the pour), or by combinations of these patterns. In this process intermixing of bentonite with concrete is conceivable, but it is more pronounced with random flow and insignificant with plug flow.

The flow motion of fresh concrete through a tremie pipe in a relatively long panel is shown in Fig. 8-5. As the initial batch is placed, it assumes an angle of repose according to the concrete slump, so that around the pipe the material travels upward faster than it does away from the pipe. Because in this case the panel is too long for one tremie pipe, more lateral movement must occur for the concrete to fill the trench at the ends, and when batch 2 is poured, it displaces batch 1 laterally and toward the ends of the panel, as shown in Fig. 8-5b. This lateral displacement continues until the entire panel is filled, as shown in Fig. 8-5d; each batch thus forms a block which is gradually squeezed to a thin column around the pipe causing the level of placed concrete to rise. The initial block is pushed to the ends whereas the last block occupies the center.

Although it is rather difficult to draw conclusions for general application, the following points should be made:

1. Plug motion (initial batch always on top) is the preferred method of placement. This benefits the final product in two ways (1) by minimizing mud trappings and inclusions and (2) by providing the sweeping action which cleans and removes bentonite from around the reinforcement bars and other vertical surfaces.

2. For relatively short panels plug motion will occur as long as the tremie pipe is sufficiently submerged in the concrete. There is some doubt what length is considered sufficient, but practical experience shows that the tip of the tremie pipe preferably should be 3 m (or 10 ft) below the concrete level. If

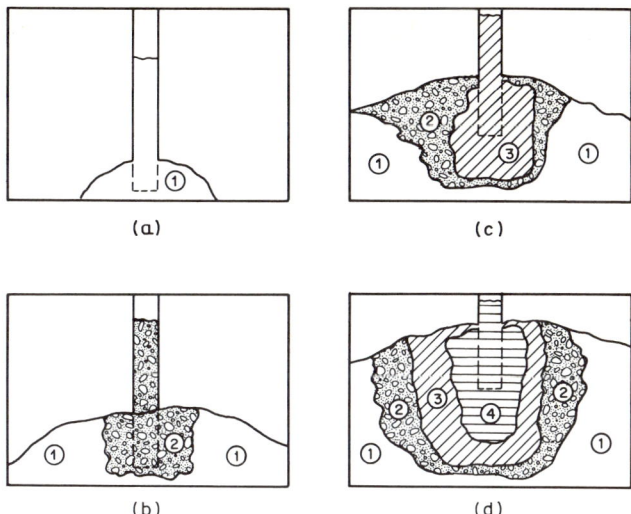

Fig. 8-5 Flow motion of concrete in long panel through single tremie pipe: (*a*) batch 1 being placed; (*b*) batch 2 being placed; (*c*) batch 3 being placed; and (*d*) batch 4 being placed.

concrete rises in a plug flow, most of the impurities in the slurry will move upward and will be collected at the top.

3. Lateral flow can cause the concrete at the top to be exchanged for new concrete. It also can cause impurities in the slurry to gradually move toward the ends of the panel, where they are likely to be trapped and mixed with the fresh concrete.

4. For long panels one tremie pipe is unlikely to produce the desired flow motion. Hence, the recommendations regarding the number and spacing of pipes should be followed.

5. If practicable the tremie pipes should be located near the construction joints. Reinforcing cages, incidental inserts, and boxing out should be arranged to hamper the flow of concrete as little as possible.

6. It is advantageous to convey the concrete to the tremie pipe at uniform speed, but toward the end of the pour the rate of placement is reduced considerably.

7. Tremie placement is more thorough and complete with workable concrete, but workability gives better results with relatively low slurry density.

8-5 STRENGTH OF CONCRETE

Diaphragm walls depend for their strength and structural integrity on the strength of concrete, i.e., compressive and bond strength; the effectiveness of integration of individual wall units, i.e., the details and strength of construction joints; and the method of load transfer to and from adjoining structural elements, i.e., connections with subpiers, columns, beams, and slabs. This section discusses the factors influencing concrete strength and its significance as a measure of general quality, based mainly on data available at present.

Flow Motion during Placement The initial batches are moved and displaced, and they may not come to final rest until the pour is completed. Since for relatively long or deep panels this may take several hours, unless the mix contains retarders, premature setting can influence strength.

Changes in the Water-Cement Ratio Mixing slurry with concrete may increase the water content of the latter with a subsequent loss of strength.

Introduction of Bentonite and Slime The final strength can be decreased as a result of intermixing with bentonite and slime from the slurry. The finished wall may also have entrapped inclusions and soft pockets of nonconcrete materials.

Curing Conditions Curing time, moisture, and temperature can affect the development of strength, mainly because of their influence on the hydration of cement. In general, because these conditions are favorable and the period of moist storage is longer and uninterrupted, the strength is greater. The curing temperature of concrete cast underground is also a favorable factor for the development of strength unless it is much lower than the temperature during casting.

Selection of Strength

In general, concrete of any specified cube strength to standard requirements can be produced. Besides the theoretical considerations affecting the selection of strength, economy and practical aspects should also govern. Usually, the requirements of excavation and concrete placement are considered first, and, as mentioned in other sections, it is advantageous to establish the wall thickness before selecting strength. For normal diaphragm-wall thickness the design strength is between 3000 and 4500 lb/in^2, or 210 and 315 kg/cm^2. In the lower range, however, the use of entrained air is more justified, and the water-cement ratio is more compatible with the requirements for high slump. The design strength can be even higher subject to special considerations for particular jobs and for unusual classes of structures.

Measured Strength

Construction records show that cores obtained from exposed walls several weeks after pouring produce strengths higher by as much as 1000 lb/in^2 (70 kg/cm^2) than the design strength. This is due mainly to the fact that concrete undergoes curing within damp soil and is not exposed to outside detrimental conditions.

The strength characteristics of concrete will be considerably affected if changes in the water-cement ratio can occur. For concrete intermixed with bentonite the probable maximum increase in the water-cement ratio is shown in Fig. 8-6 as a function of the bentonite-cement ratio. This will occur assum-

Concrete Technology and Construction 273

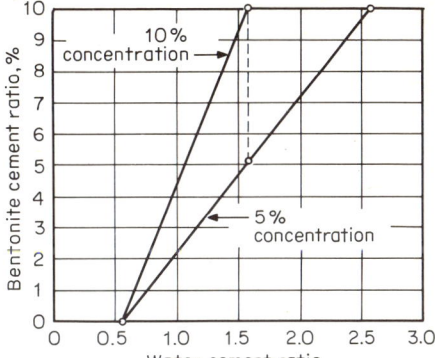

Fig. 8-6 Change in water-cement ratio for varying bentonite-cement ratios for the concrete mixes of Table 8-2. (*From Ikuta et al., 1971.*)

ing that the suspending slurry infiltrates the concrete and is completely mixed with it. For example, for a bentonite-cement ratio of 0.1 (10 percent) about 33 kg of bentonite is mixed with 1 m³ of mix. For a 10 percent slurry this means that about 330 kg of water will be taken by the fresh mix as slurry, and this will raise the initial quantity of water to 515 kg for 1 m³ (total volume adjusted) so that the new water-cement ratio is now 1.56.

Data from Activable-Tracer Test If the foregoing assumption is valid, a serious reduction in the strength of set concrete could follow. However, the measured strength of core samples shown in Fig. 8-7 clearly shows no definite relationship between the bentonite-cement ratio and strength. Further, the results are quite scattered and show no specific trend in the development of strength. This supports the conclusion that intermixing of bentonite with fresh concrete does not necessarily imply that the latter absorbs water.

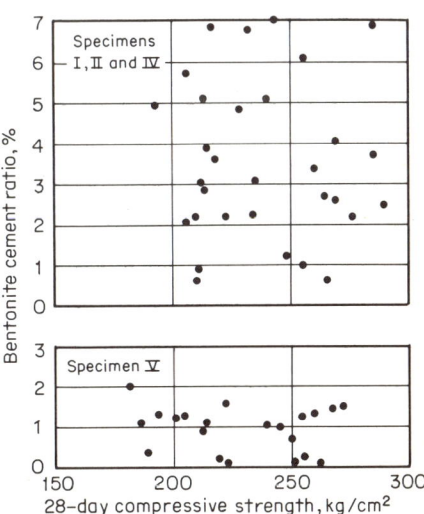

Fig. 8-7 Relationship between bentonite-cement ratio and compressive strength. Samples obtained from the specimens of the activable-tracer test. (*From Ikuta et al., 1971.*)

However, as evidenced from Fig. 8-8, the presence of bentonite in the mix affects the modulus of elasticity and its relationship to compressive strength. At some specific region both clean concrete and bentonite concrete have the same modulus, but the elasticity-strength curve for the latter is considerably steeper, meaning that bentonite concrete has a lower modulus at the same strength and therefore behaves more like a plastic material. The same behavior was observed in the plastic mixes of Chap. 7.

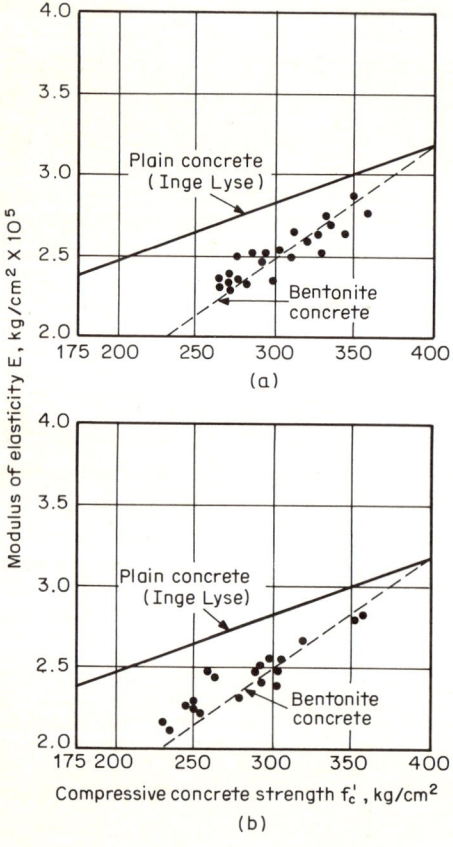

Fig. 8-8 Modulus of elasticity as a function of strength for plain and bentonite concrete: (a) specimen IV and (b) specimen V, activable tracer test. (*From Ikuta et al., 1971.*)

Results from Laboratory Tests Table 8-5 shows data from three concrete mixes. The prototype bentonite mix was tremied into four prototype panels containing 4, 7, 10, and 13 percent bentonite slurry, respectively (Japanese bentonite); the prototype dry mix was poured into an empty panel. All panels were identical in size and shape (5 m deep, 1.8 m long, and 0.45 m wide). The trench mix was tremied into a trench panel 12 m deep, 2 m long, and 0.6 m wide filled with 5 percent bentonite slurry. Before testing, all concrete was cured for 8 weeks. The prototype concrete was moist-cured under standard conditions, and the trench concrete was field-cured as nearly as practicable.

TABLE 8-5 Quantity of Materials and Properties of Mixes per Cubic Meter

Mix conditions*	Cement, kg	Water, kg	Sand, kg	Gravel, kg	Entrained air, %	Water-cement ratio	Slump, cm
Prototype dry	332	186	766	1032	3–4	0.56	20
Prototype bentonite	365	200	809	927	3–4	0.56	20
Trench	363	187	738	1048	3–4	0.515	20

*Prototype tests from *Jap. Concr. J.*, vol. 9, no. 6, 1966.

Compression tests on standard cylinders from the prototype concrete produced an average compressive strength of 360 kg/cm². These results are shown in Fig. 8-9, and evidently the strength is moderately influenced by the presence of bentonite while it shows a slight tendency to decrease with bentonite concentration higher than 7 percent. Surprisingly, core samples from the trench concrete produced average strength of 400 kg/cm² for the upper half of the structure and 450 kg/cm² for the lower half.

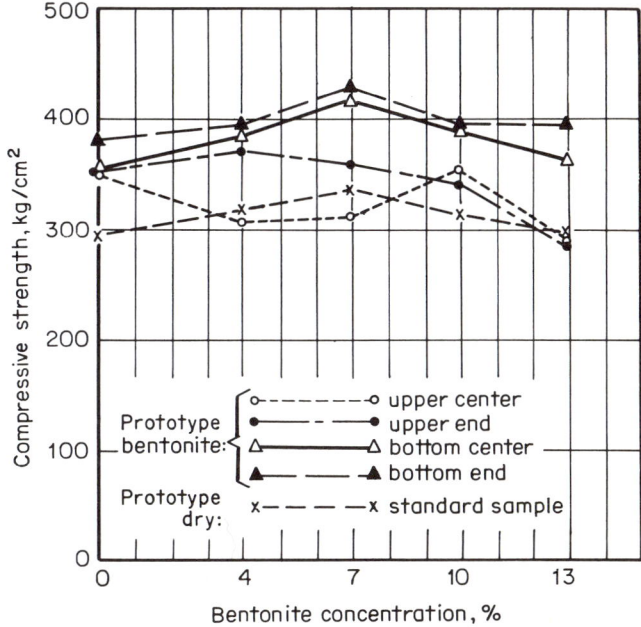

Fig. 8-9 Compressive strength of core samples for concrete tremied into bentonite-slurry prototype panels and dry panels. (*From Jap. Conc. J., vol. 9, no. 6, 1966.*)

Table 8-6 shows data for concrete used in another series of tests. The mix in this case was placed in various panels filled either with plain water or with slurry, as shown in Table 8-7. Compression tests on standard cylinders again show that the 28-day strength does not appear to be significantly affected by the presence of bentonite in the slurry.

TABLE 8-6 Quantity of Materials and Properties of Mix* (per Cubic Meter) (from Nippon Co., 1975a)

Cement, kg	Water, kg	Sand, kg	Gravel, kg	Entrained air, %	Water-cement ratio	Slump, cm
350	191	710	1033	4	0.545	20

*Used in compression and bond tests.

TABLE 8-7 Composition of Bentonite Slurry and Strength of Core Samples (Mix of Table 8-6) (from Nippon Co., 1975a)

Sample	Materials used	Proportions, %					28-day strength, kg/cm^2
		Bentonite	CMC	Telnite	FCL	Telmarch	
A	Air (dry)	0	0	0	0	0	
B	City water	0	0	0	0	0	333
C	Bentonite	8	.....	...	...	...	286
D	Bentonite, CMC, Telnite	6	0.05	0.5	...	...	311
E	Bentonite, CMC, FCL	6	0.05	...	0.5	...	310
F	Bentonite, CMC	6	0.05	...	...	...	310
G	Telmarch	...	.....	...	...	1.0	310

Results from Field Tests Table 8-8 shows data for concrete used in actual structural work at six different building sites. Tests on standard cylinders from site A showed a 28-day strength of 284 kg/cm^2 near the top and 336 kg/cm^2 near the bottom of the structure. Standard-cured samples of the same concrete showed a 28-day strength of 315 kg/cm^2. Samples at site B tested in compression showed a 28-day strength of 325 and 366 kg/cm^2 near the top and bottom, respectively, compared with an average 28-day strength of 251 kg/cm^2 for standard-cured samples. For site C the results of 15 compression tests showed a compensated strength (measured strength multiplied by a strength-correction factor) of 210 kg/cm^2, compared with a theoretical value

TABLE 8-8 Proportions and Properties of Concrete from Field Sites per Cubic Meter*

Site	Cement, kg	Water, kg	Sand, kg	Gravel, kg	Air-entraining agent, kg	Water-cement ratio	Slump, cm
A (building)	350	162	1073	754	4.3	0.46	19
B (building)	284	182	702	1191	....	0.64	20
C (subway)	320	176	775	1023	1.3	0.55	20
D-F (buildings)	340	170	†	†	†	0.50	19

*From *Jap. Concr. J.*, vol 9, no. 6, 1966.
†Not given.

of 200 kg/cm². The results of tests on samples obtained at sites D, E, and F are shown in Fig. 8-10. Although in this case the variation in strength is not pronounced, there is a slight tendency for strength to improve with depth. Thus, the record tends to show that the in situ strength of concrete placed in slurry trenches usually increases with depth, as shown in Fig. 8-11 (Ikuta et al., 1973). This generally is valid for upward flow, whereby the fresh mix is pushed up from below like a plug; it can be attributed to the consolidation effects which are intensified in the lower part of the wall as large quantities of concrete are placed at one time.

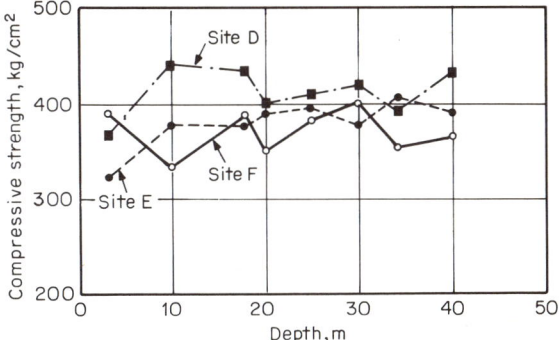

Fig. 8-10 Twenty-eight-day compressive strength of field samples. Sites D, E, and F of Table 8-8. (*From Jap. Conc. J. vol. 9, no. 6, 1966.*)

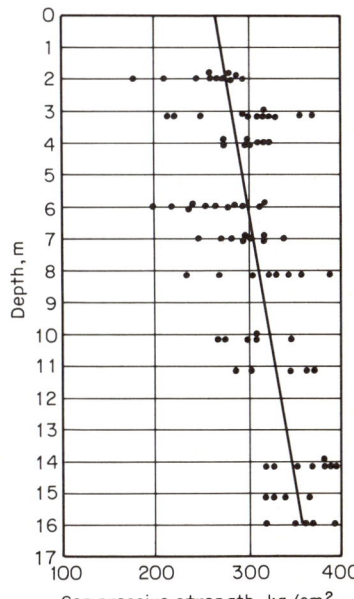

Fig. 8-11 Variation of in situ concrete strength with depth for diaphragm walls. (*From Ikuta et al., 1973.*)

Recommended Design (Working) Concrete Stress

From the present record of field-cured structural concrete it is evident that the attained strength is adequate provided the placement is carried out smoothly and within the procedure indicated for the tremie method. Further, the development of strength is aided by the moist-curing conditions which usually prevail at the site and compensate for unfavorable effects. However, interpretation of this record should take into account the actual influence of the average construction conditions on the anticipated strength, and in this respect the effect of factors limiting strength is not always controlled to the extent desired.

Thus, until authoritative bodies review the matter of strength and make more specific recommendations, it is better to take a conservative stand in choosing the working stress, particularly in permanent structures. In general, this author recommends a reduction of 15 percent in the normal working stress to compensate for effects of conditions beyond control. Accordingly, the allowable or computed stress (short members) in direct compression or the stress in extreme fiber on the compressive side of a reinforced-concrete flexural member should be

$$f_c = 0.45(0.85 f_c') = 0.38 f_c' \tag{8-2}$$

which is rounded off as shown in Table 8-9 for the most commonly used values of strength. A similar reduction is also recommended for the allowable or computed stress in concrete of an eccentrically loaded wall or column. For well-supervised jobs f_c can be taken as $0.40 f_c'$.

TABLE 8-9 Allowable Stress for Various Values of Strength

f_c', lb/in^2	2500	3000	3500	3750	4000	4500	5000
f_c, lb/in^2	950	1150	1350	1450	1500	1700	1900

8-6 BOND STRENGTH AND BOND STRESS

Special Considerations Affecting Bond

For plain smooth bars bond resistance is caused by a maximum bond stress over a short length where adhesion is about to fail and a lower friction drag over the length where adhesion has failed. Thus the bond strength depends largely on adhesion, but even after adhesion is broken, friction between the steel and concrete continues to provide bond resistance.

With deformed bars this behavioral pattern is changed. Adhesion and friction still assist the interaction of steel bars and concrete, but most of the bond resistance is provided by the interlocks of the reinforcing bars. The bond strength in this case depends primarily on the bearing of lugs on the concrete and on the shear strength of concrete between lugs (ACI, 1966).

The immersion of steel bars in bentonite slurry can influence the development of bond resistance in the following ways: (1) the slurry first may coat the steel surfaces with a slippery film, partially destroying the adhesion and frictional resistance between the two materials; (2) bentonite and impurities from the excavation may be trapped between and under the lugs; and (3) the concrete may fail to surround the bars completely in certain sections. These effects are governed mainly by the conditions of placement and by the characteristics of slurry, including its gel strength and density.

It is quite unlikely that when the steel cage is immersed in bentonite slurry the latter will deposit a filter cake on the steel surface since this would require a filtration process. Instead it is conceivable that any adhesion between the steel and the bentonite coating is of the order of the gel strength of slurry, which is comparatively low in relation to the shearing stress induced by the rising concrete. Owing to its granular composition and inherent friction, as the concrete moves upward it performs a sweeping action which tends to remove the bentonite coating from around the bars. In normal hydration this coating is replaced by the cement particles, which cover the reinforcement and create normal adhesion. At present it is not explicitly clear how and when the bentonite coating is completely swept or absorbed, except as noted in previous sections.

In a mixed flow it is possible for some bentonite to remain and continue to envelop the steel bars. If the slurry contains too much slime, some impurities will probably be squeezed and remain trapped between the lugs. If the injected concrete does not produce the thorough displacement of all residual materials, their presence in the final structure can influence the development of bond. Deformed bars can be more vulnerable than plain ones since they collect more impurities, but this should not be the basis for comparing the performance of the two bar types in bond under the slurry-trench conditions.

Bond Tests

The first systematic investigation carried out on bond was presented (CIRIA, 1967) following tests on specimens cast in conditions simulating diaphragm-wall construction. The specimens in this case were reinforced-concrete beams tested in bending to produce bond failure over a 6-in length of bar at the lower end as cast.

Results from these tests are shown in Fig. 8-12a and b for mild-steel plain round bars and for deformed bars, respectively. For the plain bars the bond-stress–bar-slip curve shows no significant differences between the conditions of injection. Bond stresses are higher for the hand-punned concrete but within the basic range of results. For all conditions and specimens there is little increase in bond stress between 0.001- and 0.01-in bar slip (0.025 and 0.25

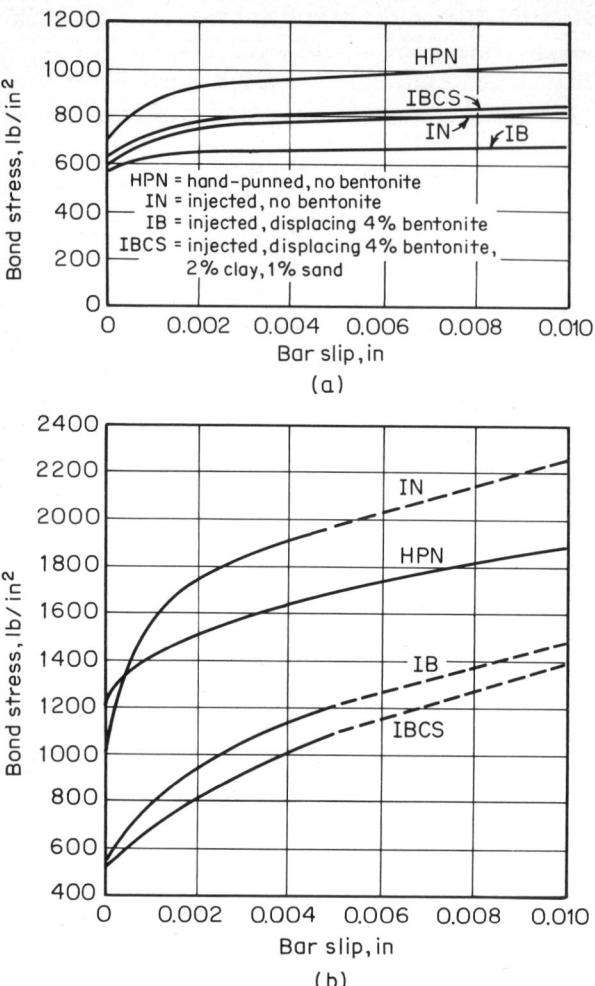

Fig. 8-12 Results of bond tests: (a) $\frac{7}{8}$-in-diameter mild-steel round bars; (b) $\frac{7}{8}$-in-diameter high-tensile-steel deformed bars. (CIRIA, 1967.)

mm), and for all practical purposes the maximum bond stress is reached at a bar slip of 0.002 in (0.05 mm).

For the deformed bars there is considerable reduction in bond stress with bentonite or bentonite-clay-sand slurry. However, there is also considerable increase in bond stress with bar slip, and this increase seems to hold for bar slip 0.01 in or higher (0.25 mm). An interesting observation is that for 0.001-in bar slip the average bond stress of plain and deformed bars in bentonite is nearly the same. However, the average bond stress of deformed bars is considerably higher than that of plain bars as bar slips approach or exceed 0.005 in.

Pullout Tests Results from pullout tests carried out according to ASTM C234 are shown in Fig. 8-13. These tests were for the concrete mix of Table 8-6 in conjunction with the slurry of Table 8-7. The bars were of the deformed type and 19 mm in diameter (similar to ASTM no. 6 bars) and were immersed in the thixotropic slurry for 24 h. Subsequently the concrete was tremied and allowed to cure for 28 days. During the pullout test the loading was continued until a slippage of 0.25 mm (0.01 in) occurred at the loaded end.

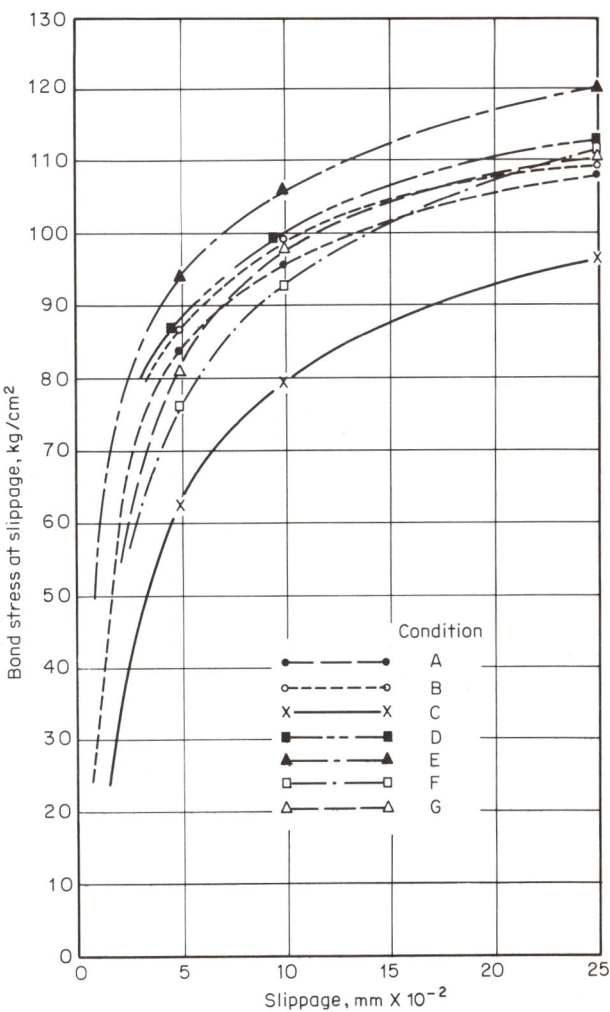

Fig. 8-13 Results of bond pullout tests (ASTM C234) for dry concrete and for concrete placed in bentonite slurry. Conditions A to G correspond to notation of Table 8-7. (*Nippon Co.,* 1975a.)

Other recent tests were conducted by the Japanese Highway Public Authority (1975) on deformed bars for concrete cast in conditions simulating plug flow. The concrete was placed in dry panel, panel containing plain water, and panels containing bentonite slurries at 2, 8, and 12 percent concentrations. The tests were carried out 8 days after placement, and at the end of that period the concrete attained an average compressive strength of 277 kg/cm². The bond-stress–bar-slip relationship is shown in Fig. 8-14. It can be concluded from the result of Figs. 8-13 and 8-14 that tremied concrete results in some loss of bond strength, and this loss is greater with increased bentonite concentration, i.e., increased gel strength. Peptizers appear to moderate these effects on the bond, probably by reducing the gel strength of the film enveloping the bars.

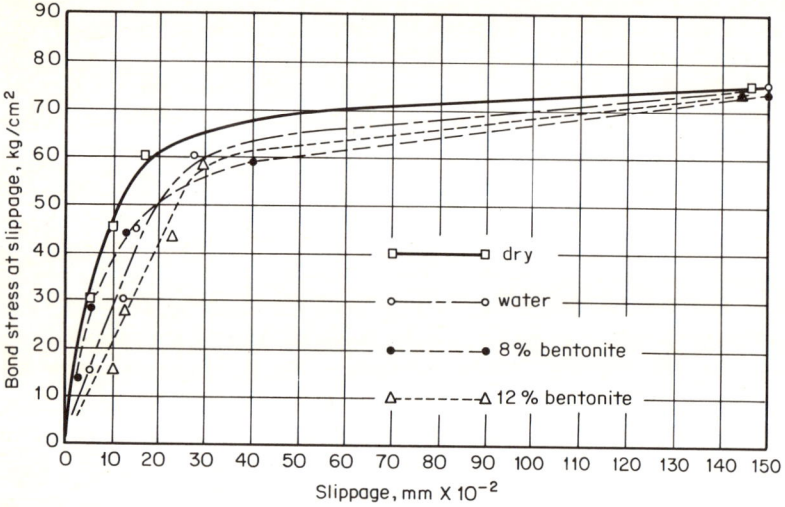

Fig. 8-14 Results of bond tests for deformed bars (19 mm diameter); dry concrete and concrete placed in plain water and bentonite slurry. (*From Japanese Highway Public Authority, 1975.*)

The bond was also shown to vary considerably with the colloid content in tests conducted on the prototype dry and bentonite concrete of Table 8-5. In this case vertical and horizontal bars were tested for slip and ultimate bond in both plain and deformed types. The results of tests for the horizontal bars are shown in Fig. 8-15. It is evident that the bond stress decreases with increasing bentonite concentration and that deformed bars have a higher average usable stress in bond resistance. These results also provide the average range of bond stress for both vertical and horizontal bars for the standard slippage of 0.25 mm (0.01 in), summarized in Table 8-10, which again shows the higher effectiveness of deformed bars. These conclusions are valid for both vertical and horizontal reinforcement.

The concrete is better consolidated above the horizontal bars than below them. For normal cast-in-place concrete the rule is that the more concrete below a bar the greater the resulting loss of bond strength, and any bleeding of

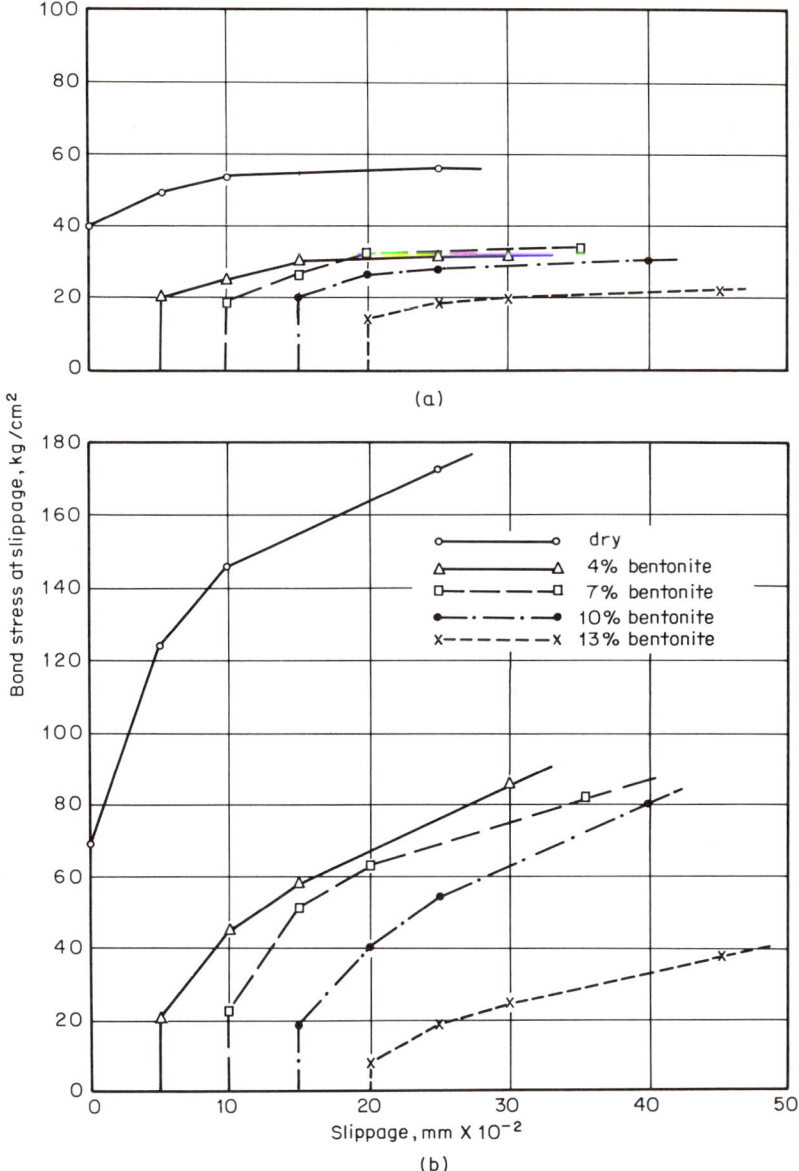

Fig. 8-15 Results of bond tests for horizontal bars (concrete of Table 8-5): (*a*) plain smooth bars; (*b*) deformed bars. (*From Jap. Conc. J., vol. 9, no. 6, 1966.*)

the mix will tend to accumulate water and air beneath the bar. According to this rule, the bond should improve somewhat for bars away from the top, and this is indeed demonstrated by Fig 8-16, which gives a crude approximation of this variation for horizontal bars.

TABLE 8-10 Bond Stress for Slippage of 0.25 mm (0.01 in), Deformed and Plain Bars
$f'_c = 300$ kg/cm²; vertical and horizontal bars, concrete of Table 8-5

	u	
Concrete sample	Deformed bars	Round plain bars, kg/cm²
Dry	$0.35f'_c$–$0.50f'_c$	45–55
Bentonite, 4%	$0.17f'_c$–$0.20f'_c$	21–22
7%	$0.13f'_c$–$0.18f'_c$	16–43
10%	$0.09f'_c$–$0.19f'_c$	20–33
13%	$0.07f'_c$–$0.18f'_c$	11–18

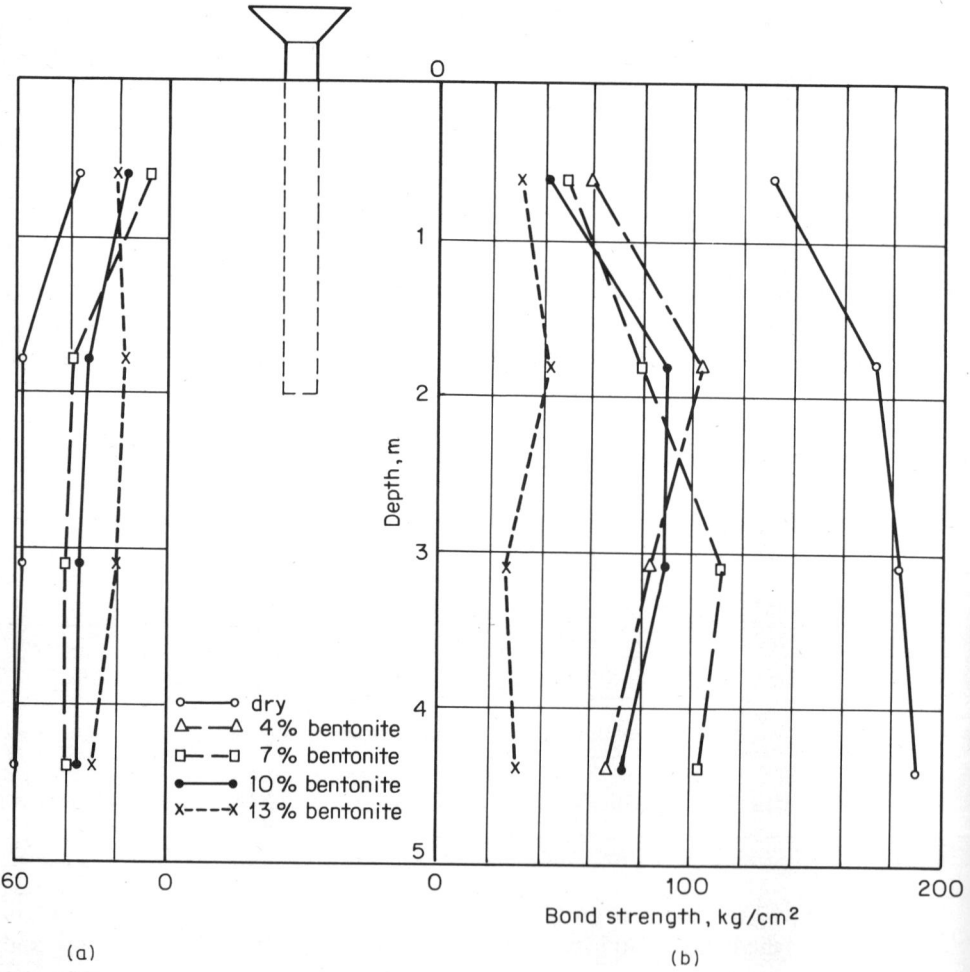

Fig. 8-16 Variation in bond strength of horizontal bars with depth: (*a*) plain round bars; (*b*) deformed bars. (*From Jap. Conc. J., vol. 9, no. 6, 1966.*)

The influence of bentonite slurry on the bond stress is also demonstrated in a report by Ikuta et al. (1973), presenting results of tests conducted to simulate field conditions. The consistency of these results shows that in certain situations the loss of bond strength can be significant for plain bars but is less serious for deformed bars. The report also presents a relationship between bond stress and concrete strength f_c', shown in Fig. 8-17 for deformed bars, and recommends an allowable bond stress 20 percent lower than the stress in conventional construction. The same report discourages the use of plain round bars, particularly in horizontal reinforcement.

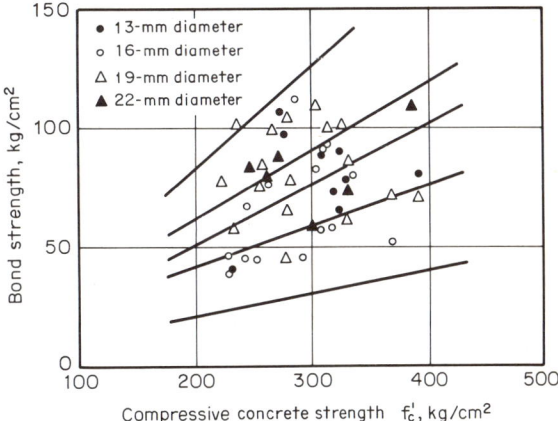

Fig. 8-17 Relationship of bond strength (0.25 mm slippage) and concrete strength. (*From Ikuta et al., 1973.*)

Recommended Bond Stress On the basis of the available data, criteria for an acceptable bond distribution cannot be established. The record thus far has been collected more by sampling than by systematic investigation, and does not permit a specific procedure to be formulated. Bond analysis should therefore be based on a conservative approach. The following guidelines are recommended: for deformed bars use a reduced allowable bond stress (80 percent) and even less if conditions warrant; if possible, avoid the use of plain bars, particularly in horizontal reinforcement; avoid splices, especially at points of maximum stress; if splices must be used, provide generous laps (1.5 to 2 times the normal splice length).

PART B: Construction

8-7 ASSEMBLY AND DETAILS OF REINFORCEMENT

Reinforcing Cages In general reinforcement bars are prefabricated and assembled in cages. If practicable, this is done in the shop, and then the cages are shipped to the site. For relatively deep and wide diaphragm-wall panels, reinforcing cages can reach formidable dimensions, in which case they must be assembled at the site. Occasionally, this can give rise to handling problems since the only practicable position for assembling a cage is the horizontal, from which it is picked up by crane and held in suspension before being inserted in the panel. During hoisting the action of its own weight can cause severe distortion of the cage. When detailing the reinforcement it is therefore necessary to decide whether to make the cage rigid (by additional bracing or by welding the bars) or flexible. In some instances the latter is favored because of economy and because a flexible cage is less liable to suffer permanent distortion during handling. If the bars are large and closely spaced, it is advantageous to make the cage rigid. A distorted cage should never be inserted in a panel.

A special lifting sling and two crane lines can be connected to the cage to prevent excessive distortion. When the cage is almost vertical, the second line is disengaged since it is only used to keep the cage straight during lifting. Alternatively, for very heavy and massive cages the assembly shown in Fig. 8-18 is used. Once in the vertical position, the cage is slowly inserted into the panel and held in place by spacer blocks. A device frequently used is the concrete roller spacer, shown in Fig. 8-19, placed over the outer bars to allow the cage to roll down the trench and also ensure adequate cover for the reinforcement.

A common practice is to place the main vertical reinforcement on the outside of the cage. This, however, will reduce the concrete cover, and in case of bond failure splitting of concrete may precede shearing by the lugs. Despite this disadvantage it is better to facilitate the unobstructed flow along the outer side so that the fresh concrete can fully surround the bars and provide full contact where it is needed most.

Bar Spacing The minimum bar spacing of vertical reinforcement should not be less than 6 in and preferably 9 in. Horizontal bars give minimum impediment to flow if they are arranged in open spacing. The spacing of horizontal bars should never be less than 12 in unless these bars constitute the main reinforcement. A generous concrete cover should be provided, especially for the earth side of the exposed wall, and the minimum clear cover should not be less than 3 in (or 7.5 cm).

(a)

(b)

Fig. 8-18 Method for handling heavy cages: (*a*) the cage is braced to withstand handling and erection stresses; (*b*) the cage is suspended vertically and slowly inserted into the panel. (*Franki.*)

288 Slurry Walls

Fig. 8-19 Concrete roller spacers placed over outer bars to guide cage during placement. (*Franki.*)

It is conceivable that the bond strength may be reduced where laps occur, particularly where several horizontal bars in the form of distribution steel, links, or ties are in contact with each other against vertical splices and where such bars are very closely spaced. In general, it is desirable to have the most open bar spacing that is practicable, and avoid bar arrangements, hooks, and shapes that could trap bentonite, cause cavitation, or otherwise impede the free flow of concrete.

Boxes and Inserts Recesses are mandatory in a wall which is to support future floors. Where reinforcement bars must be used for future mechanical connections, they commonly are placed behind a recess. Shear and bendout bars for future structural connections also require recesses, but they should be avoided at the extremities of a panel, where the flow energy of concrete may not be sufficient to displace the slurry.

Recesses are formed by using boxwork inserts attached to the reinforcing cage. The insert boxes should have nominal dimensions. If they are made too large, they are likely to restrict the free flow of concrete. If possible, they should be placed only on the outside of the reinforcing cage. Any boxwork of a size comparable to the wall thickness will probably cause inclusions of bentonite mud at the edges. No particular difficulties should be encountered in positioning insert boxes if they are robust and firmly attached to the cage, provided the cage itself is accurately positioned and firmly held.

Occasionally an insert box is displaced by the force of rising concrete or lost when it is caught on the small ridge formed where the ground changes vertically. Cages are quite heavy while they are handled, and a brief resistance to motion during insertion does not always register with the crane operator. Accordingly, adequate clearance must be provided for insert boxes after considering the width of trench, the cage dimensions, and the tolerance in the excavation. The required degree of accuracy in the vertical and lateral position of the cage can be maintained if the cage is suspended from its true center of gravity.

Materials for boxing out and recesses are standard. Long horizontal bands of wood are inserted and tied to the cage to form continuous shear keys for floor-slab connections; when the wall is exposed after excavation, the wood is chopped out and the dowels behind are bent out to form the connection. Steel brackets, plates, or similar sections are welded to the cage and provide the connection to structural-steel frames when they are exposed. Very practical and effective is the use of styrofoam planks to form keys and inserts, attached to the cage as shown in Fig. 8-20. They are burned out with a torch when the wall is exposed. Special porcupine plates are used to transmit raker shears and thrusts to the wall. The details of these connections are discussed in subsequent sections.

Splices Clearance during lifting, lifting capacity of equipment available, panel size, and similar construction considerations determine whether one,

Fig. 8-20 Reinforcing cage ready to be inserted. The white marks are styrofoam planks used to form inserts for connections.

two, or more cage lengths are to be used. When splices are necessary, the first length is left just projecting above the trench and the second cage is welded or otherwise attached to it before the entire cage is lowered into position. The splice connection of reinforcing cage shown in Fig. 8-21 consists of a U bolt fastened with two nuts and a link. This bolt can carry a load of 12 tons and is preferred by many contractors to welding the bars.

Fig. 8-21 U-bolt connection used to splice the upper and lower part of the reinforcing cage. (*Franki.*)

If it is practicable and economical to have a cage assembled in the shop but the assembly is wider than can be transported by truck, the problem is remedied by making the cage in two sections, as shown in Fig. 8-22a and b; the two sections are transported as shown in Fig. 8-22a, and when they arrive at the site they are assembled as shown in Fig. 8-22b.

8-8 COMMON TYPES OF CONSTRUCTION JOINTS

In the past years vertical construction joints in wall panels have undergone major changes and continuous modifications. In the early applications plain butt joints were commonly in use, but they were weak in transferring shear while they were conducive to moisture transfer. With the improvement of excavating techniques construction joints were given more attention, and various types were evolved and tested in order to produce connections which would offer resistance to shear and bending stresses and also provide a watertight wall.

When scraping shovels and chisels were used to remove the soil along the concrete end face of poured panels, joint details were relatively simple. They were formed in different ways with shaped forms giving half-round, tongued-and-grooved, encastre, or plain butt joints with a scarified surface. In certain cases adjacent panels were built with a separate joint of plastic concrete. Of these, the half-round joint known as the *interlocking-pipe joint* was proved effective and still is in wide use.

Basic Requirements In general, a construction joint must satisfy the following requirements:

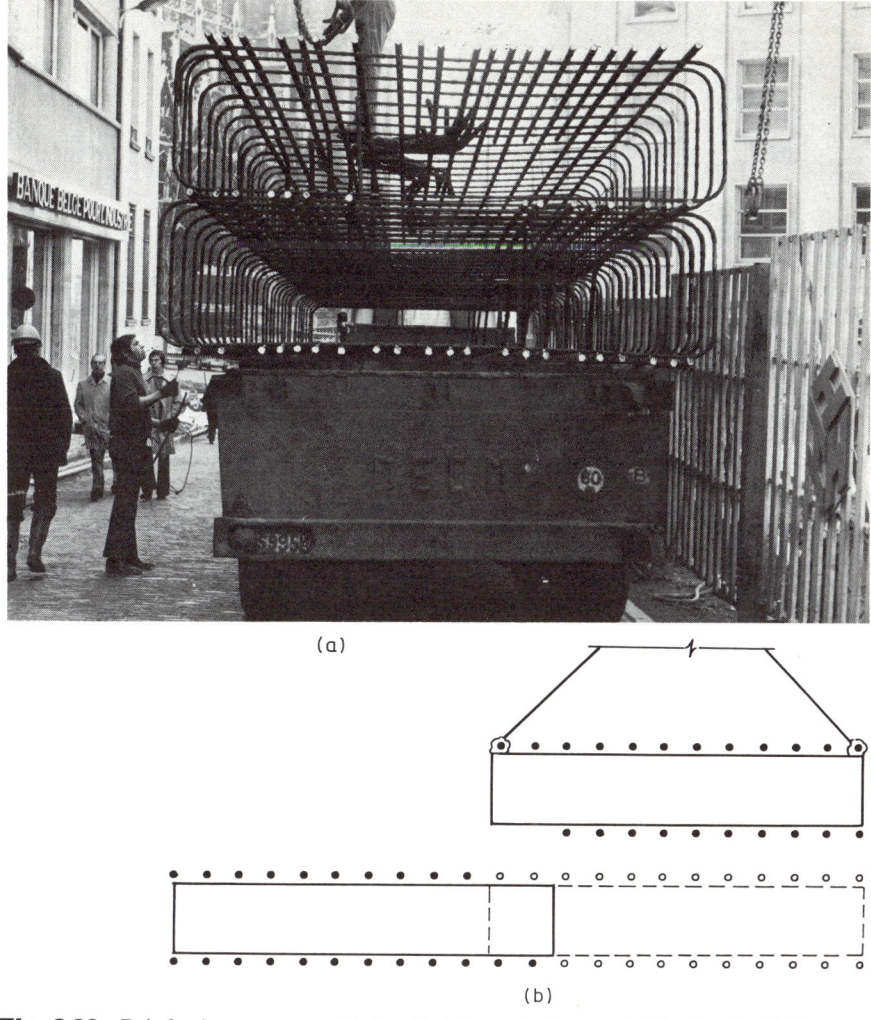

Fig. 8-22 Reinforcing cage assembled so that it can be transported by truck: (*a*) the cage upon arrival at the site; (*b*) diagram showing how the cage is spliced before it is inserted into the panel.

1. Physically the joint should not disturb the previously poured panel. On the other hand, it should accommodate the excavation of the adjacent panel, preferably without restricting the type of equipment.
2. While the joint is executed, no leakage of fresh concrete should occur.
3. The joint forms, plates, and the like should withstand the pressure of fresh concrete without undergoing too much distortion or deformation.
4. The joint should be capable, where required, of transferring shear and other stresses. On the other hand, the joint should be fairly watertight.
5. The joint should collect minimum amounts of slime and bentonite and

should facilitate the cleaning and removal of residual material from hardened concrete surfaces.

6. The construction of a joint should be feasible by simple methods and equipment.

7. Finally a joint should be executed at a cost compatible with the economics of the project.

Common Types of Joints

Round-Tube (Interlocking-Pipe) Joint This is very common and relatively simple to construct. It is a semicircular joint formed by means of a steel tube inserted at one end as a stop for the concrete. Some time after the start of the pour (usually 2 h) the pipe is given a slight rotational movement to break the bond with the fresh concrete. When the pour is completed and the concrete begins to set, the tube is slowly extracted, as shown in Fig. 8-23; initially the extraction is accomplished with the aid of two jacks (Fig. 8-23a) each applying the necessary pulling force (sometimes more than 200 tons if the pipe is left in too long), and then with the use of a crane, as shown in Fig. 8-23b. Removal of the tube leaves a half-round concrete key at the end of the unformed panel, which is used to guide the excavating tool as construction continues. Any deviation from the true vertical alignment of the completed panel is thus introduced to the next panel and is corrected in the next equipment pass. Theoretically, the outside diameter of the tube should be the width of trench; hence some difficulties may be encountered when inserting the pipe if the trench is not sufficiently straight.

The round-tube joint is practicable for excavations carried out with rotary drilling equipment and round-end clamshells. The connection is good for transferring shear, but it cannot transfer bending stresses since the reinforcement does not go through the joint.

If cavitation or overwidth excavation occurs at the end, the fresh concrete may flow around the stop-end tube, as shown in Fig. 8-24, and must be broken with chisels when excavating the adjacent section. This can be time-consuming and expensive and requires a special round chisel if the concrete has hardened excessively. If such concrete rings are not detected and removed, they can force the excavating tool out of alignment and result in a defective panel.

Steel Plate and Vinylon Sheet This detail allows the horizontal reinforcement to be extended through the joint. Usually, therefore, the joint is located near the center of the panel or some distance from the ends. A steel plate is welded to the cage, as shown in Fig. 8-25a, to provide a barrier between the concrete and the slurry section. Chemical textile sheets (usually vinylon) surround the ends of the cage, as shown in Fig. 8-25c, and also along the bottom to provide maximum protection against leakage of concrete beyond the partition plate. The horizontal reinforcement is extended through holes in the plate and is spliced with the steel cage inserted in the adjacent side. A properly stretched sheet produces a smooth face.

(a)

(b)

Fig. 8-23 Extraction of interlocking pipe by means of (a) jacks and (b) crane. (*Franki*.)

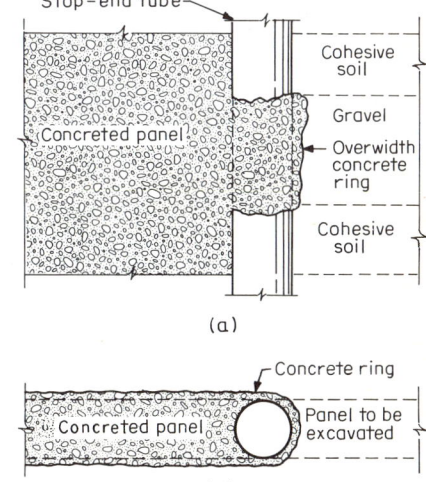

Fig. 8-24 Penetration of fresh concrete beyond the stop-end tube due to overwidth excavation: (a) partial elevation; (b) partial section through gravel layer.

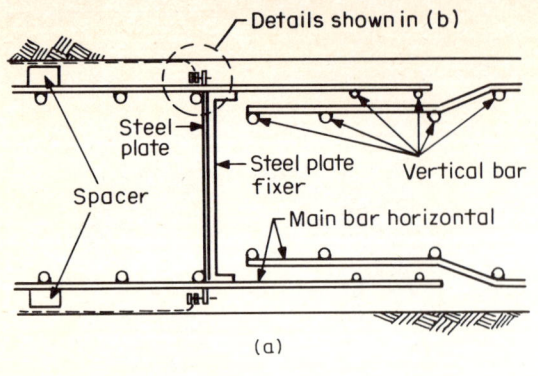

(a)

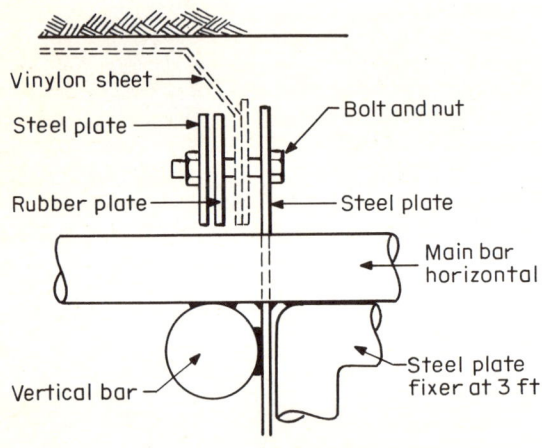

(b)

(c)

Fig. 8-25 Vinylon-sheet and steel-plate joint. (*a*) General joint detail and splicing of horizontal reinforcement. Note that the interlocking pipe is not shown. Concrete is poured against the vertical steel plate. (*b*) Detail showing the attachment of vinylon sheet to the steel plate. (*c*) The reinforcement cage is lowered into the trench; vinylon sheets are extended 3 to 4 ft to prevent concrete from leaking into the next panel.

Although this detail is relatively simple, its execution requires some previous experience and adequate field control. If the combination of the plate and the vinylon sheet does not separate the two chambers completely, the fresh mix can leak and fill the slurry section, particularly with high-slump concrete. If holes are accidentally punched on the sheet during handling of the cage and escape detection, they can create potential leakage paths. The cage should be held firmly at the guide walls or by other means since the pressure of fresh concrete on the partition plate can move it off the true position or the plate may suffer excessive distortion.

Modified Round-Tube Joint This detail is shown in Fig. 8-26. A round steel tube and a corrugated plate are attached to the cage, as shown, to prevent direct contact between concrete and the round tube so that the extraction of the tube does not depend on the hardening process and setting time of concrete. The device is useful for avoiding stuck-pipe problems. The teeth of the exterior plate provide interlocks for the two concrete sections and also increase resistance to seepage flow provided bentonite and slime do not accumulate in between.

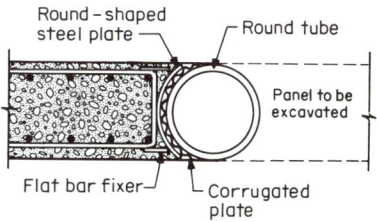

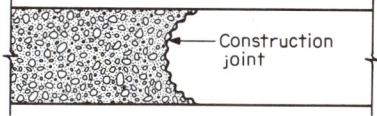

Fig. 8-26 Modified round-tube joint.

The same joint can also be executed using a V-shaped plate inserted with the cage and bearing against the round tube. This again is intended to facilitate the withdrawal of the tube provided concrete does not leak into the annular space.

I-Beam Joint This detail, shown in Fig. 8-27, is generally used in composite steel and concrete panels. It is particularly suited to trenches excavated with square-end clamshells. In composite walls the beam is used to make structural connections. When it is used in conjunction with the round-tube joint, as shown in Fig. 8-27, care should be taken that concrete does not leak on the other side of the web line.

RPT Joint This was initially developed for a special project in England by Randel, Palmer, and Tritton, but the joint can be used on a general basis. The conventional steel end tube is combined with sections of straight web piles

296 Slurry Walls

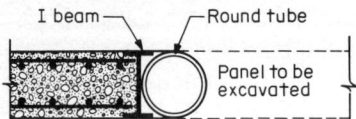

Fig. 8-27 I-beam joint details.

attached to the cage and incorporated into the concrete panel, as shown in Fig. 8-28a. After the concrete has set, the split pile provides a recess at the end of the section into which one of the clutches of the straight web pile protrudes. When the end pipe is extracted and the next panel is excavated, two sections of straight web pile are sunk, as shown in Fig. 8-28b, and linked with the web of the first panel. The connection can therefore transfer shear and axial tension.

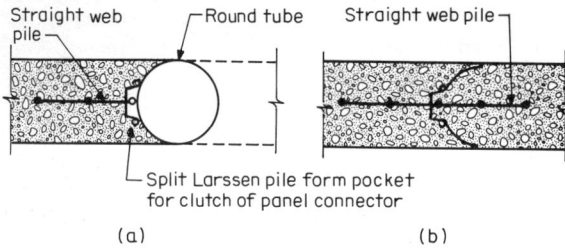

Fig. 8-28 RPT joint details.

Keyed and Water-Stop Joints These are used mainly to ensure better contact between adjoining panels and improve the watertightness of the joint. The construction involves the creation of a vertical cavity in the same alignment and between adjoining panels as shown in Fig. 8-29. One or two key

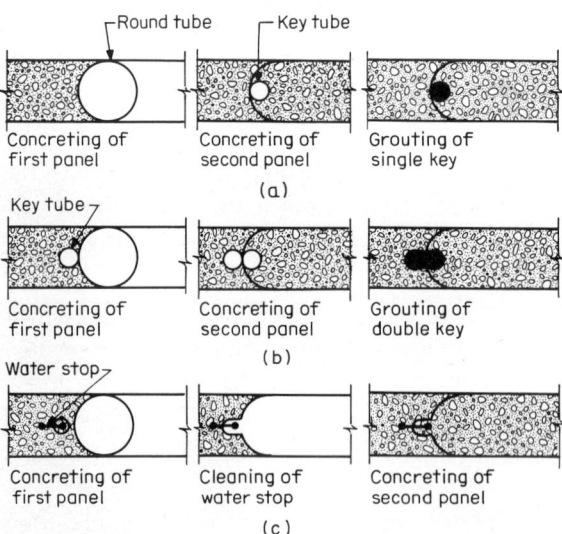

Fig. 8-29 (a) Single key joint; (b) double key joint; (c) water-stop joint.

tubes can be inserted with the main end tube. When the adjoining panels are completed and the concrete has set, the cavity formed by the key is thoroughly cleaned and grouted. A water stop may be inserted following the sequence shown in Fig. 8-29c.

8-9 SPECIAL CONSTRUCTION JOINTS

With the exception of the steel-plate joint, the details described in the foregoing section and modifications of them are used for diaphragm-wall panels in which the vertical steel is the principal reinforcement. When the wall must resist the lateral loads horizontally, however, these construction joints cannot be used since they do not provide the required structural continuity in that direction.

For such situations special joints have been developed and introduced by researchers and major specialist contractors. Their success and effectiveness depend on how well they are executed. Thus an adequately designed joint can become defective if improperly constructed and thereby fail to provide the intended transfer of load or stress. On the other hand, an improperly executed joint may not be watertight to the degree desired. Theoretically a plain keyed joint is satisfactory in resisting and transferring lateral loads of a magnitude equal to the shear strength of plain concrete at some critical section. A joint with a mechanical connector, e.g., the web extension of the RPT joint, additionally resists axial forces along the wall and transfers vertical loads; hence it is useful in preventing laminar tilting of the panels. Reinforcement extended and spliced into adjoining panels provides all these functions and also transfers bending stresses.

Despite the theoretical adequacy of a joint, its in situ performance and effectiveness are a different matter, and engineers should be cautioned that the joint is only as good and adequate as built. This should be the principle to follow when considering the special joint details described in this section.

Steel-Plate-and-Casing Joint

This joint, executed in the sequence shown in Fig. 8-30, presumably eliminates the problems sometimes caused by concrete leaking to the slurry chamber with the steel-plate and vinylon-sheet joint of Fig. 8-25. The partition steel plate and vinylon sheet are attached to the reinforcing cage as before, but in addition two specially shaped steel blocks or casings are inserted in the slurry section, as shown in Fig. 8-30a. Block A serves as barrier for concrete that tends to leak toward the slurry panel but also creates a chamber into which the horizontal bars can protrude. Block B is also a barrier that stops leaking concrete, and because it fits tight at the end of the excavated panel, it prevents the reinforcing cage from being displaced or moved out of position. When the concrete is poured, the panel is as shown in Fig. 8-30b. When the poured section has sufficiently hardened, the blocks are withdrawn and the panel is completed, as shown in Fig. 8-30c.

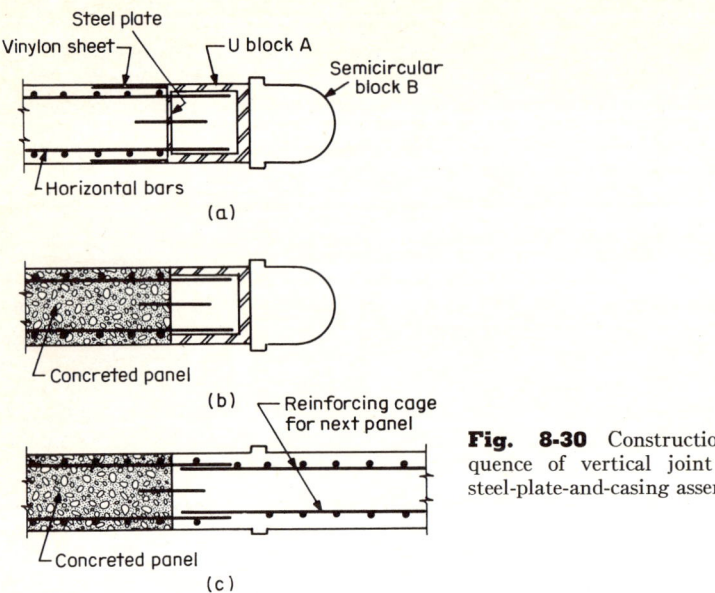

Fig. 8-30 Construction sequence of vertical joint using steel-plate-and-casing assembly.

The joint is relatively more expensive than the simple steel-plate joint because of the extra appurtenances and the longer installation time required. The blocks have greater contact area with the surrounding soil than the conventional round tube and usually are inserted with vibrohammers. The driving power should be limited to the power necessary to overcome skin friction. If the latter is unusually high, some difficulties may be encountered as the boxes are withdrawn from the panel; in this case considerable jacking forces may have to be applied. If despite all precautions and care concrete still leaks, it may be quite difficult to extract the boxes.

Casing Joint by Franki

The special steel casing joint shown in Fig. 8-31 has been developed and used by Franki in horizontally continuous reinforced-concrete diaphragm walls (Baar, 1971). Although in principle it is similar to the joint detail of Fig. 8-30, it is executed differently. The joint allows the horizontal bars to protrude into the unfinished panel but also gives them full protection from fresh concrete during pouring and from slime and bentonite impurities when the next panel is excavated. Full-scale in situ tests have shown that the bond with the concrete along the contact area and around the shear key can be broken without serious difficulties, and the casing can be withdrawn even several days after pouring by applying normal pulling force.

The casing is made to have two deep slots, or housings, that receive the horizontal bars; the depth of the slots is obviously determined by the required bar-splice length. The slots are wider at their ends than at the entry to allow a vertical spacer bar to be inserted in order to keep the horizontal bars in the

Concrete Technology and Construction 299

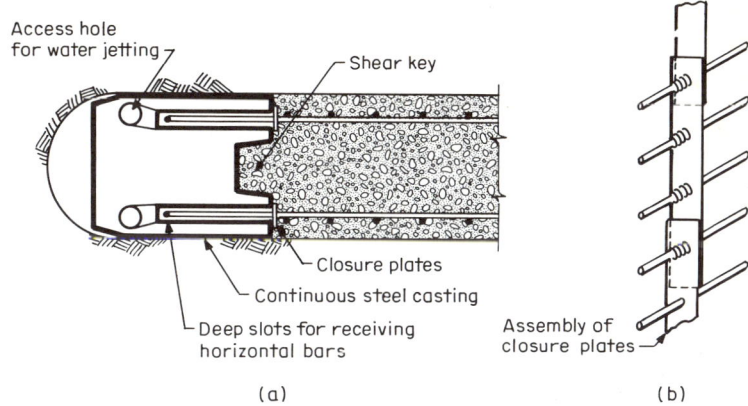

Fig. 8-31 Steel-casing joint developed by Franki. (*From Baar, 1971.*)

right position. This spacer bar also guides the next cage and facilitates its introduction into the panel without distorting or displacing the dowel bars. The earth end of the casing can be shaped as shown in order to reduce the contact area with the soil and permit extraction with less pulling force. Two holes at the ends of the slots, usually 10 cm (4 in) ID, allow water jetting to be forced in to clean the bars after concreting. The protruding bars are then kept in clean water.

The housings are separated from the fresh concrete during pouring by means of plates, as shown in Fig. 8-31a and b, perforated at intervals corresponding to the bar spacing. The plates are cut in suitable lengths, assembled, and overlapped, as shown in Fig. 8-31b. The pressure of the rising concrete pushes the steel plates one at a time firmly and tightly against the edges of the slots, and this prevents the fresh mix from entering the housings. The casing is made so that it can be left in place for several days while the next panel is excavated. Its presence affords the splice bars a protective shelter against the action of grabs, chisels, and other tools and prevents the intrusion of bentonite.

Installation The casing is installed as soon as the panel is excavated and before the reinforcing cage is inserted. After the first panel is concreted and the second panel is excavated, the casing is extracted and placed at the end of this panel. The casing should be entirely vertical and set at the correct distance from the end of the concreted panel and within the specified tolerance. The reinforcing cage should fit loose and should never be forced into the panel.

Figure 8-32 shows a construction joint after the casing has been removed. It is evident that the shear key is as accurate as in conventional formwork, and the bars appear entirely clean and free of any residual material. Baar (1971) recommends a minimum width of trench of 80 cm (about 32 in) for walls of average depth, but the detail is better executed if the wall thickness is 100 cm (40 in) or more.

300 Slurry Walls

Fig. 8-32 Construction joint executed as shown in Fig. 8-31. The joint is seen after the casing is removed; note the shear key, the horizontal splice bars, and the vertical spacer bars. (*Franki.*)

Locking Box by Takenaka

From the preceding discussion of construction joints and problems related to their execution it is evident that among the prime objectives when designing these details are the method of load transfer, particularly for the complex loading conditions that often must be considered in diaphragm walls, and the prevention of concrete leakage. Although the transmission of shear, axial load, moment, and torsion is theoretically a design problem, in reality it depends on the actual load-carrying capacity of the connection. Concrete leakage often is assumed to be the contractor's problem, but it can seriously affect the strength of the joint and therefore the structural integrity of the wall.

For permanent walls the unit lengths between joints generally should correspond closely to those established by the design. The joint connector should be placed accurately with regard to the next unit, especially in walls which are structurally continuous. During pouring every precaution and step should be taken so that no leakage or flow-around of fresh concrete occurs. Experience has demonstrated that if this happens, the penalty for the restoration of the joint usually is a high cost and a doubtful performance.

These considerations have resulted in the development of a construction joint used with the TBW method discussed in Chap. 5. A partitioning device called a *locking box* is used to form the connections in continuous walls. The process is shown schematically in Fig. 8-33. The locking box is fabricated using two steel sections, usually channels placed back to back, each receiving and holding two sealing hoses, as shown in Fig. 8-34. The assembly is lifted and held in a vertical position, from which it is slowly inserted into the joint

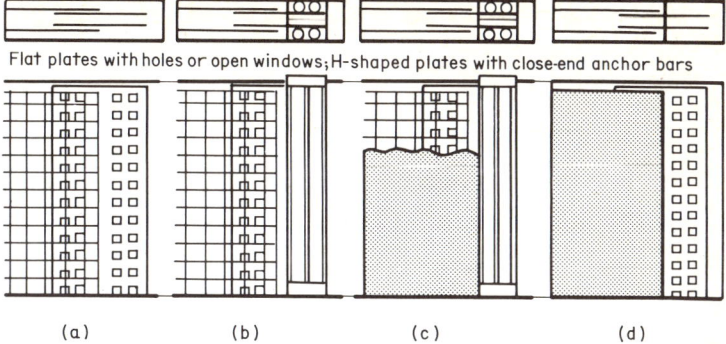

Fig. 8-33 Locking-box joint: (*a*) insertion of reinforcing cage and steel connection plates, which in this case are provided with holes or open windows; (*b*) setting of the locking box and inflating the sealing hoses; (*c*) pouring concrete; and (*d*) withdrawing the locking box.

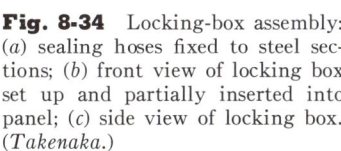

Fig. 8-34 Locking-box assembly: (*a*) sealing hoses fixed to steel sections; (*b*) front view of locking box set up and partially inserted into panel; (*c*) side view of locking box. (*Takenaka.*)

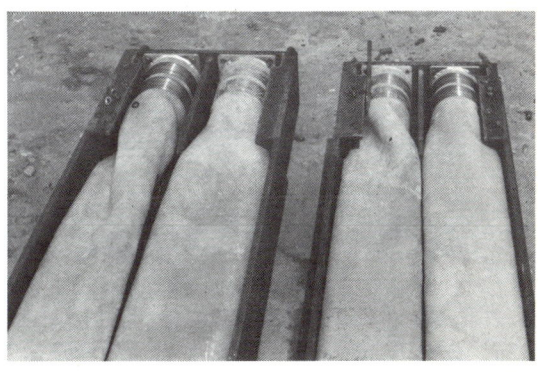

(a)

(b)

(c)

location so that the four hoses fill the entire trench width, as shown in Fig. 8-33b and c. When in place, the hoses are inflated and expand to give a tight barrier beyond the concrete line, and this is achieved even in irregular excavation.

The connection can be executed either using the perforated plates of Fig. 8-33 or with the steel I beam and bent bars shown in Fig. 8-35. Each of these details is intended to accommodate possible loading conditions of shear and axial load.

Fig. 8-35 Construction joint in continuous reinforced-concrete wall executed with steel beam and U bars. The joint is shown after the locking box has been removed. (*Takenaka.*)

8-10 EFFICIENCY OF CONSTRUCTION JOINTS

Common Joint Defects

Three common defects of the round-tube joint are shown in Fig. 8-36. In Fig. 8-36a bentonite mud and impurities are trapped between panels. In Fig. 8-36b the edge of the concreted panel was not thoroughly cleaned, and slime was left between panels, resulting in a cavity and causing a weak zone. The same problem caused the partial cavity shown in Fig. 8-36c. These defects influence the efficiency of the joint by causing more seepage than permitted through the joint, by reducing the effective cross-sectional area which resists the lateral loads and transfers shear, and by making the wall more prone to differential movement.

Bentonite mud and impurities pushed toward the joint and collected there during excavation and concrete placement are removed and swept by the rising concrete better when they are against smooth-metal surfaces than when they are against the relatively rougher concrete face. Thus, any joint executed

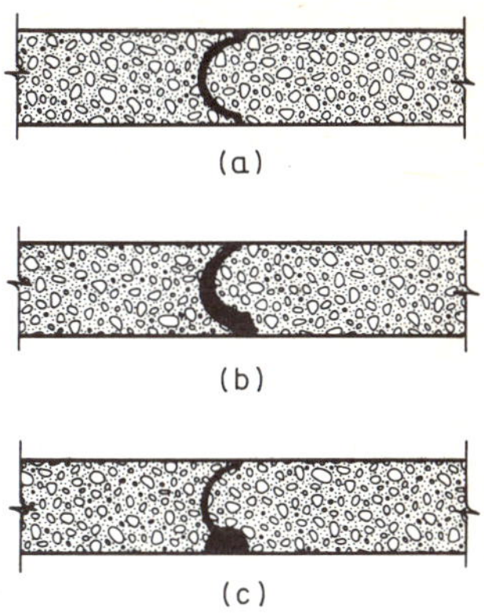

Fig. 8-36 Common defects of construction joints: (a) bentonite mud and impurities left between concrete; (b) cavities caused by soil not completely removed from the end of concrete; (c) partial cavity due to soil trapped between concrete.

with a steel plate is likely to have a lower percentage of mud trappings. On the other hand, the problem presented in Fig. 8-24 can occur despite the most elaborate precautions to avoid leakage or flow-around of fresh concrete. Any time the difference between the trench width as excavated and the diameter of the end tube is more than 3 in (7.5 cm), cement paste and even aggregate from the concrete will seep out and around the pipe.

Tests on the Transfer of Load

In certain situations the actual load-carrying capacity of a vertical connection may be in doubt, and thus the question will arise whether this would require a test. Properly devised and performed tests may serve in place of any design procedures or other recommendations, but the costs involved seldom are justified. Certain relevant tests have been carried out under research programs for both the design loads and for combinations of all sources of stress to which a joint may be subjected, and the results from these tests can be used as guidelines.

Transfer of Vertical Load Tests (University of Osaka, 1971) on diaphragm walls used as load-bearing elements show that a plain round-tube shear joint is greatly improved in its ability to transfer vertical load if it is provided with dowel anchorage. In these tests a steel ball weighing 2 t was freely dropped from a certain height onto a diaphragm-wall panel, and during this process observations were made on the decline in the transfer of energy and shock waves through the joint to the adjoining panel. For effective joints the energy was transmitted as in continuous, monolithically cast walls. Conversely, changes in shock-wave amplitude indicated that a portion of the wave energy had dissipated and was absorbed at the joint.

Figure 8-37 shows the energy transfer through three joints. The joint in Fig. 8-37a is open, i.e., filled with premolded material or bentonite and other impurities. In Fig. 8-37b since the joint has full contact with the concrete sections, it depends on certain frictional resistance. The joint in Fig. 8-37c is provided with dowels.

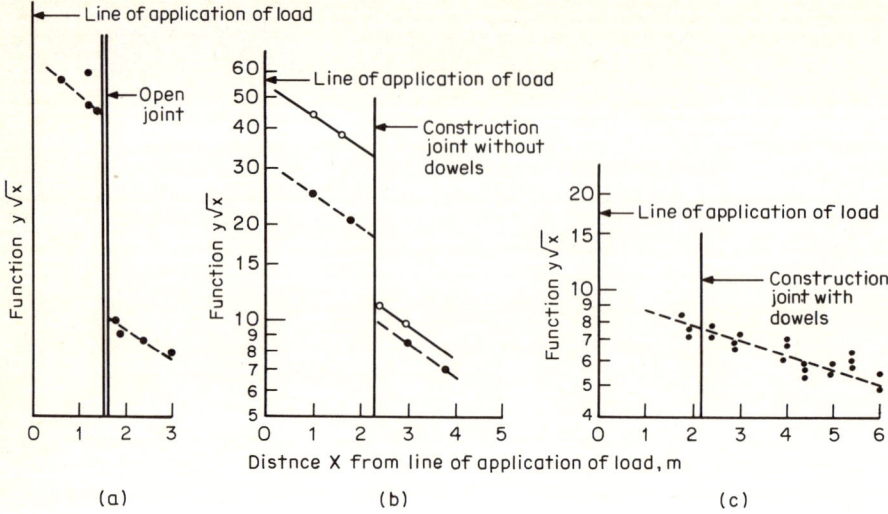

Fig. 8-37 Energy transfer across construction joints. (*From University of Osaka, 1971.*)

If the wave amplitude y is multiplied by the square root of the distance x from the point of application of load and then plotted on a logarithmic scale while the distance x is plotted on an arithmetic scale, the function will approach a straight line. Thus we see that the wave-transfer line is discontinued in Fig. 8-37a, where a serious energy drop occurs at the joint. In Fig. 8-37b a partial transfer of wave energy is attained, while a considerable fraction of energy dissipates at the joint. In Fig. 8-37c the full energy is transferred across the joint.

Transfer of Direct Axial Load (Tension) Examples of connections which allow the transfer of direct tension across the joint are the straight plate of the RPT joint and dowel bars. These devices also prevent sliding of a panel with respect to an adjoining one, thus increasing the resistance of a wall to laminar tilting.

Direct-tension tests on concrete sections taken from diaphragm-wall panels have been performed by Ikuta et al. (1969b). The connections consisted of dowel bars, plain-steel rectangular plates, tapered plates, and plates punched to provide slits. The last two types are thought to increase the ultimate resistance of the connection by direct bearing against the slits or by forcing sliding to occur along a tapered surface. The connection devices actually used in these tests are shown in Table 8-11, together with results from the tests.

If we consider type B as the reference connection, the average bond stress on the steel plate is $\tau_b = 8.33/[2(10.35)(50)] = 8.05$ kg/cm². This can be taken as the basic stress and applied to other types. For type C the total bond area (excluding the slits) is about 849 cm², and therefore the actual increase in the

TABLE 8-11 Various Types of Connection Devices and Results of Tests on the Transfer of Axial Load (from Ikuta et al., 1969b)

Type	Dimensions, mm, and shape	Average initial cracking load, t	Ultimate load, t	Comparison with A type, %
A	1000, 16-mm ⌀ bar	6.5	10.35	100
B	500, Thickness = 3.5, 100	6.3	8.33	80.4
C	25, 50, 25, 100	6.32	9.15	88.4
D	60, 100	6.4	7.77	75
E	60, 100	6.1	9.23	89.2

yield strength of the connection is $9150 - 8.05(849) = 2310$ kg, or 2.31 t. If we assume that this increase is due to the single shear of concrete across the slit area, the shear stress is $\tau_c = 2310/100 = 23.1$ kg/cm². Likewise it can be shown that for type D the increase in the yield strength due to the single shear of concrete in the tapered zone is 1050 kg, corresponding to a shear stress $\tau_c = 1050/100 = 10.5$ kg/cm².

Ikuta et al. (1969b) have provided a detailed discussion of the performance of these devices in tension. The main points are summarized below:

1. With type A (dowel bars) initial cracking appeared at the center of the connection when the applied load reached about one-fifteenth the compressive strength of concrete. Thereafter the cracking continued in a longitudinal direction, and eventually the bars underwent pullout failure before fracture could occur.

2. With the B type (plain rectangular plate) cracking first occurred wide open and across the section, although the plate showed no visual signs of

rupture. This might indicate that the bond strength of the plate was reached and exceeded before yielding or rupture, and the plate was subsequently subjected to pullout.

3. For certain C types (plates with slits) rupture was observed at the ends where the smallest effective section actually occurs. The calculated ultimate load at this section is $0.35(5)(4.19) = 7.31$ t. Therefore the effect of bond and slits is to increase this load by $9.15 - 7.31 = 1.84$ t.

4. For the D type no visual rupture was observed on the plate, although the concrete cracked. In some specimens, however, the initial longitudinal construction cracks opened simultaneously, probably due to the tapering effect which forced the concrete to split after the plate slipped.

5. The E type underwent rupture in the smallest section. Again for a calculated ultimate load of 7.31 t and an actual ultimate load of 9.23 t, the contribution from plate bond, tapering, and slitting is 1.92 t, or about 20 percent of the ultimate load.

From these and other tests Ikuta et al. (1969b) concluded that rusted or fabricated plates increase the load-carrying capacity of these connections. Although no set relationship exists between the size and shape of holes, the thickness of the plate, and the effect of flanges around the holes, it is appropriate to suggest that better results are achieved when the surface is rusted, the plates are relatively thick, and the number of holes is greater for the same hole area. The effect of flanges is similar to that of the plate thickness.

Transfer of Shear Most vertical construction joints, including the round-tube detail, generally are adequate in horizontal shear because of the greater than usual effective width of the section. However, for relatively long spans and deep panels braced at the joint the end reactions can be unusually high. In this case the resistance of concrete to shear can constitute an exceptional situation and might have to be considered separately.

A shear connector used frequently in underground structural walls for deep basements is the device shown in Fig. 8-38. Ikuta et al. (1972) have investigated the capacity of this connection in full-scale field tests on panels constructed in slurry trenches. The main conclusion from these tests is that the ultimate shear strength depends largely on the bearing pressure of concrete on the U-shaped bars, and therefore it can be taken as the crushing strength of concrete on the same area. This principle is given as criterion for designing the connection.

8-11 STRUCTURAL CONNECTIONS

Since the introduction of diaphragm walls in permanent underground structures the use of connections to transfer or receive loads from beams, columns, floors, and slabs has become common practice; but because joints and connections directly affect the integrity of a structure in which a diaphragm wall is a

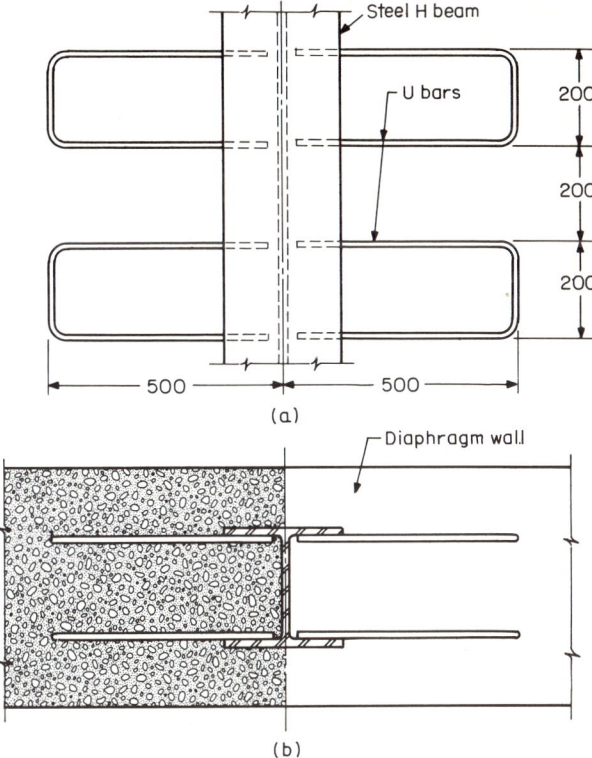

Fig. 8-38 Shear connector consisting of steel H beam and U bars; dimensions in millimeters: (*a*) elevation and (*b*) sectional plan.

part, their design and method of construction must be adequate to ensure the performance of the structure as a unit. Additionally, the strength of a partially or fully completed structure should not be governed by the strength of connections but should be measured by the strength of principal members, including the walls.

Joints and connections can be made by welding steel bars or structural-steel inserts to the main cage; by transferring tensile or compressive stress by bond or anchorage; by using steel plates and angles to prevent separation of the wall from independently supported members; by using key-type devices; or by using bonding media which affect the adherence of one member to another.

This section presents certain types and details of connections which have been used rather widely and found effective. Their load-carrying capacity and performance are indicated from field experience or by properly devised and performed tests. In selecting these connections engineers should take into account their structural characteristics, the feasibility and compatibility of the proposed method with other structural members, and the relative cost.

Transfer of Shear

The transfer of shear to a diaphragm wall can be accomplished by means of bent-out bars used as dowels combined with cast-in-place concrete placed against naturally roughened wall interfaces or by such mechanical devices as embedded plates or shapes, brackets, or other similar sections.

Transfer of Shear from Floors and Slabs For shears from nominal spans and loads the extension of bent-out bars anchored in each connecting member and with sufficient embedment to develop the full yield strength of the bar generally provides adequate means to transfer shear. These dowels require recesses in the wall formed either by horizontal bands of wood or by styrofoam planks. The continuous shear key which is thus provided and the roughened concrete face help transfer shear. No precise theoretical method or experimental basis is available for the design of this connection; thus if the bent-out bars are horizontal, resistance to the applied load is derived from the bearing of concrete on the steel bars and the shear strength of concrete along the key; if the bent-out bars are inclined into the slab, the shear resistance is a function of the tensile strength of the bars adjusted according to their angle of inclination.

Alternatively, a floor slab can be supported on a diaphragm wall, as shown in Fig. 8-39. The vertical connection plate preferably should be continuous rather than intermittent, and for better construction it should be attached to the cage. Its accurate position is therefore quite important. The connection angle is welded to the plate after the wall is exposed.

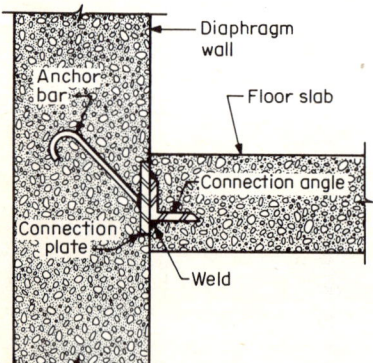

Fig. 8-39 Shear connection, floor slab to diaphragm wall.

The strength of this connection is measured by the shear strength of concrete at some critical section or by the direct bearing on the horizontal leg of the connection angle. The use of a connection plate without sufficient anchorage must be avoided. Likewise, the bearing stress along the underside of the embedded plate should be ignored because of possible accumulation of bentonite trappings at this location. The shear should therefore be transferred to the wall by means of the anchor bars and probably by some bond along the steel-plate–concrete interface.

Transfer of Shear from Beams Although brackets made by fastening suitable steel shapes to plates cast flush with the concrete face can be designed to resist and transfer shear from beams to walls, most connectors consist of steel bearing plates or angles combined with anchor bars, as shown in Fig. 8-40.

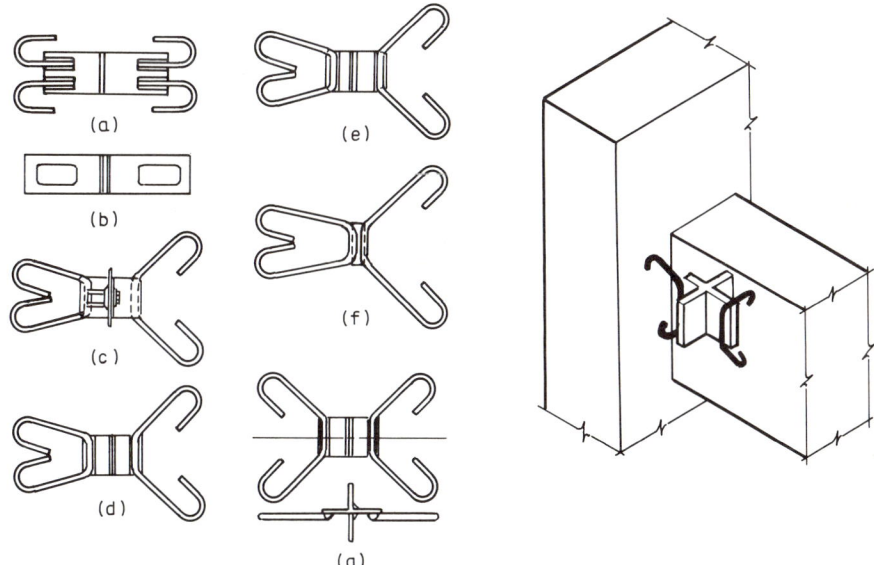

Fig. 8-40 Shear connectors of beam to diaphragm wall, consisting of bearing plates, angles, and bolts. (*From Ikuta et al., 1969.*)

Tests on these connections (Ikuta et al., 1969a) show that in the initial stage of loading slide occurs along the connection face, and as more load is applied, cracking develops in the concrete panel. With further load complicated compressive cracks occur, leading to fracture. Other important points regarding the performance of these connections are summarized as follows:

1. For types that have pressure-bearing plates identical in shape and size but different anchor bars the shear stiffness is the same. The ultimate strength of the connection depends, however, on the type of anchorage because flexure and tension act in the final stage.

2. The ultimate strength (measured by the yielding or displacement) is increased with increased bearing area of the vertical plate. For the average situations, for plate thickness of 1 cm (0.4 in) or greater the ultimate strength of the connection is practically equal to the bearing strength of concrete. Therefore, relatively thick plates are more effective.

3. For the same cross-sectional (bearing) area the use of angles results in higher initial shear stiffness and also ultimate strength.

4. All connections should be shaped and dimensioned to give minimum impediment to the flow of concrete.

Transfer of Moment

Transfer of Moment from Floors and Slabs The transfer of moment through connections between diaphragm walls and concrete slabs usually is accomplished by reinforcing steel extended as dowels. A typical moment connection is shown in Fig. 8-41. The transfer of the corresponding tension by the protruding bent-out bars can be accomplished by sufficient lap, by welding, or by other mechanical devices. Once the bars are bent to the right position and cleaned, the continuity at the joint is obtained as in conventional construction. The most difficult phase in executing the connection probably is the accurate positioning of the connection bars in the cage so that there is minimum eccentricity of force as it is transferred through the connection.

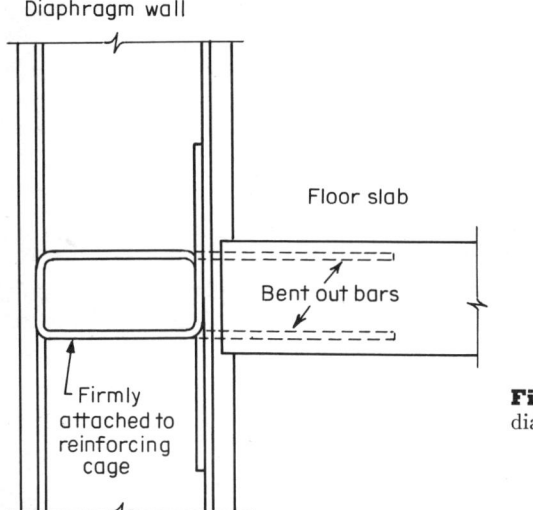

Fig. 8-41 Moment connection, diaphragm wall to floor slab.

Transfer of Moment from Beams This can be accomplished by steel bars extended as dowels or by a composite construction of steel bars, embedded plates, or other suitable steel shapes. Usually the local building codes will insist on considering the entire moment as being transferred through one type of device, although a combination of devices will actually be incorporated in the connection details. In this case the resulting number of bars may be so great that it cannot be provided without crowding the area, thus posing the risk of omissions and cavities in the set concrete and therefore unbonded bars.

Instead of risking this possibility it is better to choose a combination of bars lapped or welded to steel plates and other shapes, to analyze the transfer of moment through the entire combination, and to detail the connection so that there is a minimum eccentricity of force through the joint. If small eccentricities cannot be avoided, the disposition of the laps or welds should be made symmetrical about the centroid of the section if this is practical.

REVIEW PROBLEMS

8-1 Discuss the principal requirements of concrete tremied in slurry trenches.

8-2 Explain how the properties of fresh and set concrete are influenced by the water-cement ratio, entrained air, plasticizers, retarders, aggregate size, and ratio of fine to coarse aggregate.

8-3 In order to provide entrained air, would you use a treated cement, i.e., a cement with which the air-entraining agent has been interground, or add the agent, e.g., Vinsol resin, to the mixture?

8-4 (a) Following the ACI method of proportioning concrete mixes, design a mix using entrained air, a slump of $7\frac{1}{2}$ in, maximum aggregate size $\frac{3}{4}$ in for a compressive strength of 3500 lb/in². State your assumptions. (b) Repeat using non-air-entrained concrete. Compare the cost of the two mixes.

8-5 Explain how the strength characteristics of set concrete are influenced by the method of placement.

8-6 Give the range of compressive strength of concrete you would consider appropriate for (a) general building basement construction; (b) retaining walls; (c) subway sections and traffic underpasses; and (d) load-bearing elements.

8-7 What would your principal objections be to using more than one tremie pipe for concreting a panel? In that case, what would be your alternative?

8-8 Discuss the factors that prevent the strength of concrete in a diaphragm wall from being directly comparable to that of test specimens.

8-9 Discuss the effect of generous cover (3 in or better) on longitudinal splitting of the concrete. Would that prevent premature bond failure of deformed bars?

8-10 For bars immersed in bentonite, would the strength of a bar splice in the set concrete be improved if the spliced bars were in contact or if they were separated?

8-11 Indicate the type of construction joints you would consider suited for (a) general building basement work; (b) cantilevered retaining walls; (c) subway sections; (d) load-bearing walls; and (e) plain-concrete cutoff walls.

8-12 How can soil conditions dictate the choice of the construction-joint details?

8-13 If the coefficient of active pressure of concrete is 0.6 and the fresh mix weighs 145 lb/ft³, what should the maximum specific gravity of slurry be to allow free displacement? Disregard all other factors and assume the free surface of concrete in the tremie to be at the same elevation as the slurry level.

8-14 Explain whether and how the soil conditions can influence the choice of a vertical construction joint. For the construction joints described in Secs. 8-8 and 8-9 give examples showing where a joint may be suitable or unsuitable for the soil conditions.

8-15 Investigate the feasibility of expansion joints in slurry-wall construction and develop a joint detail that would accommodate this situation.

8-16 Summarize the criteria that should be considered in designing and detailing vertical construction joints for the transfer of vertical load, axial load, shear, and bending moment.

8-17 Explain how you would design the inclined anchor bars of the connection shown in Fig. 8-39. State your assumptions.

8-18 By reference to Figs. 8-12 to 8-17 formulate a statement regarding the effect of bentonite on bond strength for plain and deformed bars. Relate the development of bond resistance to bar slip.

REFERENCES

American Concrete Institute, 1954: Recommended Practice for Selecting Proportions for Concrete (ACI 613-54).
———, 1963: Formwork for Concrete, *Spec. Publ.* 4.
———, 1966: Bond Stress: The State of the Art, ACI Committee 408.
———, 1971: Placing Concrete by Pumping Methods, ACI Committee 304.
Baar, M., 1971: Continuous Reinforced Structures for Cast-in-Place Diaphragm Walls, *Proc. Franki Congr. Madrid.*
CIRIA, 1967: The Effect of Bentonite on the Bond between Steel Reinforcement and Concrete, *Constr. Ind. Res. Inf. Assoc. Interim Rep.* 9, London.
CUR, 1972: Underground Concrete, *Neth. Comm. Concr. Res. Rep.* 56.
DiBiagio, E., and J. A. Roti, 1972: Earth Pressure Measurements on a Braced Slurry-Trench Wall in Soft Clay, *Proc. 5th Eur. Conf. Soil Mech. Found. Eng., Madrid,* vol. 1, pp. 473–483.
Ikuta, Y., et al., 1969a: An Experimental Study on the Integration of Diaphragm Walls with Major Building Structures, *Trans. Jap. Concr. Inst.* August.
———, 1969b: Studies of the Vertical Joint Method for the Slurry Trench Basement Wall, *Takenaka Tech. Res. Rep.* 4, Tokyo, November.
———, 1971: An Experimental Study on the Flow Motion of Fresh Concrete in Slurry Trench Wall by Activable Tracer, *Takenaka Tech. Rep.* 6, Tokyo.
———, 1972: Studies of the Vertical Joint Method for the Slurry Trench Basement Wall, *Jap. Concr. J.,* vol. 10, no. 3, March.
———, 1973: Compressive and Bond Strength Characteristics of Concrete Replacing Bentonite Slurry, *Takenaka Tech. Rep.* 10, Tokyo.
Japanese Highway Public Authority, 1975: *Bond Stress Test Rep.* 2, Tokyo.
Neville, A. M., 1973: "Properties of Concrete," 2d ed., Wiley, New York.
Nippon Co., 1975a: Compression and Bond Tests, unpublished report, personal communicaton, Tokyo.
———, 1975b: *Bond Stress Test Rep.* 1, unpublished, Tokyo.
Troxell, G. E., et al., 1968: "Composition and Properties of Concrete," 2d ed., McGraw-Hill, New York.
University of Osaka, 1971: Tests on the Efficiency of Construction Joints in Diaphragm Walls, *Spec. Bull.*
U.S. Bureau of Reclamation, 1963: "Concrete Manual," 7th ed., Denver, Colo.
Xanthakos, P. P., 1974: "Underground Construction in Fluid Trenches," Colleges of Engineering, University of Illinois, Chicago.

Chapter Nine
CONSTRUCTION FUNDAMENTALS OF DIAPHRAGM WALLS

Chapter 8 discussed the requirements for assuring reinforced concrete of satisfactory quality and acceptable strength and demonstrated the effects of construction conditions and methods of concrete placement on the finished structure. Emphasis was placed on the choice of construction joints and connections to adjoining structural members as well as on the method of execution, since any related problems will generally carry to the finished construction.

For a seemingly simple but usually complex construction process it is necessary to set forth useful guidelines. The following sections review the construction fundamentals of diaphragm walls. These requirements often have a significant influence on the design and details of a project.

9-1 SITE INSPECTION AND PREPARATION

In general, a complete site inspection before construction is started and sometimes before the design is completed is necessary and helps avoid serious problems. This is important because underground construction of this type usually must be scheduled with regard to traffic maintenance, preservation or relocation of existing utilities, and frequent interference from existing structures and facilities. These are among site conditions that determine the extent to which construction can be feasible.

The site inspection usually is confined to the immediate construction area, but it is advantageous to include the surrounding neighborhood. The following points must be investigated.

Environmental Effects Noise-producing equipment and facilities may be objectionable, and often the noise level is regulated by local ordinances. In addition the method of mud disposal and treatment of used slurries can sometimes be rejected as a public nuisance.

Space Availability This can dictate the type of wall that is feasible and can influence the slurry-plant layout and location of miscellaneous facilities; working space for assembling reinforcing cages; and miscellaneous field work. Space availability thus can determine the panel length and dictate the panel sequence. Where the construction partially encroaches on adjacent property, e.g., tie backs, or falls within the street right-of-way, an easement must be obtained.

Local Traffic This can have a decisive effect on the construction conditions and priorities. Traffic can dictate when a concrete pour should be carried out but can also delay or disrupt a pour already started. On many occasions the work must proceed simultaneously at several different locations while traffic must move around and through the site.

Overhead Structures and Facilities These should be identified and located. Fire escapes, canopies, and overhead structures can interfere with the excavation and construction of a panel and may have to be removed temporarily. Restricted headroom generally limits the choice of equipment with regard to height, size, and mobility. Exceptional mobility of equipment is required under limited clearance to existing buildings and structures, particularly for corner panels.

Underground Utilities and Structures If the construction interferes with existing utilities, they must be protected or relocated. Excavation near existing sanitary sewers can result in serious trouble, e.g., leakage causing the contamination of slurry. Near defective water mains excavation can be subjected to dynamic groundwater flow, or near abandoned sewers excavation can result in sudden loss of slurry followed by loss of stability.

On the other hand, excavating machines are quite sensitive to man-made structures and hard obstacles which are difficult to handle. Thus, underground utilities, tunnels, debris, and other incidental construction parts must be identified, and a method must be selected for their removal.

Availability of Utilities Invariably fresh water and electricity are necessary for this type of work. If good water is not available or its supply is limited, it may be necessary to use special slurries, which can affect the construction as a whole.

Site Preparation and Plant Layout

Sites which are irregularly contoured or have sloping topography generally require extensive preparatory work, the cost of which raises the total cost of the project. Excavation in man-made fills or at the toe of a slope must be preceded by certain steps to ensure that the operation will not become critical from the viewpoint of trench stability.

In general trench excavation does not start until guide walls have been built at the ground level. The function and details of these walls are discussed in subsequent sections in this chapter.

The next step is to locate and set the mud-circulation and preparation plant. This consists of slurry mixers, storage tanks, the mud-plant separation units, and mud storage area. It is essential to provide adequate mixing facilities and mechanical separators for the job size and type of excavating equipment. Although this means additional cost, any compromise can result in a variety of problems during construction.

As a rule of thumb the average volume of bentonite slurry for a given panel size is estimated according to the type and permeability of soil; in fine soils of relatively low permeability the slurry volume to be available at all times may be about 1.5 times the panel volume; for excavations in gravel and relatively pervious ground the extra supply of slurry often is 100 percent of the panel volume. These are average requirements and they should be used with judgment.

All plant facilities should be set in convenient locations where they will not interfere with panel excavation, assembly of reinforcing cages, or flow of traffic. If the construction is carried out along a perimeter, as for building basements, the plant units are set at or near the center, where they remain until the work is completed. If the walls follow a long alignment, the mud plant must be moved as needed. Mud assemblies must be within the optimum range of reverse-circulation systems when they are used.

Spoil containers and storage tanks can be set either below or at ground level, where they can receive and hold soil and incidental materials from the excavation. Since it is not always possible to remove these materials as soon as they are excavated, extra storage capacity is necessary.

In fairly large and regular sites the construction can proceed by simultaneously carrying out several phases (excavation, concreting, materials removal) as shown in Fig. 9-1. In this case the operation is considerably more economical and has the benefit of substantial time savings.

9-2 PANEL DIMENSIONS

The factors dictating panel size and dimensions are related mainly to the site conditions. Thus a specified panel length often cannot be provided in the field because of restrictions on availability of construction space and time.

Fig. 9-1 Construction of a traffic underpass in Brussels. The site conditions allow several phases of the work to be carried out simultaneoulsy (excavation, concreting, materials handling, assembly of cages).

Panel Length

It generally is advantageous to have a wall processed in longer units. This reduces the number of vertical construction joints, with a corresponding saving in the cost. It also means fewer problems with the alignment of the wall and less water seepage at these locations. Despite these obvious advantages, the panel length is also governed by the following factors.

Face Stability and Slurry Loss   As mentioned in Chap. 2, short panels are better than long ones when it is necessary to resort to the arching effect for trench stability or when there is danger of slurry loss through open cavities underground. Short panels usually have a length of 4 to 8 ft (1.2 to 2.4 m).

Concreting   It is advantageous to complete the concrete pour before any significant stiffening or setting of the fresh mix occurs, and in practice the pouring should take about 4 h. For average panel lengths from 3.5 to 4.5 m (11.5 to 15 ft) and for one tremie pipe the speed of rising concrete will be similar to that shown in Fig. 9-2, which is plotted as function of depth, and lower speeds should be expected near the end of the pour. According to these data, the mean upward speed of concrete is from 8 to 15 cm/min (3 to 6 in). At this speed a panel 70 ft deep (21 m) will be poured in about 3 to $3\frac{1}{2}$ h, and will

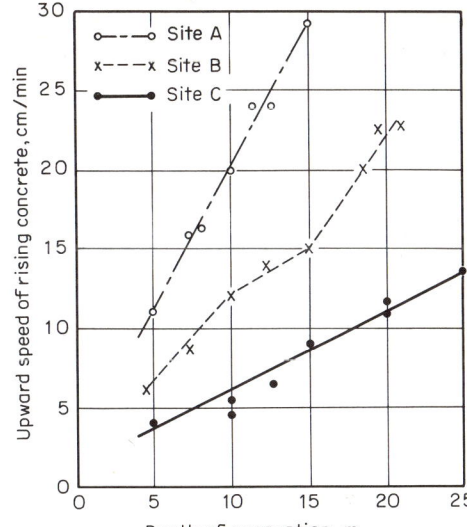

Fig. 9-2 Relationship between depth of excavation and speed of rising concrete level; panel length from 3.5 to 4.5 m (11.5 to 15 ft), and one 8-in diameter tremie pipe.

probably require 110 to 130 yd³ (84 to 100 m³) of concrete. Thus longer panels may need more than one tremie pipe.

Reinforcing Cages The panel size may have to be changed or reduced if no space is available to assemble and handle the cage. The cage must have a size and weight that is within the range of equipment that can be operated at the site.

Location of Bracing and Anchorage Panel lengths and joint locations should be coordinated with both the temporary and the permanent bracing. If this includes the use of anchors and tiebacks, the corresponding spacing will dictate the panel length.

Excavating Equipment The panel length should not be shorter than a single pass of the excavating tool. For panels that require more than one pass it is advantageous to provide the compatibility discussed in Chap. 5. Grabs work more efficiently if the load is symmetrically applied to the jaws, especially in dense ground, and this efficiency is further improved if the grab is assured of full bite.

Incidental Factors These include the site and traffic conditions as they may affect the orderly supply and availability of concrete; coordination of excavation, materials handling, and concreting operations; and water supply and plant facilities at the site.

For the average project a tentative panel length is first selected according to trench stability and concreting requirements. It then is compared with the range of the excavating systems intended for use and finally checked for any other considerations that may govern. It therefore follows that seldom is the panel length known at the design stage unless the design is precoordinated with the construction phase and sequence.

Panel Depth

For most jobs the panel depth is within the excavating range of almost any clamshell or rotary drill. Thus the base of the wall simply follows the project requirements. A wall may be founded on rock or other firm materials to transfer vertical loads; it may be extended into an impervious layer to protect the excavation from groundwater; or it may simply be embedded below excavation level for lateral stability and resistance to movement.

The top-of-wall elevation is established with regard to the details of the subgrade and superstructure, the location and elevation of adjacent footings, and the proximity of the wall to existing or future utilities.

Panel Width

The panel width cannot be less or more than the range of the available excavating machines. For clamshells the usual range is from 45 cm to 1.5 m (18 in to 5 ft) and for rotary drilling from 40 cm to 1.2 m (16 in to 4 ft). A practical minimum width is 45 cm (18 in) to allow the easy passage of tremie pipes between reinforcement bars. Walls wider than the foregoing range must be built using special grabs or by double digging. A frequent wall thickness is 60 cm (24 in).

A relatively thin wall is not necessarily the most economical design, especially if it is heavily reinforced. Furthermore, there is some evidence that a narrow trench can affect the flowability of fresh concrete. Thus, if the panel width is less than 24 in (60 cm), the tremie-pipe spacing should be reduced to 80 percent of the normal.

In view of the unavoidable tolerance in the width of the finished wall the panel thickness should also be considered in relation to the tolerance factors during excavation, particularly in very deep walls. This tolerance is discussed in subsequent sections.

Method of Detailing

For ordinary diaphragm walls with round joints the reinforcement should be detailed in relation to the stop ends. Some panels have no end tubes, most have one, and others have two; this variation can complicate the detailing of reinforcement. Sliwinski and Fleming (1974) recommend the method of detailing presented in Fig. 9-3. For simplicity the stop end or ends are considered to lie outside the panel length as shown. If the cover to the main steel longitudinally is 9 cm ($3\frac{1}{2}$ in) measured from the face of the concrete to the center of the main steel, and if the tolerance in the verticality of stop ends is according to the FPS (1974) specification, the relationship between wall thickness, panel length, reinforcing-cage length, and equivalent panel length is as shown.

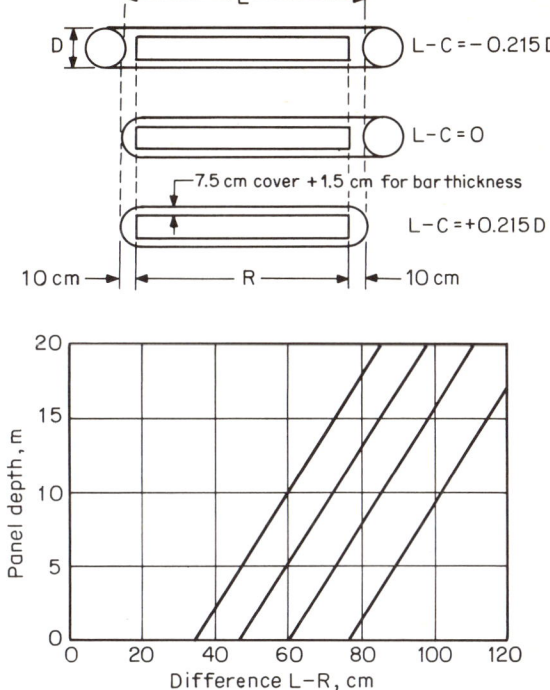

Fig. 9-3 Recommended method of detailing and relationship between panel dimensions. Verticality tolerance for stop ends = $\frac{1}{80}$. L = panel length for detailing; R = standard cage length for detailing; C = mean concrete length; D = panel width. (*From Sliwinski and Fleming, 1974.*)

Effective Wall Thickness

Some codes and specifications recommend a certain reduction in the effective wall thickness to compensate for construction imperfections in the finished wall and also in consideration of a probable decrease in the concrete strength near the face due to intermixing with bentonite. For instance, for a load-bearing wall this reduction, usually 1 in, or 2.5 cm, should be applied to both faces, whereas for a wall designed for bending moment it should be applied to the compressive side. The foregoing recommendation is primarily for permanent structures or for combined walls. On the other hand, the design of a wall should recognize the usual overexcavation and the resulting wall overwidth; hence in the author's opinion a reduction in the effective wall thickness is neither necessary nor justified.

9-3 PANEL SEQUENCE AND ARRANGEMENT

For construction along streets the panel sequence usually is determined by traffic considerations. For building construction common practice is to choose the first panel near the entrance and then work continuously along the perimeter to complete the wall. This can be changed as necessary if more than one rig is used at the site, and attention is paid to accessibility and equipment mobility.

The panel sequence should allow free and unrestricted flow of traffic in and out of the site, make all excavation possible from a fixed location of the mud plant units, and allow the simultaneous work on three successive panels, as shown in Fig. 9-4, which is generally desirable so that as one panel is excavated, a second panel is prepared and a third is concreted. On the other hand, it is important to avoid damaging panels already cast, and before deciding the panel sequence this point should receive ample consideration.

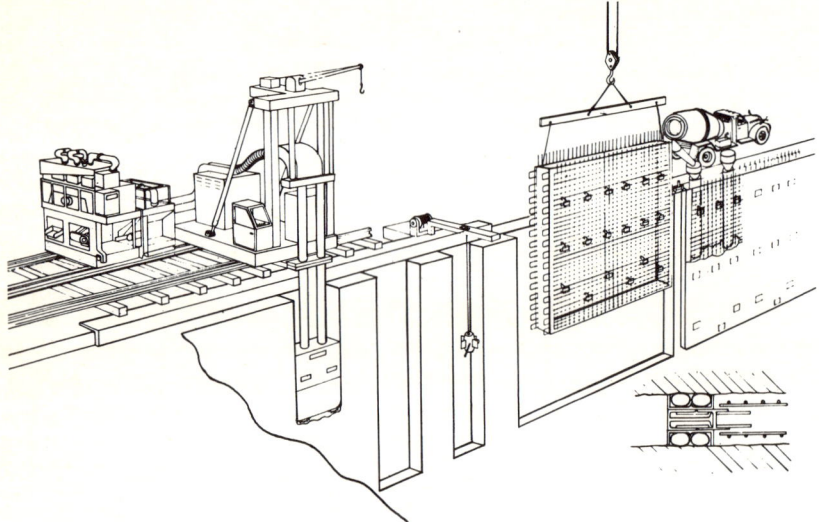

Fig. 9-4 Typical panel installation and construction sequence.

An example of panel sequence is shown in Fig. 9-5. This wall required several panel lengths and configurations, as shown. Referring to the same example, it is interesting to note that the corner panels were dimensioned so that each side corresponds to one equipment pass. This is good practice and allows the panel to be filled with one tremie pipe placed at the corner.

9-4 GUIDE WALLS

Guide walls are reinforced-concrete sections of a suitable configuration built at grade and along the alignment of the trench. They are always constructed ahead of the trenching operations. They serve only temporarily, and at least one side is removed when the structure is completed. Guide walls fill several functions: (1) they control the line and grade of the trench and therefore guide the excavation and movement of equipment along the outline of the wall; (2) they make it possible to start excavation from a level below grade when

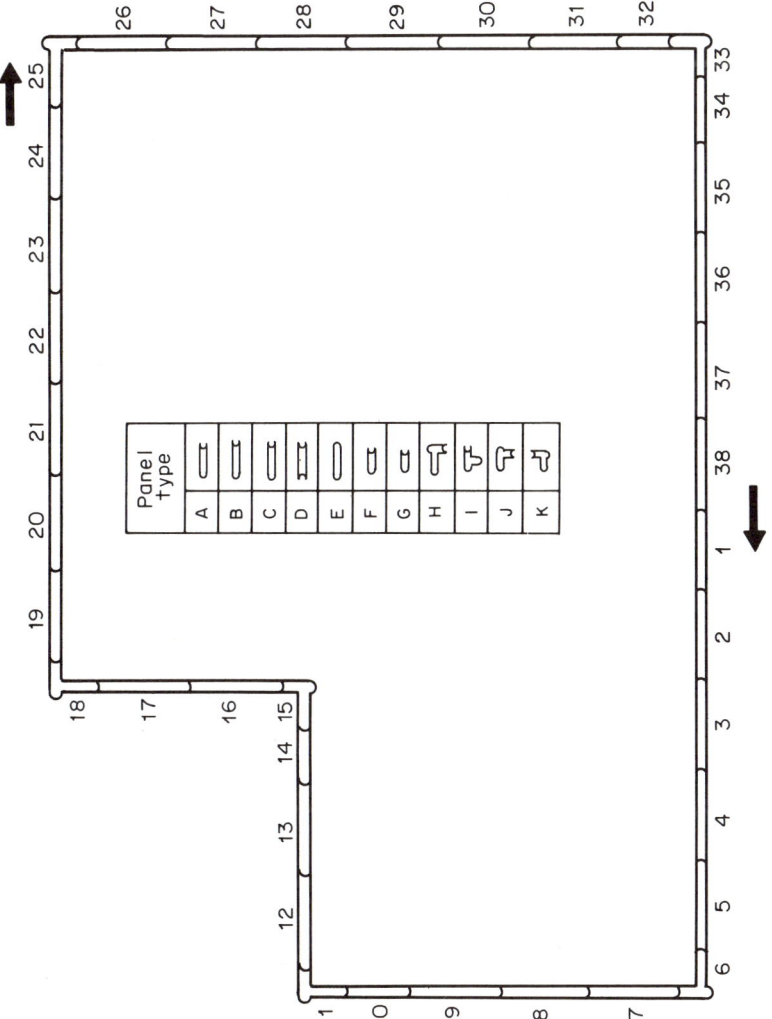

Fig. 9-5 Construction and panel sequence for a deep building basement. Note the various panel configurations and types.

man-made fills, sewers, footings and other obstructions exist; (3) they support the guide trench from heavy-construction surcharge, including the passage of equipment, and at levels below which the slurry may conceivably fall; (4) they protect the sides against turbulent action created during the up-and-down passage of equipment or with the introduction of fresh slurry; (5) they brace the face near the top by the use of timber cross struts; and (6) they act together with the guide trench as reservoir for the slurry.

The distance face to face is made greater than the nominal width of trench, usually by 2 in (5 cm), so that the equipment can move up and down without hitting the concrete face. After the panel has been excavated and cleaned, the reinforcing cage is inserted and suspended at the guide walls. As mentioned, guide walls should be braced with wood struts; otherwise they can move out of position and even fall into the trench. It is also good practice to have the guide walls structurally continuous at the construction joints.

Where shallow utilities or footings exist it is advantageous to carry the tip of guide walls down to the same level. On the other hand, the base of guide walls should be on firm and compact soil; otherwise protrusions in the finished diaphragm panel can occur just below the guide-wall base level. In very soft or saturated soil it is prudent to take special precautions to prevent washout of soil from the area below the guide walls, as shown in Fig. 9-6.

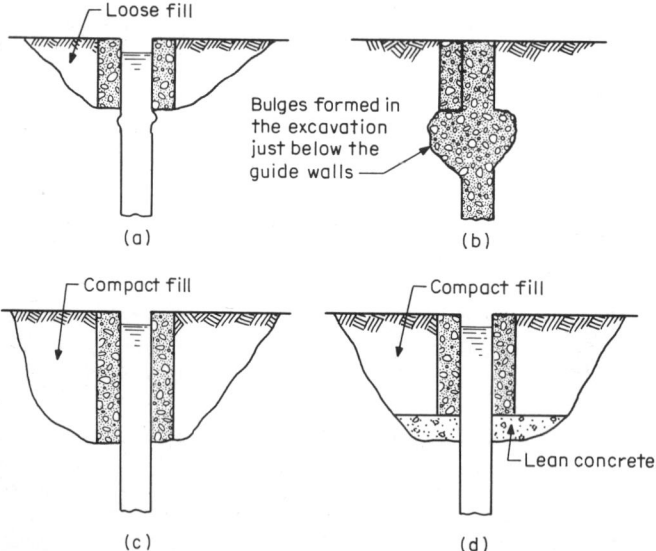

Fig. 9-6 Special guide-wall construction in soft or in caving soil: (*a*) variation in slurry level induces flow in outside soft soil and creates cavities; (*b*) shallow guide walls and loose fill behind high water table cause the formation of bulges; (*c*) preventive action provided by deep guide walls and compact fill behind; (*d*) preventive action provided by lean concrete at the base of guide walls and compact fill.

A track is laid alongside the guide walls and serves to receive the moving derrick, except when the machine is suspended from crane tractors. Since neither the weight nor the size of this assembly is known ahead of the construction, it is necessary to detail guide walls only after the excavating system is selected. In firm and stable ground portable guide walls made of steel trench sheets, wood blocks, and the like can be used in lieu of concrete guide walls. Steel forms are preferred by many contractors for long continuous walls.

Construction Details Guide walls are detailed and built as shown in Fig. 9-7. The ground side usually is cast against the soil, whereas the inner face is formed. If a rail is used, the top should be finished smooth and truly horizontal. It is evident from Fig. 9-7 that certain details are intended to accommodate the site and ground conditions rather than the excavating system; these details should be specified ahead of the main construction.

9-5 CONSTRUCTION ACCURACY AND TOLERANCE

A finished diaphragm wall may show three basic deviations from the specified accuracy. The first relates to the true vertical alignment; the second involves the alignment of the wall in plan; and the third involves irregularities and protrusions from the average wall face. The incidence and extent of these deviations depend on several factors but mainly on the excavating equipment and the skills of its operator; the method of supporting and guiding the rig, including the guide-wall construction; and the type of soil encountered and the soil-slurry interaction. In the field the construction control that is practicable and attainable should recognize the importance of early detection of any deviation. Control, however, is difficult to maintain in ground with boulders and erratic or sloping hard layers, even if sophisticated detection devices are used.

The horizontal accuracy is better maintained by building the guide walls to true line and by monitoring and checking their position during excavation and application of surcharge loads. Concrete which has bypassed the stop-end tube, either through an existing cavity or because of overwidth excavation, will act as an obstruction while the adjacent panel is being excavated, especially if there is long interruption between successive excavations. The result is the angular distortion of the wall in plan shown in Fig. 9-8.

The usually accepted tolerance on verticality is between 1/100 and 1/200. For some excavating systems and in ideal ground conditions a tolerance of 1/500 has been possible, whereas in some instances the tolerance could not be better than 1/50. The FPS specification recommends an accuracy of vertical alignment 1/80, measured along the wall face at the ends of panels. All these limits are academic, however, since in some situations they are unattainable. There is an observed tendency for the vertical construction accuracy to be better in cohesive than in granular soil. In any case, an actual deviation in

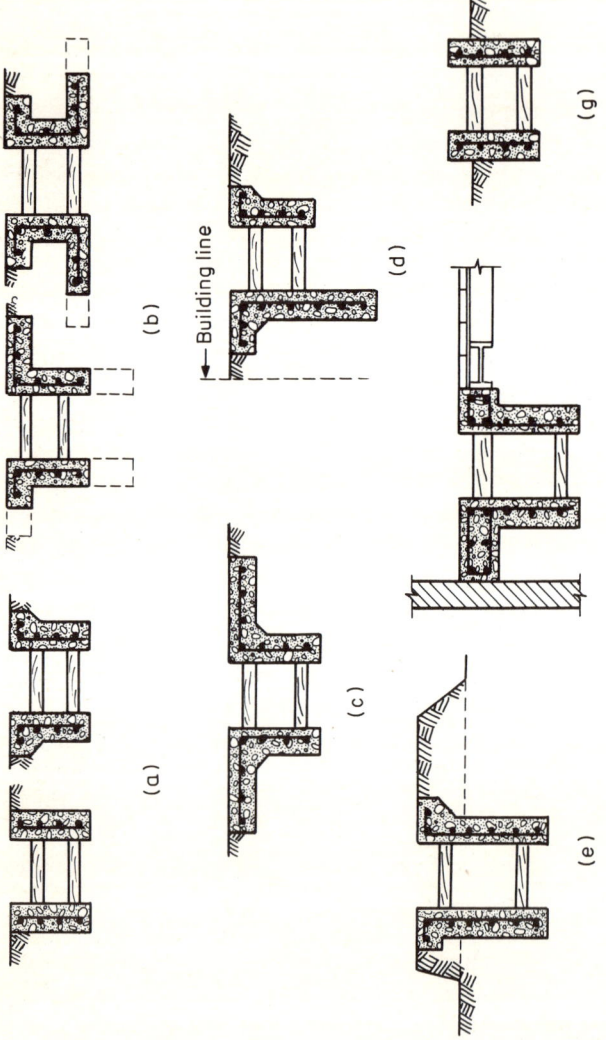

Fig. 9-7 Guide-wall details: (*a*) stable ground conditions and normal-weight excavating machines; (*b*) unstable ground conditions or heavy-surcharge loads; (*c*) guide walls for machines requiring a platform on one or on both sides; (*d*) guide walls near existing buildings or structures; (*e*) guide walls in man-made fills; (*f*) guide walls built as part of the main bracing system; and (*g*) guide walls built above existing grade to raise the slurry level.

Construction Fundamentals of Diaphragm Walls 325

(a)

Fig. 9-8 Angular deviation from true alignment caused by concrete penetrating beyond the stop-end tube: (*a*) deviation from verticality; (*b*) deviation shown in plan.

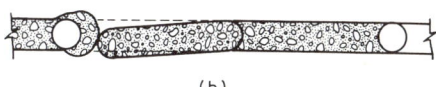

(b)

verticality of more than 1/80 might indicate a serious condition even if there is no structural continuity between panels. For example, it could result in construction joints which are separated and thus create ideal seepage paths, or it could mean greater eccentricity of the applied vertical loads. If the minimum vertical accuracy cannot be ensured, it may be advisable to limit the wall depth.

Within the specified limits of vertical tolerance a finished panel may show an angular deviation at any level when viewed in plan. This occurrence depends on the depth and can become objectionable with respect to the exposed face of the wall. Tolerances for such angular deviations are not always necessary, except in special cases when they must be established according to panel length, position, and relation to site conditions.

Special Devices The device shown in Fig. 9-9 is used in Japan to check the verticality and unevenness of slurry-trench panels and boreholes. It can measure deviations for depths up to 50 m (about 165 ft) by means of supersonic waves emitted against the face of the trench from a detector trembler. The trembler is freely suspended like a plumb bob and measures the distance to the face from time differences between emitted and reflected waves received and recorded on paper. The device provides simultaneous measurements for both faces, and the wall unevenness or any wavy surfaces are continuously recorded.

Protrusions, Wall Finish, and Thickness

For each job and project it is necessary to specify the tolerance to be allowed for normal protrusions resulting from irregularities beyond the general face as the trench is excavated. An overbreak of some 3 in (7.5 cm) has been allowed on many walls for average soil conditions. The usual tolerance for protrusions

326   Slurry Walls

Fig. 9-9 Device used to check the verticality and unevenness of slurry trenches. (*Takenaka.*)

on the finished wall is from 3 to 4 in ($7\frac{1}{2}$ to 10 cm), but this applies to homogeneous clays or to dense sand and gravel formations where face stability is better controlled. In badly fissured clay formations and in loose sand and gravel layers or where there are underground obstructions and pockets of soft ground interbedded with hard formations the tolerance should be increased accordingly. Since some structures show a certain extension of irregularities beyond the average wall face, provision should be made for trimming the wall when it is exposed. The general appearance of a wall may also be disrupted by various irregular sections which should be finished by chipping and grinding or by applying cement grout in order to make the final surface conform to the permitted variations.

The surface finish of a concrete wall cast against a vertical earth cut depends to great extend on the nature and texture of soil. Since some of the filter cake along the soil face remains during concreting to be broken out later, it acts as a plaster to produce a relatively smooth face. The same effect is also noticed at the ends of panels, where vinylon sheet is used. Despite these effects the general wall finish is influenced by the roughness of the soil worked through, varying from good in clay to rough in gravel, and thus face treatment is sometimes necessary. The wall finish is improved usually by guniting, masonry blocks or face brick, precast panels, or by building an inner form-finished decoration wall.

The actual wall thickness usually is greater than the design or theoretical width, and this may be due to several reasons: (1) initially the excavator is likely to trim off soil beyond the theoretical lines as it moves up and down the trench; (2) some individual soil grains are lost from the face by sloughing and peeling off until the filter cake can exert its plastering effect; and (3) in very soft silty soil the face can be displaced inward under the lateral action of the fresh concrete, as explained in previous sections. Practical experience shows that the actual wall thickness exceeds the design by as much as 3 to 6 percent in clay, 4 to 8 percent in sand, and 7 to 10 percent in gravel.

Top Concrete Layer

When the fresh concrete is taken up to ground level, it usually is overpoured at the end so that the top layer of intermixed slurry and concrete is skimmed off until clean concrete emerges. When concrete is terminated below the slurry level, the uppermost layer (usually 12 in or 30 cm) must be broken out and replaced by new concrete after the wall is exposed.

As pointed out in previous sections, the flow motion of fresh concrete through tremie pipes produces a top which is seldom level, and this complicates the control of tolerance. With a central tremie position the highest point at the rising surface is around the tremie pipe. This surface slopes gently away from the pipe according to the angle of repose, and the more flowable the concrete the flatter the slope. Since it is possible for the concrete around the tremie pipe to reach the top while the concrete level at the ends is much lower, some hand tamping may be necessary to finish the wall.

Walls Terminated below Guide Walls

For walls which must be terminated below ground level certain problems can arise with respect to the final top of wall elevation. The top surface must be as nearly level as possible, and this sometimes is achieved only by having the fresh concrete well rammed with heavy flat-faced or grid-type tools at the end of the pour. In addition to hand tamping, immersion vibrators can also be inserted and used to consolidate the fresh mix around the tremie pipe and spread it evenly in the panel. Since this is done through the slurry, and since there can be no visual inspection of this operation, the methods can produce a reasonably smooth and level top, but they can also cause some mixing of fresh concrete with bentonite.

Some difficulties can be experienced when concreting the adajcent panels for walls terminated below ground level. In this case it may be necessary to backfill the previous panel with granular material above the concreted portion and use an end barrier to prevent the backfill from entering the new excavation or to use a lean concrete mix as backfill above the top of wall.

If the upper part of the wall is to receive soldier piles with lagging, their installation should be correlated with the concrete placement and the method of securing a relatively level top concrete surface, since this will affect the actual embedment of the piles into sound concrete. These details should be studied before construction since they involve the inclusion of appropriate items in the specifications and in the bill of material, and also result in construction tolerances according to the final arrangement and the difficulties involved.

If the wall must be terminated below ground level, it is necessary to indicate a lowest elevation for good, durable concrete.

9-6 WATERTIGHTNESS OF DIAPHRAGM WALLS

The watertightness of diaphragm walls often is as important as strength. This is true for a permanent wall used in a deep building basement, a subway, a traffic underpass, or a cutoff structure. Concern for the means by which watertightness can be achieved warrants examination of all pertinent factors. Thus, consideration must be given primarily to the pore structure of set concrete and the joint details and how they have been executed. In practice the watertightness of a wall must be secured even though its back cannot be treated with bituminous coating or with other suitable sprayed layers of waterproofing materials. Fortunately, it is possible to obtain a structure which can be sufficiently watertight for all practical purposes.

Watertightness of Concrete

For normal concrete the better the quality of constituent materials the less permeable the finished wall. In this case other factors influencing its permeability are the degree of compaction actually attained, the lack of any special surface finish, and the subsequent treatment and exposure to favorable curing conditions.

Cement and Water For workable mixes used in diaphragm walls the permeability increases quite rapidly with the water-cement ratio provided the latter exceeds 0.6, or about 7 gal of water per sack of cement. For water-cement ratios up to 0.6 the permeability k_c of hardened concrete is of the order of 10^{-11} cm/s but increases with wetter mixes, so that for water-cement ratios of 0.7, k_c is close to 10^{-10} cm/s (Powers et al., 1955, 1959).

Permeability decreases as the cement-voids ratio increases. This relationship appears to be more confirmed (Troxell et al., 1968). On the other hand, greater fineness of the cement generally improves watertightness as it improves strength and durability.

Aggregate Type and Grade The greater the maximum size of aggregate for a given water-cement ratio the greater the flow, owing to the larger water voids developed on the underside of the coarser particles (Valenta, 1968). Aggregates should be of low porosity and permeability. As a means of comparison, a well-graded aggregate is even more important from the standpoint of watertightness than in terms of flowability and strength since it produces denser concrete. For a dense mix with maximum aggregate size $\frac{3}{4}$ in (2 cm) and a water-cement ratio of 0.6, the permeability is not likely to exceed 10^{-10} cm/s. Values of this order have been established by testing samples taken from large-diameter piles placed under bentonite slurry (Sliwinski and Fleming, 1974).

Admixtures The use of the so-called waterproof cements or special admixtures to improve the watertightness of concrete may not always produce the desired effect. In general, except for pozzolans, the use of extra cement is more effective.

The permeability of concrete containing entrained air is not increased for the percentages ordinarily used in diaphragm walls, provided, however, the water-cement ratio does not exceed the limits set forth in the previous sections. On the contrary, entrained air benefits watertightness by reducing bleeding and segregation; the danger of demixing during the long transport and placing time is overcome, and while in place the capillary system is inactivated by the presence of a limited amount of air.

Placing and Curing The watertightness of diaphragm walls is markedly influenced by minor construction defects or by nonhomogeneous conditions. Major leaks can be caused either by such defects as cracks in the finished structure or by void spaces left in the concrete as a result of honeycombing or segregation rather than by the inherent porosity of the cement paste and of the constituent aggregates. The lack of vibratory compaction can be serious unless it is compensated for by a flowable mix and by proper methods of placement. Bentonite or mud trappings in the finished wall are more critical in terms of concrete strength since they are practically impermeable, but cavities and openings around inserts which the fresh concrete failed to reach can be particularly harmful to the watertightness of the finished wall.

The generally humid conditions of drying mean that shrinkage cracks are almost eliminated. Since panel lengths are relatively short, the remaining (long-term) shrinkage is in practice accommodated at the vertical joints so that intermediate cracks need not be developed. Under the usual conditions of underground curing the continued hydration of the cement can result in gel development, which reduces the size of voids and has a beneficial effect on watertightness.

Expected Watertightness of Diaphragm Walls From the foregoing discussion it follows that it is not difficult to produce walls with permeability of the order of 10^{-10} cm/s. Suppose a wall is 60 cm (24 in) thick, retains a hydraulic head of 10 m (33 ft), and has a porosity of 15 percent; if we take into account a suction pressure of 1 atm to assist water flow, the quantity of water percolating through the wall is close to 0.3 L for 1000 m² of wall surface over a period of 24 h (Sliwinski and Fleming, 1974).

Watertightness of Construction Joints

The simple butt connection or the round-end tube joint cannot be considered waterproof since they cannot completely stop seepage from entering. Experience shows, however, that significant leakage seldom occurs, probably because of bentonite trappings across the joint and the impregnation of soil by slurry in the immediate vicinity of the joint. Whenever some leakage has occurred, it has been attributed to differential deflection and movement between adjoining panels, and this situation is more pronounced near corners. It is common experience that excessive differential deflection and movement between such panels typically is followed by a corresponding leakage at the construction joint. This problem must be solved in a structural sense, and in this respect it involves a number of variables, namely the panel configuration in plan, wall height, soil conditions, type and uniformity of lateral loads, method of excavation, and bracing. If it is solved structurally, the construction joints will be sufficiently watertight.

The usual and probably the most economical way to correct damp joints is to allow leakage to occur until most of the differential movement has been completed and then remedy the problem by grouting up from ground level or from within the excavation after the wall is exposed. Alternatively, a steel or a suitable plate bedded on epoxy-resin mortar can be bolted to the concrete over the interior face of the joint.

If a waterproof joint is specified by the design, the water-stop details described in Chap. 8 can be used in conjunction with a round-tube joint provided differential movement is controlled. The use of shear-transfer devices between panels basically is indicated for structural reasons but also reduces differential movement and therefore produces watertight joints. However, the possibility of causing cracks elsewhere in the wall and away from the joints should not be ignored, particularly for panels which are only lightly reinforced.

Waterproofing Underground Structures

The durability of concrete for underground walls, subway galleries, and traffic underpasses depends, among other factors, on the water or seepage infiltration. Besides affecting appearance and causing some safety problems, this results in icicle growth during winter, leading to gradual degradation of the structure. It is therefore necessary to ensure that the method of diaphragm-wall construction will not limit the waterproofing which could normally be attained. In many instances waterproofing is decided as a compromise between watertightness and economy after all merits and relative cost are known. The same principle is also used to determine the size and location of the areas to be waterproofed; thus, if a dry roadway or a dry railbed is all that is required and lateral water infiltration is not significant, it usually is sufficient to waterproof the roof of the box section.

Diaphragm Walls in Building Basements As already mentioned, the watertightness as a whole is initially improved by paying proper attention to joint construction, including joint spacing for the site conditions (average climate, temperature variations, wind action, concrete drying, and the like). Premolded water stops anchored in the concrete on both sides of the joint will improve watertightness.

Common cause of leakage in deep basement walls is panel separation caused by uneven and variable lateral loads and poor excavation and bracing procedures. These effects are considerably reduced by the use of shear-transfer devices to keep panels together, but these devices are sometimes troublesome and difficult to install and therefore result in unnecessary costs.

Walls for Transit Structures The leakage of tunnel structures usually is attributed to poor construction. The junction between the roof slab and the side walls is the most critical and should be detailed taking the tunnel length and the associated shrinkage effects into account (see also Sec. 12-10).

At present there are two waterproofing methods. One is to treat the roof of the section externally with hot bituminous materials and overlap the joint between tunnel walls and roof. By contrast the second method shows complete reliance on fully watertight concrete and joints and is used with other forms of cut-and-cover construction and with the completely mechanized tunnel boring and lining systems.

A review of waterproofing methods for underground structures given by Otter (1974) emphasizes tunnels and gallery construction.

9-7 PROBLEMS CAUSED BY UNDERGROUND STRUCTURES AND UTILITIES

Certain difficulties should be expected because of the presence of utilities, sewers, water mains, and house connections; existing subbasements; and freight tunnels and other underground openings. The severity of this problem often is ignored and often is exaggerated. From the construction standpoint the situation can be remedied by special work or combination of methods and by the use of special equipment.

Sewers, Water Mains, and Utilities Sewers and water mains can make the construction troublesome or even impossible, particularly at crossings where they occur as a dense network. Examples of the most common problems are slurry contamination by sewage, stability difficulties due to broken mains, and slurry escaping to existing openings. Underground cables for power and communications are commonly protected or relocated before trenching.

Where sewers and water mains are at relatively shallow depth, it is good practice to have the walls terminated below that level. If the ground support must be extended to ground level, the solution is to install soldier piles with lagging in the uppermost part. Many city ordinances prohibit wall construction in the first 8 to 10 ft (about 3 m) below grade to accommodate future installation or relocation of utilities. If the utilities are still shallow but too numerous to deal with, it is better to start construction by pretrenching to just below the utility line and carry the guide walls to the same depth.

Despite the foregoing measures it still may be impossible to position the equipment so that it can excavate continuously under the utility network and install the reinforcing cage. In the extreme case it may thus be necessary either to abandon the method or combine it with other ground-engineering techniques. In cases where only a few isolated sewers intercept a panel, the excavation can be started at ground level under slurry as in normal conditions and then special diggers can be attached to equipment of high mobility, e.g., backhoes, to scrape and remove the soil under a sewer pipe, as shown in Fig. 9-10a. When enough depth is reached, a low-head clamshell is lowered and moved laterally to bring the cable as far out as the utility permits so that the reach of the bucket is beyond the sewer line, as shown in Fig. 9-10b.

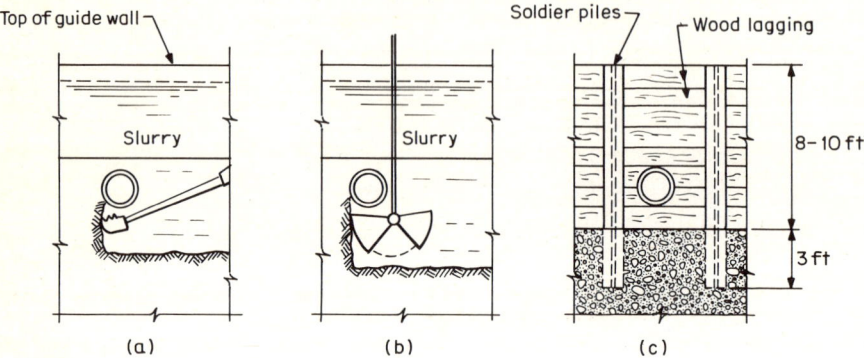

Fig. 9-10 Excavation around and under a sewer: (a) scraping the soil around and under the pipe using special tools attached to a backhoe; (b) lowering a low-head bucket close to and under the pipe; (c) setting soldier piles into fresh concrete and installing wood lagging.

If isolated utilities exist at considerable depth, the foregoing construction should be avoided since it can cause damage to the sewer. In this depth range (over 20 ft, or 6 m) special diggers become less effective and more difficult to operate. The construction is better carried out by bypassing the utility, as shown in Fig. 9-11, and by pregrouting the zone around the pipe before trenching. The portion of the wall left unfinished around and beneath the sewer is completed in sections after general excavation. If the utilities are too many and erratically located, diaphragm-wall panels can be combined with bored piles.

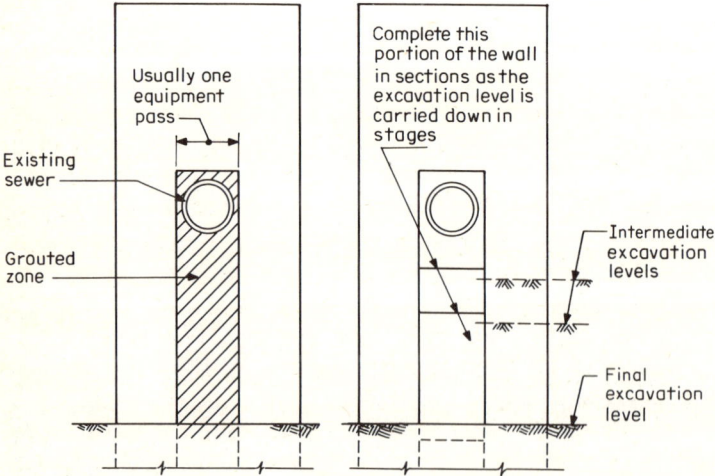

Fig. 9-11 Construction around and under an existing sewer located at relatively great depth.

Subbasements This group includes junction chambers, manholes, and various small underground openings. If a subbasement is long and within the alignment of the slurry trench, it can be used as working platform, functioning also as guide wall. Quite often a subbasement is an obstruction in the path of excavation and interferes with new construction. The usual procedure in this case is to provide a closure wall to protect adjacent usable space from construction operations, break and remove a strip of the base slab to allow trenching, backfill with granular material, and start excavation from the ground level, as shown in Fig. 9-12a. Alternatively, temporary shoring is placed to strengthen the roof if necessary, and the subbasement is flooded with slurry, as shown in Fig. 9-12b.

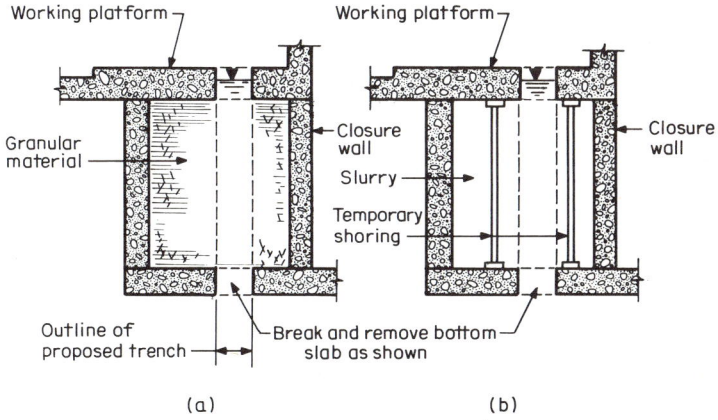

Fig. 9-12 Construction through an existing subbasement: (a) break slab as shown, backfill with granular material, and start trenching from the existing ground level; (b) support subbasement temporarily as shown, fill with slurry if necessary, and pour wall to basement level.

Abandoned Tunnels Since abandoned tunnels are usually found at considerable depth, their removal before construction is not practicable. If abandoned tunnels, galleries, or similar openings cross a proposed wall, the construction can proceed as shown in Fig. 9-11. If the structure is in good condition and accessible from within, it may be better to plug the sections on either side of the trench and then break and remove the portion that interfaces with the excavation. If neither solution is possible, the next alternative is to use chisels and other percussive tools to break and consolidate the concrete, but this is difficult and expensive, especially in heavily reinforced structures.

9-8 PROBLEMS CAUSED BY BOULDERS AND OBSTRUCTIONS

Large boulders or man-made obstructions always slow down excavation. Isolated obstructions and boulders which occur occasionally do not necessarily require the conversion of the excavating equipment to percussive tools, and their removal can be achieved by conventional grabs or by special extractors. It is quite difficult to determine before construction exactly how many boulders per panel would justify the use of chisels and percussive equipment in place of conventional grabs. This decision usually is made by the contractor and is based on experience with similar jobs.

If unusually large boulders exist at shallow depths and above the groundwater table, it may be better to dig out the obstructions, as shown in Fig. 9-13, backfill with silty or sandy soil to satisfactory compaction, rebuild the guide walls, and repeat the excavation several days later.

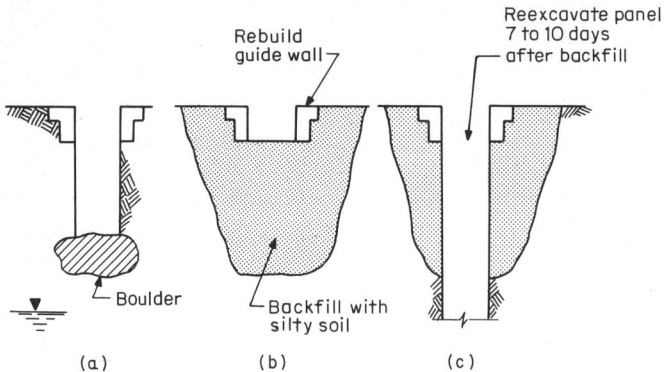

Fig. 9-13 Method of removing boulders above the groundwater table: (*a*) boulder is found at relatively shallow depth; (*b*) obstruction is dug out and the area is backfilled; (*c*) panel is reexcavated.

If very large boulders are encountered at relatively deep levels or below groundwater table, which will make the foregoing method impracticable, it still may be better to avoid chiseling, bypass the obstruction after making a pass to the top of the boulder, and pour concrete, as shown in Fig. 9-14*b*.

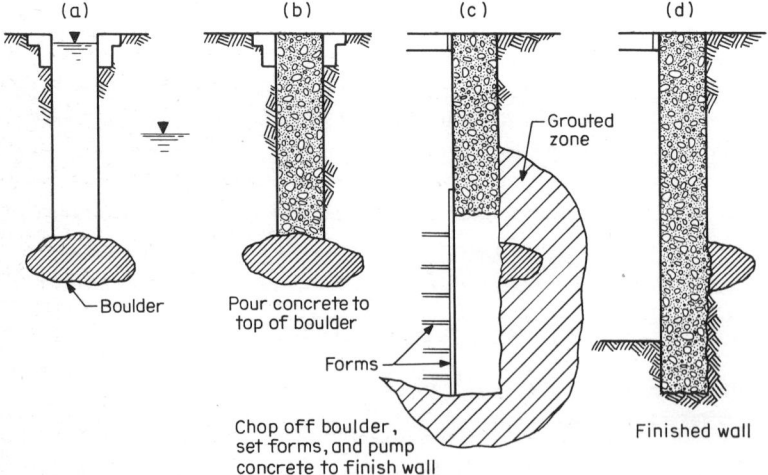

Fig. 9-14 Method of removing boulders and finishing panels when boulders exist below the groundwater table or in deep strata.

During general excavation it may be necessary to grout the soil around the portion omitted. When the boulder is exposed, it is chopped off, and the missing portion of the wall is concreted in normal formwork, as shown in Fig. 9-14c and d.

As mentioned earlier, the construction tolerance in ground with boulders should be established accordingly since removal of these obstructions will always affect the accuracy attained in verticality, particularly under prolonged use of chisels and in erratic formations. A corresponding problem is the inevitable protrusions on the exposed wall. In ground interbedded with harder soil or boulders extending beyond the face of the wall frequent protrusions are certain to appear on the exposed face, and provision should be made for trimming material and finished portions beyond the tolerance line.

REFERENCES

FPS, 1974: Specifications for Cast-in-Place Diaphragm Walling, Federation of Piling Specialists, London.

Otter, M. K., 1974: Waterproofing Tunnel and Gallery Construction, *Spec. Rep.*, Dames and Moore, Middlesex, N.J.

Powers, T. C., et al., 1955: Permeability of Portland Cement Paste, *Portland Cement Assoc. Res. Bull.* 53, Chicago.

———, 1959: The Flow of Water in Hardened Portland Cement Paste, *Portland Cement Assoc. Res. Bull.* 106, Chicago.

Sliwinski, Z., and W. G. K. Fleming, 1974: Practical Considerations Affecting the Construction of Diaphragm Walls, *Proc. Diaphragm Walls Anchorages, Inst. Civ. Eng., Lond.*

Troxell, G. E., et al., 1968: "Composition and Properties of Concrete," 2d ed., McGraw-Hill, New York.

Valenta, O., 1968: Durability of Concrete, *Proc. 5th Int. Symp. Chem. Cement, Tokyo*, pt. III, p. 193.

Chapter Ten

VARIOUS WALL SYSTEMS

10-1 PREFABRICATED DIAPHRAGM WALLS

Prefabricated concrete panels installed in slurry trenches to form structural retaining walls were introduced first in 1970. This method can be looked upon as the combined result of two foundation types, the prefabricated interlocking pile and the sheet-pile curtain wall. The procedure again consists in the excavation of a trench under slurry, but the in situ tremie placement of concrete is replaced by the insertion of precast panels.

Guide walls are required before excavation. Their detailing must be preceded by exact estimates of the anticipated panel size and weight of the precast sections, since the guide walls hold and temporarily support these panels until the grout has gained enough strength.

Grout Systems

A conventional grout or lean concrete is not used because its limited flowability will not ensure the effective sealing of the joints between panels. A seal is best obtained by the use of flowable cement-bentonite mixes to which suitable agents are added to retard premature stiffening.

Single Grout This consists of water, bentonite, cement, and some additives and is essentially similar to the cement-bentonite grouts described in Sec. 7-2. Initially it is introduced to support and seal the face of the trench. Since the presence of cement in relatively high ratios will generally cause flocculation, appreciable slurry loss should be expected before a seal is obtained. Since the initial slurry is used as final grout, excessive soil retention from the excavation must be avoided and therefore the slurry must be checked regularly and recycled if necessary to remove the soil fraction.

The twofold function requires conversion of the slurry from a colloidal system into a self-hardening material. Such a slurry usually is referred to as *coulis*. Figure 10-1 shows the development of strength of a typical coulis as a function of time after mixing. Evidently the setting process must be controlled within a relatively narrow range; the slurry must remain flowable until the excavation is completed and the precast panel is inserted and positioned, but it must rapidly gain strength thereafter and develop resistance to deformation in order to allow the next panel to be excavated. With time the grout becomes solid and blends with the panels at the construction joints. Its ultimate strength, however, is much lower than that of the precast panels, and this allows the grout to be handled and removed while cleaning the joints.

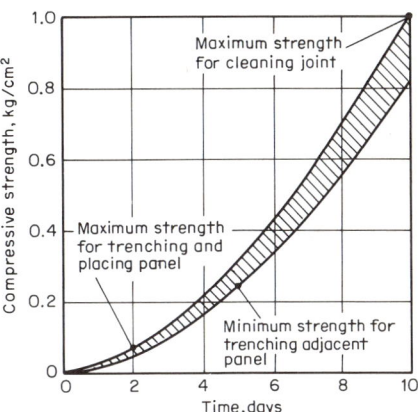

Fig. 10-1 Increase in grout strength with time for installing prefabricated panels. (Some contractors have successfully cleaned the joints when the grout attained twice the maximum strength shown.)

The installation depends on successful controls over the process so that the grout can progressively stiffen around the precast panels and finally set to a hardened material. The strength of this material is influenced primarily by the cement content and usually is in the range of 7 to 15 kg/cm² (100 to 200 lb/in²). Its density, however, is only 25 to 30 percent higher than the density of plain water.

Displacement Grout This has been developed and used by Bachy in France, and, as the name indicates, it requires two different materials. Initially, bentonite slurry is used during excavation for face support, but it is replaced by a suitable bonding grout just before the precast sections are placed. The mechanical properties and strength of the bonding grout are therefore less dependent upon the requirements of excavation, and can be specified with emphasis on the finished wall. This grout can develop higher strength and resistance to deformation.

Compared with the single grout, the two-grout application offers the following advantages: (1) it eliminates strict adherence to schedule and thus allows better timing; (2) it accommodates the development of strength of the set material over a wider range; (3) contamination of grout with materials from the excavation is no longer a problem; and (4) since the slurry is not intermixed with cement, the excavation is completed in stable conditions. The substitution of slurry is accomplished in a displacement process, however, and this requires utmost care (Des Francs, 1974).

Usually the grout is prepared from the initial slurry merely by adding cement and probably more bentonite. In order to ensure the correct proportions of the constituent materials it is necessary to check the composition and flow properties of the slurry before preparing the grout and adjust them if necessary. The final strength of the cement-bentonite grout should be checked considering the possibility of soil inclusion in situ. Laboratory results should not be the only indicator of grout behavior, since low-strength grouts of this type are known to be anomalous systems showing wide and erratic variations in their properties.

The effect of variations in viscosity, due for example to variations in bentonite content, on the compressive strength of set grout is less pronounced for water-cement systems (Des Francs, 1974). Therefore, when the final grout is prepared from the initial bentonite slurry, it is more important to deal with the water-cement ratio than with any other factor. Next, it is necessary to recycle the slurry until the sand content is less than 8 percent, the specific gravity is less than 1.10, and the cone viscosity is less than 40 s. These adjustments are necessary not for control of strength but because slurries which are too dense or too viscous will not readily be displaced by the grout.

A trial grout sample is first prepared at the batching plant and checked for strength-time behavior under the effect of all constituent materials. When sufficient quantities are available in the field, the grout is pumped into the trench using a special spreader, which is lowered near the bottom so that the displacement begins from the bottom up. When the substitution is complete, routine tests are necessary on the grout density. After the precast sections are inserted, control tests are mandatory on samples obtained from preselected panels to monitor the development of grout strength.

Example of Displacement Grout Table 10-1 shows a displacement grout. A trial batch was prepared using the following materials per cubic meter of grout: water, 870 kg; cement, 310 kg; and bentonite, 62 kg. The 90-day compressive strength of the laboratory sample was 2 MPa (about 285 lb/in^2), and this was adequate for the construction. The recycled slurry had a specific gravity of 1.08, sand content of 2 percent, and a viscosity of 36 s before displacement. The grout at the site was prepared from another batch of recy-

TABLE 10-1 Properties of Grout Used with Prefabricated Wall Panels (Data from Des Francs, 1974)

	Location			
			With panels in place	
Property	Mixing plant	When pumped into trench	At surface	4 m from surface
Specific gravity	1.29	1.29 at bottom 1.27 at 2 m from bottom 1.25 at 4 m from bottom	1.25	1.25
Funnel viscosity, s	20			
Strength, lb/in^2,				
7-day	45	38 (at bottom)	31	34
28-day	305	272 (at bottom)	214	245
90-day	460	415 (at bottom)	355	385

cled slurry (specific gravity 1.10, sand content 3.4 percent, and cone viscosity 39 s) by adding cement until the cement-water ratio reached the value 0.4.

Panel Types and Configuration

At present prefabricated panels are of reinforced concrete. Prestressing has been introduced on a trial basis but is not yet available for general construction. Variations are conceivable with the use of lightweight aggregate or with hollow panels. A practical limit in the use of prefabricated walls is imposed by the weight of individual units. For the usual crane and hoisting capabilities available at the site the practical maximum weight for a single panel is about 40 t, and the practical maximum panel length can accommodate a four- or five-story basement.

Panel configurations and installation details have been standardized. Figure 10-2 shows a wall made from typical panels installed so that they slot together. This interlocking is possible because the tongue of each succeeding panel fits into the groove of the preceding panel and provides high accuracy and a smooth structure. A second type, shown in Fig. 10-3, consists of beam-and-slab-sections and is therefore equivalent to the conventional soldier piles with lagging. The beams usually are made twice as thick as the slabs; a standard thickness is 50 cm (20 in) for the beams and 25 cm (10 in) for the slabs.

Other configurations are possible and can be used to fit the job conditions. Examples and variations are shown in Fig. 10-4. In Fig. 10-4*a* the wall is hybrid construction, in which the prefabricated elements are I sections and are used as joint beams for cast-in-place panels. The wall shown in Fig. 10-4*b* is similar to steel soldier beams with lagging. The wall shown in Fig. 10-4*c* presumably allows the splicing of horizontal bars and provides continuity in that direction. The prefabricated panels (uprights) are installed in slurry-filled holes and are provided with anchor bars bent as shown to fit. Cast-in-place panels are installed in between and receive the dowel bars. Experience shows that it is quite difficult to bend the anchor bars for overlapping since

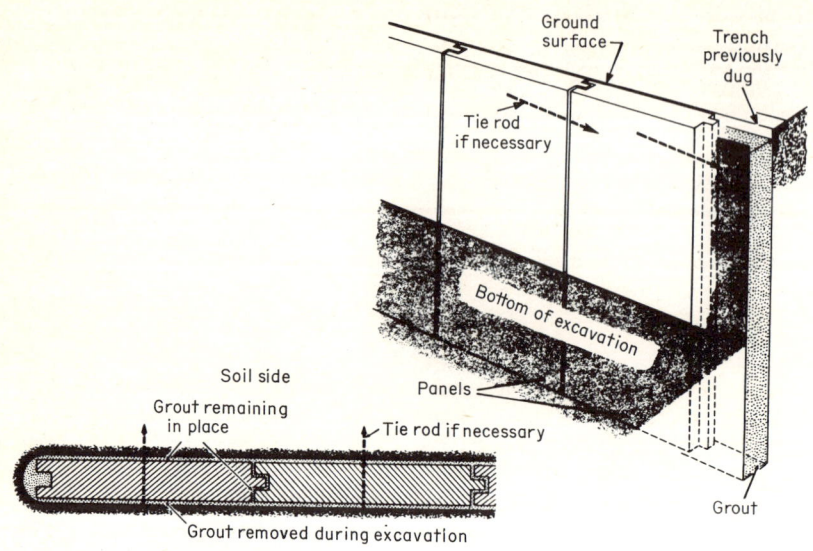

Fig. 10-2 Prefabricated wall with identical panels.

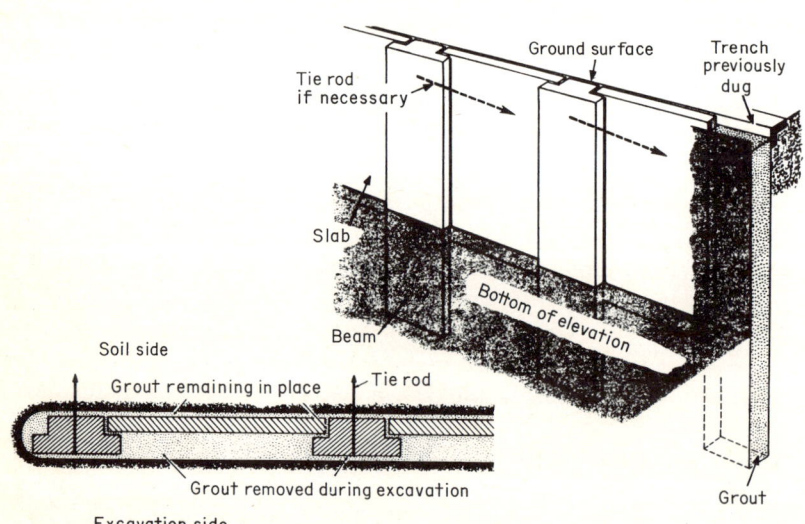

Fig. 10-3 Prefabricated wall with beam-and-slab panels.

340

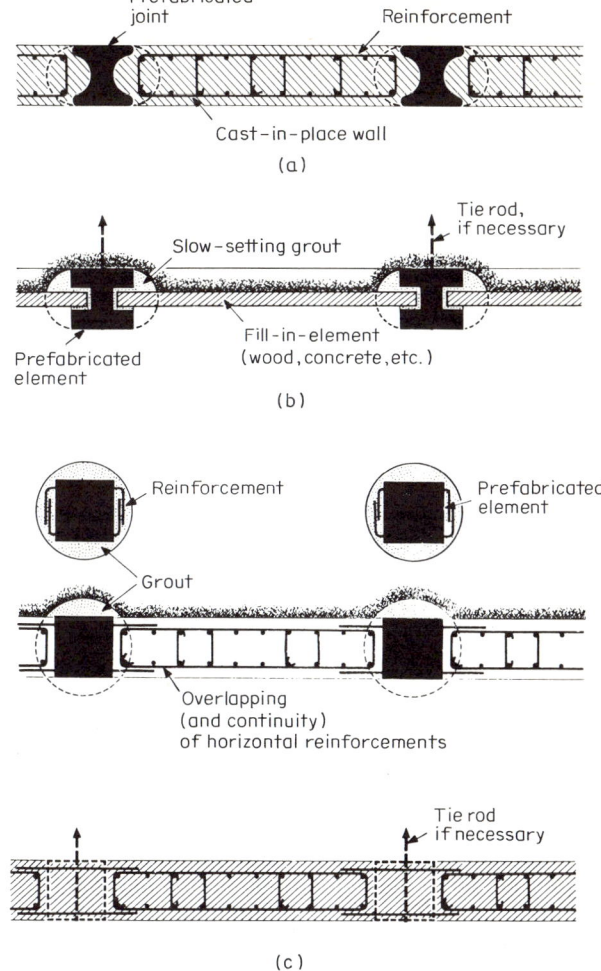

Fig. 10-4 Prefabricated sections for hybrid-wall construction.

this must be done from the surface (contractors report that trained divers often must be lowered into the trench to bend the dowels).

Prefabricated Stop Ends The type shown in Fig. 10-5a is very suitable for use as stop end. In place the prefabricated stop ends are as shown in Fig. 10-5b. Although their use may raise the direct cost of the wall, this is offset by practical and technical advantages. The stop ends remain in place, and thus it is not necessary to maintain lifting equipment and men at the site once the panel is concreted. About 4 h after a pour is completed it is possible to start excavating the next panel. Besides acting as barriers for the fresh concrete, properly reinforced prefabricated ends can be used as structural members to resist lateral loads.

Fig. 10-5 Prefabricated stop ends: (*a*) sections ready for installation; (*b*) panel plan showing how stop ends provide a barrier between concreted and unexcavated area.

Installation

A practicable maximum length is of the order of 15 m (50 ft) for a panel width of 2 m (6.5 ft). This is by no means a fixed limit, and it can be increased or reduced depending on the available headroom and other restrictions at the site. Standard details include fitting and lifting hooks for handling and positioning. The interior face is coated with a special compound to facilitate removal of any grout that may adhere to the concrete.

It is advantageous to provide space near or at the construction site as casting yard. The facilities should accommodate at least the number of panels expected to be installed per day. If the sections must be transported to the site from a central plant, undue costs and delays will be incurred, particularly with long hauling distances and heavy traffic conditions. For average jobs the rule of thumb is to schedule the installation of three panels per day using one excavating machine and one crane. The slurry is prepared at a slurry plant that has mixing, storage, and recycling units, and the grout is mixed at a batching plant. Slurry or grout displaced by the panels usually is disposed of.

The panels are lifted and held vertically by cranes, and in congested sites this means strict compliance with Occupational Safety and Health Act (OSHA) requirements. The panels are lowered slowly into the trench, and the insertion is facilitated if the trench is slightly wider. This overwidth allows also the grout to surround the precast sections fully and provide a continuous seal (Fig. 10-6). The sections are checked for alignment, and as soon as they are disengaged from the crane, they are held in position by special devices which bear on the guide walls. This suspension is continued until the grout has hardened. It is advantageous to stop the tip of the sections slightly above the bottom of the trench to allow the grout to remain underneath and level uneven spots at the base.

Fig. 10-6 Setting a prefabricated section in place. Note that the trench is made wider than the concrete section.

Some contractors have developed special joints for rectangular panels. For example Bachy uses a joint that has a slot in the end, as shown in Fig. 10-7, and this improves watertightness by controlling differential movement between panels. The double recess is regrouted with the insertion of a water stop. A special locking device at the lower end serves to align the sections and keep them together, as shown in Fig. 10-8. The hook at the bottom of the next panel engages in the locking bar of the previous panel.

With the use of displacement grout it is sometimes desirable to graduate the strength characteristics of the bonding material. The transition and variation in the composition and strength of the grout is feasible even at different levels within the same panel. For example, it is possible to have high-strength material underneath the panel where heavy loads must be transferred and a relatively weak but plastic grout in the upper portion of the wall to resist differential lateral movement without cracking.

Advantages and Disadvantages

Prefabricated walls have the merits inherent in precasting. The general appearance of the exposed wall is entirely satisfactory, the face is smooth and clean, and further treatment therefore is not necessary. The concrete quality

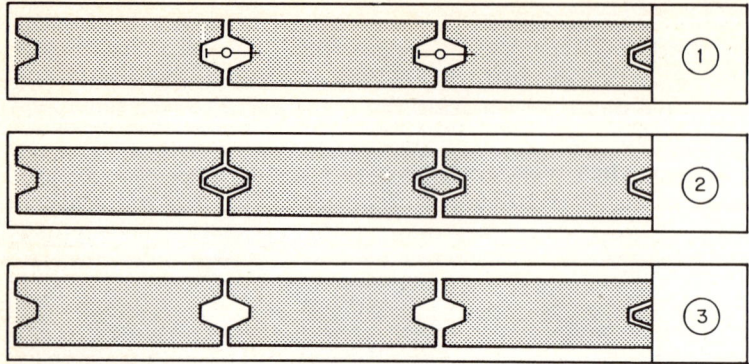

Fig. 10-7 Use of slots to make a watertight joint: (1) water-stop detail; (2) reinforced-concrete key; (3) sealing grout only.

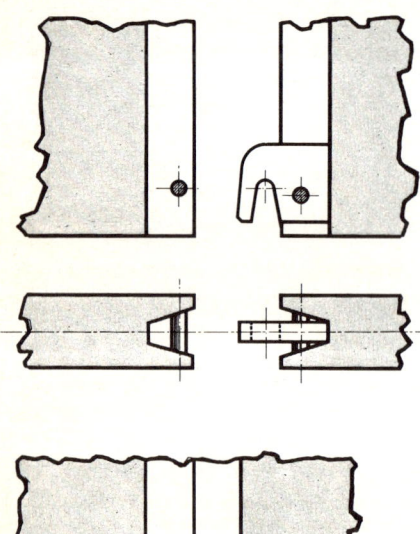

Fig. 10-8 Special coupling device fitted at the bottom of prefabricated sections.

control available with precasting invariably gives better assurance of specified strength and allows better accuracy in positioning the reinforcement. Material savings are possible since prefabricated walls generally are thinner than cast-in-place walls. The final structure is built to finer tolerances, and openings and miscellaneous inserts for connections are more accurately positioned.

The continuous grout on the back face improves watertightness, and if necessary, a waterproofing compound can be applied to the exterior face after fabrication. If exposed concrete is not desired on the interior face, a special finish or face treatment is applied before the panels are installed. Finally, the many practical problems associated with cast-in-place walls and tremie placement of concrete are eliminated.

On the other hand, these favorable considerations must be balanced against the conditions of the job, e.g., the careful planning of the operation and strict adherence to schedule; the minimum job size necessary to offset certain fixed costs inherent in precasting; bad soil conditions causing differential panel movement, which forces the grout to crack at the joints; and the lack of structural continuity across panels. A prefabricated wall is shown in Fig. 10-9.

Fig. 10-9 A prefabricated diaphragm wall consisting of beam-and-slab sections shown after general excavation; the wall is laterally braced with tiebacks.

10-2 BORED-PILE WALLS

A bored-pile wall can be used as ground support. A great advantage in this case is that the installation can be carried out in almost any site and ground conditions. The presence of utilities or difficult soil, such as caving, sloughing, or water-bearing formations, offers no serious impediment to the construction operations. On congested sites the increasing use of auger rigs and the availability of reverse-circulation rotary drills for both small- and large-diameter piles allow both speed and flexibility in construction. Furthermore, bored-pile walls can be built to be reasonably watertight, and if necessary they can be capped with a reinforced-concrete beam to distribute vertical loads.

Configuration and Size of Piles

A bored-pile wall can be made contiguous, i.e., with the piles in contact, or adjoining; interlocking, also called *secant piles*; or intermittent, i.e., with a spacing exceeding the pile diameter if the ground is fairly stable.

Usually the piles are designed to resist both lateral and vertical loads, and these considerations can influence the pile spacing and diameter. A circular section generally requires more main steel per unit length of wall for the same moment capacity than a diaphragm wall of constant section. This is illustrated in Fig. 10-10, which compares the lever arm of reinforced bored piles and a diaphragm wall of thickness equal to the diameter of the piles. The usual practice is to increase the moment capacity of the piles by providing more steel on the tension side if spacing permits and if there is no stress reversal. When an unusually large number of bars are required and the congestion may hinder the flow of concrete, it is better to use steel sections such as beams or columns (North-Lewis and Lyons, 1974).

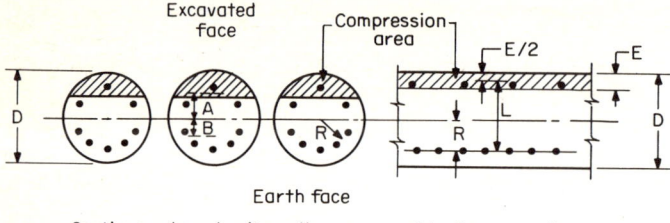

Fig. 10-10 Plan sections through contiguous bored-pile and diaphragm wall. Comparison of effective section and lever arms for circular and constant section. Lever arm for bored piles = $A + B$. Lever arm for diaphragm wall = $L = D/2 + R - E/2$. Given $D = 600$ mm and $R = 200$ mm, it is estimated that the lever arm for the bored pile is 267 mm and for the diaphragm wall is 433 mm.

When it is practicable from the standpoint of hole stability, bored piles can be installed inclined to the vertical to better balance the overturning moment of the lateral loads and thus increase the arm of a free-standing cantilever wall. For vertical piles and average soil conditions this arm is about 8 m (26 ft), but at this excavation depth it is advantageous to brace the piles laterally, usually by means of ground anchors. In this manner the wall movement is limited, and a considerable reduction in the amount of reinforcing steel is possible. If anchors are provided, a reinforced-concrete waling beam usually is cast in place as the excavation is carried down.

The configuration of the piles is decided according to the loads, the ground conditions, and the proposed construction methods, including the availability of equipment. Various configurations are shown in Figs. 10-11 and 10-12. When the soil is competent to stand at least temporarily and groundwater presents no problem, it is advantageous and economical to install only one row of piles and select a pile spacing exceeding the pile diameter. Precut lagging can be inserted either in preformed grooves or wedged against the piles, as shown in Fig. 10-11a and b. However, the location of lagging can influence the magnitude and distribution of lateral stresses. The intermittent pile wall can be combined with a face wall, as shown in Fig. 10-11c.

In caving soil or in soil with water contiguous bored piles usually have the

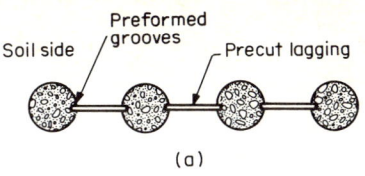

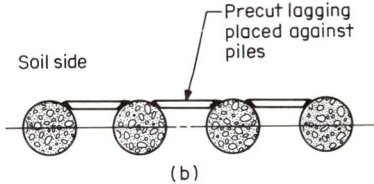

Fig. 10-11 Various types of bored-pile walls: (*a*) wall with lagging inserted in preformed grooves (made by positioning metal forms or foamed-plastic strips); (*b*) wall with precut lagging placed against piles as the excavation is carried down; (*c*) bored piles with a separate concrete-face wall.

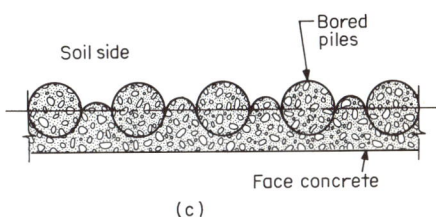

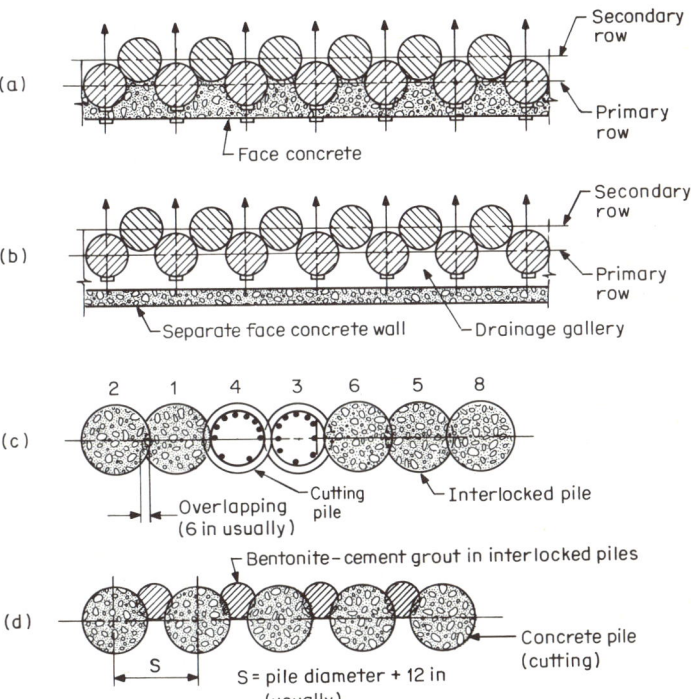

Fig. 10-12 Contiguous bored-pile walls: (*a*) tangent wall with a concrete face cast against the piles; (*b*) tangent wall with a separate concrete facing for seepage and drainage control; (*c*) secant wall formed with equal-diameter piles; (*d*) secant piles of different materials.

347

configuration of Fig. 10-12a and b. The outer row is constructed first, with the piles spaced less than two diameters apart. The piles of the secondary row are located as shown and are displaced toward the outside of the excavation. Since the piles are touching each other, the wall is also called a *tangent wall*. A concrete face may be cast against the piles, as shown in Fig. 10-12a, or it may be a separate wall, as shown in Fig. 10-12b, to provide a drainage gallery. In unfavorable ground conditions, e.g., very loose soil, or where complete watertightness is desired it may be better to use the secant-pile method as shown in Fig. 10-12c and d.

Construction and Installation

Bored-pile walls are processed using auger-type rigs, supplemented in difficult ground or hard rock by core barrels, chisels, jackhammers, and oscillators. Vibratory methods complement mudding or percussive techniques for inserting casing to support the pile hole. On a regional basis and for certain soil types rotary drilling with slurry muds is fast and economical, e.g., in Great Britain and the southern United States. Rotary drilling under slurry protection is particularly indicated where the overburden consists of sand, often in a loose state, and has a high water table.

Excavation with Temporary Casing With auger-type tools it usually is necessary to use temporary casing, not only to support the face of the borehole through cohesionless soil but also to ensure the correct pile spacing by presetting a number of casings in advance of the excavation. Whether the casing is telescoping or is installed in successive lengths by lugs and pins, gaps are formed between piles. With casings 19 mm ($\frac{3}{4}$ in) thick, the pile spacing for a contiguous wall usually is from $7\frac{1}{2}$ to 10 cm (3 to 4 in) greater than the pile diameter to keep the shaft within the normally accepted tolerance.

With auger boring machines the accuracy of the installation, both vertically and in plan, can be improved by having the kelly bar guided instead of suspended. The plumbness of the hole should be checked regularly, especially in hard ground. This is important not only in terms of the structural requirements but also because any deviation from plumbness will make sealing of the joints more difficult.

Problems can arise when attempting to extract the casing, even where moderate casing lengths are used. A situation very likely to cause trouble is a sealed-off zone with artesian pressure above the initial bottom seal of the casing (Woodward et al., 1972). These conditions indicate the probability of deviations in the size of the concreted pile and in its verticality and may result in either intrusion of soil or protrusion of concrete beyond the theoretical line. Hence, the configuration of the wall, pile spacing, and tolerance must be decided accordingly.

Installation under Slurry Support Preboring with bentonite slurry can be used to facilitate the installation and extraction of temporary casing or as an independent method in conjunction with rotary drills and reverse circulation. The construction of a bored-pile wall close to an existing building and in soil overlain by dense sand and gravel may not be possible without the risk of damage to the building; the ground may be too compact for driving or for extracting casings. In this case preboring using slurry down to the bottom of the gravel layer is indicated. The casing is then lowered through the slurry and sealed into the underlying clay. At this stage the slurry can be pumped out, and the excavation is continued in the dry. The concrete is pumped or lowered with buckets.

When the upper formations consist of clay, the hole can initially be bored in the dry. Slurry is introduced as soon as troublesome layers and water-bearing formations are reached. However, in some instances unless the machine is handled along a sufficiently controlled alignment, a temporary casing can provide a useful function in setting the correct spacing and line. In this case the process involves the following steps: (1) prebore and insert collar casing; (2) drill with auger to top of sand layer in dry conditions; (3) fill hole with bentonite slurry; (4) drill to foundation level using drilling and cleaning buckets; and (5) concrete the hole using tremie pipe. If it is necessary to use bentonite slurry for the entire depth of the hole, it is advantageous to use machines equipped with reverse circulation to facilitate materials handling.

Installation under slurry support offers certain advantages:

1. Even in clay formations a dry hole cannot always be guaranteed. Furthermore, delays in concreting can jeopardize a pile since the stability of the shaft cannot be relied on for a long time.

2. The thixotropic slurry can keep loose sand in suspension, minimizing the amount of unconsolidated materials settled at the base of the hole. This is important if the pile must transfer vertical loads.

3. It is often desirable to have a head of bentonite available to oppose heave at the base level.

4. The concrete can be tremied instead of lowered.

Other considerations that influence installation under slurry are deviations from the true vertical line in both the longitudinal and lateral direction; hole enlargement caused by factors discussed in previous sections; and slurry-mud communication between holes in the same vicinity when the excavation is carried out in pervious ground.

Installation under slurry support often requires a hit-and-miss process. If the piles must be made contiguous, the sequence is as shown in Fig. 10-13. In Fig. 10-13a the pile spacing exceeds the pile diameter by the tolerance, usually 15 to 20 cm (6 to 8 in). The odd-numbered holes are first bored and concreted, followed by the even-numbered holes. This is intended to eliminate the problems discussed in the preceding paragraph and also to prevent disturbance to concrete that has not set when an adjacent hole is bored. In the finished series the piles may or may not touch each other, so that both gaps and overbreaks are noticed between the piles when the wall is exposed.

If the pile spacing and sequence is not specified and must be established at

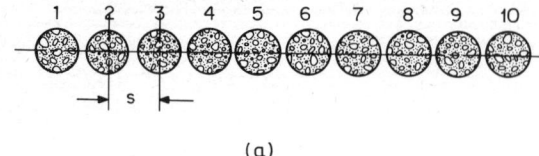
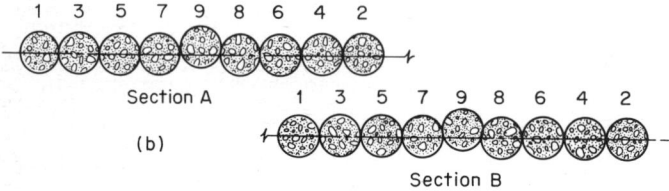

Fig. 10-13 Construction sequence of contiguous bored-pile walls in slurry-filled holes: (a) adjoining piles; S = pile diameter + tolerance; (b) "touching piles" built by the hit-and-miss method.

the site, the construction can proceed as shown in Fig. 10-13b. Production is improved if the drilling rig works in more than one section at one time, but this requires extra mobility. The holes in each section are bored and concreted in the sequence shown, and this prevents slurry-mud communication or other interference between adjacent holes. Each hole is located by the hit-and-miss method, so that in the finished wall all piles are along the same alignment except perhaps the last (center) pile in each section, which may have to be displaced slightly in order to fit.

Sealing of Joints

In water-bearing ground it is necessary to seal the joints, usually by various grouting techniques. Grouting, however, is not always certain to provide a seal which is entirely watertight (North-Lewis and Lyons, 1974). Under normal conditions this sealing is adequate, in that it allows the general excavation to proceed with minor interference and problems for the average water head.

The grouting is carried out behind the piles and before any major excavation begins. The sealing may be less effective near the ground level unless it is accompanied by additional soil cover or surcharge. A higher degree of watertightness is accomplished by guniting or by raking and sealing the defective joints as a supplement to the initial operation and when the face is exposed. Cementation Co. uses a $2\frac{1}{2}$-in-thick (about 6.5-cm) facing of reinforced gunite to make bored-pile walls watertight. Satisfactory results are also obtained with a separate concrete face or a drainage gallery.

Current sealing techniques include the presetting of grout tubes in the shaft during construction. Another method provides the wall with a continuous concrete membrane or face, but this can be used only in good cohesive strata.

Uses and Limitations

Bored-pile walls built with auger-type tools supplemented with casings are suitable in cohesive soils or where the upper crust of the ground consists of water-bearing formations of limited depth. On the other hand, contiguous walls built in slurry-filled boreholes have only a minor dependence on the type and conditions of soil, and the most serious impediment to their use is probably the presence of hard rock and similar obstructions.

The walls provide a positive construction for either shallow or deep retaining structures and can also be used as load-bearing elements. Depending on the design requirements, such walls may be single structures or stepped where excessively high single walls are undesirable for technical or economic reasons. When the walls are permanent, they are provided with helical steel cages for the entire depth. If the piles are to serve only temporarily, the reinforcing cages may be replaced by rolled-steel joists, which can be recovered when the permanent construction is completed.

The use of bored piles as retaining walls is largely a matter of economics and ground conditions. The method has an inherent flexibility, particularly with regard to existing shallow footings and underground utilities, and in congested or confined sites. When the walls are built as individual piles, it is usually possible to chisel through hard objects or obstructions, and to bore close to and around footings with a minimum risk of damage and in a minimum of time. When squeezing or caving ground with water is encountered, the use of slurries and especially mounted pier-drilling machines with reverse circulation will prevent hole instability and allow completion and concreting before appreciable soil movement occurs. In certain cases lack of working space, headroom, accessibility, or working time can preclude the use of other ground supports, including diaphragm walls, on the basis of construction efficiency and in spite of favorable soil conditions.

However, bored piles have certain limitations and disadvantages. Excessive overbreak can occur in unstable ground while the casing is withdrawn, or an oversized hole can result with the use of slurry, especially in granular soil. Despite the measures taken to alleviate the consequences of these problems, seepage and leakage through the joints cannot always be stopped to the degree desired, particularly with a high water table and deeper than normal excavations. Finally, structural connections are more difficult, especially with horizontal members such as slabs and beams, and require elaborate details.

Secant-Pile Walls

Secant-pile walls, shown in Fig. 10-12c, can be regarded as the predecessors of diaphragm walls. Indeed, the development of machines that could excavate linear slots made the conversion from interlocked holes to straight walls possible. Like contiguous piles, secant piles are used as ground support. However, they exhibit the flexibility inherent in bored-pile walls without the undesirable characteristics.

The minimum practical wall thickness is about 45 cm (18 in), but this depends on the overlapping dimension, which usually is from 10 to 15 cm (4 to 6 in). The resulting advantage of this interlocking is a nearly watertight joint provided the specified maximum allowable tolerance is satisfied to prevent out-of-plumb piles from becoming unlapped.

Initially, only alternate piles are cast (numbered 1, 3, 5, etc.), using conventional auger rigs, and concreted with or without reinforcement. The cutting of the reinforced piles (marked 2, 4, 6, etc.) into the previously cast piles usually is done 1 day later so that the concrete has set but is not too hard. This operation is carried out by means of a hydraulically actuated casing fitted with a special cutting edge. The casing is guided at two points on a heavy boring rig.

Since all piles in a secant wall resist the lateral loads equally, it is advantageous to provide reinforcement in every pile; some contractors object to this procedure, however, on the grounds that it may create difficulties when cutting into the adjacent piles. Having reinforcement in every pile is particularly desired with large diameters, where more steel is placed on the tension side. For the cutting piles the bars are arranged along a spiral, whereas for the interlocked piles they are arranged within square ties. Secant piles generally are more expensive than other types of bored-pile walls.

This author has developed a variation on the conventional secant piles, shown in Fig. 10-12d. The interlocked holes can be made with a diameter one-half that of the cutting holes, and the holes are filled with a cement-bentonite grout of the type described in previous sections. Normally, this grout attains a strength of 100 to 200 lb/in^2 (7 to 15 kg/cm^2). The resulting advantages are a considerable saving in materials by transforming the initial slurry into a construction material and more flexibility in the use of available equipment for cutting the piles.

The Benoto Rig This special equipment, made for secant-pile walls, can travel under its own power up to 300 m. For greater distances it usually is necessary to lower the mast and fit suitable road wheels. The rig moves on a rail track mounted on a steel plate. It lifts itself up on four hydraulic legs, so that the rail track hangs on the machine and can be moved backward or forward. The machine is then lowered onto the track and continues its travel. Because of the large contact area, the bearing pressure is low when the machine sits on its track; hence no platforms or guide walls are necessary to distribute the load.

The attendant plant with a rig consists of a light crane for handling the casings and the reinforcement and a small traxcavator for handling materials and leveling the ground for the guide timbers. When two rigs work in tandem, they can be serviced by one set of attendant plants.

A Benoto rig is shown in Fig. 10-14. During operation of the piling rig a steel casing is forced into the ground by means of hydraulic rams, while a secondary transverse ram imparts a twisting motion to the casing. A grab

Various Wall Systems   353

Fig. 10-14 A Benoto piling rig working on forming a subway tunnel wall close to the boundary of residential buildings. (*Courtesy of David Jobling.*)

removes earth material from inside the casing as sinking progresses, and the casing is always kept ahead of the grab. When the excavation is completed, the steel cage is inserted into the casing and the concrete is poured. As the concrete level rises, the casing is gradually withdrawn with the same twisting motion, which leaves a distinctive finish on the concrete face.

10-3 COMPOSITE WALLS

Concrete diaphragms cast in slurry panels can be combined with other structural elements to produce a composite ground support or foundation wall. Successful combinations are obtained with drilled piers or steel I beams. These variations are worked out from special design and construction requirements, but frequently they are a matter of custom and regional trend.

Steel-and-Concrete Panels

Soldier Piles and Reinforced Concrete The two most usual variations of this type are shown in Figs. 10-15 and 10-16, and clearly their difference is only in the method of construction. In the United States this wall type was developed regionally on the West Coast and had its first applications in San Francisco. Difficult ground, interbedded with soft bay muds, loose sand, recent rubble fill, abandoned timber piles, debris from old sailing ships, and the like, required the use of relatively short panels. The adaptation of steel fram-

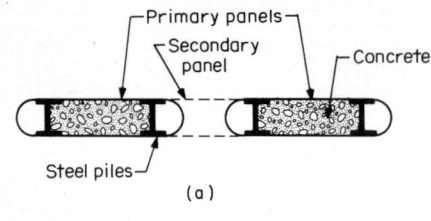

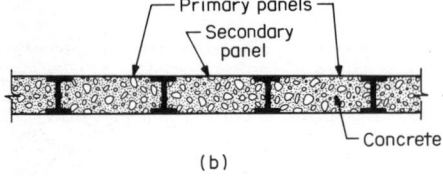

Fig. 10-15 Typical composite wall: (*a*) outline of excavated panels; (*b*) finished wall.

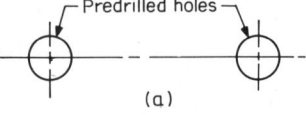

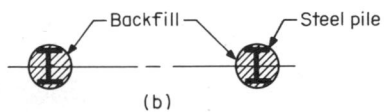

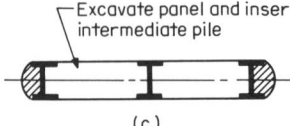

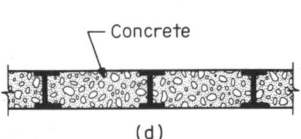

Fig. 10-16 Alternate construction method of composite wall.

ing for the underground structure dictated the utilization of steel members in the wall to make direct structural connections and thus resulted in the composite wall.

Where site and ground conditions permit, the installation proceeds as shown in Fig. 10-15. First, the primary elements are built in alternate panels, using either round- or square-end machines to carry out the slot excavation. Each panel must correspond to one equipment pass. The steel I beams are assembled with the reinforcing cage and inserted as one piece. Each panel is concreted between the webs of beams, as shown in Fig. 10-15*a*, and the wall is completed by the casting of the secondary panels, as shown in Fig. 10-15*b*. The joint detail can be executed as shown in Fig. 8-27. A round tube may be needed together with the I beam in poor ground or where the construction accuracy is low and overbreak is expected.

Alternatively, a composite wall must be built as shown in Fig. 10-16. The steel beams are set in predrilled holes, as shown in Fig. 10-16a and b, along the proposed perimeter of the structure, and this usually is done by preboring under slurry support. After the beams are set, the holes are backfilled with granular material to prevent concrete leakage. The panels between beams are excavated with square-end clamshells, usually in three passes, the intermediate beam is inserted as one assembly with the reinforcing cage, and the panel is concreted.

Of the two methods the construction shown in Fig. 10-15 has the advantage of simplicity. The installation does not require special equipment like auger rigs or pile-driving machines. The lifting and handling of the beams and cages is done in one operation while the beams serve to brace and stiffen the cages.

The usual beam spacing is from 2 to 3 m (6 to 10 ft), but it can be increased if the range of equipment permits. The wall thickness is from 60 to 120 cm (24 to 48 in).

Advantages and Disadvantages The wall types shown in Figs. 10-15 and 10-16 are more expensive than all-concrete walls for the usual loading, but in many instances the difference in direct cost is balanced by other considerations. Thus composite walls are advantageous (1) in connection with short panels to reduce the risk of caving of face collapse; (2) where it is better to resist the lateral loads by steel sections because of their greater flexural rigidity; (3) where it is desirable to use interior bracing consisting of steel framing and make permanent connections with the wall; and (4) where limited space and time preclude all-concrete walls.

Concern is often expressed with regard to the durability of I beams in earth. In spite of the generally satisfactory performance and durability of I beams and steel piles in foundation work, some codes specify mandatory protective measures, which either detract from the economy of the installation or influence its feasibility. Low pH, resistivity, and the presence of chlorides and sulfides in the soil are considered to constitute a corrosive environment against which the steel beams must be protected. Under these conditions the installation becomes more expensive.

Studies by Romanoff (1961, 1969) indicate that the type and amount of corrosion observed on steel piles driven into undisturbed natural soil is not sufficient to affect their strength or useful life significantly, and this is true regardless of soil characteristics and properties. Undisturbed soils are so deficient in oxygen at levels a few feet below ground surface or below the water table that the steel sections are seldom liable to corrosion. Consequently, such soil properties as type, drainage, resistivity, pH, and chemical composition have little meaning in the corrosiveness of soils toward steel sections. Evidently, the method of trenching under slurry should not be expected to alter the oxygen level, since the excavation is isolated from the earth environment and the presence of slurry precludes communication with the outside air.

Soldier Piles and Plain (Unreinforced) Concrete Although this wall type is comparatively recent, it has been tried and used widely in Japan. The

356 Slurry Walls

wall is built in a series of rectangular panels, each containing three or four H piles. Since the piles resist the lateral loads equally, they are of the same flexural rigidity. They are spaced close enough to each other to make reinforcing bars unnecessary in the concrete. An angle or flat bar is welded to the flanges of the end beams to act as barrier to the flow of fresh concrete, and a spacer box is used in the end chamber in lieu of the interlocking tube.

The excavation can be carried out with any type of slurry-trench equipment, including round and square ends. However, the accuracy in verticality can affect the operation. If the fin of the end piles is too short, it may leave openings in the trench and thus provide leakage paths for the fresh mix. If the fin is very long, to ensure sufficient penetration, driving the end pile will require considerable force, causing some sloughing at this location.

The construction is staged according to the sequence shown in Fig. 10-17 and requires the following steps:

1. Excavate primary (alternate) panels as shown in Fig. 10-17a. The panel size depends on the maximum beam spacing (without reinforcement) and on the number of tremie pipes that can be used.

2. Install H piles. The intermediate piles are lowered, but the fin piles usually must be driven. The pile width is the same as the width of the trench.

3. Install spacer box, as shown in Fig. 10-17c.

4. Place concrete using tremie pipes, as shown in Fig. 10-17d. All chambers in the same panel must be filled simultaneously, or the intermediate piles can be distorted excessively or even displaced by unequal pressure from either side.

5. Excavate secondary panels, extract spacer boxes, and scrape steel face to remove residual material for a clean joint, as shown in Fig. 10-17e and f.

6. Construct secondary panels, as shown in Fig. 10-17g amd h. The piles are kept in position better if they penetrate the soil below the base of the trench and if their tops are held firm to give resistance to displacement while the fresh mix is tremied. No special difficulties should be encountered as the spacer box is extracted as long as it is not in contact with the concrete. The box can be extracted any time after initial concrete setting.

The concrete section is assumed to act in bending between the steel piles, as shown in Fig. 10-18, and therefore failure will occur in the concrete when the tensile strength is exceeded. Recent tests have shown, however, that the ultimate resistance of concrete is many times the strength of plain concrete in bending, and this is attributed mainly to the greater ratio of wall thickness to span length which results in a very different type of failure (see also Sec. 13-6).

In Japan this wall type is frequently more economical than the continuous concrete diaphragm or bored-pile wall of the same flexural rigidity. This is due to the savings of materials through a reduction in the wall thickness and to the favorable economics of using factory-produced steel vs. the high cost of on-the-job assembly of reinforcement.

Various Wall Systems

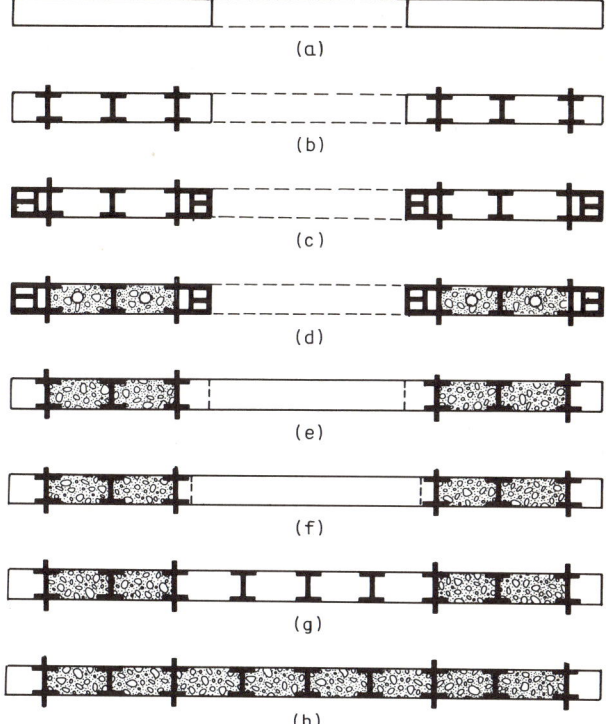

Fig. 10-17 Construction sequence of steel and plain (unreinforced) concrete composite wall: (*a*) excavation of primary (alternate) panels, (*b*) installation of steel piles, (*c*) installation of spacer boxes, (*d*) placement of tremie concrete, (*e*) excavation of secondary panel, (*f*) cleaning steel faces to remove residual material, (*g*) installation of steel piles in secondary panel, and (*h*) completion of wall.

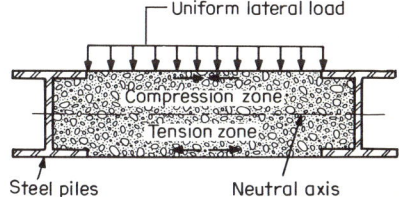

Fig. 10-18 Assumed bending action of plain (unreinforced) concrete panel.

Walls Interlocked with Bored Piles

A second example of composite walls, adopted primarily in the United States, is the diaphragm wall interlocked with large-diameter bored piles. The possibility of this combination is obvious in the underground portion of a tall building and is exemplified by a basement enclosure which must provide high load-bearing capacity although this load transfer is not possible at the tip of the wall because of insufficient soil strength there.

The installation of the bored piles usually precedes the construction of the main wall, and in some jobs it is stipulated under a separate contract. The engineer must therefore work out the particulars of the construction as a whole with emphasis on the wall-to-pile connection. Although the pile spacing is dictated by the main features of the superstructure, such as columns, bays, and the like, it cannot exceed the practicable maximum for a horizontally spanning wall; otherwise an unusually thick or heavily reinforced wall will be needed.

For the construction shown in Fig. 10-19 the casing is used only for the excavation. The reinforcing cage is provided with a slot formed by a Styrofoam filler, which is intended to receive the panel and act as shear joint. The casing is withdrawn as the hole is concreted, but the filler remains. The linear panel is usually excavated with square-end clamshells, but suitable scraping tools are necessary to remove the Styrofoam and clean the concrete surface for good connection. Despite protrusions and overbreak that may occur as the casing is pulled out, the key remains intact and smooth, and the only requirement is to position and center the cage accurately. If all residual material is removed, the joint is fairly watertight.

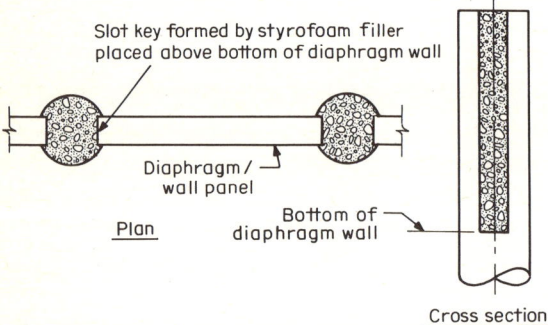

Fig. 10-19 Diaphragm wall interlocking with drilled piers.

If the steel casing must be left in place, a mechanical connector can be provided by welding sections of a straight web pile or a flat bar to the casing, as shown in Fig. 10-20. The bored hole must be large enough to receive the entire assembly; otherwise the insertion must be supplemented by driving. The annular space is then backfilled with fine gravel or granular material as shown in Fig. 10-20. The use of grout to fill this space should be avoided because it is extremely difficult, if not impossible, to scrape off and remove the grout for the wall connection. Panel excavation around the pile casing requires special chisels guided by a kelly to embrace and trim the soil from around the web attachment because this cannot be reached by the main equipment. Since this operation can be difficult and troublesome, it deserves appropriate inclusion in the specifications.

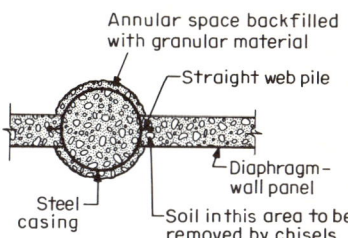

Fig. 10-20 Connection detail of cased piers and diaphragm-wall panels.

10-4 CIRCULAR AND POLYGONAL ENCLOSURES

General Considerations

Circular enclosures can be built along guide walls which are truly circular in plan. In this case the distance between guide walls is increased to include, besides the normal tolerance, the chord deflection for one panel, usually one equipment pass. Although the resulting structure is not perfectly circular, it will resist the lateral loads mainly as a circular wall, particularly for large diameters. For relatively small diameters the enclosure is built as a polygon, in which case the guide walls follow a similar configuration.

The resulting advantages can be significant. Under appropriate circumstances it may be possible, for example, to omit lateral bracing in the wall or reduce the wall embedment below excavation level. Functionally circular enclosures are most suitable for construction and access shafts, underground storage tanks, hydraulic and power installations, and underground parking where the conversion to a circular plan is not in conflict with the layout of the superstructure.

Underground circular structures are self-supported walls, i.e., in most instances they need no internal bracing. The concrete is utilized under stresses to which it is particularly suited, i.e., compression. Finally, if the base of the excavation is stable and groundwater poses no problem, the wall can be stopped just below this level.

For the same plan area a circular shape requires less perimeter than a square structure, and this saving in wall length is close to 12 percent. A circular wall resists the lateral stresses in such a way that it is certain to undergo inward movement of smaller, and often inconsequential, magnitude. Some yielding is likely to occur where one panel meets another because of the tendency to squeeze, consolidate, and compress any bentonite and impurities trapped in between, and this continues until full bearing is mobilized between the two concrete surfaces in contact. Movement due to shrinkage and creep is seldom significant, generally because of the favorable curing conditions.

The lateral wall displacement due to yielding at the construction joints can be estimated as follows. Let the wall have an initial mean radius r_i and perimeter s_i and n construction joints and panels. If each construction joint yields tangentially by an amount δ, the initial perimeter will be shorter and become s for a final radius r, as shown in Fig. 10-21. The following relations hold:

$$r_i = \frac{s_i}{2\pi} \qquad r = \frac{s}{2\pi} \qquad \text{and} \qquad s = s_i - n\delta$$

When we note that the inward movement is $y = r_i - r$, it follows that

$$y = \frac{n\delta}{2\pi} \tag{10-1}$$

which shows that the lateral inward wall movement is independent of the radius and depends mainly on the number of construction joints and the compressibility of any soft material left there. For example, for a wall with a radius of 50 ft and 20 panels, the lateral movement for a yield of $\frac{1}{8}$ in at the joints is 20(0.125)/6.28 = 0.4 in.

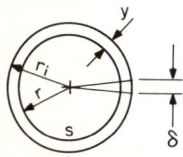

Fig. 10-21
Movement of a circular wall toward the excavation caused by yielding at the construction joints.

Polygonal Shapes For underground openings of relatively small diameter (up to 10 m, or 35 ft) the shape approaches a polygonal configuration. Examples are ventilation, access, drop, and mining shafts. The polygon is evolved as shown in Fig. 10-22. If the interior clear distance between faces and a tentative wall thickness are known, a basic inscribed circle is drawn as shown having as diameter the inside clear distance plus the selected wall thickness. A suitable polygon is drawn next so that its sides are tangent to the basic circle, and this provides the configuration of the wall along the centerline. The usual side length is from 1.5 to 2.5 m (about 5 to 8 ft) and is excavated with one equipment pass; hence the panel range of the excavating implement will determine the number of sides as well as the side length.

For unreinforced concrete polygons the usual panel extends from corner to corner and may involve two or three sides. One or two tremie pipes are placed at each interior corner, and the construction joints are executed using the round-tube detail shown in Fig. 10-22b. In addition to overwidth excavation at these ends, a further problem is the encroachment of adjacent sides; thus unless a barrier is provided, the fresh concrete in the shaded area is part of the next panel and must be broken by chisels to allow a smooth excavation.

If the structure must be reinforced, the construction joint is better executed if it is placed away from the corners, preferably at the center of the side. Since the reinforcement must be continuous horizontally, the joint must be given

Various Wall Systems

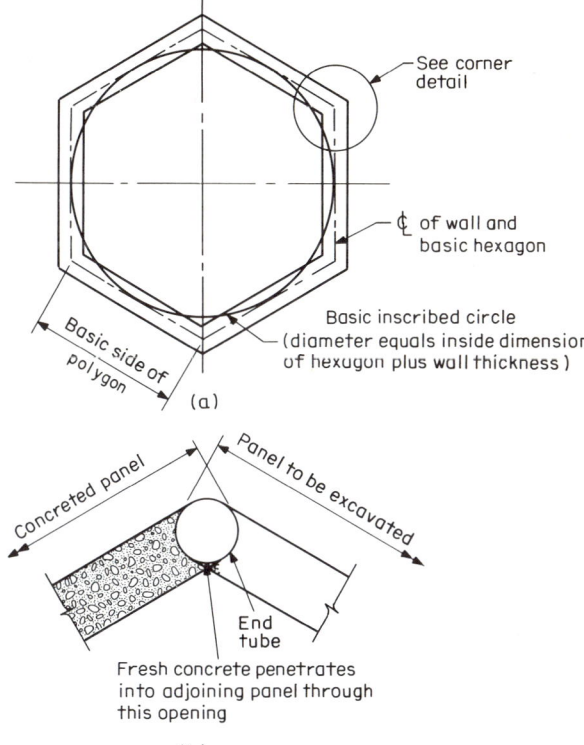

Fig. 10-22 (a) Development of a hexagonal shape from a basic circle; (b) corner detail.

the details described in Chap. 8. The panel length for concreting can consist of one or two full sides and two half sides, and this may require two or three tremie pipes located at the corners.

Construction Requirements

The usual wall thickness is from 24 to 30 in (60 to 75 cm). The following construction requirements can influence the design: the wall must have sufficient bearing at its base; the ground must be stable against base failure at excavation level; and the lateral forces combined with surcharge loads and unloading caused by excavation must give a resultant load which is substantially uniform around the structure; otherwise the wall can suffer local distortion (see also Sec. 13-5).

Because the wall is expected to perform as a compression ring, the construction-joint details must be considered accordingly. Square-end joints are more effective in transferring compression-ring stresses especially if the horizontal reinforcement is extended through. Workmanship and manner of execution can determine the actual yield of panels at the joints. Processing the enclosure in longer units will reduce the number of joints, but this also means more tremie pipes and enhances the possibility of mud trappings and inclusions at the joint.

A cap beam cast separately after the wall is completed serves as an upper ring to tie individual panels and distribute the vertical loads (see Fig. 10-23). More internal circular braces are provided, if necessary, and they can be either steel rings or cast-in-place sections. Experience shows that in circular enclosures of relatively large diameter (50 m, or 160 ft) local distortion or excessive lateral movement sometimes is unavoidable unless the wall is braced with ground anchors. For intermediate-diameter walls it is possible to eliminate bracing rings as well as tiebacks.

Fig. 10-23 Circular enclosure for a power station, Isle of Grain, Kent (England). (*ICOS*.)

The compression force on a cross-sectional area 1 ft deep is

$$P = r\sigma_h \qquad (10\text{-}2)$$

where r is the mean radius and σ_h is the lateral earth stress per square foot at the point under consideration. This force increases with depth and becomes a maximum at the base of the wall. The wall thickness must therefore be selected to accommodate the compression stresses at this level and kept constant for the entire height since it is never practicable to change the width of the trench as the excavation is deepened. It thus may be expedient to select a relatively thinner section and increase its capacity at the base by providing compression reinforcement. On the other hand, it is sometimes possible to eliminate the internal bracing of relatively large circular enclosures by improving their elastic stability through an increase in the wall thickness. For example, a circular structure, 49 m outside diameter (160 ft) and 20 m deep (65 ft), was constructed as a free-standing wall without internal rings or tiebacks merely by increasing the wall thickness from the initial 60 cm (24 in) to 80 cm (32 in).

If the wall is constructed without bracing, it is advantageous to monitor its movement during unloading of the interior area. Earth moving should be done in stages to allow distortions to be completed, and the removal of earth materials should be as uniform as possible. Frequent measurements of wall movement will disclose irregular distortion of the structure, and if this is observed, it may be necessary to install supplementary bracing before continuing the excavation.

10-5 POSTTENSIONED DIAPHRAGM WALLS

The prestressing of concrete can be applied to diaphragm walls to extend their effective structural length; it creates a compressive stress within the concrete which partially or wholly balances the tensile stress expected to occur in service. Furthermore, the accompanying reduction in the amount of reinforcement is substantial and can remedy certain problems of steel shortages. One method is to posttension high-strength steel-wire strands after the concrete has cured in the panel. A great advantage in this case is the increase in the stiffness of the section and the subsequent decrease in the elastic deflection, which makes extension of the unbraced excavation depth possible. For instance, a wall 75 to 90 cm thick (30 to 36 in) will probably have a maximum unsupported cantilever height of about 7.5 m (25 ft). With posttensioning this can be increased by more than 50 percent. This improved capacity is even more significant if a single bracing can be provided at the top. Braun (1972) has reported that a prestressed structure 80 cm thick (32 in) supported an excavation 16 m (53 ft) deep without bracing other than a single row of tiebacks installed 4.5 m below the ground surface.

The possibility of utilizing the advantages of unbraced walls through posttensioning becomes obvious with certain classes of building construction where lateral support by intermediate floors cannot be provided at all in the final stage. Such buildings are, for example, certain underground garages for which a continuous ventilation gap is required by fire regulations between the earth retaining walls and parking floors as access for pumping water and foam directly from the ground level. In this case, unless the exterior walls can be self-supported, the construction is not feasible at all.

Construction and Installation

Prestressing is not a fixed state of stress and deformation but is a time-dependent process whereby both concrete and steel deform plastically under continuing stress. For this application the provision of high-strength concrete which has low creep, shrinkage, and thermal response is very much under the influence of the construction technique. Because of the wall confinement the conditions are rather favorable for prestressing. The process is further benefited by a concrete strength which usually is higher than normal since it is developed in better curing conditions. Other favorable factors to be mentioned are the ample supply of dependable high-strength-steel strands, the availability of equipment for posttensioning, and the current progress in the state of the art.

Theoretically, the only common requirement between high-strength concrete, which is commonly used for prestressing, and high-slump concrete, which is necessary in slurry-trench construction, is the small size of the aggregate fraction; the water-cement ratio is exactly the opposite. In spite of this conflict, concrete for prestressing can easily be provided in the range of 4000 to 4500 lb/in^2 (275 to 310 kg/cm^2) without sacrificing the high slump or making the application too expensive.

At present the posttensioning of diaphragm walls is highly specialized. According to recent applications, it is possible to arrange conventional prestressing strands and ducts so that they will not be affected by submersion in bentonite slurry. This is facilitated by spacing the vertical bars farther apart than usual. With the tremie pipes properly located, the presence of ducts should not hinder the flow of fresh concrete.

Very significant are the results of field tests on prototype panels showing that the posttensioning of strands at nominal eccentricity will not necessarily cause distortion of the panel in the ground. In fact, it may be impossible to introduce tension at the extreme concrete fiber even if the strands are overstressed. This important point is discussed in some detail in Chap. 13, but an obvious conclusion drawn from experience is that the soil mass surrounding a posttensioned wall provides an ideal stressing bed of infinite strength.

Locating and Stressing the Tendons The prestressing strands are arranged in posttensioning units within the reinforcing cage, as shown in Fig. 10-24. The assembly usually represents a compromise between steel economy and handling difficulties and requires the use of temporary stiffeners attached to the cage to prevent distortion. These reusable attachments and tackling aids can be devised according to the panel size and ensure that neither the prestressing strands nor the stressing heads will be displaced during handling and lowering of the cage.

The tendon loops are cased in metal ducts. Bottom anchorages are avoided by placing the cables in a U shape. The upper ends are tensioned and anchored using a suitable device. The method of looping the cables generates the tendency for stress concentration in the bottom zone of the anchorage, but, according to the record, no loss of tension due to yielding of concrete has occurred. The cables usually are fully tensioned in one stage, this being possible because of the resistance offered by the surrounding soil, which prevents deflection of the wall and therefore unacceptable tension in the concrete.

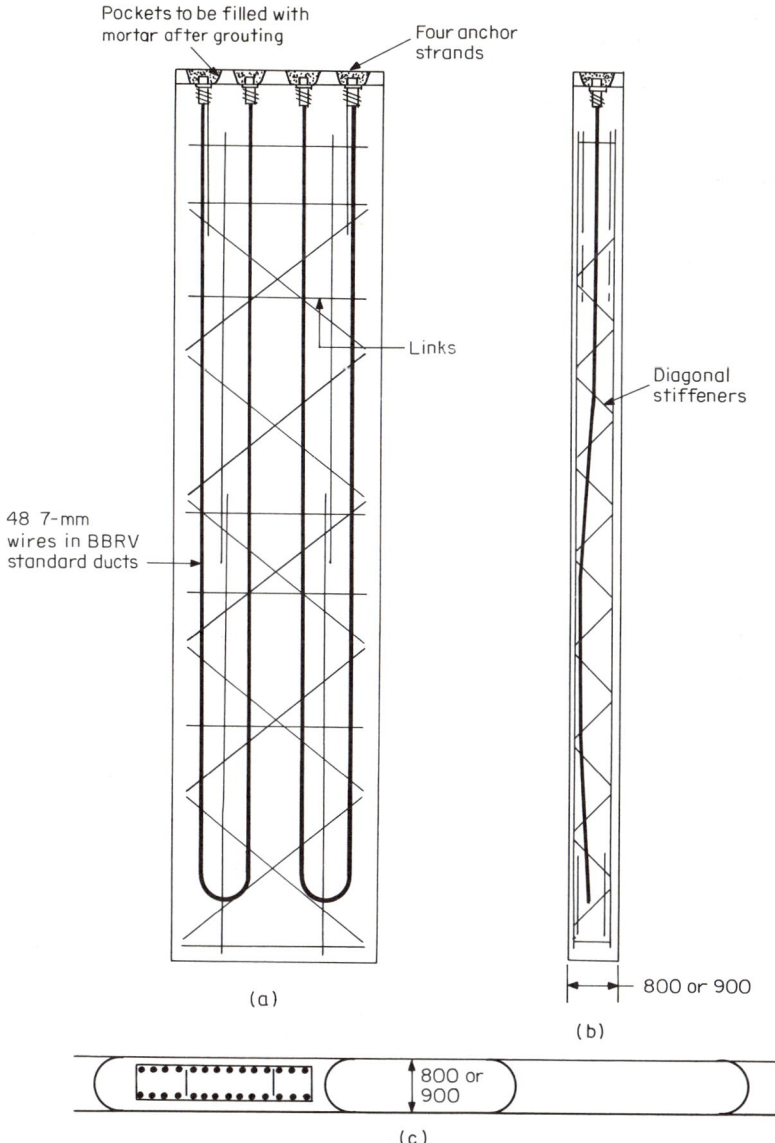

Fig. 10-24 Arrangement of strands in posttensioned diaphragm walls: (*a*) elevation, (*b*) vertical section, and (*c*) horizontal section. Plan of the German Embassy in London; dimensions in millimeters.

For nominal panels two sets of metal ducts are sufficient, as shown in Fig. 10-24*a*. The ends together with the stressing heads (see Fig. 10-25) are projected into the top capping beam and must be precisely located along the centerline of the wall. Similar precision must ensure the eccentricity of the ducts elsewhere in the panel, and spot welding the supporting rods provides the exact fixing positions within the cage. The eccentricity can reach a

Fig. 10-25 Details of stressing anchor heads in reinforcing cage; construction of the German Embassy in London. (*ICOS, Great Britain.*)

maximum of 30 to 40 cm (12 to 16 in) within the panel zone where maximum cantilever moment is anticipated to occur upon excavation.

When the wall is completed, a separate capping beam is constructed in conventional formwork using normal-slump concrete. When the concrete in the capping beams has reached at least 80 percent of its ultimate strength, stressing is applied simultaneously by means of two jacks, one at each end of the loop (see Fig. 10-26). If there is doubt about the stiffness and response of

Fig. 10-26 Posttensioning of diaphragm walls; construction of the German Embassy in London. (*ICOS, Great Britain.*)

soil, the stressing force should be applied in two or three stages. Stressing is carried out progressively by working along the anchorages of each capping beam. The operation is completed by injecting grout into one arm of the duct system until it emerges from the end of the opposite arm.

The installation can take advantage of the apparent rigidity of the panel in the original ground by allowing temporary overstresses in concrete and steel. Self-adjustment will gradually take place, due to a normal prestress loss and the application of the cantilever moment upon excavation.

A promising further development is the use of concentric tendons in conjunction with multianchored or multipropped retaining diaphragm walls.

10-6 BUTTRESSED WALLS, CELLS, AND ARCHED STRUCTURES

The wall types described in the preceding sections generally are single vertical ground supports and resist lateral earth stresses and vertical loads in one plane. Free-standing cantilever walls are feasible for relatively shallow excavations and where ground movement can be limited by the strength of the soil. Walls braced with ground anchors are feasible in favorable ground, provided it is possible to control the legal aspects associated with their construction. Situations can arise, however, where these solutions are not acceptable, e.g., in waterfront installations involving landfills behind the retaining structures. In this case diaphragm panels in the form of buttressed walls, cells, and arched structures can be used instead of cellular cofferdams, concrete monoliths, and open-quay structures. The stability concept of these structures ranges from essentially stiffened cantilevers to gravity-type structures (see also Sec. 13-5).

Buttressed or T Walls

The continuous series of T sections like that shown in Fig. 10-27 will usually be analyzed as a stiffened cantilver wall and not necessarily as a gravity wall. Although a great advantage is the resulting rigidity, which allows greater unbraced lengths, the wall derives its stability by sufficient embedment below excavation level. The efficiency of the wall is greatly improved if the structure is oriented with respect to the lateral loads as shown. Most of the flange of the T section and probably a portion of the stem are in compression, and the reinforcement is more effective if placed at the end of the stem according to the principles of T beams.

The assumption of side shear (friction or adhesion) at the interface with the surrounding soil has considerable influence on the analysis of stability, since it means that the weight of the soil column between stems must be taken into consideration in tabulating the forces acting against the wall (see also Chap. 11). Where this is justified, it will result in wall deflections of a relatively smaller order.

If the stem and flange lengths are of the order of 3 m or less (about 10 ft), it

368 Slurry Walls

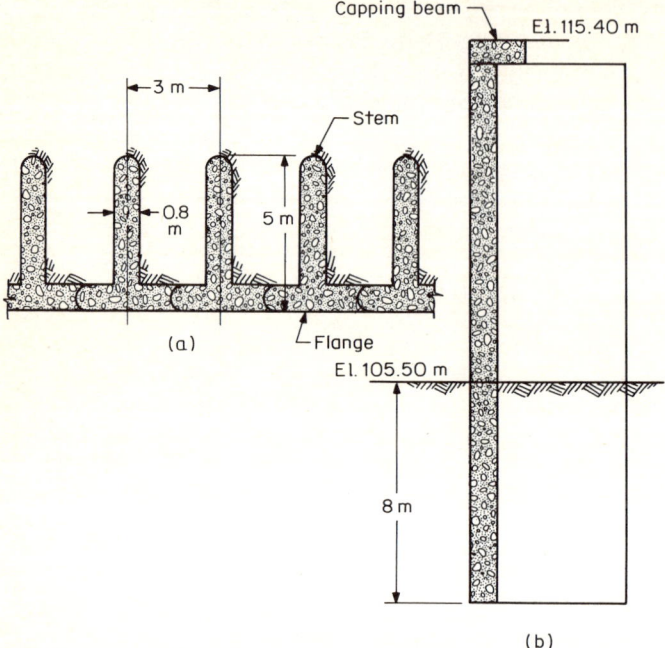

Fig. 10-27 T-shaped diaphragm wall built as perimeter wall for the Harrow-on-the-Hill reservoir, England: (*a*) sectional plan and (*b*) cross section. (*From Fisher, 1974.*)

is possible to fill the excavation with one tremie pipe placed at the junction. For longer fin sections the use of a second tremie pipe at the end of the stem is quite helpful to the operation in view of the congestion which is likely with the main steel bars there. The construction joints usually are located in the flange, as shown, and consist of a simple round tube. Because of the T shape the panels should be excavated with a relatively mobile equipment, and it is advantageous to have each leg completed in one pass. The reinforcement for one panel is assembled and installed in a single cage.

Arch-Type Walls

An example of arch-type wall, in this case provided with a closed face, is shown in Figs. 10-28 to 10-30. The ground conditions at the site are favorable for this type of structure, the upper deposits being dense sand layers over a considerable bed of boulder clay overlying sandstone layers. In this condition and with a proper geometric configuration the wall can be designed as a gravity structure, so that all lateral loads are converted into direct bearing at the base. For this reason the wall of Fig. 10-28 has its front corners constructed as bored piles to increase the contact area along the shaft and at the base. The long back-fin wall improves resistance to overturning not only by adding weight and by increasing the arm of the resisting moment but also because this section will have to be drawn and extracted like a tooth from the

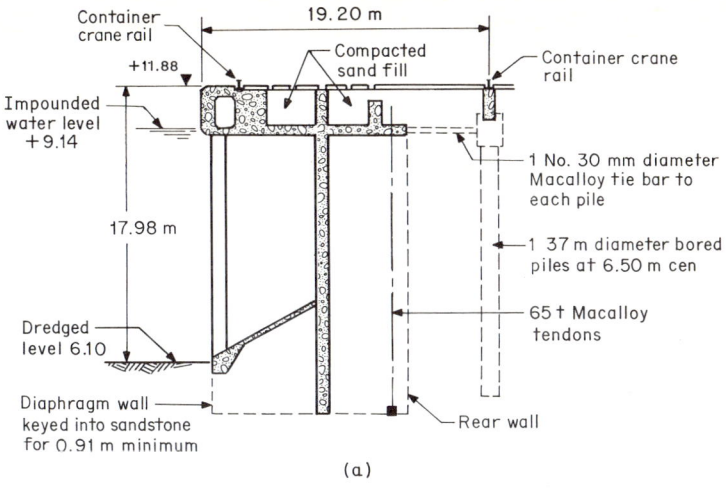

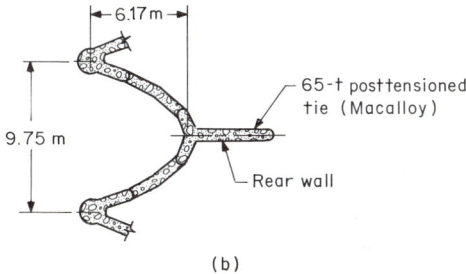

Fig. 10-28 Details of Seaforth dock wall, Liverpool: (*a*) typical cross section; (*b*) plan. (*From Agar and Irwin-Childs, 1973.*)

Fig. 10-29 Guide-wall construction for the arch wall of Seaforth dock, Liverpool.

Fig. 10-30 View of Seaforth dock arch wall, Liverpool.

surrounding soil before the wall can rotate about its toe. Stability against overturning can be further increased by rock anchors connected to the back fins to tie the structure to the ground, but results of work on prototype panels indicate that in fairly good soil this is not necessary (Agar and Irwin-Childs, 1973). More serious is laminar displacement of the wall, but this can be avoided by rigidly connecting the top of the structure to a substantial deck and by keying the base into solid unyielding materials.

A postconstruction problem that can cause considerable trouble is sliding of the wall along its base. This situation manifests itself in the absence of a base footing that normally supports a substantial earth-column load, thus adding to the frictional resistance along a plane at this level. The possibility of this occurrence warrants ample subsoil investigations. If, for example, a thin seam of plastic clay exists below the base of the wall and remains undetected, it can cause the wall to move laterally, although tilting is not necessarily involved. Thus, the decision to rely on friction along the base, penetrate into the rock, or otherwise provide embedment below the excavation level must be made after considering the lateral forces in relation to the configuration of the wall. If sliding occurs, it may continue and even accelerate with time until the failure propagates into the earth behind the wall.

The wall of Fig. 10-28 was given the arch shape to initiate progressive arching within the soil behind the completed structure and thus reduce the resultant lateral force. Resistance to overturning was increased by the heavy load of the superstructure. The most severe loading conditions occurred after the construction of the wall, when the channel was excavated but not filled with seawater.

Diaphragm-Wall Cells

Closed Cells Closed diaphragm-wall cells can replace steel-sheet cellular cofferdams in situations involving very poor soil, where the presence of hard materials can make pile driving impracticable or where the excavation is too deep to allow the piles to be driven without distortion deep enough to allow for future dredging. Reinforced-concrete monoliths or open-quay structures are good alternatives for waterfront installations, but in many instances they will cost more.

The cellular construction of such walls is based on the use of concrete diaphragms for the construction of cells. An example is the structure shown in Fig. 10-31. The principles of double-walled structures and cell-type cofferdams are applied to this case; two rows of diaphragm walls are built across from each other and connected by cross walls to form a closed cell. When steel is used, the arches are in tension, which in this application is impracticable. Therefore the first consideration in adapting the cell design to concrete is to reverse the front arches, as shown, so that both front and back walls are in compression. Simple diaphragm panels in the transverse or cross walls act as buttresses for the rear arches but also take in tension the thrust of the front arches and therefore require connection at the joints.

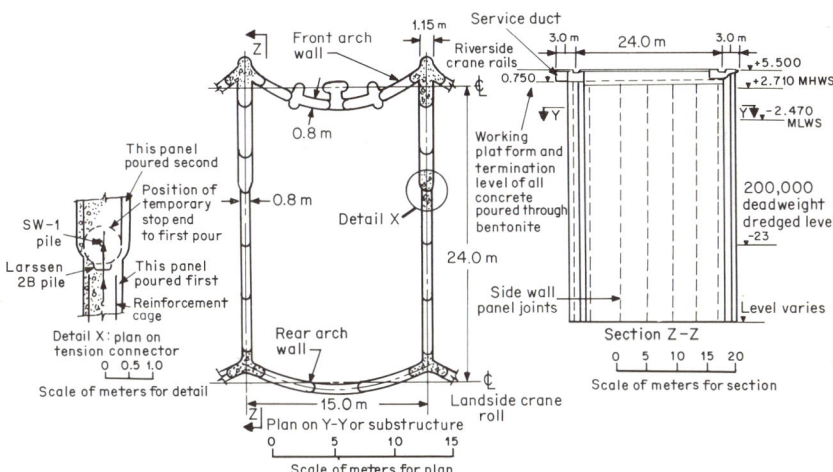

Fig. 10-31 Diaphragm-wall cell details for Redcar ore terminal, England. (MHWS = mean high-water surface; MLWS = mean low-water surface) (*From Fisher, 1974.*)

The complete elimination of bracing, which is the main characteristic of this construction, requires each cell to be stable against overturning, sliding, and tension at the joints. Each row acts as a single structure; nevertheless the cells must resist not only lateral stresses exerted by the earth between them but also overturning when one side is lowered or unwatered. A quick condition can be prevented by selecting the width of cells in relation to the height of the finished structure and the depth of penetration. If the construction is in clay, the analysis of stability must also include the safety against shearing rupture along a curved surface for the entire mass of soil behind the structure. Resistance to overturning is aided considerably by soil friction and is fully developed by preventing laminar tilting of the cross walls so as to integrate these panels into a single structural unit. The cross walls must also be analyzed for flexural resistance when a difference in the total lateral pressure exists in the adjoining cells, either from the earth or from pore water. Failure to give these details due consideration will make the cells sensitive to structural damage, and the cross walls are apt to deflect and distort excessively under nonuniform conditions during construction and service.

The front arch is stiffened against bending after dredging by means of vertical ribs or T heads located at suitable points. The completion of the structure can be achieved by the placement of a rigid top slab or by a series of peripheral beams. For good bearing the front arches and the cross walls must penetrate into bedrock or a comparable stratum.

Panel Connectors The requirement that the cross walls restrain the front arches is satisfied if laminar tilting of the panels is prevented. This means that axial forces and shear must be transferred across the half-round joints. The connection device for the wall of Fig. 10-31 consists of the RPT joint detail shown in Fig. 8-28 and was incorporated in the construction of the wall in the absence of a rigid top slab.

Since more bearing is required under the cross walls near the front arches, the wall thickness must be varied accordingly. Figure 10-31 shows how this transition is made; the thinner wall section is built first, and the stop-end tube is inserted and extracted as usual. When the thicker section is excavated, the tool will not fit exactly into the preformed half-round end; hence any soil adhering to the concrete surface must be trimmed off by a chisel.

Open Cells Open-cell construction is shown in Fig. 10-32. This structure consists of a single wall (steel sheet piles) and cross walls (concrete diaphragms) and forms a dock in Bristol, England (Fisher, 1974). The continuous transverse diaphragm shown in Fig. 10-32a is in poor soil conditions, whereas the intermittent diaphragm shown in Fig. 10-32b is adaptable to firm soil or rock.

These bents form the legs of a continuous portal frame. Unlike stiffened cantilever walls, open-type cells derive their stability from their action as gravity structures. Hence, stability is improved if more weight is placed upon the cells, e.g., a heavy deck, and if there is adequate bearing at the base.

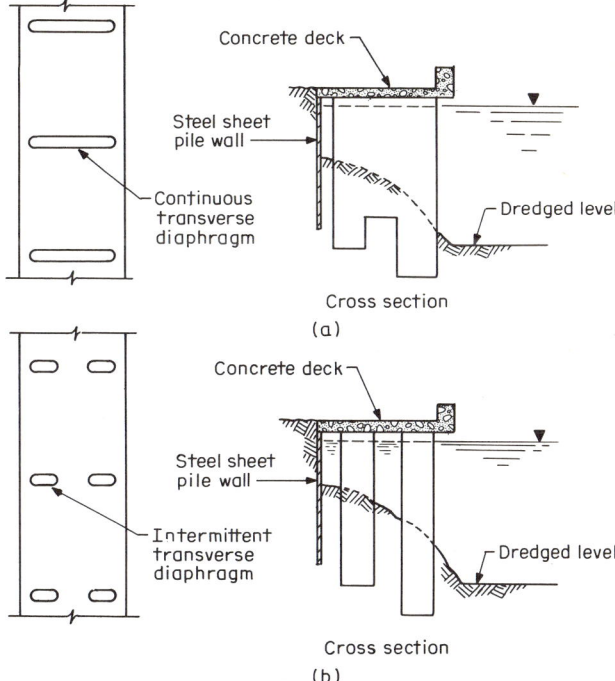

Fig. 10-32 Open-cell construction for a dock in Bristol, England: (a) construction in poor soil; (b) construction in firm soil or rock. (*From Fisher, 1974.*)

REVIEW PROBLEMS

10-1 Describe the single grout and displacement grout used in prefabricated walls. List situations where the latter is more desirable.

10-2 Prepare a tentative specification including control limits for a bentonite-cement slurry to be used as final grout for a prefabricated diaphragm wall. State your assumptions.

10-3 Establish the probable size and weight limitations for prefabricated wall panels in a congested city site.

10-4 How would you check the stability of a slurry-filled borehole for a battered pile?

10-5 Give examples for which you would recommend the various types of bored-pile walls discussed in Sec. 10-2. Explain the effect of such factors as soil conditions, geometry of structure, functional requirements, site conditions, temporary or permanent character of installation, etc.

10-6 Compare the secant-pile method with the interlocking-element method discussed in Chap 6. List similarities and differences.

10-7 Explain in some detail how the ground conditions can influence the construction of prefabricated walls.

10-8 (a) List situations where the use of beam-and-slab prefabricated walls is not indicated; (b) explain whether you could use prefabricated panels for wall cells.

10-9 Discuss the factors that might influence your decision to use (*a*) a composite wall with steel and unreinforced-concrete panels and (*b*) a composite wall with reinforced-concrete panels.

10-10 A vertical shaft is 80 ft deep, and its inside diameter is 21 ft. Develop a suitable polygon assuming a wall thickness of 30 in and a side length of 6 to 6.5 ft. Select a panel length, panel joint, and tremie pipes. The wall is of plain concrete.

10-11 Discuss the conditions under which a circular wall can be built without lateral bracing. Also explain how you would select suitable construction-joint details.

10-12 Explain the difficulties and the differences between prestressing and posttensioning prefabricated wall panels.

10-13 Make a comparison for the technical and constructional requirements between a sheet-pile cellular cofferdam and a diaphragm-wall cell.

REFERENCES

Agar, M., and F. Irwin-Childs, 1973: Seaforth Dock, Liverpool: Planning and Design, *Proc. Inst. Civ. Eng., Lond.*, 1, p. 54.

Braun, W. M., 1972: Post-tensioning Diaphragm Walls in Milan, *Ground Eng.* London, March.

Des Francs, E. C., 1974: Prefasif Prefabricated Diaphragm Walls, *Proc. Diaphragm Walls Anchorages, Instit. Civ. Eng., Lond.*

Fisher, F. A., 1974: Diaphragm Wall Projects at Seaforth, Redcar, Bristol and Harrow, *Proc. Diaphragm Walls Anchorages, Inst. Civ. Eng., Lond.*

North-Lewis, J. P., and G. H. A. Lyons, 1974: Contiguous Bored Piles, *Proc. Diaphragm Walls Anchorages, Inst. Civ. Eng., Lond.*

Rigden, W. J., and P. W. Rowe, 1974: Model Performance of an Unreinforced Diaphragm Wall, *Proc. Diaphragm Walls Anchorages, Inst. Civ. Eng., Lond.*

Romanoff, M., 1961: Corrosion of Steel Piling in Soils, *Nat. Bur. Stand. Monogr.* 58, Washington, D.C.

———, 1969: Performance of Steel Piling in Soils, *Proc. 25th Conf. Nat. Assoc. Corrosion Eng.*

Woodward, R. J., W. S. Gardner, and D. M. Greer, 1972: "Drilled Pier Foundations," McGraw-Hill, New York.

Chapter Eleven

LOAD-BEARING ELEMENTS AND FOUNDATIONS

11-1 THE USE OF SLURRIES IN LARGE-DIAMETER PILES

As mentioned in Chap 1, the first significant applications of slurries occurred in connection with rotary drilling for large-diameter piles. Initially the slurry was used to remove sand and cuttings from the excavation, but it soon became evident that it could increase the stability of the face, lubricate the tools, and thus avoid stuck-pipe problems, prevent formation fluids from entering the excavation, and keep the cuttings in suspension if pumping was stopped. With the extension of the foundation market slurries were used in large-diameter holes, mine and access shafts, visual-exploration holes, and eventually in load-bearing elements of prismatic and linear shapes.

Advancing the Hole with Conventional Tools and Casing Representative types of equipment were mentioned in Sec. 10-2 and Chap. 5. These machines can be truck-mounted for increased mobility, or they can be carried on crawler cranes or tractor trailers. Often the basic digging tool is an earth auger or a drilling bucket supplemented by special tools where it is necessary to deal with unusual soil conditions. Various mud buckets, sand pumps, and bailers are typical accessories often used to clean out water and slurry in the shaft for concrete placement in the dry.

When a casing is used, it is fabricated to suitable lengths so that it will cover the hole from the bottom to well above the slurry level. In firm cohesive soil an open hole can be completed before cavitation occurs, but in dry noncohesive soil or in water-bearing ground the hole almost invariably will not stand open, and in this case the use of slurry is mandatory.

Advancing the Hole with Slurry In this case the slurry is used to help drill the hole and is removed upon completion of the excavation. Because its functions are limited, the stability controls are simplified accordingly, thus allowing a better utilization of materials and time. Slurry-mixing facilities are seldom necessary, and although a substantial part of bentonite is wasted, this hardly justifies the time and equipment necessary for slurry mixing and testing. Visual observations substitute for physical tests and provide the basis for slurry quality control. If sodium bentonite is not available locally, commercial calcium bentonite or local clays mixed with CMC provide good substitutes.

Two methods are available to advance the hole (McKinney and Gray, 1963), depending on the type of ground. In predominantly loose sands the hole is advanced by stirring the soil with an earth auger, adding bentonite and water as the excavation becomes deeper. A surface casing is placed from the ground level to just below the unstable formation. Water and bentonite are continuously added as the auger is rotated, and the bit is lifted and churned to mix all excavated soil with the slurry. If mixing equipment is not available, the dry bentonite is merely dumped into the slurry. When the intended depth has been reached, the remaining casing is lowered to the bottom through the processed slurry. The casing is properly seated in the final bed by driving or rotating before the slurry is bailed from the hole.

The second method is used in ground consisting predominantly of silts and clays. In this case the excavation is carried out with a drilling bucket under bentonite slurry. A surface hole is filled with slurry, which is either premixed on the ground or is prepared by churning bentonite with water in the hole. It obviously is not desirable to have stiff cohesive clays processed into a complete slurry since this will produce an unworkable mix. The drilling bucket must therefore excavate and remove the clay in bulk beneath the advancing bentonite slurry. As the hole is deepened, the drilling bucket must be raised to discharge excavated materials. For this it is advantageous to provide a relief opening through the bucket, otherwise as the latter is withdrawn, excessive swabbing can occur, causing the wall to be drawn into the hole. The relief opening also acts to reclaim any excess slurry and return it to the hole as the bucket is withdrawn. Boulders and heavy gravels are removed as they are encountered, broken up, or loosened by special tools.

Large-diameter piles designed as load-bearing elements often have their base resting on rock or other firm formation (hardpan). In these conditions the entire slurry is bailed out of the hole, leaving a free casing so that it is possible to underream the hole; inspect and test the bottom visually; and place the concrete in the dry if this is desirable.

Concreting the Hole under Slurry If the bottom of the pile is not in impervious ground, the slurry cannot be removed upon completion of the excavation because of the danger of blow-in and base failure at the lower portion of the hole. Alternatively, certain building codes require the hole to be drilled under slurry protection for the reasons mentioned in Sec. 10-2. In this case the slurry control limits are established according to the requirements of tremie concrete. Very helpful is the use of special circular drills equipped with reverse circulation to remove excavated materials.

For the usual hole diameters (6 to 12 ft) one tremie pipe located at the center ensures good flow and complete displacement of bentonite. Occasionally, contractors use a concrete pump of the piston-displacement type coupled to a 3-in flexible rubber hose of sufficient length attached to a rigid pipe. After the pipe and the hose are inserted into the hole so that the end touches the bottom, a plastic plug is inserted in the line to separate the initial batch of concrete from the slurry, as is done in a tremie pipe. As the fresh concrete is discharged, it pushes the plug down and out of the pipe and begins to fill the hole; the plug is recovered as it floats to the top.

11-2 THE USE OF SLURRIES IN PRISMATIC AND LINEAR ELEMENTS

The development of machines which perform slot excavations has led to the adaptation of specially shaped foundation elements. These may provide structural possibilities and solutions which are not always available with circular piles. Several common configurations are shown in Fig. 11-1. Variations of them can be worked out for unusual combinations and magnitude of loads or for special classes of structures. In projects that include conventional diaphragm-wall construction, the adaptation of prismatic foundations can

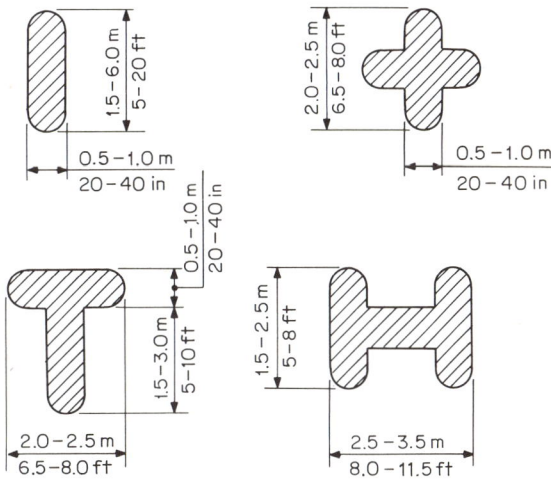

Fig. 11-1 Cross sections of typical load-bearing elements.

save money and time since it allows the use of the same equipment and incidental plant facilities.

Linear or prismatic diaphragm-wall elements may replace a group of cast-in-place piles. This has merits where heavy loads require unusually large monolithic structures; where the foundation is exceptionally deep or the installation is in very difficult ground; and where the foundation must resist, besides vertical loads, bending moments and lateral thrusts. The flexibility with which prismatic elements can be arranged regarding size, shape, and plan makes them suitable for any type and magnitude of loads, and this is true in spite of the presence of existing foundations or other site limitations.

Preparing the Base of the Trench

If an element is intended for bottom bearing, the base of the excavation must be cleaned out before the concrete is placed. Since it is not possible to have visual inspection of the bottom, special care is necessary to provide a clean base and thus ensure a firm foundation.

It is essential to understand that the shape of the base depends largely on the equipment performing the excavation. In some cases the bottom will be nearly flat and level in every direction, but depending on the features of the machine, it may be given an irregular or rounded profile, as shown in Fig. 11-2. At present there is no evidence to indicate that the shape of the bottom influences the transfer of load, at least from the theoretical standpoint; hence there should be no objection to the use of machines which do not provide flat or square bottoms. However, practical experience confirms that a smooth square base is better cleaned at the end of excavation. Such a base is obtained by passing a square-end clamshell at the end or by shifting the long drill somewhat from the initial position to level the hills.

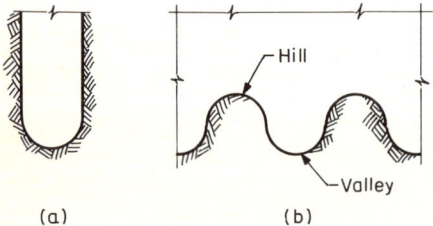

Fig. 11-2 Different bottom shapes: (*a*) round cross section cut by a round-end clamshell; (*b*) irregular bottom excavated by a rotary long drill.

Soft materials accumulate at the base or are kept in suspension near the base. Their presence is detected by taking samples of bentonite. If these have unusually high density, the slurry should be recycled or replaced by fresh solution. Figure 11-3 shows a device commonly used to obtain samples for testing purposes. This device is relatively simple, and the sampling point can be anywhere in the trench.

There should be no compromise in the matter of bottom cleanliness. In most instances it is possible to obtain a satisfactory base by passing an air lift. A small amount of dry or plastic cuttings left at the bottom will make little

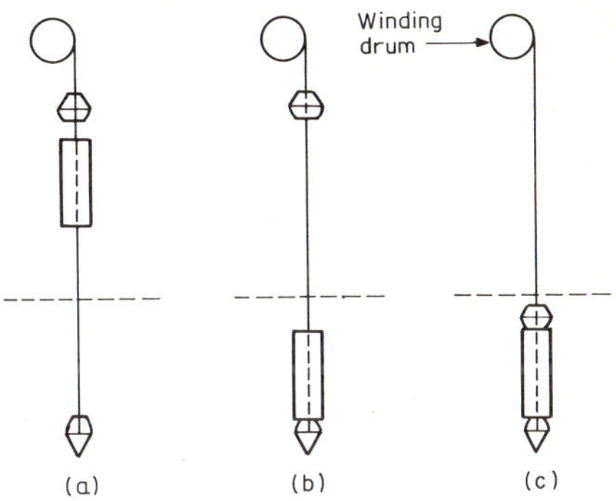

Fig. 11-3 Sampling device for checking the bentonite density at the base of a hole: (*a*) bottom weight lowered to the sampling level; (*b*) sample tube lowered down; (*c*) top cover lowered down and sampler ready for lifting. (*From Sliwinski and Fleming, 1974.*)

difference in bearing capacity, but if 1 in (2.5 cm) or more of soft mud is accumulated at the base, it can affect the transfer of vertical load. This mud is not displaced by the fresh mix, and unless it is completely consolidated by the weight of the concrete, it can cause unacceptable settlement when the structure is loaded. Clay cuttings and bentonite trapped under the fresh concrete as its level rises are not completely compressed in the short time required for the mix to be placed and hardened, partly because the concrete may not exert its full weight and partly because this weight may be resisted by friction along the shaft. After the foundation is loaded, these soft materials will constitute a zone of postconstruction settlement.

In some instances a bottom softening due to seepage and soil impregnation by slurry of the zone just below the excavation is likely to occur. This penetration is stopped when the filter cake is formed at the final excavation level but can recur with the passage of tools to level or to clean the bottom. Although it raises the moisture content of the soil for a few inches below the base, it has no real influence on the compressibility or bearing capacity of the foundation.

Concreting

The requirements for the placement of concrete in a prepared prismatic trench are essentially the same as in conventional diaphragm walls. Additionally, the shape and configuration of the trench must be considered in choosing the location and the number of tremie pipes. A T section of nominal dimensions can be filled with one tremie pipe placed at the junction of the stem and the flange. An I or a square section will probably require one tremie pipe at each opposite corner. Fresh concrete will flow in any direction from the discharge point but will not change direction easily when it confronts a stationary barrier.

A great advantage is the elimination of construction joints, round-end tubes, casings, and all the associated appurtenances. In general, this allows more efficient scheduling of the installation phases. A single element will probably be excavated and cleaned on the same day and reinforced and concreted the following day. If an air lift must be applied, it should be done just before the placement of concrete in order to reduce the possibility of cavitation and sloughing which exists in a panel left open for a prolonged period.

11-3 USUAL DEFECTS AND REPAIRS

Possible defects of the finished elements and the conditions under which they occur can be summarized as follows: (1) overstressing of soil beneath the foundation due to insufficient bearing area or to unconsolidated materials at the bottom; (2) improperly tremied concrete, resulting in voids and cavities within the set structure; (3) structural discontinuities and deviations from the true vertical line, causing local overstressing; and (4) excessive mixing with bentonite, affecting the development of concrete strength.

Unlike belled subpiers, where bearing-capacity failure is improbable or exceptionally rare, insufficient load transfer at the base of prismatic elements is a possible cause of settlement during service, particularly where the transfer of load is primarily through direct bearing. This may be due to limited contact area at the base or to the presence of soft materials at this location. Regardless of the cause, it can occur in spite of an otherwise sound and structurally adequate concrete structure. Because the cross-sectional area of the shaft usually is more than enough, the concrete stresses are lower than the theoretical design strength and therefore some localized weakening due to the above-mentioned reasons is insignificant.

Methods for Checking and Repairing Defective Elements For settlement-sensitive structures and where heavy loads are carried, a post-construction investigation may be requested to check the continuity and integrity of the finished structure. Usually, however, it is customary to request tests that can confirm the assumed load transfer, either by base bearing or by skin friction.

If there is indication that the set concrete is defective in one of the ways mentioned, a good check on its quality and soundness is by means of diamond coring (Baker and Khan, 1971). The larger the diameter of the core the more reliable the results but the more expensive the test. Diamond coring can be supplemented by other test procedures such as caliper logging, inclinometer readings, seismic-wave and velocity measurements, and three-dimensional logging. The damaged zones can be repaired by grouting, but even with the best technique and materials available the repair is carried out on a speculative basis. Efforts to repair local defects have failed more often than have succeeded. Baker and Khan (1971) report that grouting has been found effective where a clear void exists not filled with earth materials provided enough pressure can be developed between two grout holes. For prismatic elements these defects are far less serious because of the larger than necessary cross-sectional area of the shaft.

Load Tests The best procedure for confirming the assumed load capacity (base bearing and shaft resistance) is by means of load tests. If these disclose the possibility of excessive settlement, a usual remedy is to construct a second element adjacent to the first structure and connect them by means of a top concrete cap.

Most problems associated with the transfer of load can be avoided by monitoring the final stages of the excavation, particularly the rate of drilling and the type of excavated materials in the last few feet of penetration, and by cleaning the bottom as specified. Hard clay, very dense sand, or other firm materials will slow the drilling bits down considerably. On the other hand a test panel excavated and concreted in situ is quite useful and will provide information that can be used as a basis for a semiempirical design. Test panels are recommended for all major projects, and their usefulness will generally justify the cost.

11-4 THE TRANSFER OF LOAD BY BASE BEARING AND SHAFT RESISTANCE

For the general case of foundation elements built in slurry trenches or holes the transfer of load involves base bearing or base resistance and skin friction or side adhesion which together make the shaft resistance. Besides the vertical loads it is conceivable that lateral forces may act on the structure. However, the following sections pertain primarily to the transfer of vertical loads.

Base Bearing

The transfer of load by direct base bearing of an axially loaded compressible element is influenced primarily by the depth and size of the element, the soil characteristics, disturbance of the base of the excavation, and the contact between the concrete and the supporting soil underneath. These factors generally determine the resistance of soil to penetration.

Loosening of the base, which sometimes occurs with piles drilled by casing in saturated soils, is limited and rather unlikely in prismatic panels excavated under slurry protection. Good care is essential, however, to avoid accumulation of sedimented soil particles and bentonite mud at the base. Comparing the base resistance of piles built with and without slurry, Chadeisson (1961) found larger initial settlement for the latter, but this surprising fact was reconciled after compaction of the loosened soil by the first loading and both piles behaved similarly during the second load cycle. Reese et al. (1973) have examined test piles drilled under slurry protection, and in some instances the base resistance was not as high as the theoretical; this was attributed mainly to the presence of soft materials settled to the bottom and not removed because of poor cleaning procedures.

For well-supervised and well-constructed projects the method of excavation and concrete placement causes minor, if any, disturbance to the soil, so that the conditions are satisfactory for the development of base resistance.

Shaft Resistance

For working loads the transfer of load for round or prismatic elements generally begins with skin friction or side adhesion and is completed by base bearing when sufficient displacement has occurred vertically. Even the best methods for analyzing this situation are semiempirical and relate shaft resistance to a friction factor for sands and to the undrained shear strength for clays. With the slurry process failure to displace the bentonite from the interface can influence the development of shaft resistance. The questions arise, therefore, when and how the bentonite mud is really swept or absorbed by the rising fresh mix and when and to what extent any bentonite left at the interface will affect skin friction or side adhesion.

Some suspicion that slurries may adversely affect the development of shear resistance stems from their use as lubricants in caisson and tunnel work (see also Chap 1). However, there is a significant difference between the two cases. In the lubricating process the slurry remains unchanged, and the only factor resisting sliding is its thixotropic strength. During the placement of tremie concrete in a slurry trench or hole the fresh mix displaces all the bentonite, and as it rises, it exerts a sweeping action along the interface (see also Chap. 8). Any bentonite left there interacts with the fresh concrete and is absorbed by the mix through reaction with calcium from the cement, so that no slurry is left to act as lubricant.

Although these facts are confirmed in practice, the type, porosity, and permeability of the soil around the shaft will determine the associated earth-structure interaction in the final position. For example, in impervious clays neither filtration nor soil impregnation by slurry takes place, and during the concrete placement all bentonite should be expected to be removed or swept from the interface. Sand and gravel formations, on the other hand, are prone to deep filtration or rheological blocking according to the void size and distribution, the hydraulic gradient, and the shear strength of the slurry. When fresh concrete rises against such soils, it generally displaces all free bentonite at the interface, but the sweeping action does not extend beyond this zone. Despite the presence of colloid matter between the bulk of the vertical earth face and the hardened concrete in the final position, however, practical experience and field tests indicate that substantial friction still is available.

A hole in clays is thus equivalent to a cased hole, whereas a slurry hole in pervious granular soil should be looked upon as a process causing a zone of interaction in the immediate vicinity of the face (see also Chap. 3). The practicality of these phenomena can be determined by the performance of piles and prismatic sections built under slurry and tested in both cohesive and cohesionless soils, and results from such tests provide a basis for establishing safe load limits for the available friction and/or adhesion.

Results of load tests on three instrumented deep piles at Bidston Moss, reported by Sliwinski and Fleming (1974), show satisfactory load transfer in both sands and clays. A model pile test in sand revealed that a filter cake 5 mm thick around the shaft could have resulted in a reduction of shaft resistance by about 10 percent. In other tests two similar piles were constructed in drilled holes, one with bentonite slurry and the other dry, in the same site and ground conditions. The results are shown in Fig. 11-4, and clearly there is no major difference in the load-transfer characteristics of the piles to indicate any adverse effect of bentonite on shaft resistance.

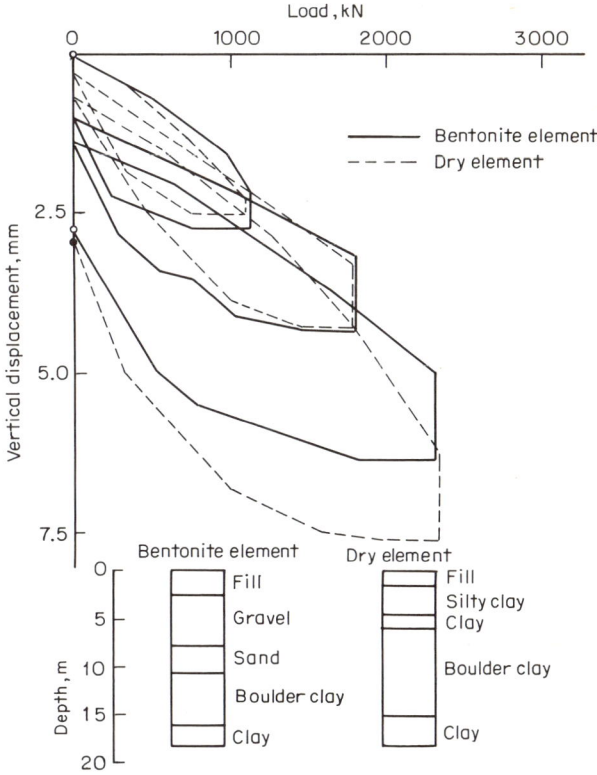

Fig. 11-4 Load-test comparison of two 600-mm-diameter piles on the same site, one under bentonite and the other dry. (*From Sliwinski and Fleming, 1974.*)

Effect of General Excavation

Where a retaining diaphragm wall must also act as foundation, two basic factors should be recognized in estimating wall resistance due to friction or adhesion. The first is the installation of the wall before general excavation, which creates load-transfer conditions similar to those for a pile. However, during and after general excavation significant stress changes are conceivable beneath and around the wall, and these will continue until equilibrium is resstablished according to the final loading, geometry, and groundwater conditions (Henkel, 1970). An analysis based on the undrained shear strength of the soil does not include the effect of these changes, so that it is better to predict the load-carrying capacity of the wall in terms of effective stress. This problem remains complex, and at present no rational procedure is available. The effective-stress approach partially explains the effects of excavation on wall friction, and considerable judgment is therefore necessary to evaluate the applicability and the limitations of this analysis.

For the diaphragm wall of the basement at Kensington and Chelsea Town Hall, London, an analysis based on the foregoing principles indicated that probable wall friction at the time of final load application was unlikely to be less than that available before excavation, but this should not be taken as general conclusion (Corbett et al., 1974).

11-5 BASIC CONCEPTS OF LOAD TRANSFER

For the transfer of load by base bearing and shaft resistance it is necessary to consider the following factors:

1. The development of these reactions as a function of the vertical displacement
2. Dimensions of the element (length, width, or diameter), shape and configuration, and relative confinement in the soil mass, e.g., whether a single element or group action
3. The stiffness of concrete with respect to the compressibility of the supporting soil
4. The soil properties and mainly the shear strength

Regardless of how desirable it is to provide a structure free of settlement, some downward displacement will take place in spite of the method of excavation, face support, cleanout practice, and concrete placement, as is true with almost any type of foundation. This downward displacement helps transfer the load, first by mobilizing the shaft resistance and then the base bearing. Because side-wall shear is developed at much smaller vertical displacements, appreciable shaft resistance (sometimes all of it) is mobilized before any load can be transferred by base bearing.

These facts have been confirmed in full-scale field tests. Whitaker and Cooke (1965) have reported the results of instrumented tests on bored piles with and without bells in stiff London clay. These show that the ultimate load (considered the sum of shaft resistance and base bearing) was reached at a vertical displacement which in some instances was of the order of 10 percent of the base diameter, although the shaft resistance was fully mobilized at a vertical displacement between 0.5 and 1 percent of the shaft diameter.

Reese and O'Neill (1969) have found that in stiff Beaumont clay a vertical displacement of 5 mm (about 0.2 in) is enough to mobilize full shear resistance, but more displacement is generally needed to develop the same resistance in sand. Vesic (1967) has reported displacements as much as 10 mm (0.4 in) for bored piles in granular soils. Regardless of the type of soil these displacements usually are assumed to be independent of the shaft diameter, so that the vertical movement corresponding to full shaft resistance is interpreted merely as shear displacement.

The amount of displacement that must occur vertically before appreciable base resistance is available depends not only on the conditions, the type, and the confinement of the bearing materials but also on the size of the base for both circular and prismatic elements; the larger the base the smaller the displacement.

Effects of Shape, Dimensions, and Stiffness The load-bearing elements of Fig. 11-5 illustrate how the shape and configuration can influence the load transfer. The circular element in Fig. 11-5a will generally develop full shear along the entire perimeter, and the same assumption can safely be made for the elongated I section shown in Fig. 11-5c. The I section in Fig. 11-5b, however, will most likely develop its shear resistance along a modified "effective" perimeter as shown, and this is less than the actual contact area. Evidently the weight of the soil fillet between flanges should be included with the total load.

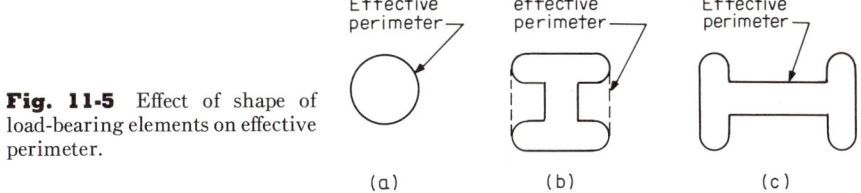

Fig. 11-5 Effect of shape of load-bearing elements on effective perimeter.

The length-to-diameter or length-to-width ratio may influence the division of load between shaft and base, but it is not clear under what conditions this will be true. From a theoretical investigation of piers with enlarged bases, Poulos and Davis (1968) have concluded that underreamed piers may not offer any appreciable advantages over straight-shaft piers of the same shaft diameter and length when the length-to-diameter ratio is greater than about 25 unless the stratum at the base is considerably stiffer than the overlying materials.

In general, a higher modulus of the base-supporting materials compared with the modulus of the shaft-supporting materials and a smaller length-to-diameter or length-to-width radio should indicate that a greater proportion of load is transferred to the base. DeMello (1969) explains this in terms of the interference of the shaft length and the relationship of the stiffness of the pile material to the deformability of the soil, as well as in terms of the sensitivity of the stress-strain curve of the soil adhesion.

11-6 TESTS ON BORED PILES INSTALLED BY SLURRY DISPLACEMENT

Reese et al. (1973) have conducted instrumented load tests on three bored piles varying in diameter from 76 to 92 cm (30 to 37 in) and from 13 to 22 m deep (43 to 72 ft). The ground at the test site consists of stiff clay in the upper crust underlain by a stratum of water-bearing sand. Soil profiles and sketches of the test piles are shown in Fig. 11-15.

The piles were installed according to the following sequence: (1) the hole was augered without slurry until a caving layer was encountered, at which time bentonite slurry was introduced in the hole and maintained at a level close to the surface while drilling was continued under mud; (2) drilling was completed either with an auger or with a special drilling bucket cleaning the bottom just before inserting the reinforcing cage; and (3) concrete was poured using a 10-in-diameter tremie pipe.

One pile was tested 16 days after casting, and the other two 1 month after casting. A quick-test procedure was used whereby load increments of one-thirtieth of the estimated ultimate load were applied at $2\frac{1}{2}$-min intervals, and readings were taken using a high-speed system. After testing, the piles were extracted for examination and visual inspection.

Test Results Figures 11-6 to 11-8 show load-distribution curves obtained by plotting the measured load at various depths of the piles. For pile G_1 a load up to 3000 kN (675,000 lb) is carried almost entirely by shaft resistance (side-wall shear). For greater loads the ultimate shaft resistance is developed along the entire length of the shaft, and additional loads are resisted by base bearing, as shown by the nearly parallel curves. The load at the tip is estimated either directly or by extrapolation, and the corresponding settlement at the base is found by subtracting the computed elastic deflection from the observed displacement at the top of the pile. Load-settlement curves for the top and the tip of each pile are shown in Figs. 11-9 to 11-11.

The load-transfer curves shown in Figs. 11-12 to 11-14 are obtained at the various depths by plotting the shear stress developed at a given point, calculated by differentiating the load-distribution curves vs. the displacement of the same point with respect to its original position. A relationship between shear resistance at a point along the shaft and shaft movement is thus obtained.

The side-load transfer is related to the shear strength of the soil by a reduction or adhesion factor α, which is the ratio of the average unit load transfer in a stratum to the average shear strength of the stratum. The concept of the α factor is extended to sands, where α expresses the portion of drained strength actually mobilized by skin friction so that the effects of the unknown earth-stress coefficient and the adhesion factor between concrete and sand are lumped into the α factor. Values of α for the three piles are summarized in Table 11-1, which also includes data from other tests on bored piles cast in

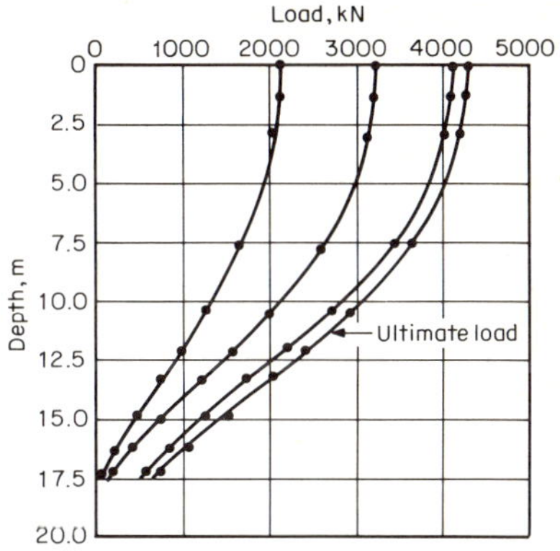

Fig. 11-6 Load-distribution curves from instrumented load tests; pile G_1. (*From Reese et al., 1973.*)

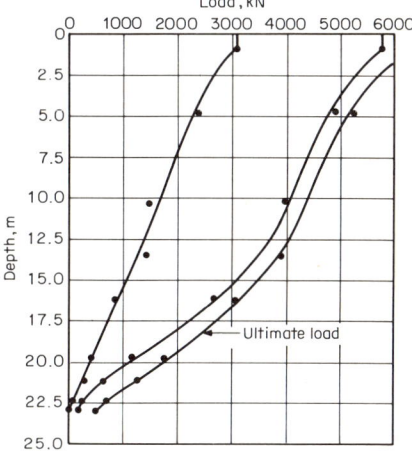

Fig. 11-7 Load-distribution curves from instrumented load test; pile G_2. (*From Reese et al., 1973.*)

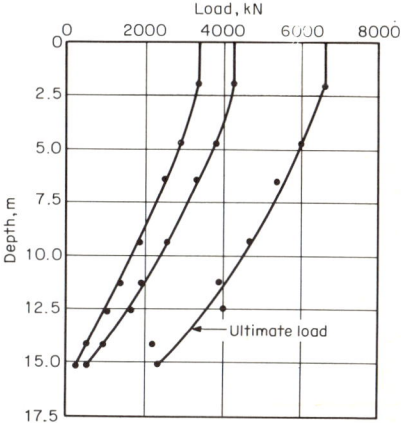

Fig. 11-8 Load-distribution curves from instrumented load tests; pile BB. (*From Reese et al., 1973.*)

387

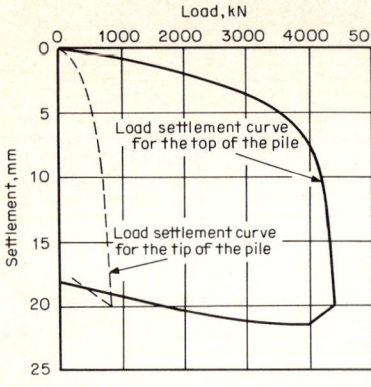

Fig. 11-9 Load-settlement curves from instrumented load tests; pile G_1. (*From Reese et al., 1973.*)

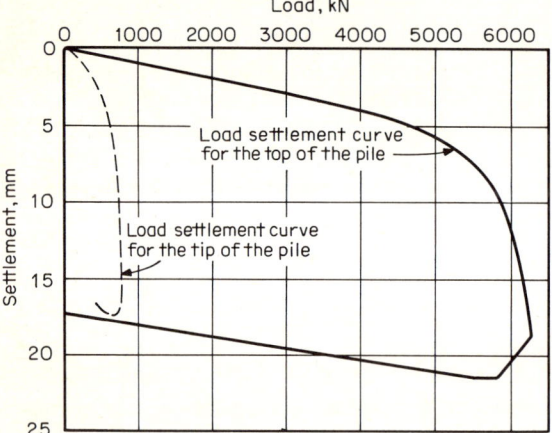

Fig. 11-10 Load-settlement curves from instrumented load tests; pile G_2. (*From Reese et al., 1973.*)

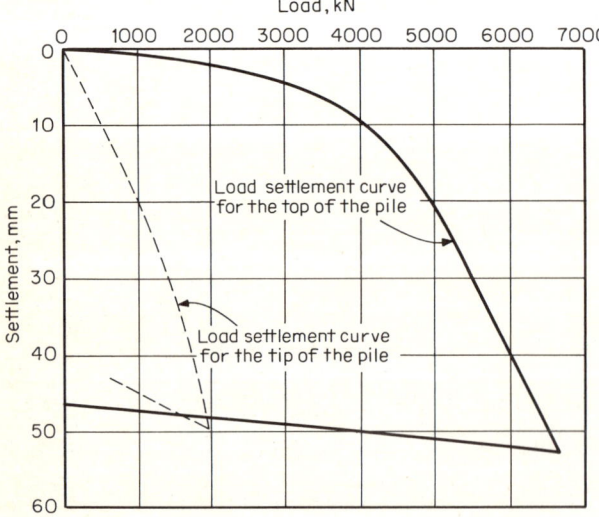

Fig. 11-11 Load-settlement curves from instrumented load tests; pile BB. (*From Reese et al., 1973.*)

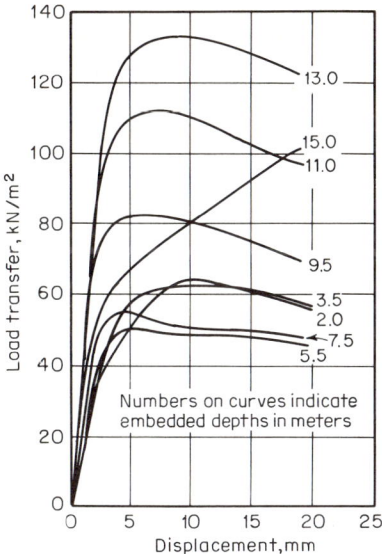

Fig. 11-12 Load-transfer curves from instrumented load tests; pile G_1. (*From Reese et al., 1973.*)

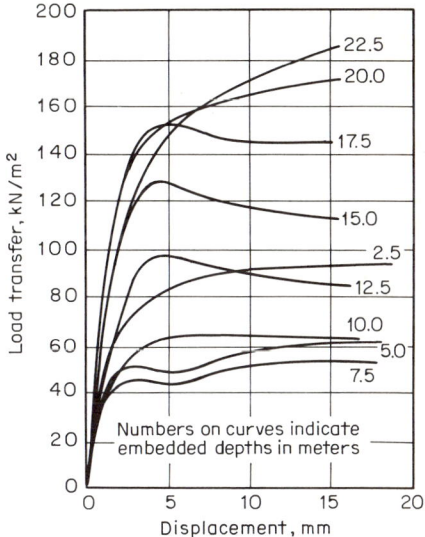

Fig. 11-13 Load-transfer curves from instrumented load tests; pile G_2. (*From Reese et al., 1973.*)

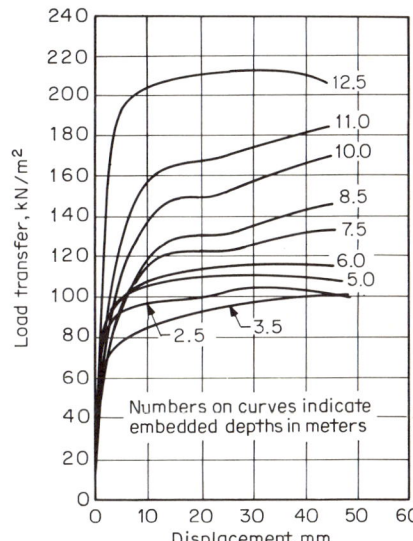

Fig. 11-14 Load-transfer curves from instrumented load tests; pile BB. (*From Reese et al., 1973.*)

stiff fissured clay but without slurry or casing (Reese and O'Neill, 1969). The latter piles are marked with an asterisk.

Interpretation of Results The load-settlement curves for piles G_1 and G_2 (Figs. 11-9 and 11-10, respectively) reveal a sudden vertical displacement when the ultimate load is reached, indicating lack of high base resistance and

TABLE 11-1 Summary of Results from Instrumented Load Tests (from Reese et al., 1973)

Pile	Soil	Average peak load transfer, kN/m²	α_{av}	Ultimate tip resistance, kN/m²
G_1	Clay	52	0.65	
	Sand	103	0.70	1140
G_2	Clay 1	74	0.75	
	Clay 2	159	1.08	
	Sand	176	0.67	1450
BB	Clay	92	0.25	
	Sand	150	0.75	4000
S_1T_1*	Clay	49	0.44	1090
S_2T_1*	Clay	59	0.53	970
S_3T_1*	Clay	62	0.54	935

*Bored piles cast in stiff fissured clay without slurry or casing.

a high degree of sensitivity of the clay. For pile BB (Fig. 11-11) the load continues to increase with displacement, which would normally be expected if the tip of the pile were in very dense sand. In reality the curves reflect the cleanout procedures of the bottom. For pile BB the bottom was left clean, and very little accumulation of soft materials was noticed. The cleanout of holes G_1 and G_2 was less thorough, and evidently a layer of soft materials remained at the bottom.

Regardless of the differences in the method of installation and bottom cleanout, the significant conclusion is that peak load transfer in stiff clays is reached at a relative shaft movement (displacement of a point with respect to its original position) of the order of 4 to 5 mm (0.16 to 0.2 in), but for sands the relative displacement varies from 5 to 10 mm (0.2 to 0.4 in). This is evident from the load-transfer curves and from the soil profile and shaft locations shown in Fig. 11-15.

An important observation from Table 11-1 is that the factor α is higher for piles with slurry than without, but before accepting this some uncertainties must be recognized, e.g., the difficulty of measuring the undrained shear strength of fissured clays and the relative reliability of shear-strength tests. Accordingly it is unsafe to recommend higher values of α for piles cast in slurry.

For pile G_2 the soil layer shown as clay 2 in Fig. 11-15 has an average value of α much higher than the one for clay 1. Clay 2 is sensitive, nonfissured organic clay, and clay 1 is stiff, fissured clay. Visual examination of the soil-concrete interface in clay 2 after the pile was extracted showed that shear failure occurred in the natural soil about 1 cm (0.4 in) from the concrete surface. This could indicate that in such clays the load transfer is substantially equivalent to the undisturbed shear strength.

The relatively low average value of α in the hard clay around pile BB is consistent with the results of other tests (Reese et al., 1973), showing that in

Load-Bearing Elements and Foundations 391

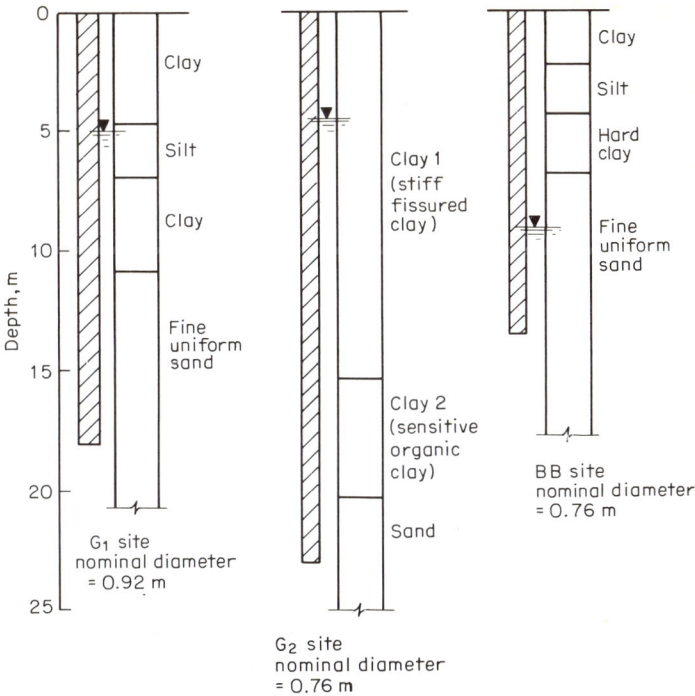

Fig. 11-15 Soil profiles and sketches of shaft location; instrumented load tests. (*From Reese et al., 1973.*)

hard soil the shear resistance around the shaft is not as dependent on the soil strength as it is on the soil-slurry interaction. In other words, the load-transfer capacity is nearly the same, irrespective of soil strength, and roughly equal to the shear strength of a thin fillet of earth mixed with some bentonite immediately around the shaft. For pile BB this strength in hard clay was 92 kN/m² (1900 lb/ft²).

Effect of Construction Techniques When the load transfer depends appreciably on base bearing and the installation is under slurry support, consideration should be given to the possibility of reduced base resistance due to incomplete bottom cleanout. Very often the deposition of soft or loose materials from sloughing and peel-off occurs just before concreting. If the excavation is in caving soil and the bottom is cleaned several hours before concreting, it will be necessary to protect the uppermost part of the excavation with a temporary casing since this is the section most liable to sloughing and peel-off.

The initial batch of concrete should not be expected to displace soft materials from the bottom, and the washing action of the flowing mix may be reduced by poor placement methods. Visual inspection of the tips of the piles of Fig. 11-15 disclosed, for example, the presence of soft materials at the bottom. This was particularly true for piles G_1 and G_2, whereas pile BB had much less soft material at its bottom, evidently because of better cleanout techniques.

The drilling bucket used for pile BB effected a more thorough cleanout of the bottom than the auger used for piles G_1 and G_2.

A hard-coating of mud, clay, and sand was observed for most of the extracted length for all three piles. Examination of the concrete-soil interface showed that the failure surface was in the soil mass rather than along the interface, except for the hard clay in pile BB, where the failure occurred at the interface.

11-7 TESTS ON SKIN FRICTION AND WALL ADHESION

Laboratory Tests on Concrete-Sand Interface

Using the apparatus shown in Fig. 11-16, Farmer et al. (1971) investigated the effect of bentonite on skin friction for concrete-sand interfaces. The sand samples had a bulk density of 1660 kg/m³ (104 lb/ft³) and a relatively low degree of saturation to allow the formation of a stable cylindrical opening at the center for the introduction of bentonite slurry. The slurry was introduced into the opening and kept there until an interaction zone 5 mm thick (0.2 in) was formed at the interface. Then the slurry was replaced by fresh concrete (6-in slump) poured through a 75-mm-diameter tube from the bottom up to simulate the upward flow of tremie concrete. A filter cake, relatively unscoured, was observed upon exposing the interface.

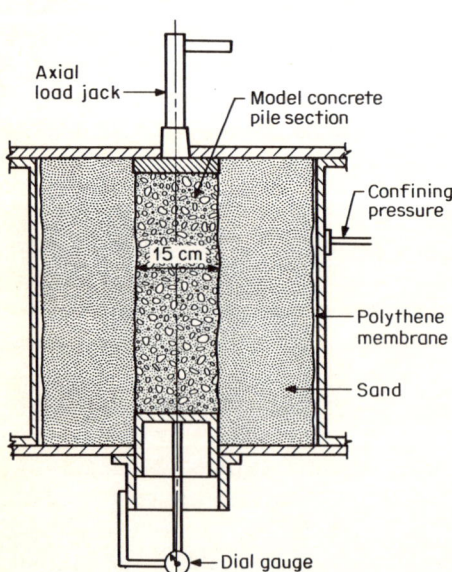

Fig. 11-16 Triaxial shear apparatus used to investigate the effect of bentonite on skin friction. (*From Farmer et al., 1971.*)

After the concrete was allowed to set, a confining pressure was applied through the polythene membrane for 5 h, inducing horizontal compression while constraining vertical compression against the top and bottom plates. The vertical displacement of the pile was next measured upon the application of incremental loads. The test was repeated at a confining pressure of 0.07, 0.14, 0.21, 0.28 and 0.35 N/mm², and for each set the load application was continued until excessive displacement occurred.

Test Results The average skin-friction–confining-pressure ratio for the five values mentioned is plotted vs. the vertical displacement in Fig. 11-17 for concrete-sand and for concrete-bentonite-sand interfaces. For low displacements the skin friction is somewhat greater when bentonite is present, but with larger displacements it is greater without bentonite. The difference, however, is rather small and less than 10 percent.

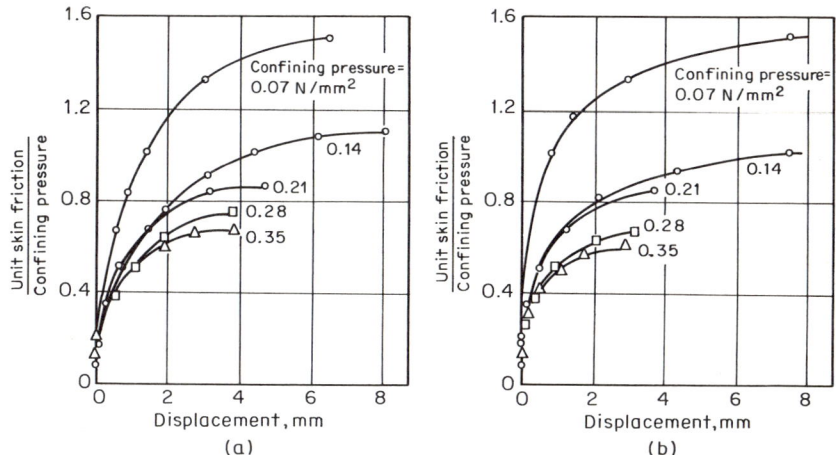

Fig. 11-17 Friction-displacement characteristics of model pile: (a) concrete-sand interface, (b) concrete-bentonite-sand interface. (*From Farmer et al., 1971.*)

Skin friction increases with increasing displacement and attains peak values at displacements from 5 to 10 mm (0.2 to 0.4 in), depending on the confining pressure, which is in good aggreement with the results reported by Reese et al. (1973). Although the fully developed load transfer increases with confining pressure, the corresponding relationship is nonlinear, primarily in the range of low confining pressures (Fig. 11-18). For a concrete-sand interface the coneefficient of wall friction $\tan \delta$ at maximum load transfer decreases from a value in excess of unity at low confining pressures to about 0.60 at 0.35 N/mm² confining pressure. For the concrete-bentonite-sand interface this coefficient attains a nearly constant value more rapidly, indicating that the friction and any adhesion in the impregnated zone were rapidly mobilized even at small displacements. For these samples the soil-slurry interaction

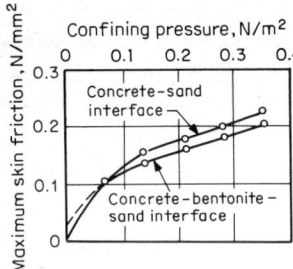

Fig. 11-18 Skin friction vs. confining pressure; fully developed load transfer. (*From Farmer et al., 1971.*)

evidently occurred as deep filtration and resulted in the formation of a cake of sand and bentonite 10 mm (0.4 in) thick, confirmed by observation.

This behavior is in good agreement with results from other investigations (Potyondy, 1961) and shear-adhesion tests on cohesive mixes prepared by blending clay with cohesionless sand. For the soil-slurry interaction discussed in Chap. 3 the effect of impregnation by bentonite is an increase of the shear resistance of the sand by imparting some cohesion to the soil, although in certain instances a small reduction of the friction angle is also conceivable.

Field Tests on Cast-in-Place Piles in Mixed Soil

Field tests on cast-in-place large-diameter piles placed by slurry displacement have revealed a high level of load transfer, corresponding to an adhesion factor higher than or close to 1. This, however, should not be taken as a general guideline since it probably means an increase in wall shear due to structural irregularities and accentuated wall roughness caused in many instances by the method of excavation and construction.

Figure 11-19 shows three test piles for the foundation of the approach viaduct to the Second Mersey Tunnel (England). The tunnel was constructed in firm to stiff sandy clays intermixed with sand beds. The tips of the piles are in sandstone bedrock of variable depth between 15 and 60 m (50 and 197 ft). The holes were bored using 6 percent bentonite slurry (Fulbent 570) and have an average diameter of 1.22 m (about 4 ft).

The piles were loaded in 50-t increments until a working load of 550 t was reached. This was maintained for 24 h, and then the piles were unloaded in 100-t increments and left unloaded for 24 h. They were again reloaded in increments until a load equal to $1\frac{1}{2}$ times the working load was reached and were kept under this loading for another 24 h before unloading. Settlement of the pile caps was observed every hour for the duration of the tests and at each load increment during the loading and unloading cycles.

Comparison of Computed and Actual Settlement Results of these tests are shown in Figs. 11-20 and 11-21 (Farmer et al., 1971). For the load-settlement curves the presence of rock at the pile tip is considered by introducing an appropriate modulus of elasticity for this material ($E = 10^4$ N/mm²). The probable vertical displacement at which full skin friction has been

Load-Bearing Elements and Foundations 395

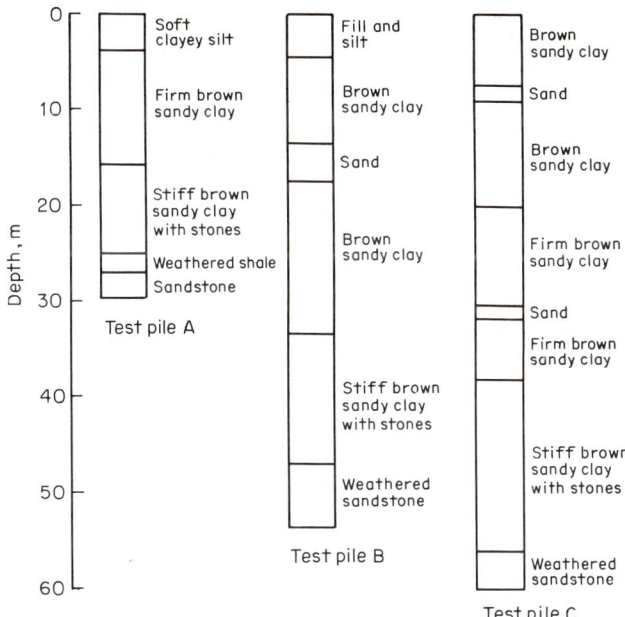

Fig. 11-19 Soil profiles, test piles A, B, and C. (*From Farmer et al., 1971.*)

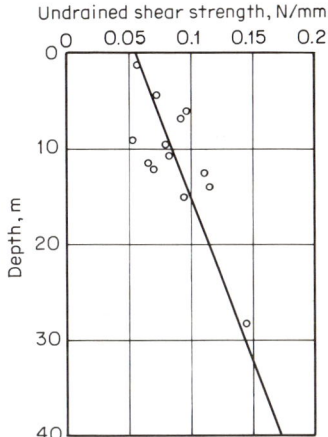

Fig. 11-20 Shear strength vs. depth; test piles A, B, and C. (*From Farmer et al., 1971.*)

mobilized is taken as 6 mm (0.24 in). For simplification zero load transfer is assumed for the upper end, which was cased, and full load transfer for the lower part. The bulk of the pile length is assumed to be in contact with clay and/or sand, and the undrained shear strength is extrapolated from tests on undisturbed samples (Fig. 11-20). Load-settlement curves are shown in Fig. 11-21 for adhesion-factor values α of 0.2, 0.5, and 1.0. These curves can be compared with the actual load-settlement curves. There is clearly a uniform

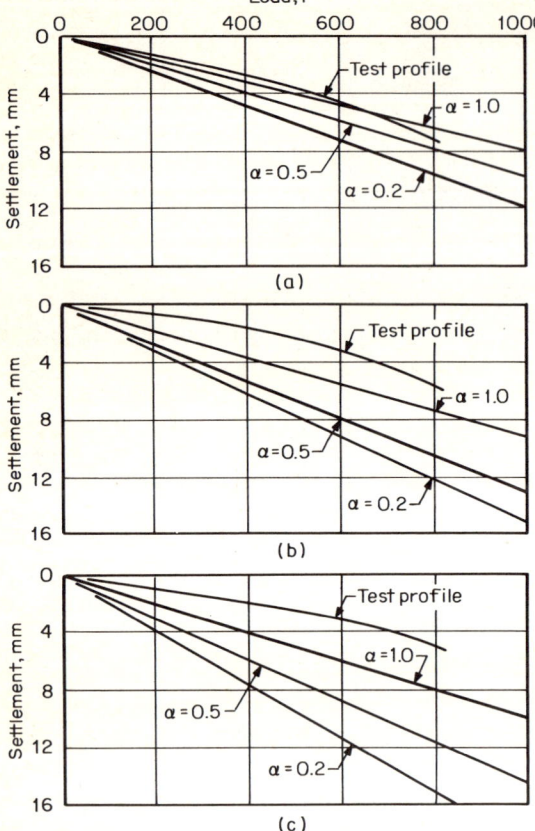

Fig. 11-21 Comparison of computed and actual load-settlement curves: (*a*) test pile A; (*b*) test pile B; (*c*) test pile C. (*From Farmer et al., 1971.*)

pattern of nearly linear settlement with increasing load, usually observed with compressible friction piles at small tip displacements, and invariably the actual load transfer is higher than predicted.

11-8 TESTS ON DIAPHRAGM-WALL PANELS

UNO Building in Vienna Kienberger (1974) has reported results of tests on circular and linear slurry panels used in the foundation of this building. Pertinent data are shown in Table 11-2. The ground formations at the site consist of gravel, silty clay, and fine sand, generally stiff or dense.

Load-settlement curves are shown in Fig. 11-22. Although panels I and II have the same depth, a direct comparison of load transfer is not possible because of differences in the shape and cross section; thus the base cross sections have a ratio of 0.9 whereas the ratio of the shaft surfaces is 0.8.

It is possible, however, to correlate the load-transfer characteristics if we note that in both cases the base bearing is about 10 kg/cm² (100 t/m²). The

TABLE 11-2 Test Elements for the UNO Building in Vienna (from Kienberger, 1974)

Element	Dimensions and shape	Test load, t	Depth, m
I	50- by 150-cm diaphragm-wall panel	500	13
II	90-cm-diameter pile	500	13
III	50- by 150-cm diaphragm-wall panel	1000	24

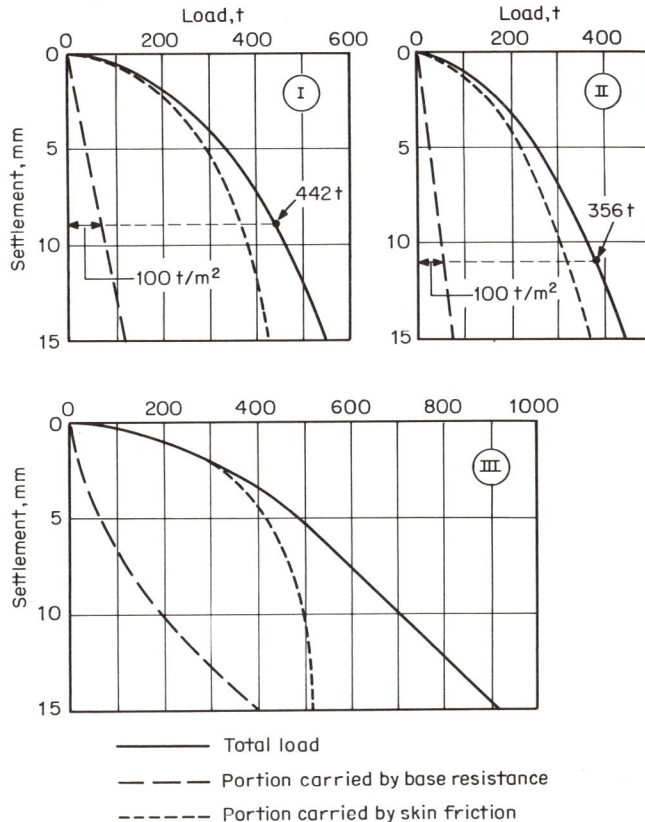

————— Total load

— — — Portion carried by base resistance

– – – – – Portion carried by skin friction

Fig. 11-22 Load-settlement behavior of test elements for the UNO Building in Vienna. (*From Kienberger, 1974.*)

total load carried by the circular element is 356 t, of which 293 t is resisted by skin friction and 63 t by base bearing, so that the average side-wall shear is about 7.9 t/m². For the linear panel the total load is 442 t, and skin friction amounts to 372 t, corresponding to an average side-wall shear of 8.0 t/m², almost the same as in the circular panel. The corresponding vertical displacements are 9 mm (0.36 in) and 11.4 mm (0.45 in) for the linear and circular panel, respectively. The slight difference in settlement is due largely to differences in the method of cleaning and installation. For the linear panel a premixed and carefully controlled bentonite slurry protected the excavation and prevented sedimentation, whereas for the circular panel bentonite and water were mixed and added as the hole was deepened. Although the base of the hole was cleaned before concrete was poured, some sedimentation probably occurred, leaving loose material at the base.

Pullout tests showed that practically the same skin friction was developed for elements I and II at the same top displacement. For the two linear panels a displacement of 15 mm (0.6 in) mobilized an average shear resistance of 4.2 t/m^2 in the zones of silty clay and fine sand. The shear resistance was rather uncertain in the gravel and fill layers but higher than in sand and clay.

Surface deformations between and around the panels were uniform. Although the three elements were spaced far enough apart (15 m, or 50 ft) to prevent mutual interference, their zone of influence diminished in a shorter distance. This zone appeared to cover an area radially from the face about $1\frac{1}{2}$ times the pile diameter or the diameter of a corresponding circle for the wall panels. For the two linear panels the zone of influence was virtually the same, despite the difference in depth.

Test Panels in London Clay Burland (1963) has reported tests on two diaphragm-wall panels in London clay. Both panels are 120 cm long, 50 cm wide, and 40 ft deep. One panel was excavated dry, whereas the other was excavated with a clamshell bucket under bentonite slurry. Both bases were cleaned before concreting.

After the concrete was allowed to cure for 3 weeks, each element was subjected to two load cycles. The first consisted of incremental loads (one-fourth the working load) until $1\frac{1}{2}$ times the working load was reached. The increments were maintained until the rate of settlement was reduced to 0.002 in per 30 min. The maximum load was applied for 24 h, after which the panels were unloaded and the rebound was observed.

The ultimate loads were estimated using adhesion factors 0.45 and 0.25 for the dry and bentonite element, respectively. The value of 0.45 is often used for straight bored piles in London clay, whereas the value of 0.25 was arbitrarily selected for the bentonite panel to account for possible effects of slurry. For both panels a bearing-capacity factor of 8.2 was used for base bearing. From these values, the calculated ultimate loads are 289 tons for the dry and 196 tons for the bentonite panel. One-half of these values was taken as the working load.

The results of the incremental load tests are shown in Fig. 11-23. In the range of working loads both elements have a similar performance, although the bentonite panel tends to have a vertical displacement slightly higher than the dry element. Figure 11-24 shows results from the second load cycle, consisting of a constant-rate penetration test to failure. The resistance to penetration is nearly the same for both panels, although the actual ultimate load for the dry element is 340 tons, or 15 percent higher than the estimated. Peak side-wall shear is reached at a vertical displacement of 0.2 in (5 mm).

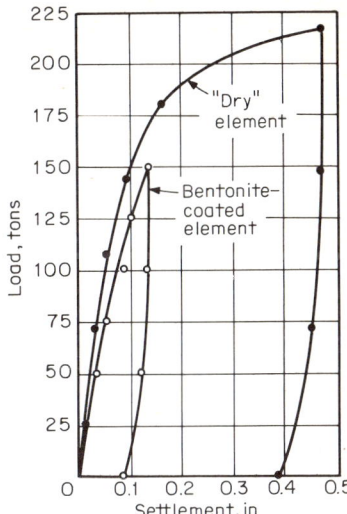

Fig. 11-23 Load-settlement curves for incremental-load tests on diaphragm-wall panels. (*From Burland, 1963.*)

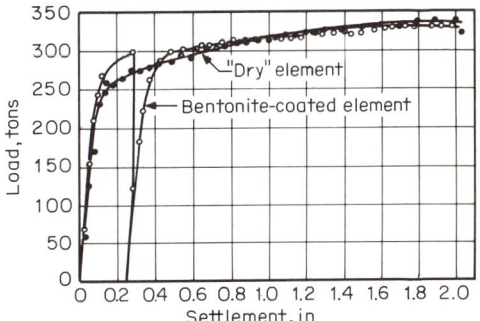

Fig. 11-24 Load-settlement curves for constant-rate penetration tests on diaphragm-wall panels. (*From Burland, 1963.*)

The variation in the moisture content several months after the load tests is shown in Fig. 11-25. This is for samples obtained at depths of 30 and 35 ft. The lateral extent of moisture migration is from 2 to 3 in (5 to 7.5 cm), and the increase of moisture content toward the concrete face varies from 3 to 4 percent.

Kensington and Chelsea Town Hall, London Comparatively recent tests have been reported by Corbett et al. (1974). The test panels measure 1.2 by 0.5 m in plan and are 14.4 m deep. The excavation was completed with a cable-operated grab, and before placing the concrete the slurry was replaced with a fresh solution. When the pour was completed, the observed overbreak was 8.5 percent, indicating some cavitation and overexcavation in the granular layers. The ground characteristics at the test site are typical for London, with a soil profile consisting of successive layers of sandy clay (brick earth), sand and gravel, and stiff gray fissured clay.

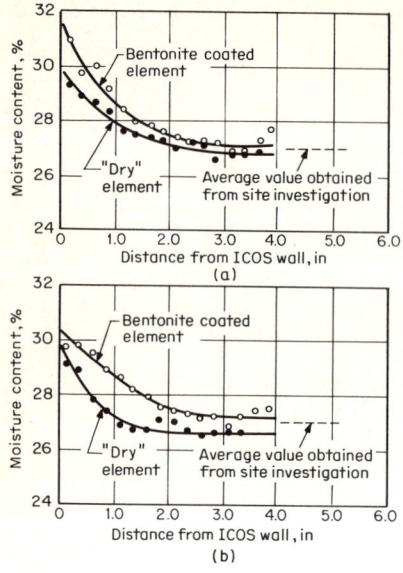

Fig. 11-25 Variation of moisture content with distance from diaphragm-wall test panels: (a) 30 ft below ground surface; (b) 35 ft below ground surface. (*From Burland, 1963.*)

For an estimated ultimate capacity of 2930 kN (obtained by using adhesion factors 0.5 for clay, a friction factor 0.7 for sand, and a bearing-capacity factor 9.0) the ultimate shaft resistance is 2190 kN and the ultimate base bearing 740 kN.

The panels were subjected to six cycles of loading, the first five being incremental and the sixth cycle consisting of a load application at a constant rate of penetration. For the latter a maximum safe reaction of 4000 kN was provided. The vertical displacement was sufficient to mobilize full shaft friction. Thus at 4000 kN the shaft carried 3650 kN, leaving only 350 kN for base bearing, or less than one-half the estimated base bearing.

The actual ultimate shaft resistance for this case is 67 percent higher than estimated. It is therefore conceivable that the adhesion and friction factors (based on dry conditions and a relatively smooth face) were underestimated and that incidental construction procedures usually ignored in the design, e.g., overbreak and face irregularities, contributed considerably to the increased ultimate capacity.

11-9 GUIDELINES FOR THE SELECTION OF LOAD-BEARING ELEMENTS

Choice of Type and Section The selection of the size, shape, and group arrangement will be governed by the type and magnitude of the applied loads, the site and ground conditions, and the structural features of the superstructure. Alternatively, local availability of equipment and regional construction trends dictate the choice in many instances.

Round cross sections offer a good solution where vertical loads are combined with bending moments in varying directions, e.g., forces from wind loads, and where the zone of influence of individual members is limited by site constraints or by the geometry of the foundation. A single slot section will resist vertical loads and bending moments in one direction, and for nominal dimensions (usually one equipment pass) it can replace an equivalent row of piles, thus giving a monolithic structure which is independent of such fixed tolerances as pile spacing. A T section has the same advantages as an equivalent T beam and offers compatible solutions near or along the perimeter of a foundation. The choice of I or H shapes is suitable to stress reversal. Finally an X section is indicated where it is necessary to resist predominantly concentrated vertical loads.

Recommendations for the Distribution of Load The usual approach to the load-distribution problem involves the application of limit theory in connection with a semiempirical choice of parameters based on regional experience. There is considerable variance in how this experience is expressed, as can be judged from the following statements. In Europe, for example, Sliwinski and Fleming (1974) recommend no reduction in the normal adhesion factor α for impervious clays but suggest a reduction of 10 to 30 percent for the friction factor of granular soils. Farmer et al. (1971) have suggested that in sand a reduced coefficient of wall friction $\tan \delta$ is appropriate with increasing confining stress, and this is in good agreement with results obtained by other investigators for various interfaces of sand-concrete materials. Reese et al. (1973) have correlated the average load transfer to the drained shear strength of sands $\tau = \sigma'_v \tan \phi'$, where σ'_v is the effective overburden stress and ϕ' is the friction angle for effective stress. The ratio α of the actual average load transfer to the average τ is plotted in Fig. 11-26 vs. the average pile settlement for the tests described in Sec. 11-6. In this case the apparent difference between the ultimate value of α for the three bored piles is due to the varying influence

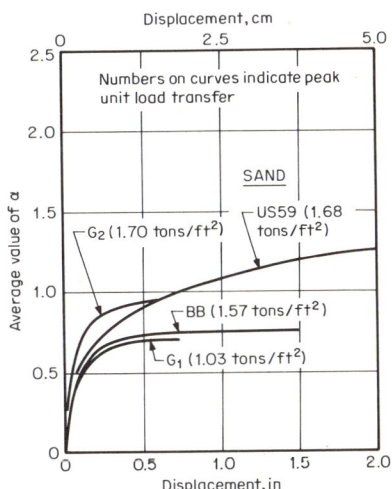

Fig. 11-26 Curves of load transfer versus displacement. (*From Reese et al., 1973.*)

on the load transfer of the upper clay layer. The minimum value of α is about 0.7, and this is recommended if the penetration is 25 ft (7.6 m) or less. For greater penetrations the factor could conceivably decrease, and thus reduced values are recommended in this case (Touma and Reese, 1974).

In some instances local building codes require the shaft resistance to be ignored completely if a foundation bears on a stratum which is relatively rigid compared with the overburden materials. At best the practice is to design the distribution of load for both shaft resistance and base support. Alternatively, load-bearing elements of uniform section built in clays and shales may sometimes have to be checked for shaft resistance alone, considering base bearing an added safety margin.

Semiempirical Criteria and Interpretation of Tests

Conventional or specially instrumented prototype load tests on panels built in average construction conditions can provide correlation between an assumed distribution of load and the in situ capacity (Reese et al., 1968), and this can be done with little or no interference with the construction operations. However, in some instances the cost of such load tests is not justified, or the results thus obtained may not always be fairly representative of the average site conditions. It follows, then, that in some cases the best available choice is to resort to a conservative design to compensate for all the unknown factors.

The interpretation of results from tests like those described in the foregoing sections can be useful provided the engineer understands that these tests also reflect the relative scarcity of construction sites with uniform and homogeneous soil conditions; the difficulty of providing the same control of the excavation process for both the regular and the test panels; and the prohibitive cost of full-scale investigations for the average project. The engineer should also exercise great caution in interpreting comparative tests due to the difficulty of simulating the side-wall configuration, obtained using the bentonite slurry process, by any form of dry construction in soils where the slurry method would normally be used.

For a semiempirical design the following guidelines are useful:

1. It is advantageous to excavate the panel, place the reinforcing cage, clean the bottom, and pour the concrete as soon as possible. The shorter the time a panel is open the less the soil disturbance, cavitation, sloughing, and other related effects.

2. Invariably the method of excavation and bottom cleaning will influence the development of side-wall shear and base resistance. The length of exposure of the open panel, the condition of the base, the roughness of the concrete-soil interface, the moisture sensitivity of the supporting earth materials, the slurry-displacement conditions, and changes in the earth-stress environment due to construction are among the factors influencing the load transfer. Since it is seldom possible to measure these effects directly, they can only be approximated and lumped into a common parameter.

3. Machines which excavate a relatively smooth face are in this case no

better than those which give a rough interface, but in practice it is impossible to distinguish between the two on a quantitative basis. On the other hand, machines which remove the excavated soil instantaneously (such as reverse circulation) should be preferred to those which do not.

    4. At present there is no evidence that the load-transfer characteristics of slurry panels (either circular or prismatic) differ materially from those in dry construction. The indication is that only certain empirical parameters are likely to vary. Thus ultimate side-wall shear is reached practically at the same vertical displacement for both slurry and dry elements. This displacement, expressing the relative movement of a point with respect to its initial position, can be taken as 5 mm (0.2 in) for firm clays. In sands it varies from 5 to 10 mm (0.2 to 0.4 in), depending on the confining stress, and the lower this stress the greater the displacement necessary to mobilize full shear.

    5. The assumption that in cohesive soil shaft resistance can be expressed as a percentage of the undrained shear strength, independent of other factors, usually corresponds to reality but can also be an oversimplification or lead to incorrect predictions.

    6. The average shear value of shaft resistance for circular and prismatic panels of the same depth and in similar soil conditions is the same, irrespective of the base-to-shaft surface ratio.

    7. Because side-wall shear is developed at smaller vertical displacements than base bearing, the entire shaft resistance is sometimes mobilized before any load can be transferred by base bearing.

    8. In soft clays shear failure normally occurs in the natural soil a small distance from the face of concrete, probably because this is the weakest shear zone. In this case the load transfer is equivalent to the undisturbed shear strength, and the adhesion factor α is unity. For stiff and hard clays the shear factor α, although less than unity, is the same for dry and for bentonite elements.

    9. In hard clays shear failure occurs along the concrete interface since this is the weakest zone, and in this case the wall adhesion is merely the shear strength of the hardened bentonite coating. For practical purposes this strength can be taken as 90 kN/m² (about 1900 lb/ft²).

    10. In sands the shaft resistance is not necessarily a linear function of the overburden stress. Furthermore, the direct shear tests described in the foregoing sections show that the coefficient of wall friction was reduced with increasing confining stress and reached a value of 0.6 at 0.35 N/mm² (about 50 lb/in²). The nonlinear function of side shear and depth also reflects the influence of arching near the tip (Touma and Reese, 1974), and the continuing growth of the arching zone with settlement results in larger displacements at failure. These phenomena are observed primarily in dense and very dense sands, where the shear transfer in the arching zone accounts for a large portion of the total shear transfer.

    11. In sands appreciable vertical movement of the tip is necessary to mobilize sufficient base bearing. Base bearing is significant with respect to the total capacity only in dense and very dense material. Since sandy soils are

more prone to sloughing and peel-off, both causing some deposition of loose materials at the bottom, it is safe to expect an actual tip resistance smaller than the theoretical.

12. For any soil conditions and soil type a small vertical displacement due to the presence of some soft sediment at the base is beneficial to the development of side-wall resistance. On the other hand, excessive accumulation of unconsolidated soil can prevent the development of base bearing and also cause unacceptable settlement.

11-10 DESIGN FOR VERTICAL LOADS, PRISMATIC OR CIRCULAR ELEMENTS

Estimation of Working Load

For the general case this requires a semiempirical prediction of the peak-load transfer, i.e., the ultimate load. A working load is then obtained by dividing this by an appropriate factor of safety.

The ultimate load P_u is the sum of the ultimate base resistance P_b (direct bearing) and the ultimate shaft resistance P_s (wall friction and/or adhesion). Since the ultimate shaft and base resistance are reached at different vertical displacements, two different factors of safety must be used. Accordingly, the working load P_w is

$$P_w = \frac{P_b}{F_b} + \frac{P_s}{F_s} \qquad (11\text{-}1)$$

in which F_b and F_s are the factors of safety for base and shaft resistance, respectively. If F is the factor of safety relating P_u and P_w ($P_w = P_u/F$), the relation $P_u = P_b + P_s$ combined with Eq. (11-1) yields

$$F = \frac{F_s F_b (1 + R)}{F_s + F_b R} \qquad (11\text{-}2)$$

in which $R = P_s/P_b$.

It is customary to omit the weight of the element and the overburden term from the bearing-capacity calculations, the assumption being that they balance approximately. If we accept this assumption, the actual load to be transferred to the soil is the load applied externally to the element. This, however, is not recommended for retaining diaphragm walls functioning as load-bearing elements, and in this case the weight of the wall should be part of the load.

Estimation of Shaft Resistance

Cohesive Soil If S denotes the effective perimeter of an element, the ultimate shaft resistance P_s of a given segment length ΔL is expressed in terms of the undrained shear strength s_u and the reduction factor α. This leads to the expression

$$P_s = \Sigma S \, \Delta L \, \alpha s_u \qquad (11\text{-}3)$$

Average values of α are recommended for three main consistencies of clays, namely soft, stiff, and hard. For soft clays ($q_u < 0.5$ ton/ft²) the factor α can be taken as 1, since failure should be expected to occur in the soil some distance away from the face. For stiff clays (q_u from 1.0 to 2.0 tons/ft²) α can be expected to vary from 0.45 to about 0.60. The value 0.45 is commonly used for the stiff London clay and is close to the average shear strength obtained in triaxial tests. The value 0.6 is appropriate for the lower limit of this strength range. For hard clays ($q_u > 4$ tons/ft²) the shaft resistance is roughly equal to the shear strength of a hard coating at the interface. This can be taken as 2000 lb/ft² (96 kN/m²), but some variations should be expected.

Cohesionless Material Ultimate shaft resistance is expressed in terms of the effective overburden stress σ_v', the coefficent tan δ, and a reduction coefficient K_s. This relationship is expressed as

$$P_s = \Sigma S \, \Delta L \, K_s \sigma_v' \tan \delta \qquad (11\text{-}4)$$

Another expression is (Touma and Reese, 1974)

$$P_s = \Sigma S \, \Delta L \, \alpha \sigma_v' \tan \phi' \qquad (11\text{-}5)$$

in which α is the reduction factor and ϕ' the effective friction angle. Equations (11-4) and (11-5) are equivalent as long as the parameters K_s, α, δ, and ϕ' can be correlated.

For concrete-soil interface and dry elements ϕ' approaches δ closely (Potyondy, 1961). For most practical purposes the same assumption can be made for slurry panels, and any difference is thus lumped into the coefficient K_s and α. The assumption that the shear resistance along the shaft in granular soils is a linear function of the effective overburden stress is valid only for limited depths (Kerisel, 1964; Vesic, 1968). At present, this author recommends computing either K_s or α from the expressions

$$K_s = \begin{cases} \alpha = 0.7 & \text{for depths} < 25 \text{ ft} \\ \alpha = 0.5 & \text{for depths} > 25 \text{ ft} \end{cases} \qquad (11\text{-}6)$$

Estimation of Ultimate Base Bearing

The analysis of base bearing of deep foundations is based largely on the work of Terzaghi (1943), Meyerhof (1951), and Berezantsev et al. (1961). The following paragraphs will supplement the investigation of base resistance.

Cohesive Soil The ultimate base resistance on the underside of an element is expressed in terms of s_u as

$$P_b = A_b N_c s_u \qquad (11\text{-}7)$$

in which A_b is the area of the base and N_c is a bearing-capacity factor. Whitaker and Cooke (1965) recommend N_c values of the order of 6.5 when s_u is taken as the average shear strength from soil tests, and this is intended to compensate for sample disturbance and other incidental effects. N_c can be taken as 9 in connection with the $\phi = 0$ condition and when s_u is closer to the lower limit of the shear strength range under consideration. The value of 9 also is appropriate for fairly undisturbed samples of nonfissured clays.

Cohesionless Material For this case P_b likewise is expressed as

$$P_b = A_b N_q \sigma'_v \qquad (11\text{-}8)$$

in which the bearing capacity factor N_q is estimated as a function of the effective friction angle ϕ' and the σ'_v term is not necessarily always equal to the overburden pressure.

Correlation of Settlement with Shaft and Base Resistance

There is considerable practical significance in the observed vertical displacement at which ultimate shaft resistance is mobilized. As mentioned, for clays this can be taken as 5 mm (0.2 in), but for sands it is closer to 10 mm (0.4 in). Unless an element is founded on rock or on fairly incompressible materials, the normal tolerance of the construction methods and bottom cleanout procedures will result in settlements of that magnitude. It is therefore expedient to take into consideration the full shaft resistance.

The much larger and often uncertain settlement associated with ultimate tip resistance will in many cases make base bearing only a small supplement to the working load. It is interesting to note that for either cohesive or cohesionless soil, once the bulk of tip resistance is reached (and this usually occurs at relatively small tip movements), the load-settlement curve becomes progressively flatter and reaches an ultimate value at movements which vary considerably. Thus, whether the settlement at peak base support is 8 or 10 percent or more of the base diameter or mean width is purely academic since settlement of that order cannot be accepted.

At present, a procedure for estimating base bearing as a function of the permissible settlement is as follows.

Cohesionless Soil Touma and Reese (1974), dealing primarily with bored piles, suggest defining the failure load as the tip resistance corresponding to 1 in (25 mm) of vertical tip movement. This resistance has been determined empirically and is given by

$$P_{b1} = \frac{A_b}{0.6D} q_t \qquad (11\text{-}9)$$

where A_b = area of base, ft²
 D = diameter, ft (or, by extension, equivalent diameter of prismatic element of same area)
 q_t = average tip resistance at tip movement not exceeding 5 percent of base diameter

The value of q_t can be taken as zero for loose sand; 32,000 lb/ft² (1530 kN/m²) for medium dense sand; and 80,000 lb/ft² (3830 kN/m²) for very dense sand. These values are for diameters greater than 2 ft and for depths greater than 10 diameters. Although they have been derived empirically, they give fairly good predictions for the usual bearing-capacity problems.

Cohesive Soil The same concept can be applied to cohesive soils provided it is possible to correlate the ultimate base resistance with the fraction mobilized at 1-in (25-mm) tip movement. For soft clays it is safe to ignore base bearing completely. For stiff and hard clays examination of results from a large number of field tests indicates that at a tip movement of 1 in the base resistance is from 0.5 to 0.7 times the ultimate base resistance as expressed by Eq. (11-7). Hence, it is proposed to approximate this base resistance by the relation

$$P_{b1} = 0.6P_b = 0.6A_b N_c s_u \qquad (11\text{-}10)$$

which generally should be applied to fairly large diameters or widths.

For tip movements less than 1 in it is safe to assume a linear progression for both cohesive and cohesionless soils. Within this range the base load is obtained by straight-line interpolation.

Settlement Analysis

With the exception of long-term consolidation effects, the initial settlement is influenced by the elastic shortening, the load-transfer characteristics, and the resistance to penetration of the soil at the tip. Since the last depends largely on the construction methods, a settlement analysis is useful only as an approximation of the in situ conditions.

According to some theories, the influence of shaft resistance on tip movement is considered by dividing the shaft load into a series of segmental loads. In the simplest form this procedure is carried out as suggested by Farmer et al. (1971). Predictive techniques giving rigorous solutions have been put forward by Poulos and Davis (1968).

A computational approach suitable to digital computers analyzes friction or end-bearing elastic elements embedded in uniform and in stratified elastic media (Thurman and D'Appolonia, 1965). The accuracy of the computed results for a particular problem depends in this case on the accuracy of the selected coefficient of lateral earth stress, the modulus of elasticity of the soil, and the actual elastic or elastoplastic movement of the tip.

Factors of Safety

Once the ultimate shaft resistance P_s and the base bearing P_{b1} corresponding to a 1-in movement are estimated, a working load is obtained by applying to these values a factor of safety F_s and F_b, respectively. The selection of these factors may be arbitrary, or it may have a displacement-compatibility basis if data from load tests are available.

For the usual problem F_b may be taken the same as F_s, in which case they should be at least 2 and preferably 2.5. Where base bearing is rather uncertain compared with shaft resistance, or where the tip resistance constitutes the larger fraction of the total load capacity of the member, a greater factor of safety F_b is recommended, usually 3. In this case the factor of safety F for combined shaft and base load should be not less than 2.5.

A more rational approach to the safety-factor problem is possible if the shaft and base deformation compatiblity is known (Whitaker and Cooke, 1965). An example based on an experimentally derived relationship between shaft and base settlement is shown in Fig. 11-27. The mean curves of F_b and F_s are obtained from load-settlement curves and are plotted vs. the percentage settlement expressed in terms of the shaft and base diameter. The two curves are so far apart because ultimate shaft resistance is reached at much smaller displacements. For example, for s_c/d greater than 0.7 percent F_s approaches unity, indicating that ultimate shaft resistance is exceeded. This does not mean that the shaft resistance is now inoperative but shows that further settlement under more load is governed by the response of the base.

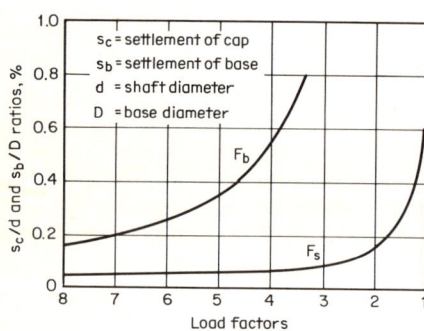

Fig. 11-27 Examples of curves relating load factors with s_c/d and s_b/D for use in design. (*From Whitaker and Cooke, 1965.*)

In order to use similar curves in design, estimates are made first of the ultimate shaft and base resistance. Assuming that the settlement of the cap (top) and the base are equal to the allowable settlement, a value of F_s and an approximate value of F_b are found from the curves. Working values of shaft and base load are then determined, the shaft compression is calculated, and the base settlement is thus corrected, leading to a second value of F_b from the curve. By successive approximations a more accurate value of F_b is obtained, allowing estimation of the base working load.

Since the foregoing procedure requires an empirical relationship which applies to the test site only and reflects the actual duration of the test, it does not include long-term consolidation effects. Evidently, a displacement-compatibility determination of the working load results in a greater factor of safety for base resistance than for shaft resistance.

REVIEW PROBLEMS

11-1 Prepare a tentative specification and the control limits for a slurry which is to be used in conjunction with a casing. In this case the slurry is to hold in suspension excavated materials and is pumped out before concreting.

11-2 Referring to Fig. 11-5b and c establish criteria for estimating the effective perimeter of shaft resistance.

11-3 Prepare a statement about the prequalifications of the supervising engineer to ensure adequate supervision according to the construction requirements of load-bearing elements.

11-4 Describe situations where a hole for a foundation element would not stand open and stable without slurry.

11-5 Describe the requirements and the differences for boring a hole under slurry (a) in loose cohesionless material and (b) in cohesive clays.

11-6 Discuss the factors to be considered in the design and construction of load-bearing elements adjacent or close to existing basements and foundations.

11-7 Using an effective-stress approach, establish a procedure to predict the load-carrying capacity of the diaphragm wall shown in Fig. P11-7. State your assumptions.

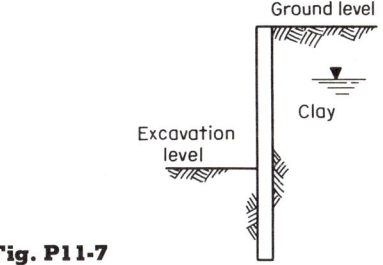

Fig. P11-7

11-8 Discuss how you would evaluate the side resistance of the underground structure shown in Fig. P11-8. The structure is built according to the under-the-roof method.

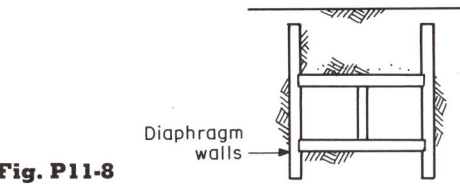

Fig. P11-8

11-9 The load-bearing element shown in Fig. P11-9 is 50 ft deep. The soil is stiff clay with $q_u = 1.5$ tons/ft^2, $N_c = 8$, and $\alpha = 0.535$. Estimate the safe working load using $F_s = 2.5$ and $F_b = 3.0$. What is the factor of safety F?

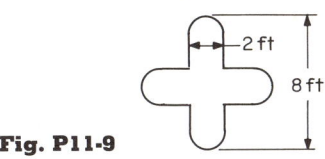

Fig. P11-9

11-10 Repeat Prob. 11-9 if the last 10 ft consist predominantly of hard clay ($q_u = 4$ tons/ft^2) assuming that in this region the shear strength at the concrete-soil interface is 1500 lb/ft^2. Use the same factors of safety as before.

11-11 Determine the effective perimeter for shaft resistance and the effective base area for tip support for the load-bearing element shown in Fig. P11-11.

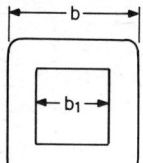

Fig. P11-11

11-12 Develop an approximate method for estimating the elastic shortening (change in length due to a variable load along the shaft) for a load-bearing element. Assume the surrounding soil to be elastic material.

11-13 A 36-in-diameter 40-ft-long bored pile is cast in slurry. The soil is sand with $\phi' = 30°$, $\gamma = 120$ lb/ft^3, and the water table is near the ground level. The pile tip rests on very dense sand. Estimate the working load using $F_s = 3.0$ and $F_b = 3.5$.

11-14 Explain how you would design prefabricated wall panels to act as load-bearing elements. Analyze both side resistance and base bearing.

11-15 With the help of Fig. 11-17 explain how the fully developed load transfer (side shear) in sand changes with confining pressure and derive the data shown in Fig. 11-18.

11-16 Compare the experimental finding indicating that the coefficient of wall friction in sand is reduced with increasing confining stress and the recommendation to reduce the friction factor α (load-transfer–shear-strength ratio) for depths exceeding the normal range of 25 ft. Discuss the similarities and the differences between the two approaches.

REFERENCES

Baker, C. W., and F. Khan, 1971: Caisson Construction Problems and Corrections in Chicago, *J. ASCE Soil Mech. Found. Div.*, February, pp. 417–439.

Berezantsev, V. G., et al., 1961: Load Bearing Capacity and Deformation of Pile Foundations, *Proc. 5th Int. Conf. Soil Mech. Found. Eng., Paris*, vol. 2.

Burland, J. B., 1963: Discussion, Session 4, Symposium of Grouts and Drilling Muds, in "Grouts and Drilling Muds in Engineering Practice," Butterworths, London.

Chadeisson, R., 1961: Influence of the Boring Methods on the Behavior of Cast-in-Place Bored Piles, *Proc. 5th Int. Conf. Soil Mech. Found. Eng., Paris*, vol. 2.

Corbett, B. O., et al., 1974: A Load Bearing Wall at Kensington and Chelsea Town Hall, London, *Proc. Diaphragm Walls Anchorages, Inst. Civ. Eng., Lond.*

DeMello, V. F. B., 1969: Foundations of Buildings in Clay, *Proc. 7th Int. Conf. Soil Mech. Found. Eng., Mexico City.*

Farmer, I. W., et al., 1971: The Effect of Bentonite on the Skin Friction of Cast-in-Place Piles, *Proc. Behavior Piles, Inst. Civ. Eng., Lond.*

Henkel, D. J., 1970: Geotechnical Considerations of Lateral Stress, *Proc. ASCE Spec. Conf., Lateral Earth Stresses Design Earth Retaining Struct., Cornell Univ.*, June.

Kerisel, J., 1964: Deep Foundations: Basic Experimental Facts, *Proc. Deep Found. Conf., Mexico City.*

Kienberger, H., 1974: Diaphragm Walls as Load Bearing Foundations, *Proc. Diaphragm Walls Anchorages, Inst. Civ. Eng., Lond.*

McKinney, J. R., and G. R. Gray, 1963: The Use of Drilling Mud in Large Diameter Construction Borings, in "Grouts and Drilling Muds in Engineering Practice," Butterworths, London.

Meyerhof, G. G., 1951: The Ultimate Bearing Capacity of Foundations, *Geotechnique*, vol. 2, no. 4.

Potyondy, J., 1961: Skin Friction between Various Soils and Construction Materials, *Geotechnique*, **11:** 343–353.

Poulos, H. G., and E. H. Davis, 1968: The Settlement Behavior of Single Axially-loaded Piles and Piers, *Geotechnique*, vol. 18.

Reese, L. C. et al., 1968: Instrumentation for Measurements of Lateral Earth Pressure in Drilled Shafts, *Univ. Tex. Res. Rep.* 89-2, Austin.
——— and M. W. O'Neill, 1969: Field Tests on Bored Piles in Beaumont Clay, *ASCE Annu. Meet., Chicago, Repr.* 1008.
———, ———, and F. T. Touma, 1973: Bored Piles Installed by Slurry Displacement, *Proc. 8th Int. Conf. Soil Mech. Found. Eng., Moskow.*
Sliwinski, Z., and W. G. K. Fleming, 1974: Practical Considerations Affecting the Construction of Diaphragm Walls, *Proc. Diaphragm Walls Anchorages, Inst. Civ. Eng., Lond.*
Terzaghi, K., 1943: "Theoretical Soil Mechanics," Wiley, New York.
Thurman, A. G., and E. D'Appolonia, 1965: Computed Movement of Friction and End-bearing Piles Embedded in Uniform and Stratified Soils, *Proc. 6th Int. Conf. Soil Mech. Found. Eng., Montreal,* vol. II, pp. 323-327.
Touma, F. T., and L. C. Reese, 1974: Behavior of Bored Piles in Sand, *J. ASCE Geotech. Div.,* **100:** 749-761.
Vesic, A. S., 1967: Ultimate Loads and Settlement of Deep Foundations in Sand, *Proc. Symp. Bearing Capacity Settlement Found., Duke Univ., Durham N.C.*
———, 1968: Load Transfer, Lateral Loads and Group Action of Deep Foundations, *ASTM Spec. Tech. Publ.* 444.
Whitaker, T., and R. W. Cooke, 1965: Bored Piles with Enlarged Bases in London Clay, *Proc. 6th Int. Conf. Soil Mech. Found. Eng., Montreal.*

Chapter Twelve
USES AND APPLICATIONS

Although engineers generally regard the use of subsurface space as an attractive alternative to the congestion of city sites and the usually high land values, in many instances underground construction has less appeal to the public because of the frequently high cost of underground work and the often undesirable interaction with the natural or the man-made environment. Examples are the high cost of new subways in some cities, particularly in North America, and the potentially deleterious effects of deep excavations on surroundings. The result has been a growing concern and lack of faith in the effectiveness of certain construction techniques and some skepticism about the benefits associated with the use of subsurface space.

Some of the key problem areas appear to be influenced by factors which in strict terms are not engineering issues. We can mention, for example, the market and site conditions of a job; the constructional competence and performance as a regional record; the allocation and sharing of risks; and miscellaneous items such as insurance and bonding.

However, to the engineer of an underground construction more important is the ability to do the work in terms of technical resources and at a minimum cost. The choice, whatever it may be, is not always easy, and very often the engineer must exclude certain methods merely because of their adverse effect on the urban environment. In a broad sense this may be damage to the physical surroundings, disruption of surface functions, or interference with the living habits of people. In the final analysis the choice of method depends on technical as well as nontechnical considerations (see also Chap. 1), but more often it is dictated by the presence of buildings and installations either above or below ground, some of which are liable to be adversely affected by work below ground level; traffic and pedestrian provisions, including traffic operational requirements; utility interference, especially the abundance of utilities that must be terminated, replaced, diverted, supported, or protected; the soil and groundwater conditions, especially the water table, and the possibility of base failure during general excavation and uplift after construction; and certain constraints, e.g., special working hours and limitations on time schedule or equipment clearance from existing buildings and facades imposed by the risk or risk-damage factor.

This chapter demonstrates the uses and applications of the slurry-wall technique in various underground works and deals in a practical manner with the problems of underground construction. It also presents some special solutions which are feasible in conjunction with the site and ground conditions most typically encountered in urban situations.

12-1 UNDERPINNING OF STRUCTURES

In this context underpinning does not mean the insertion of a new foundation or support below an existing one to transfer the load to a lower level. This transfer is often necessary because of adjacent work, such as excavation at a lower level, which can affect the strength of the supporting ground or generate movement. In a broader sense (as used here) underpinning can also be the lateral protection of the foundation or the strengthening of the ground beneath.

Consider, for example, the structures shown in Fig. 12-1. In Fig. 12-1a the need for underpinning and/or protection is obvious; in Fig. 12-1b although the excavation level is above the existing footing, the removal of the overburden can render the supporting ground weaker and thus decrease the load-carrying capacity of the foundation. Instead of constructing a deeper foundation, a diaphragm wall can be used to protect the open excavation. It is conceivable, however, that ground movement will still occur and that the settlement profile behind the wall will determine the associated effects. If this settlement is large, it will affect any building in the zone of influence and cause differential stretch and shear. Even if a structure rests on deep foundations, e.g., piles or the footing shown in Fig. 12-1b, it still may suffer damage due to horizontal extension even though it does not settle appreciably.

From the foregoing it is evident that a diaphragm wall can be used in lieu of underpinning provided that large ground movement is prevented and the construction is supplemented by ground strengthening where necessary. The latter may involve systematic ground treatment by a suitable process to form

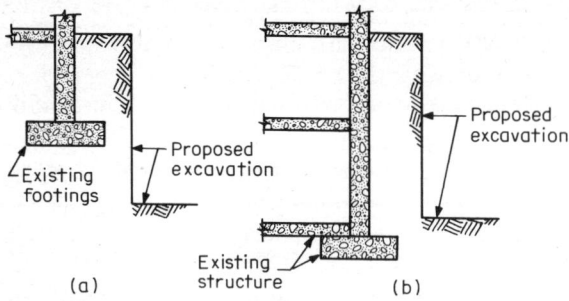

Fig. 12-1 Excavation adjacent to existing structures: (*a*) base of excavation below existing footing; (*b*) base of excavation above existing footing.

an extension of the load-bearing foundation. This improvement can be accomplished by freezing if it is temporary only or by grout injection if it must be permanent. If the ground is groutable, injection is safer before trenching, and this also reduces the risk of caving during trenching.

An example of this type of underpinning is shown in Fig. 12-2. In this case the protection was necessary to secure the structure of a metro in Paris, the bottom of which is 6 m (20 ft) from ground level and well above the base of the new excavation. After grout injection of the ground beneath the structure, protection was provided by two parallel walls bearing on hard strata. The walls were constructed in short panels and were braced laterally with four rows of permanent prestressed tiebacks.

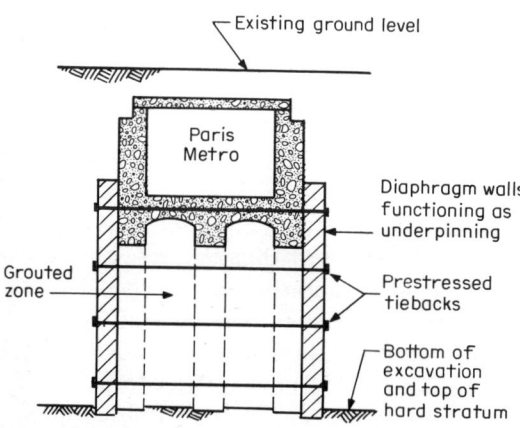

Fig. 12-2 Use of diaphragm walls in conjunction with grouting to underpin an existing subway structure.

12-2 DIAPHRAGM WALLS WITH TIEBACKS

The use of tiebacks and ground anchors is becoming widespread for two reasons: (1) the substantial progress in the anchoring techniques and the remarkable increase in the capacity of ground anchors; and (2) the advantage of free excavation and the absence of interior obstructions which permit mechanical earth moving and greatly improve the construction conditions of interior supports and intermediate floors, especially on irregular or congested sites.

The analysis, design, and installation of ground anchors is highly specialized (see also Chap 13). Anchoring usually involves driving a casing through the retaining soil mass using rotary, rotary-percussive, or vibrodriving techniques. The drilling method should not result in loss of ground beneath existing buildings, and grouting of tiebacks should not cause heave or transmission of high direct or induced pressures to the back of the diaphragm wall being supported. It is also desirable to keep the wall away from such operations as pile driving, blasting, and freezing.

Tiebacks cannot be installed if major obstructions exist, e.g., utility lines and incidental underground structures. In some instances there may be specific statutory requirements, or the adjoining owner may have the right of support and refuse an easement. The record shows, however, that although complex, the legal, ownership, and regulatory problems can be solved. More difficulties should be expected in relatively deep cuts, where it is necessary to install the tiebacks under several properties. In this case if the installation is permanent, it should be made under the implied guarantee that future construction will not impair, damage, or otherwise interfere with the structural integrity of the system.

Tied-back diaphragm walls are quite common in Europe, where the combination has enhanced the possibilities of deep excavations and has made deeper and larger basements possible. They are gradually becoming common in the United States because of the availability of space for their installation and the generally favorable position toward the legal and contractual aspects. In Japan tied-back walls are used only rarely, due to lack of space and the congestion of building sites.

12-3 DIAPHRAGM WALLS ON STILTS, PILES, SUBPIERS, AND ROCK

In permanent construction diaphragm walls may have to transfer quite heavy loads to the supporting soil. These loads are likely to include reactions from the top and intermediate floor levels for the underground portion of the structure, but sometimes even superstructure loads are transferred to the perimeter wall. Under these conditions if we assume that the normal working stress of concrete is utilized in compression, a wall of nominal thickness (60 cm or 24 in) can have a load-carrying capacity in excess of 200,000 lb/ft (linear) (about 300 t/m).

Where the available soil strength at the tip of such a wall (which tip is determined according to lateral-stability requirements) cannot accommodate the transfer of load, instead of extending the entire structure deeper until sufficient bearing is available, it may be cheaper to support the wall in one of the following ways.

Support on Stilts A wall on stilts is shown in Fig. 12-3. For each wall panel a certain portion (usually one equipment pass) is extended down to firm bearing, so that its tip is considerably lower than excavation level. Usually one or two stilts are sufficient for each panel, depending on the panel dimensions and the loads to be transferred. The arrangement and construction of the panels is governed by the following considerations: the panels should be provided with dowels to form shear joints; the reinforcing cage should be assembled and installed as one section, including the stilt portion; the average load transfer of each stilt should be nearly the same from one panel to another; the stilt spacing should be compatible with the recommended maximum spacing of tremie pipes; and the panel length and stilt depth should give a gross volume according to the maximum practicable concrete pour for the conditions involved.

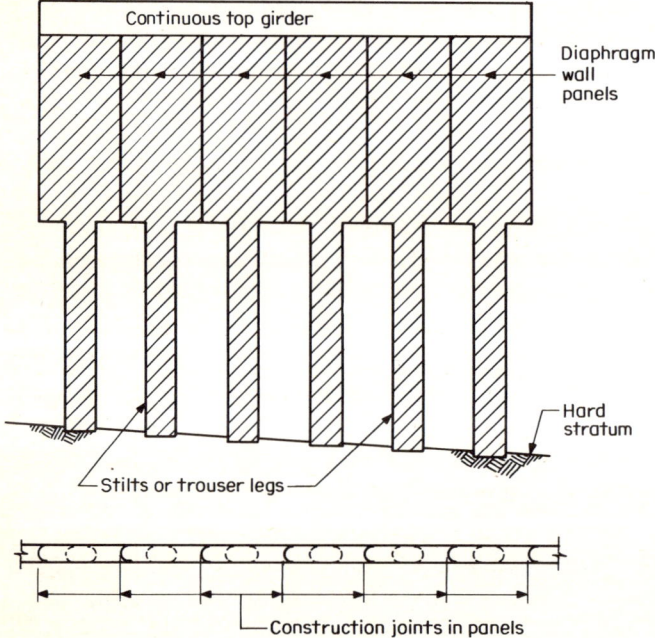

Fig. 12-3 Example of a diaphragm wall supported on stilts.

The load-carrying capacity of the wall should be based on a displacement compatibility. Side shear may be included for both the wall and the stilts, but the base resistance should be estimated considering only the bearing at the tip of the stilts.

Support on Piles Walls on piles are sometimes used if the distance from the general excavation level to stiff or hard formations is appreciable. The wall may have an embedment below excavation level, in which case it does not have to be braced at this location, or it may just be extended 2 to 3 ft into the ground, in which case bracing at or near excavation level should be provided. The choice depends on relative cost, lateral stability, and groundwater conditions.

Steel piles can be driven through the excavated panels, and alternatively steel casings can be installed in the panels to allow the piles to be rammed after the wall is constructed. Each method has advantages and disadvantages.

Driving the piles through open panels is preferred by many contractors and by the author. The pile installation is in this case independent of the panel sequence and size and offers no impediment to the phase of the excavation and concrete placement since it is carried out between these two phases. The piles have their tops projecting about 3 ft (1 m) into the concrete. A common problem resulting from pile driving is some sloughing and peel-off due to vibrations and slurry disturbance, but this is not any more serious than the disturbance caused by chiseling to break obstructions and consolidate hard materials. The construction sequence, however, requires the pile contractor to be on the job until the entire installation is completed.

Driving the piles in preformed holes after the wall is constructed allows better use of pile-drilling equipment and labor. The difficulty in this case is how to arrange the pile casings, the tremie pipes, and the construction joints without blocking the flow of fresh concrete (both lateral and upward). On the other hand, it is possible to use a different pile size after the wall is built if this becomes necessary and also to extend the piles to ground level for direct connection with superstructure columns.

Support on Subpiers This arrangement is different from the composite walls mentioned in previous sections in that the wall is not interlocked with the subpiers but is directly supported on them. The two usual methods of installation are shown in Fig. 12-4. In Fig. 12-4a the holes are bored from the ground level, and the subpiers are concreted to a level which is the base of the wall. The casing, if used, is then withdrawn, and the hole is backfilled with granular material above the concreted subpiers. The diaphragm wall is constructed as a separate structure and usually is arranged so that there is a vertical panel joint at each subpier in order to eliminate any differential movement between adjoining panels.

The wall shown in Fig. 12-4b is cast as an integral unit between panel joints; the hole is first excavated from the surface using a circular drill or an auger-type equipment under slurry, and the excavation of the panel is completed with clamshells or slot-type machines. The reinforcing cage is assembled and installed as one unit for the entire panel, and the placement of concrete begins by setting the tremie pipe in the hole. The main advantage of the construction shown in Fig. 12-4b is the elimination of the horizontal construction joint between the wall and the subpiers, which is sometimes difficult to execute, especially in troublesome soils.

Walls on Hard Stratum When rock or other hard formations exist at relatively shallow depth so that it is economically feasible to carry the tip of the

418 Slurry Walls

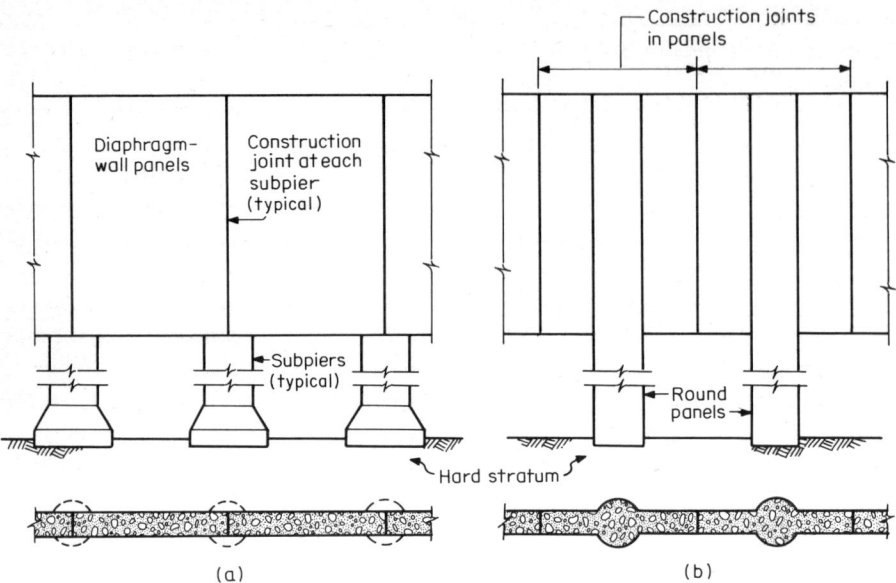

Fig. 12-4 Diaphragm walls on subpiers: (a) subpiers drilled from ground level and concreted to bottom of wall; (b) round panels excavated simultaneously with wall panels.

wall down to the same depth, the transfer of load can be accomplished in two ways: by terminating the diaphragm wall just at the top of the hard formation and using tiebacks if necessary near this level for lateral stability or by penetrating into the hard stratum for stability if this construction is more economical. As mentioned in previous sections, rock excavation for wall extension is particularly difficult and costly since it means conversion from soft ground equipment to percussive tools and rotary drills; where possible it should be avoided.

12-4 PREFOUNDED COLUMNS

Heavy column loads, usually transferred from tall buildings or from massive structures, often require special or deep foundations extended to firm bearing or rock. Commonly the construction of these foundations is carried out from the base of the general excavation, and a usual objection to this stems from the standpoint of efficiency since the size of rigs, movement and mobility of machines, and traffic through the access ramps are not provided for in the most productive way.

Alternatively, deep foundations of this type are installed before general excavation and before earth moving is begun, using the ground level as working platform. The top of the foundation is terminated at the level of the raft or at the base of the proposed excavation. However, when earth moving begins, the ground beneath the base of the excavation is unloaded. If the formations include clayey layers and a natural water table, when unloaded these beds will probably swell upward, subjecting the deep foundations to considerable tension along the shaft, applied as negative friction. Subpiers subjected to this condition have been known to fail by fracture if they are not reinforced.

It is possible to avoid these problems by adapting a joint design and construction method for the columns of the structure and their foundation portion known as *prefounded columns*. The operation consists of constructing from the ground level and in a single phase the foundation itself and the basement part of the column, as shown in Fig. 12-5. In the example shown a casing is sunk to just below the basement level as the hole is bored. The excavation of the hole is continued below this level under bentonite slurry. When the final excavation level is reached and the bottom is cleaned, the steel column is lowered into the casing, positioned, and firmly held from the top. Concrete is tremied to a level which is the underside of the excavation, so that the top portion provides an embedment for the column and seals the base. When this operation is completed, the casing is withdrawn and the annular space is backfilled with granular material.

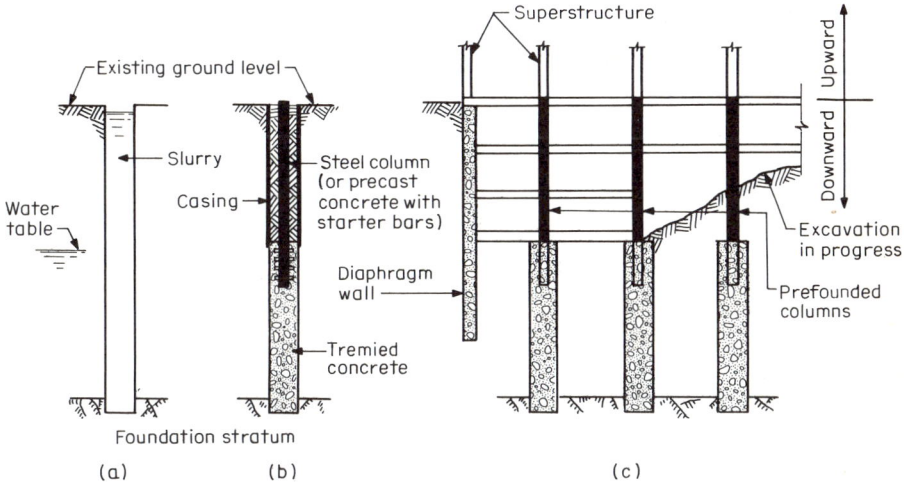

Fig. 12-5 Prefounded columns and use in downward construction: (*a*) drilling hole under slurry; (*b*) lowering and positioning column and concreting the foundation element; (*c*) simultaneous construction of superstructure and substructure by upward and downward process.

With both the foundation and the column in place it is possible to begin erection of the superstructure immediately and simultaneously construct the basement part of the building, as shown in Fig. 12-5c. A difficulty inherent in this method is that it requires considerable construction accuracy to attain the specified verticality, and the positioning of the column both before and during concreting must be checked for accuracy in plan location, plumbness, and correct level.

12-5 STRIP PANELS

For wall connections with intermediate floor slabs, beams, etc., it is necessary to design and detail suitable construction joints using bent bars, as shown in

Chap. 8. Instead of direct structural connections with the diaphragm wall, the interaction can be achieved with pillars cast against the wall as the latter is exposed and as the excavation is carried down in steps, as shown in Fig. 12-6a. The pillars, or counterforts, are built in sections, using starter bars where the intermediate floor slabs are to serve as lateral bracing, or monolithically at the end of excavation if temporary bracing is used.

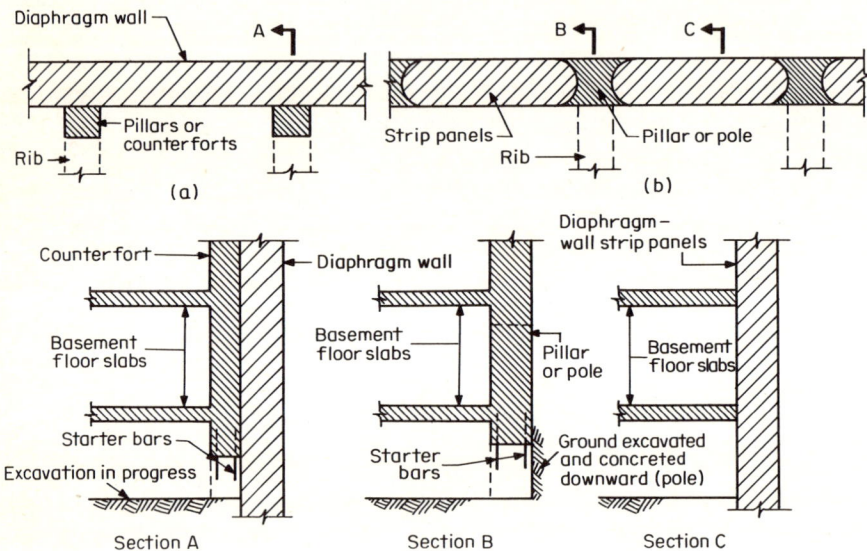

Fig. 12-6 Construction of diaphragm walls: (*a*) conventional method; (*b*) strip-panel method.

If the natural water level is below the bottom of the excavation and the ground has some strength to stay temporarily stable, it may be advantageous to build the diaphragm wall in strip panels, as shown in Fig. 12-6b. The wall is initially processed in individual slot sections without end tubes or vertical construction joints. As the excavation is deepened, the intermediate tongue is filled in sections, using again starter bars, and receives structural connections from intermediate floor slabs or frame ribs. In many instances this may yield appreciable economy, resulting primarily from the elimination of vertical panel joints and wall-to-beam and wall-to-slab connections and from a reduction in the diaphragm-wall area. Excluding water problems, the strip-panel method can also provide a self-supporting wall if used in connection with tiebacks installed through the tongues.

A different example of strip-panel wall is shown in Fig. 12-7. This construction is a joint development within the right-of-way of the Keiyo Transit Line in Tokyo. The strip panels provide the foundation structure for a continuous mat cast at ground level and used to support miscellaneous facilities for sanitary works and several warehouses for the nearby Haneda airport. Each strip panel was excavated and concreted on the same day. A 1-m (about

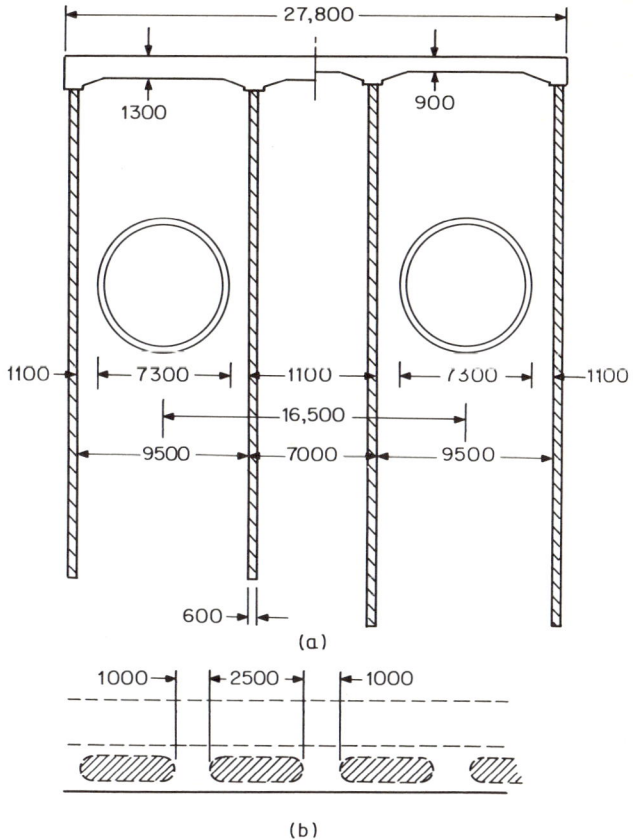

Fig. 12-7 Use of strip panels for the construction of subway tunnels at Haneda Airport in Tokyo: (*a*) typical section and (*b*) partial plan; all dimensions in millimeters.

3.3-ft) solid earth tongue was left between strips. This arrangement ensured trench stability in predominantly unstable ground acquired through landfills and where groundwater is practically at sea level.

Following this phase, the two subway tunnels shown in Fig. 12-7 were constructed in steel tubes using the shield process. Tunneling primarily through flowing ground was in this case feasible under conditions which resulted in considerably less serious effects on surroundings. Ground loss and movement, which might have occurred with the advance of the face and the installation of the liner, was primarily confined to the zone between the two walls, so that buildings and structures outside this zone were laterally protected and withstood the tunneling operations without damage.

A more conventional plan would have been to construct continuous diaphragm walls as ground support for the exterior bays, build two concrete boxes for the subway lines, backfill, and cast the mat for the ground facilities. In this case the completion of public works and all private construction at ground level would have depended on the completion of the subway construction. The latter was carried out as a separate contract and at a later time, so that it precluded the adoption of this plan.

12-6 PROTECTION OF THE BASE OF AN EXCAVATION

Protection of the base of an excavation is often indicated in anticipation of groundwater effects, uplift pressures, and bottom heave failure in soft plastic clays. These situations can conceivably arise as temporary or permanent conditions.

Temporary Condition Temporary protection of an excavation from groundwater usually means keeping the bottom dry. At best the site may be underlain by a naturally impermeable layer at a reasonable distance from the intended excavation base, and clearly the solution is to carry the slurry-trench wall down and let it penetrate into the impervious bed, although this can mean that the wall is deeper than necessary for its own stability. Below the minimum required structural embedment it usually is sufficient to fill the trench with an essentially watertight but less expensive material, e.g., plastic concrete, and omit reinforcement in this portion of the wall. This operation should be carried out in two steps: (1) fill the bottom part with plastic concrete and allow it to set and then (2) tremie the structural concrete. Bentonite trappings at the junction of the two pours are likely to accumulate, but they are unlikely to affect the strength or the watertightness of the wall.

If it is feasible and economical to lower the groundwater level (considering, for example, the effects on surroundings and the local hydrology), the excavation is protected merely by dealing with any water and seepage emerging from the bottom. Water control by lowering the natural level may also be indicated if there is an impervious bed some distance below the base of excavation that is too thin to resist uplift pressures and blowout failure; in this case the water level is carried down far enough to reduce uplift to a safe limit.

Where a naturally impervious layer does not exist at or close to the bottom of the excavation and the ground is too pervious to allow pumping from sump pits, the diaphragm perimeter wall must be combined with grouting to form a relatively impermeable bed below the base. The location of this bed should be established so that the uplift pressure at the base of the grouted zone does not exceed the weight of ground above it, and the thickness of the grouted zone should be sufficient to prevent blowout failure of the materials used. However, an artificially made impervious layer beneath the base is only relatively watertight. On the other hand, absolute watertightness is neither possible nor necessary, and it is sufficient to provide a zone of lower permeability so that any residual quantity of water can be handled by normal pumping operations from sumps installed below the base.

Figure 12-8 shows an excavation protected by diaphragm walls and grout-

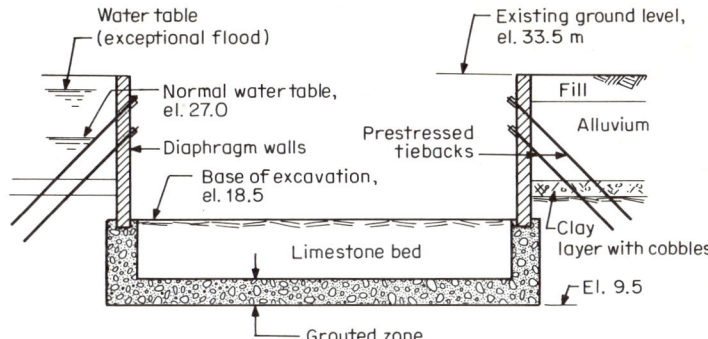

Fig. 12-8 Protection of excavation from groundwater. Construction adjacent to the Seine River, Paris. (*From Fenoux, 1971.*)

ing. The construction was adjacent to the Seine River in Paris and involved a basement 15 m (50 ft) deep to a bed of permeable limestone under a water head of almost 14 m (about 46 ft) caused by the rising water in the river. The walls were laterally braced by two rows of prestressed tiebacks, and they just penetrated the limestone bed. Beneath the walls lateral protection against seepage flow was provided by extending a vertical grout curtain to the underside of the structure as shown.

Permanent Condition Permanent protection during service of an underground structure usually requires fairly watertight perimeter walls and protection of the base against uplift pressures. The construction of deeper basements often creates the condition of permanent uplift. In some instances this situation can arise in connection with a water level which is not necessarily the maximum flood for which the local building code requires protection. A common solution is to install a thick mat or raft to balance uplift by the weight of the concrete, but this is expensive and means additional earth moving in unused space.

A less conventional and often cheaper solution is to use a thinner mat and keep it stable by vertical ground anchors. For relatively small uplift pressures (provided the soil is suitable for the method of installation) simple anchoring by sealed bars is sufficient without prestressing since the displacement necessary to mobilize tension in the bars is small and within the elastic range.

With large uplift pressures it is necessary to use high-capacity prestressed anchors, especially if a good anchorage zone is not available beneath the mat. During prestressing the reaction applied to the raft will be transmitted to the ground beneath and thus tend to consolidate it. This consolidation can cause settlement and subsequently reduce the prestressing force by somewhat decreasing the elastic extension of the cables until this process is stopped and equilibrium is reached. Since the initial prestressing force is associated with the magnitude and type of settlement thus produced, it follows that the state of equilibrium invariably represents a partial loss of prestressing, which should be recognized in analyzing the capacity of ground anchors.

If the direction and source of seepage flow are known, it may be possible to reduce uplift by extending the diaphragm wall sufficiently into the substratum below the base, but in many instances this can only reduce the amount of flow toward the ground beneath the raft. A more positive control measure lends itself to the short-term condition, e.g., seasonal or temporary rise in the water table, where a perimeter enclosure wall and a relatively impervious base (either natural or by grouting) already exist or where the watertight vertical barrier is deep enough to reduce the flow. In this case the problem of uplift pressure and base protection can be handled as shown in Fig. 12-9. Within the excavated area the pumping carried out during construction is extended to a level below the base, as shown, and this constitutes a permanent groundwater-level lowering. A drainage layer is inserted below the raft or the floor slab, and this slab can be relatively thin since it is not designed for uplift pressure. Any water infiltrating underneath is conveyed to a deep screen well, from which it is pumped out. In case of multiple breakdowns or power failure, relief holes in the floor permit excess flow to enter the area and flood it temporarily without structural damage.

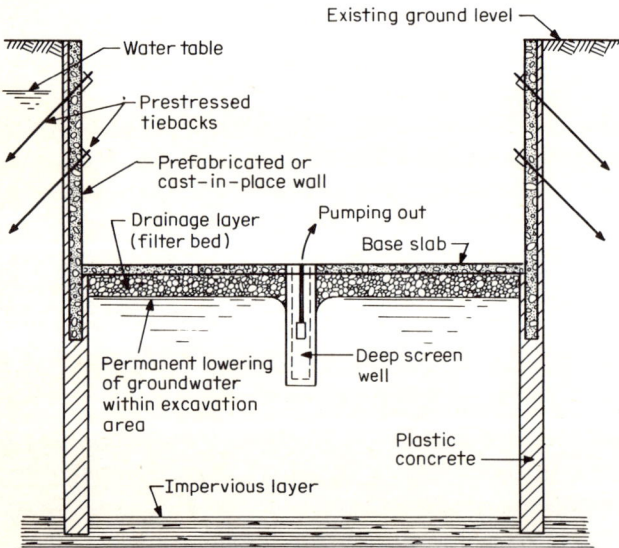

Fig. 12-9 Protection of excavation from groundwater by lowering the water table permanently within the excavation area.

Base Failure due to Heave In soft clays a major problem is stability against bottom heave failure. Heave is associated with inability of soft clay to sustain the weight of the overburden when a portion of the ground is unloaded. Diaphragm walls can be used either by themselves or combined with cross-walls to control heave. In either case the solution depends on the adhesion (side resistance) mobilized on the embedded portion, but heave is also a function of the depth-to-width or the width-to-length ratio of the excavation.

Heave is better controlled in relatively narrow excavations carried out in sections, whereas it may persist in very long or very large excavations (see also Chap. 13).

12-7 THE DOWNWARD CONSTRUCTION METHOD

In this process the objective is to build part or all of the underground portion of a structure or building from the top down rather from the bottom up; hence it also provides progressive underpinning. The method is particularly suited to deep excavations and eliminates the need for temporary strutting or bracing berms.

The first step is to construct a perimeter diaphragm wall as permanent ground support, making provision for connections with floor slabs, beams, and the like. Next, the top slab is cast at ground level and acts as uppermost lateral support for the wall. Excavation is carried out beneath this slab to the level of the first basement, and the second slab is cast, which also acts as a second bracing level. The process is continued and repeated for the remaining basement levels.

An obvious disadvantage of this construction is the difficulty of earth moving compared with the freedom and speed of conventional excavation. Under these conditions excavation should be expected to be more expensive than conventional earth moving, and the added cost will depend on the type and size of equipment that can be used, the distance traveled underground, the number and location of access openings, and the type of soil to be excavated. The most serious problem results from restrictions on equipment type and size. In spite of that, excavation under a permanent roof has certain advantages. Under a street it saves the cost of temporary decking and traffic control and allows the early reestablishment of surface activities. Under a building and in conjunction with prefabricated columns it makes concurrent upward and downward construction from ground level feasible.

The technical requirements of the downward process do not differ from conventional construction. The location and design of slabs, beams, and columns depends largely on superstructure loads and the function of the basement and is seldom affected by the excavation process. For large basements the construction is supplemented by intermediate prefounded columns or strip panels. Since the general excavation is done gradually, so that for every level excavated there is a corresponding bracing tier, the process constitutes progressive underpinning.

An example of downward construction is the multilevel underground parking structure shown in Fig. 12-10. The site is surrounded by streets and existing buildings along all sides, and the groundwater level is close to the base of the sixth basement. Diaphragm walls, continuous or strip panels, provided the exterior supports, interior bearing walls and partitions, and supports for the access ramp. The simultaneous construction of superstructure

426 Slurry Walls

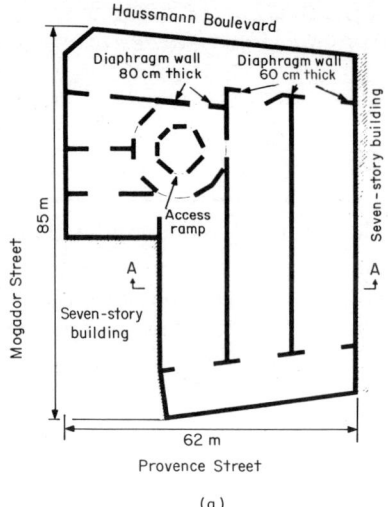

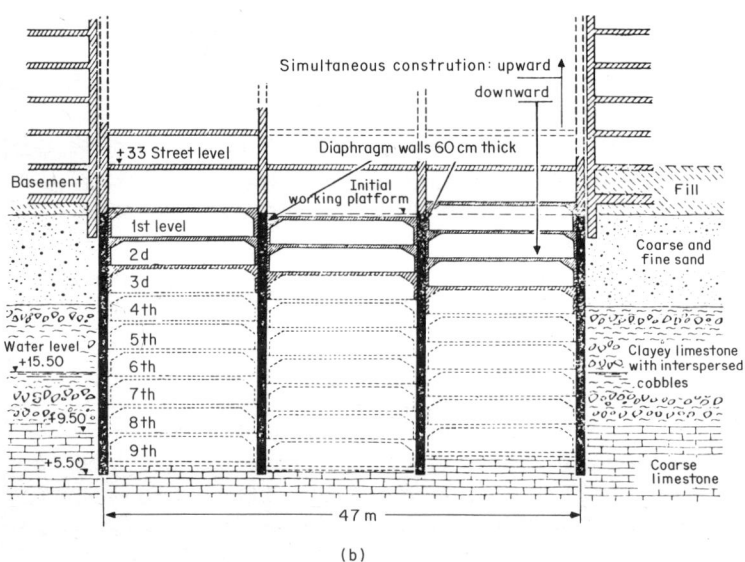

Fig. 12-10 Example of downward construction showing a multilevel structure for underground parking in Paris: (*a*) plan and (*b*) section *A-A*. (*From Fenoux, 1971.*)

and basement yielded in this case time savings, and the contractor was able to complete the upper parking levels and release them for service before the lower levels were excavated.

Figure 12-11 shows a cross section of a basement construction in London (Hodgson, 1974), comprising a two-block development from an existing basement level shown in dashed line. A third basement provided for part of the total area extends about 7 m (23 ft) from the second basement or some 14

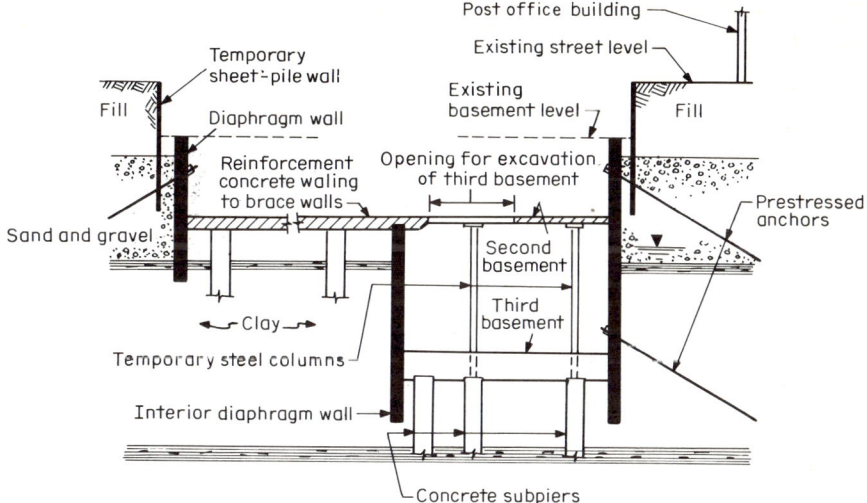

Fig. 12-11 Basement construction on Victoria Street, London. (*From Hodgson, 1974.*)

m (46 ft) from the street level. Because of the close proximity of nearby buildings the exterior diaphragm walls were braced by a concrete waling slab at the second basement level according to the downward process. The interior diaphragm wall and the concrete subpiers were built in boreholes or slots using the second level as a working platform. The subpiers have a cutoff level at the underside of the third basement raft. The temporary steel columns, necessary to carry heavy construction loads, were not preset in the holes, but the steel casing was left in place and enabled workmen to descend to the base to grout, position, and anchor the columns. Excavation of the third basement was carried out essentially as mining operation through a large opening left in the waling slab.

In more restricted sites it is better to cast only the structural frame of the underground portion to brace the walls, leaving larger and more than one opening for vertical access. Alternatively, steel lattice columns are built to rest on small temporary bases and support waling slabs inserted as excavation proceeds downward, leaving a central core open for earth moving. When the final excavation level is reached, the column loads are transmitted to a raft and the center core is built from the bottom up.

12-8 USE OF PREFABRICATED PANELS

Use as Retaining Walls A prefabricated wall with identical panels, like the structure shown in Fig. 10-2, is suitable for ground consisting mainly of stiff or dense formations close to the base of excavation so that the wall can derive lateral stability with minimum embedment, and most of its depth is thus usable in enclosing the basement levels. In relatively loose or soft ground the required embedment is from 20 to 40 percent of the overall wall depth, resulting in a considerable waste of the usable wall height. In these conditions it may be cheaper and more practicable to build the wall with beam-and-slab panels, like those shown in Fig. 10-3.

A structure of this type is similar to soldier piles with lagging. Because the beam sections are relatively narrower than the slab panels, they can be made longer without exceeding the maximum practicable weight for handling and lifting. They are embedded far enough below the bottom of excavation and resist active and passive stresses below this level, at the same time transferring vertical loads to the surrounding soil. Since the slab sections do not resist passive pressure, they have only minimum embedment below excavation level, usually 0.5 to 1 m (1.5 to 3 ft). Thus, in plan the wall appears as a series of T shapes or as a wall on stilts.

The installation follows the sequence shown in Fig. 12-12. Primary panels comprising two beam sections and one slab are inserted in alternate trenches numbered 1 and 2. The secondary panel numbered 3 is installed between panels 1 and 2 and consists of two slab sections and one beam. The number of units in each panel is thus three, which is the optimum daily schedule for one excavating machine and one crane. With the three panels in place, the installation continues by alternating primary and secondary panels according to the sequence shown, so that the grout in any panel becomes self-standing as the adjacent panel is excavated. This adherence to schedule is not necessary where a displacement-type grout is used.

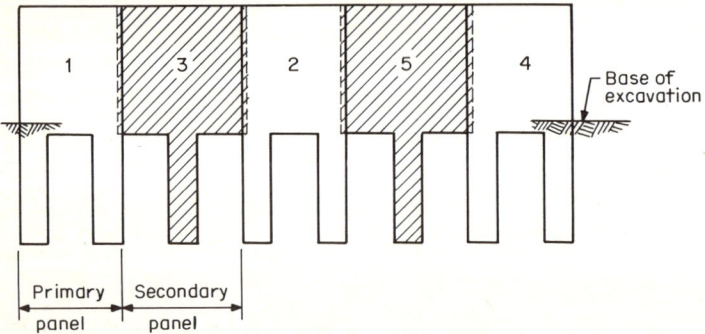

Fig. 12-12 Installation of a prefabricated wall consisting of beam and slab sections: one primary panel = two beam sections and one slab section; one secondary panel = two slab sections and one beam section.

Walls for Structural Support and Groundwater Control The construction method of such a prefabricated wall (Fig. 12-13) is similar to the protective techniques discussed in Sec. 12-6; the upper part of the wall is the ground support, whereas the lower provides a barrier for the groundwater. In the permanent condition the water table in the interior of the enclosure is lowered in conjunction with the installation of relief wells connected to a sump for collecting and pumping out any underseepage. Protection is afforded first by trenching down to the impervious substratum using a self-hardening coulis and then by placing prefabricated sections in the upper part where ground

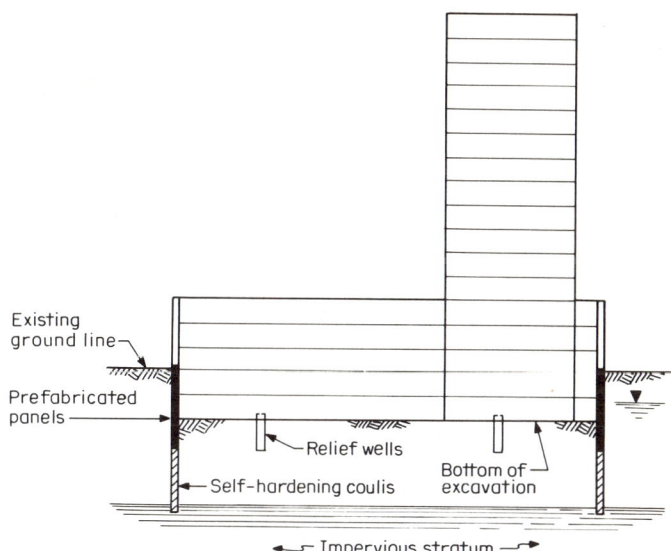

Fig. 12-13 Use of prefabricated walls for ground support and protection of the excavation from groundwater.

support is needed. Besides providing a seal, the coulis can also transfer considerable load since its compressive strength usually is adequate for this purpose. Where the transfer of load is desired, the coulis should be checked for the probable elastic shortening; if the depth of the wall is appreciable, elastic settlement due to the low modulus of elasticity can be excessive and more than the structure can tolerate.

If the loads are greater than can be transferred by a coulis, the wall should be made with beam-and-slab sections in order to make the load transfer at a deeper stratum. Supporting a prefabricated wall on cast-in-place stilts has been tried but has certain difficulties, one being how to secure firm contact between the cast-in-place and precast portions of the wall for full bearing. For a precast panel to bear on tremied concrete it is necessary to grout the contact surface between the two materials using a suitable device, e.g., the one shown in Fig. 12-14. Three vertical pipes are inserted into the panel before precasting, and one of them is a central tube large enough to allow a concrete vibrator to be lowered to the level of tremied concrete. The others are two side tubes for future grouting of the contact area. Grouting can be avoided, however, if the prefabricated panel is placed while the tremied concrete is still fresh, in which case good bearing is obtained merely by allowing the tip of the panel to penetrate the fresh mix.

12-9 CONSTRUCTION FROM A LOWER LEVEL

In some instances it is necessary either to start or to continue construction from a lower level. This may involve extension of an existing basement or the

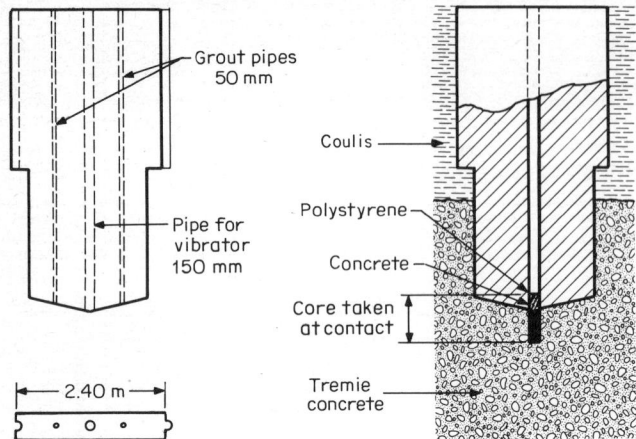

Fig. 12-14 Device used to grout the contact surface between cast-in-place concrete and precast panels.

need may be irrespective of ownership. Thus, if a subway is proposed beneath the underground portion of a building, it may be necessary to plan the new construction using the existing basement level as working platform.

An example of this type of work is shown in Fig. 12-15 for a new subway under a six-level underground parking garage. In this case the construction level was the base of the parking center, about 15 m (49 ft), from existing ground line. Construction in these conditions is complicated since it must overcome the following difficulties:

1. Problems regarding trench stability, particularly the difficulty of balancing the earth stresses and pore-water pressures by the thrust of the slurry alone without lowering the water table. It should be noted that for the construction of Fig. 12-15 it was necessary to lower the water table as shown.

2. Problems of equipment mobility when limited clearance and low headroom confine the choice to crawler-type machines and specially made rigs that can work in these conditions. For the excavation of Fig. 12-15 the contractor had to use a lightweight machine with a hydraulic bucket attached to a short boom. Further problems arise from the assembly and the placement of reinforcing cages, which often requires some daily acrobatics, and from the handling of end tubes. All these must be installed in short sections and then spliced.

3. Lack of working space and the effect on the handling of excavated materials, accessibility to the construction area from outside, and the delivery of fresh concrete to the working level.

4. Noise and ventilation problems caused by exhaust gases from working engines.

In spite of these difficulties, however, the foregoing technique is the only way to construct an underground facility beneath an existing building; it

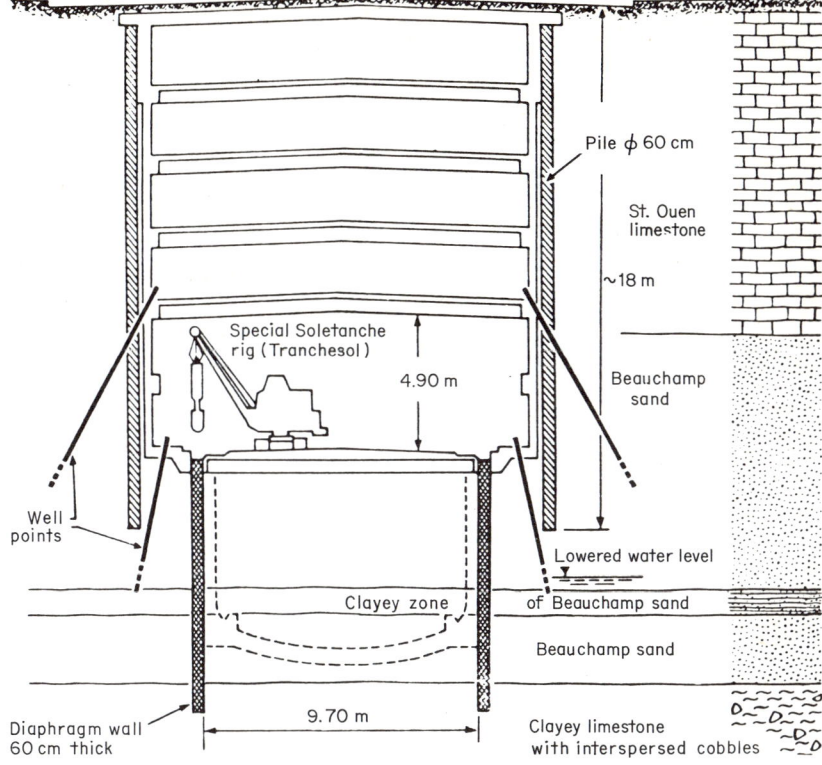

Fig. 12-15 Construction of a subway section in Paris from an existing six-level underground parking garage.

therefore can have unexpected applications. A favorable factor is the independence of construction from weather conditions, which allows better scheduling.

Construction under Temporary Decking Diaphragm walls are occasionally built from a lower level in conjunction with a modified cut-and-cover construction that involves a temporary decking carried on soldier piles with lagging. When excavation under this temporary decking is taken down to a level that provides adequate headroom for the operations, a suitable platform is provided and used as working level. The main advantage is that the street does not have to be used as a construction site for the walls.

12-10 APPLICATIONS FOR SUBWAY CONSTRUCTION

Basic Considerations

A main problem in new subway construction with cut-and-cover techniques is competition for space where space hardly exists. The scarcity of vacant sites frequently implies unusually high property values, and this excludes the opportunity of expanding the public right-of-way. Construction in these conditions becomes further complicated due to streets, buildings, and utilities above and below grade.

Underground utilities in particular can bring prolonged delays which will increase the cost accordingly. Utilities usually are found within 10 to 11 ft (3 m) below ground surface and often occupy 30 to 40 percent of the street width. They typically require some form of rearrangement such as temporary support, diversion, relocation, or complete reconstruction. Utility preparatory work often extends for many months. Because of these requirements the best practice is to adapt construction techniques and solutions which give minimum disturbance to utilities and deal only with the problem of their temporary support or minor deversion.

In many instances the foregoing considerations will preclude the use of a single method and favor instead adoption of different types of ground support supplemented by other forms of ground engineering. Without knowing the details and characteristics of the project at hand it is impossible to establish general guidelines, except for the following comments, which have wide applicability.

1. For long projects utility rearrangement will necessitate (Jobling, 1975) simultaneous preconstruction of shorter sections, e.g., at intersections. The use of diaphragm walls at these locations will require frequent dismantling, moving, and resetting of plant facilities, rigs, and equipment, and this can become a major operation. It is therefore better to use fairly simple and mobile rigs such as bored-pile machines.

2. An extensive utility system at an intersection can bring construction to sudden halt. In order to resume activity it will be necessary to locate such obstructions and carry out preexcavation to divert them. Most of these problems are avoided if utilities are left in place and supported and a flexible ground support is chosen such as intermittent bored piles or strip panels, even though this will require supplemental work and grouting.

3. For long sections or where time is of the essence secant-pile walls pose their own problems. For an average excavation depth of 40 ft (12 m) a single rig will produce five to six piles daily, which is equivalent to about 5 m (16 ft) of wall. Unless several rigs work at the site, the construction period may be too long.

4. Difficult ground and hydrologic conditions usually can be handled with diaphragm walls where other techniques may fail. With complex underground untilities the only choice sometimes is to plan a flexible construction and recognize the possibility of changes in the field.

5. In stiff, firm clayey soil bored piles can be used as side walls of a transit tunnel and be braced at the roof and base-slab levels for a nominal pile diameter. In soft clay the same piles will pose problems and require intermediate

bracing, which is impracticable in the permanent condition, leaving no choice but to increase the pile diameter considerably.

6. Street congestion and traffic maintenance can force a diaphragm-wall contractor to operate on one side only; avoid the use of rigid-arm excavators; and concentrate within a narrow enclosure to locate plant facilities and assemble reinforcing cages. The result is a significant loss of productivity.

7. A single, fairly deep subway section is not necessarily more economically built with diaphragm walls. A multilevel section satisfies the goal of an economical construction, and this may be true regardless of depth. There are no fixed limits indicating the cost range of the application, but this should be judged considering all the pertinent factors.

8. Exposed permanent diaphragm walls are acceptable and functional in subway construction unless they are within station limits.

9. Diaphragm walls are successfully combined with bored piles, sheet piles, soldier piles and lagging, and ground strengthening more often than expected to remedy the problems of underground construction.

10. For construction along streets problems related to trench stability, slurry loss, spilage of slurry mud, and disposal of used bentonite should not be underestimated or exaggerated.

Construction under the Roof for Single Tunnels

As already mentioned, the main advantage of construction under the roof is the reduced disruption of surface activities. If the work is to be executed with temporary decking, a typical construction sequence and phasing is as shown in Fig. 12-16, which is self-explanatory. In this example the upper part of the excavation (usually 10 to 12 ft or 3 m) is braced by soldier piles with lagging inserted in the top of the wall. This arrangement allows the portion below ground surface to remain unobstructed for future installation of utilities and often is requested by local authorities.

If there are not future utility requirements, the construction shown in Fig. 12-17 is executed. Provision is made only for existing utilities, which are accommodated by suitable gaps and omissions in the walls. If the construction does not involve temporary decking, it can be executed as shown in Fig. 12-18. The walls can be extended to guide-wall level if necessary, but more frequently they are terminated just above the roof slab and temporary bracing is used above this level.

Besides the reasons mentioned in Sec. 12-9, a temporary decking is also required where the existing utilities must be supported and protected rather than diverted or relocated during construction and this can be done with a temporary decking; where the sequence requires casting a separate concrete box against the walls and carrying out the construction from the bottom up; where the excavation is braced with dissimilar wall systems, like those of Fig. 12-20; where the walls of Fig. 12-18 cannot carry the loads imposed upon

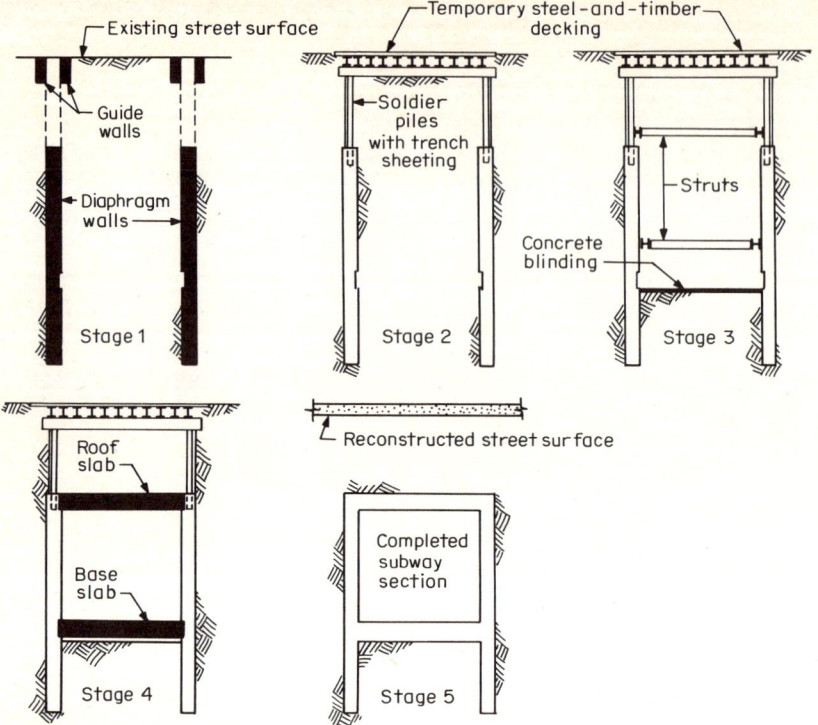

Fig. 12-16 Construction under the roof for a single tunnel; temporary decking supported on soldier piles.

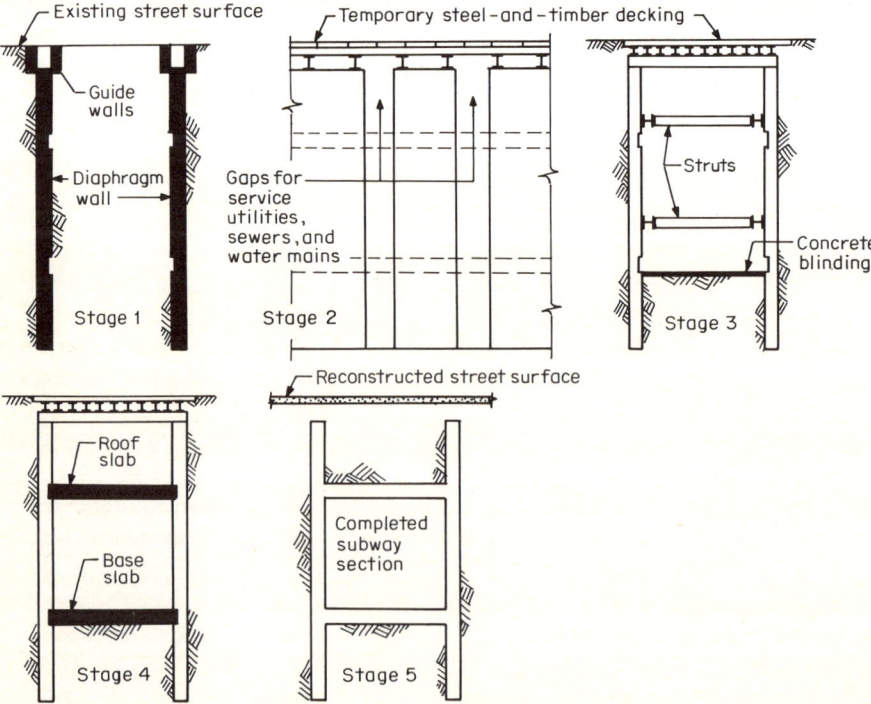

Fig. 12-17 Construction under the roof for a single tunnel; temporary decking supported on diaphragm walls.

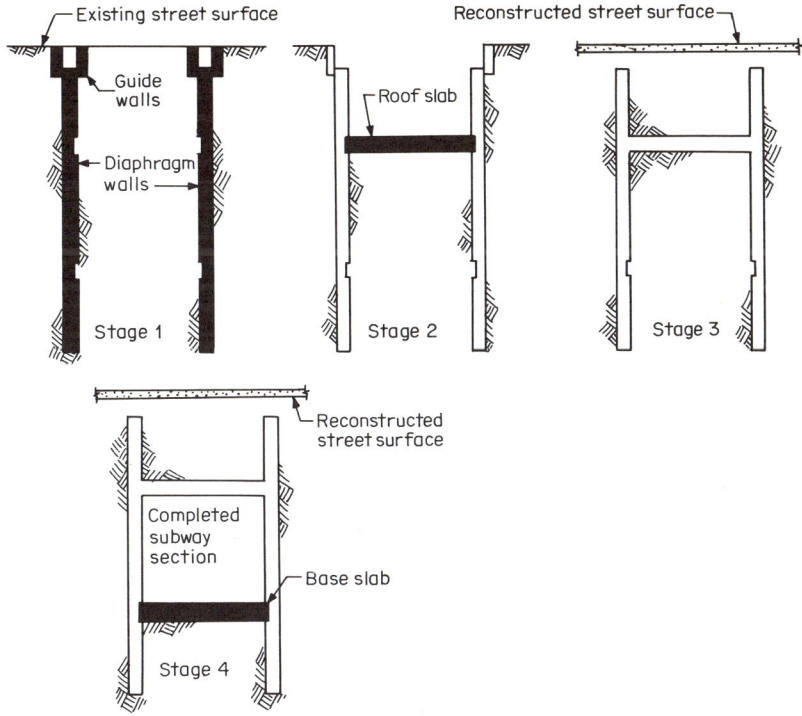

Fig. 12-18 Construction under the roof without temporary decking; single subway tunnel.

them from stage 3 to stage 4; where stages 2 and 3 of Fig. 12-18 cannot be coordinated with the time period for which street disruption is allowed in the phasing schedule; and where the subway is relatively deep, or its roof is at considerable distance from street level, making the excavation period too long.

Construction under the Roof for Multiple Tunnels

Multiple tunnels involve more than one trackway on split alignment or on different levels. Their construction is basically similar except that it may require center walls. An example is shown in Fig. 12-19, taken from a subway construction in Tokyo. In this case the exterior walls are built first, whereas the center wall cannot be installed without closing one side of the street to traffic. It is advantageous to build the upper (left) tunnel first in order to give some relief to the bracing requirements of the lower (right) tunnel section. When the difference between the two subway levels is nominal, the roof slab can be common to both sections and ventilation openings can be left in the center wall.

The construction shown in Fig. 12-20 is for a multiple tunnel and is carried out under a temporary decking. The operation must maintain and therefore

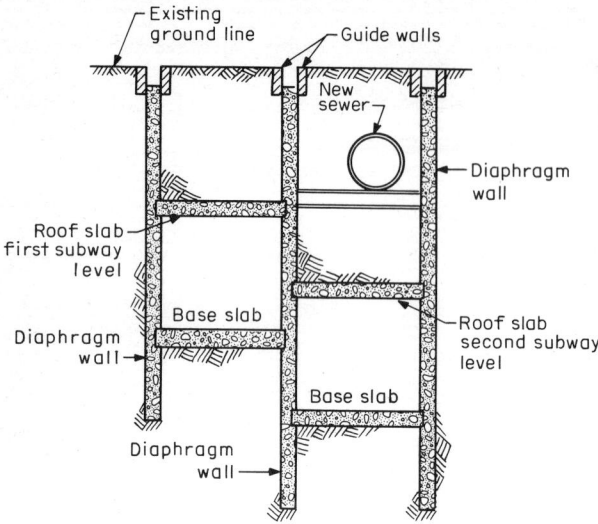

Fig. 12-19 Construction of a subway tunnel with split alignment.

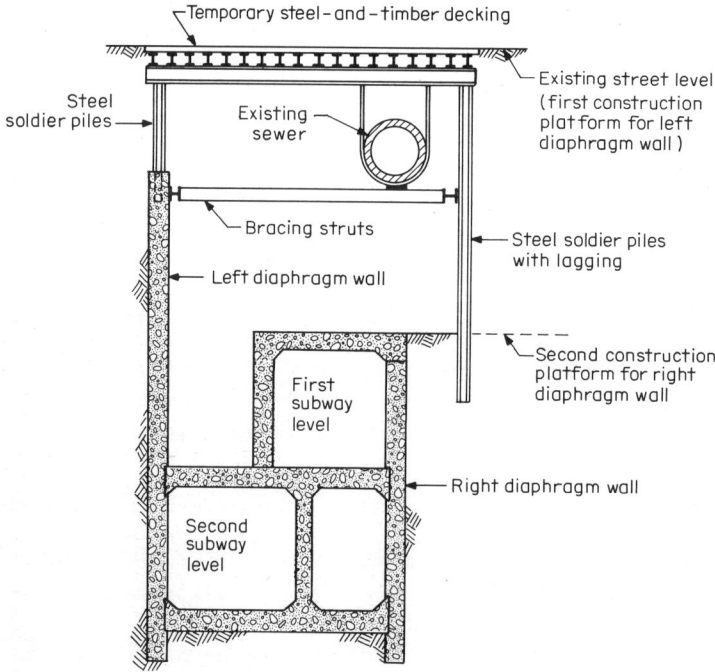

Fig. 12-20 Combination of diaphragm walls and steel soldier piles for the construction of a subway.

protect the existing main sewer rather than divert it; accommodate a split and irregular subway alignment with one exterior wall located almost directly under the sewer; keep the area just below surface unobstructed for future utility installations; and deal with a relatively low groundwater table. From these considerations the following excavation sequence is evolved: (1) construct the left diaphragm wall from street level and provide soldier piles with lagging in the upper part; (2) install soldier piles as right support instead of diaphragm walls, since the latter cannot be used as permanent structure; (3) place temporary decking; (4) excavate to second construction platform, placing intermediate bracing and supporting the sewer as shown; (5) construct the right diaphragm wall from this level, using low-head equipment and lowering the water table as needed; (6) build the subway section as shown; and (7) backfill, dismantle the temporary decking, and restore the street. The small clearance from the second construction level to the underside of the sewer (some 4.7 m, or 15.5 ft) is accommodated by the use of a long drill supported on a frame along the guide trench.

Miscellaneous Details

Wall-Roof Joints The usual connection between roof slab and side walls is made by shear keys. A different connection, shown in Fig. 12-21, is fairly watertight and satisfactory where frame action is desired. It is possible when the wall is terminated just below the roof or consists of lean concrete which can be removed to the joint level. A blinding is first placed to make the base for the roof slab. The construction joint in the wall is then cleaned and finished to give a roughened connection. The protruding bars are bent according to standard practice, and the roof slab is cast to the outside line of the wall. A concrete fillet in the slab will improve the frame action.

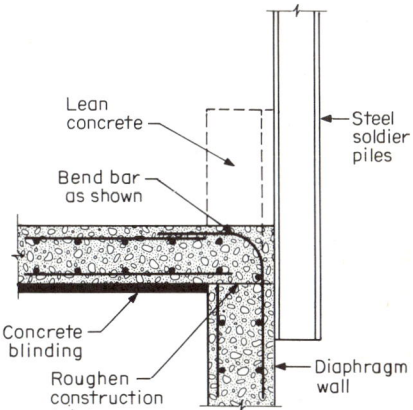

Fig. 12-21 Roof-slab and wall connection for a subway section.

Waterproofing Despite the wide variations in local standards for waterproofing, waterproofing details usually are worked out as a compromise between complete watertightness and economy. In most instances leakage through tunnel sections must be attributed to poor construction methods and execution rather than to poor details. Among the critical areas, special attention should be paid to the junction between roof slab and side walls and shrinkage effects, so that it is necessary to correlate construction joints with tunnel length and then detail the joints in order to prevent the formation of shrinkage cracks.

At present there are two waterproofing methods. One is to treat the roof of the section externally with bituminous waterproofing materials and overlap the joint between tunnel walls and roof. Since a small amount of seepage is unavoidable, drainage channels should be provided alongside the walls leading to sump pits, where the seepage is collected. Leaking joints can also be corrected from outside.

By contrast, the second method shows complete reliance on waterproof concrete and joints. Since few tunnels exist that do not leak, regardless of method, the former method is more economical and corresponds to the realities of any underground construction.

Excavation and Ventilation The most usual complaint about the under-the-roof method is that the excavation is delayed and costs more due to lack of space to move and maneuver equipment. For example, a single-track tunnel may have a distance face to face of walls about 12 to 14 ft (or 4 m) and a headroom from the underside of the roof to excavation level of 18 ft (5.5 m). For conventional equipment the vertical clearance is enough to complete the mucking out of the section, but there is room for only one access ramp and hardly any room for turnarounds.

Slowdown in the excavation can also give rise to ventilation problems from the exhast fumes of excavators and engines. For a single-track tunnel this problem becomes serious when the length exceeds about 500 ft (about 150 m), and for longer subpavement sections it is therefore necessary to modify the roof-casting method in order to provide improved natural ventilation. A practical solution is the use of a shuttering system placed some 3 ft (1 m) below the underside of the roof slab to provide an airspace. The necessary formwork supports only the wet concrete during placing and compaction. The reuse of the shuttering components can become part of the tunnel roof-construction routine.

Occasionally contractors prefer to install a belt-conveyor system at the pit head, which is extended in short lengths and reaches a maximum about 350 m (almost 1100 ft). This is intended to remove the muck on the lower belt, while excavation at the face is done with an excavator equipped with a side-tipping bucket. Again, some trouble will be experienced with fumes afrom the excavator.

Transfer of Loads In most of the foregoing examples the weight of the roof slab, the new fill on top, and any surcharge from the street or from moving loads are transferred to the diaphragm walls. This requires that sufficient load-carrying capacity be available, either by side resistance or by base bearing. Since the base slab is cast after these loads have been applied and transferred to the ground, it can be designed to carry only the subway loads and any hydrostatic uplift. The construction procedure does not always warrant making the base slab a foundation mat for the entire section, since this would require a displacement compatibility which may not be available in the final soil-structure interaction.

Special Conditions

Temporary and Permanent Walls Although in a properly supervised job the quality of diaphragm walls is good and warrants their use as permanent structures, this is not always technically feasible or economically justified, as can be understood by reference to Fig. 12-22. The scheme shown in Fig. 12-22a involves walls which are used in the temporary as well as in the permanent condition. This is feasible when the walls can carry all the loads mentioned in the foregoing section and before the base slab is in place; when approximately the same wall thickness is required during construction and excavation with temporary bracing as with the finished section; when the tunnel is not too deep considering the cost of ground support; and when no concrete cover is considered necessary on the inside face for architectural treatment.

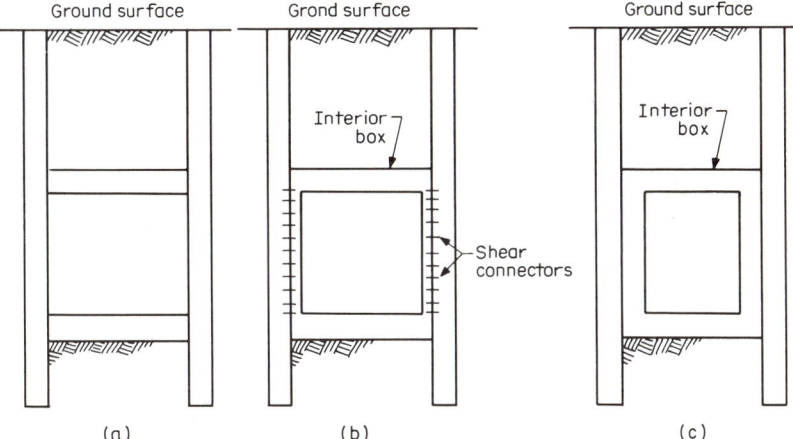

Fig. 12-22 Construction of a subway section using diaphragm walls: (a) single walls used both as temporary and as permanent support; (b) composite walls made by connecting the interior box to the exterior walls; (c) walls separated completely from the interior box.

The solution shown in Fig. 12-22b is indicated when the wall thickness of the final structure exceeds the thickness required during excavation and bracing. This configuration is used widely in Japan, where it is found economical under appropriate circumstances. It is a composite section, and major Japanese contractors such as Kajima and Zenitaka have developed special details for connecting the two wall portions. The construction is better executed in conjunction with a temporary decking.

The configuration shown in Fig. 12-22c has a separate box section built conventionally after excavation reaches the final level. The construction normally requires a temporary decking, and the walls serve temporarily only. This solution is indicated when the walls cannot carry all the loads imposed during construction and obstructions and other restrictions necessitate a different alignment in plan for the temporary and the permanent support.

The main factors influencing the choice of section are therefore the depth of the final subway section from ground level, the configuration of the section and mainly its uniformity in plan, the soil conditions and strength, and the site conditions.

Construction at Crossings From the construction standpoint the major problem at crossings, e.g., street intersections, is the maze of existing underground utilities. Practical experience indicates that in most instances it is better to avoid diversion or relocation of utilities, and instead concentrate on a construction sequence according to the hit-and-miss method shown in Fig. 12-23.

The intermittent wall consists of individual linear panels, bored piles, and the like installed from street level and arranged to miss the underground

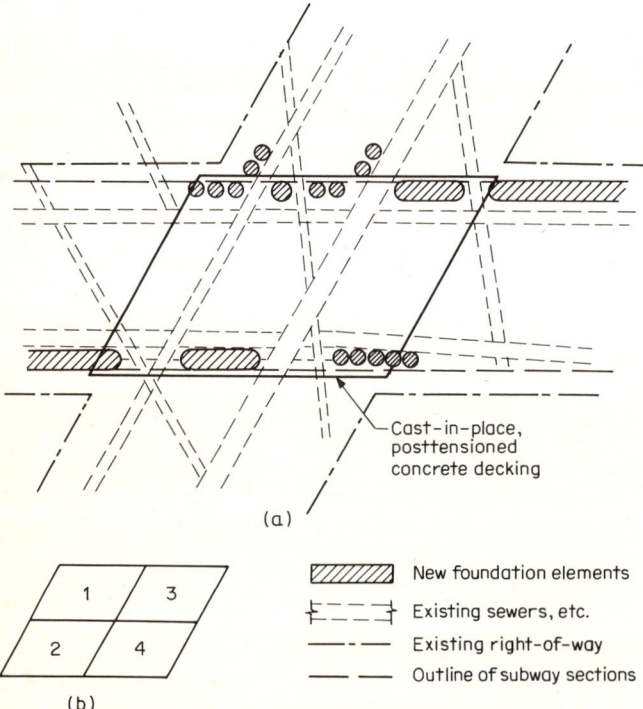

Fig. 12-23 Construction at crossings and street intersections: (a) arrangement of wall panels, bored piles, and the like by a hit-and-miss method; (b) construction of concrete decking at street level in four quarters to avoid complete closing of intersection to traffic.

obstacles. Once the basic support framing is in place, the concrete mat is cast at ground level in the sequence shown in Fig. 12-23b. This enables at least half the street to remain open at the interchange. In many instances the slab has been prestressed (posttensioned) after all four quarters are cast.

This basic layout brings full restoration to street activities and allows the excavation to proceed under an umbrella type of cover. In unstable or water-bearing ground it is necessary either to fill the gaps between support elements as the excavation is taken down or use supplementary ground-strengthening techniques. Once enough clearance is available below the utilities, the construction can be completed from a lower level.

Construction of Stations

Modern subway stations in central business districts or in high-density residential areas must provide for efficient vertical travel and interline transfer and communications and also house auxiliary facilities, relay rooms, control offices, amenities, and equipment. Thus they commonly comprise at least two floors and often three or four.

A relatively simple station, shown in Fig. 12-24, demonstrates the use of diaphragm walls and prefounded columns in a downward process. The construction is executed in five stages. Stage 1 requires partial street disruption to install the bored piles and the prefounded columns, whereas in stages 2 and 3 one side of the street is closed to traffic. Street restoration is attained in stage 4, which involves the simultaneous excavation under the roof to construct the mezzanine floor. Stage 5 consists of excavating to the underside of the track level and constructing the base slab and the main platform.

Construction with Temporary Decking The station of Fig. 12-24 can be built with diaphragm walls and temporary decking, as shown in Fig. 12-25, where the construction is represented by three stages for simplicity. In this instance the entire excavation is carried down to the track level using temporary bracing, and the interior structure is built from the bottom up.

Construction of Multilevel Sections Multilevel stations usually contain, besides the subway facility, commercial establishments and joint developments. Where four or more floors are involved, it has been found advantageous to execute the construction from two different levels, as mentioned before. This provides the opportunity to complete the construction by proceeding upward and downward simultaneously. The main requirement is to advance the excavation to a level providing adequate headroom and deal effectively with the groundwater.

The construction of a deep station is exemplified in Fig. 12-26. Evidently the upper part of the excavation is braced with soldier piles and lagging to the second construction level, which in this case is level B-2. Diaphragm walls and prefounded columns are built from this platform, and thereafter the construction proceeds upward and downward simultaneously as shown in Fig. 12-26b and c.

442 Slurry Walls

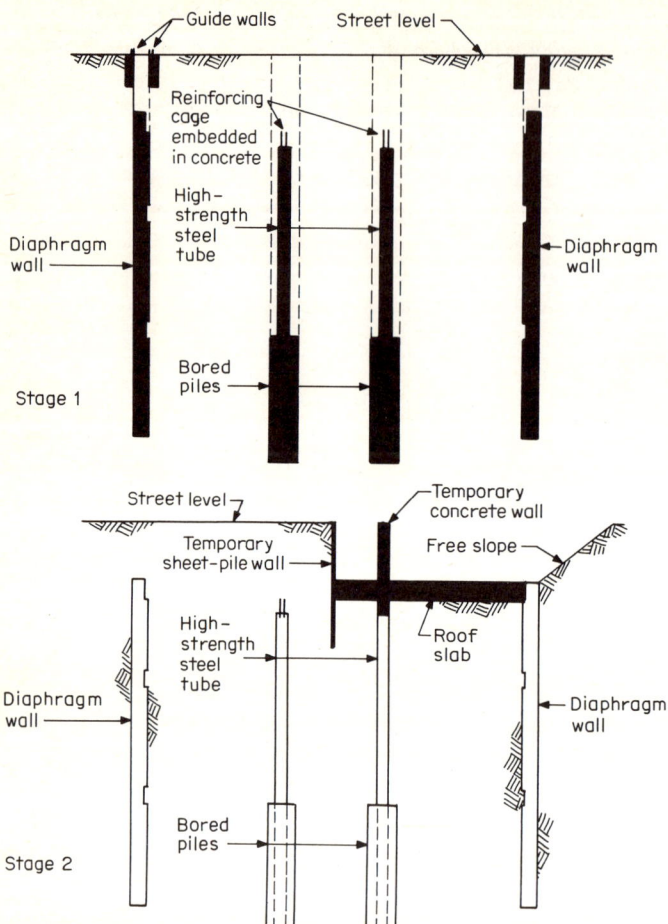

Fig. 12-24 Station construction by diaphragm-wall method; downward process.

Bottom Bracing of Long Narrow Cuts

Earth moving in the downward construction is primarily a braced excavation; since a cut is braced as it is excavated from the top down to the bottom, for relatively deep excavations the process can cause large wall movement. If it is possible to brace the lower part of the wall, especially the embedded portion, and if this can be done before any excavation begins, wall movement should be relatively small and related to the elastic deformation of the unbraced wall depth at the various stages of excavation.

Such a bottom bracing can be provided by digging transverse trenches at suitable intervals along a long narrow cut. Suitable locations are, for example, vertical panel joints. The transverse trenches are excavated usually from ground level and to the same depth as the long side walls, but the concrete in

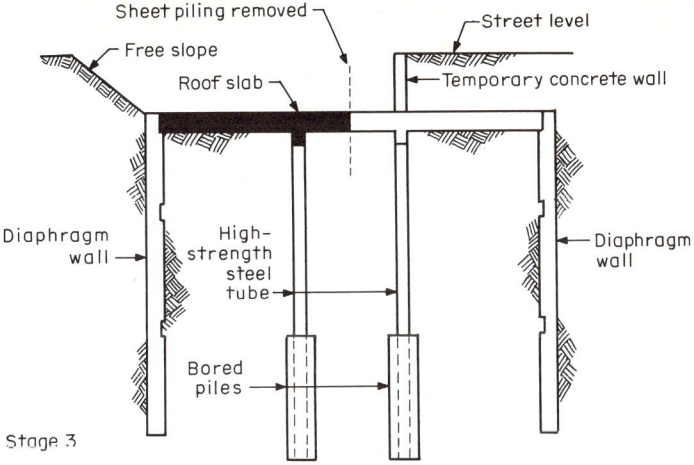

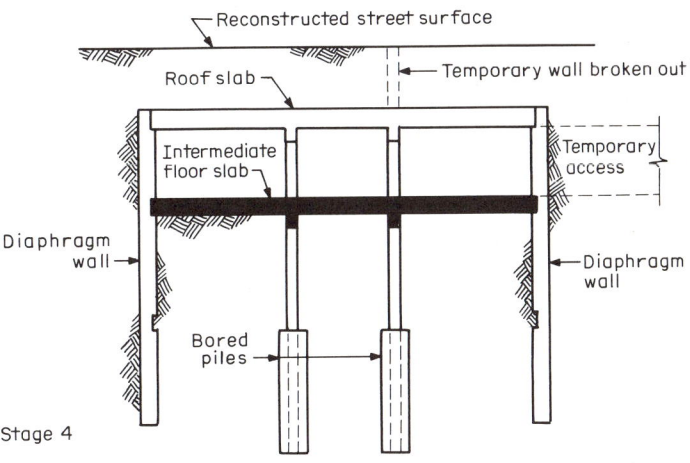

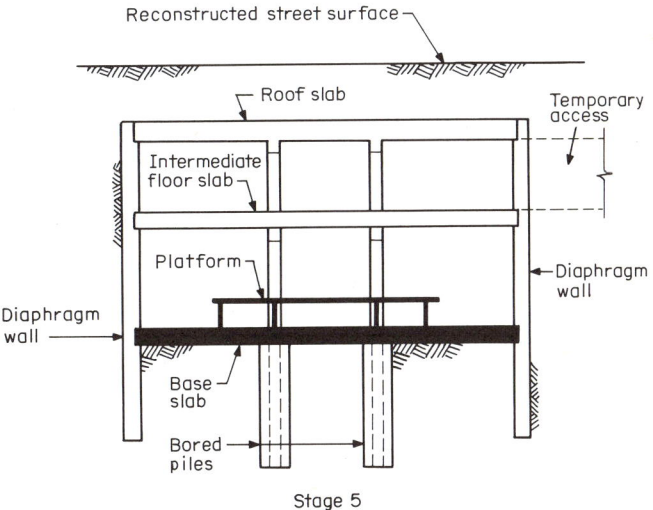

Fig. 12-24 (Continued)

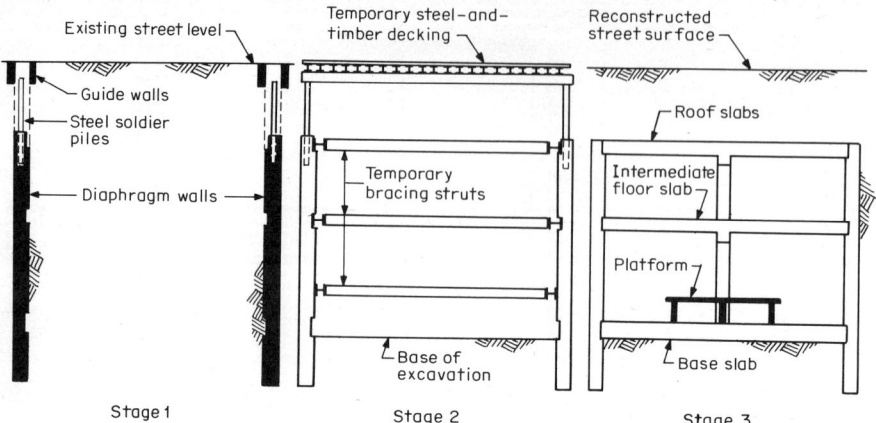

Fig. 12-25 Station construction by diaphragm-wall method and temporary decking; upward process.

the cross walls is cast only to the underside of the tunnel base. Since the cross walls are not to be exposed, their construction accuracy and tolerance is not critical. Thus guide walls can be omitted, and since the cross walls act mainly in compression, they need only nominal reinforcement. Other functions of the cross walls are to support the base slab, form a closed system with the side walls to resist uplift, or protect the base against bottom heave.

Figure 12-27 shows cross-wall bracing for diaphragm walls in soft clay in Oslo (Eide et al., 1972). The cross walls are placed at 4.5-m intervals (about 15 ft). Bracing at other levels is provided by permanent floor slabs cast downward. Extra bracing was required for this job under the middle deck. For this excavation the predicted maximum wall movement at midheight was 4 mm (0.16 in).

A frequent difficulty encountered in the construction of cross walls is trench stability. This condition will arise, for example, when the first trench is excavated but not backfilled and the next trench is to be started. To avoid this problem it is necessary to excavate alternate trenches, omitting one or two as required and using heavier than normal slurry, and backfill the cross-wall trench above the concreted portion before the intermediate cross trenches are started.

Examples of Subway Construction

Subway Station for BART Project The steel-and-concrete walls described in Sec. 10-3 have been found advantageous in poor soil conditions, for construction close to buildings, and where the interior framing of the finished structure will consist of steel members. If it is not practicable to lower the water table for a deep excavation close to buildings, the resulting pore pressure, earth stress, and surcharge load may still be counteracted by an effective wall system which must also control ground movement. However, when the

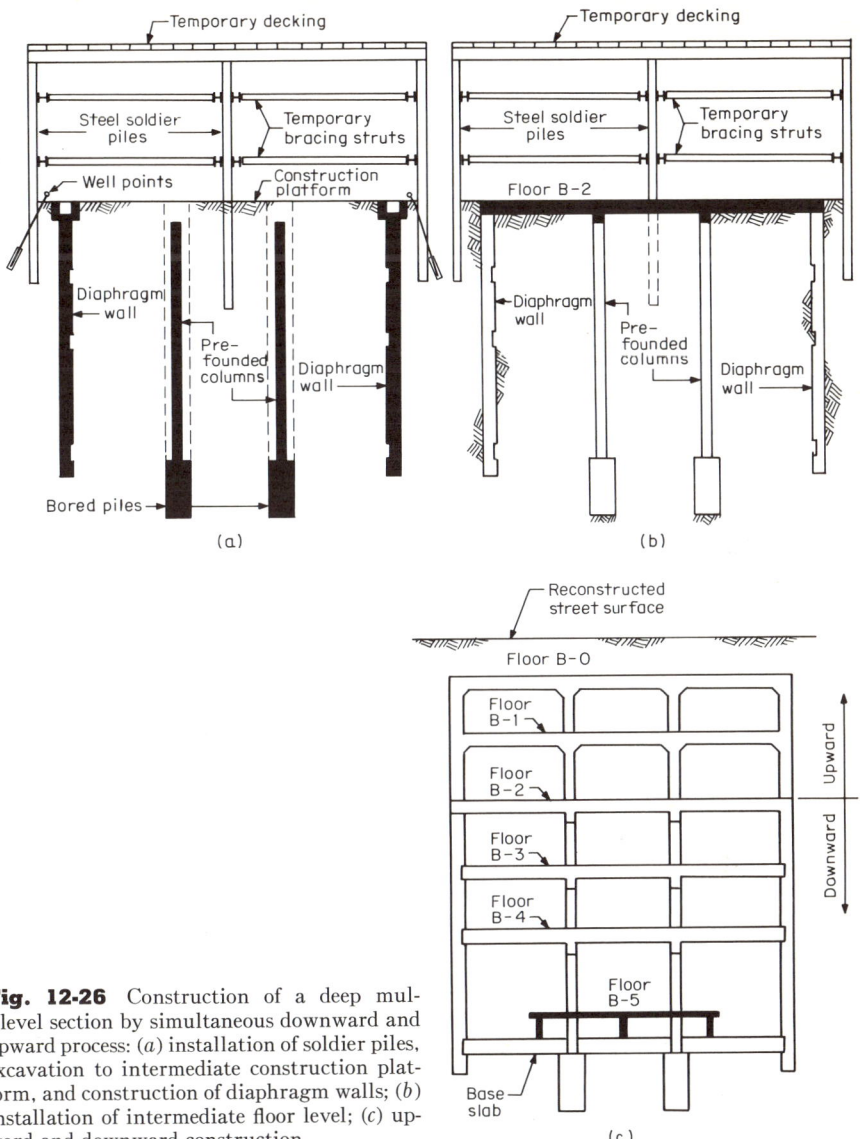

Fig. 12-26 Construction of a deep multilevel section by simultaneous downward and upward process: (*a*) installation of soldier piles, excavation to intermediate construction platform, and construction of diaphragm walls; (*b*) installation of intermediate floor level; (*c*) upward and downward construction.

excavation is unusually deep (say 80 ft or more) and is carried out inside a watertight wall without lowering the outside water table, the combined earth and full hydrostatic pressure will result in very heavy bracing loads, making temporary bracing impracticable and costly. In principle, then, the advantage of the downward construction method is the ability to use all permanent floor levels to brace the walls and thus confine additional temporary bracing to certain locations only.

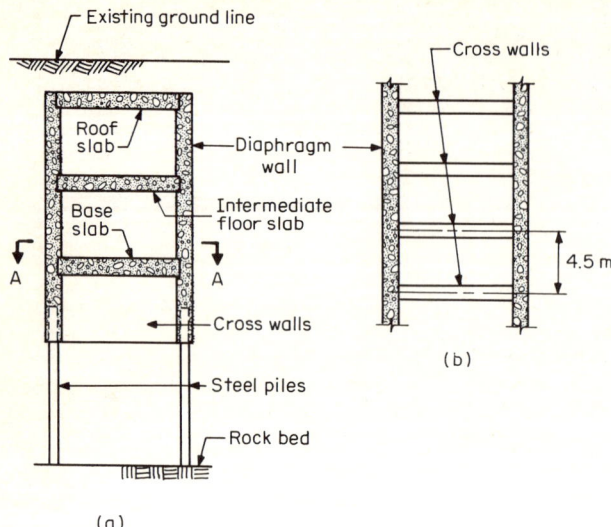

Fig. 12-27 A two-level subway tunnel with cross walls for bottom bracing: (*a*) subway cross section and (*b*) sectional plan A-A. (*From Eide et al., 1972.*)

The foregoing considerations are exemplified in the project shown in Fig. 12-28, taken from the construction of the Civic Center Station on Market Street in San Francisco (Thon and Harlan, 1971). The finished station is shown in Fig. 12-28a. The construction required first the installation of two exterior steel-and-concrete walls extending just below the mezzanine level and then an inverted process using two interior steel-and-concrete walls installed from the mezzanine level. The latter walls were also load-carrying members for the structure. The permanent structural-steel framing was connected to the steel beams of the walls, and the required construction accuracy was attained by setting and positioning the vertical beams in predrilled holes at the beginning of construction.

As shown in Fig. 12-28b, three of the required five bracing levels were provided by the permanent roof beams and the box girders of the floor steel framing. The 36-in soldier beams were set in predrilled holes filled with weak grout. Before the main operations it was necessary to undertake preparations for relocating the utilities in the vicinity of the walls and to dig a trench approximately 12 ft (3.6 m) deep to clear obstructions and retain the slurry.

Edmonton Subway Figure 12-29 shows a typical cross section for the Churchill Station of the Edmonton subway, constructed recently. The soil profile ie rather typical for this area and consists of an upper sediment (silty material with fissures and sand beds, fairly stiff) underlain by local till containing some cobbles and boulders up to about 50 ft; a formation of dense

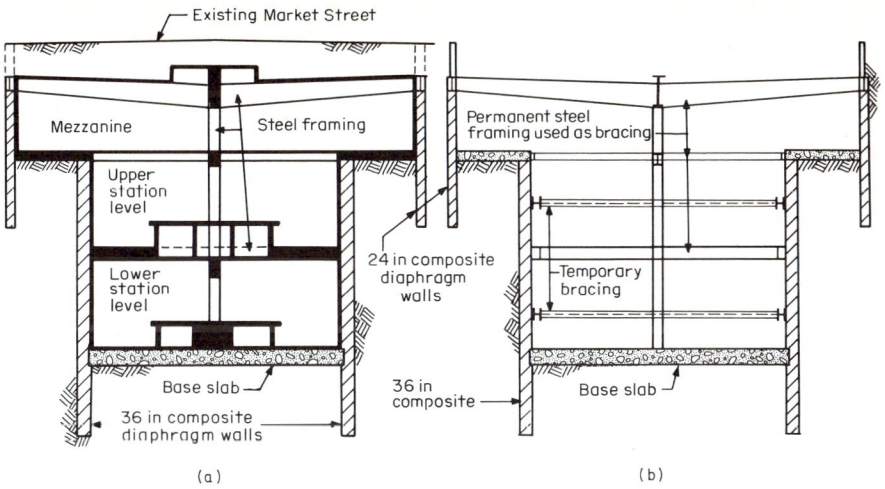

Fig. 12-28 Civic Center Station, Market Street, BART San Francisco; (a) finished section; (b) bracing system during excavation.

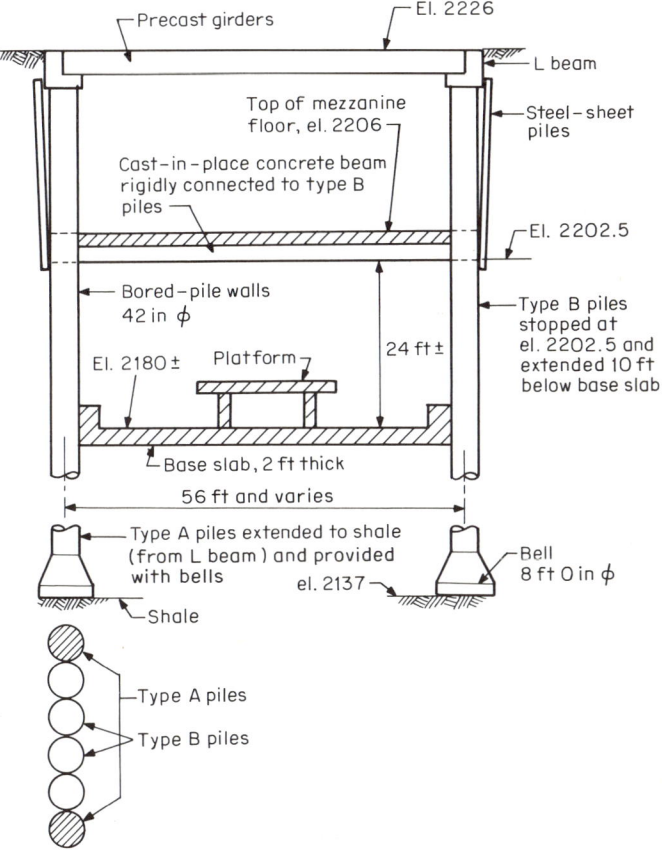

Fig. 12-29 Typical cross section of the Churchill Station for the Edmonton subway.

447

moist sand to approximately 90 ft below grade; and clay shale below that level, with average bearing capacity of 20 k/ft².

The structure comprises two floors, requiring excavation to approximately 50 ft below existing grade. Bored-pile walls were used as ground support, both temporary and permanent. The construction phasing consisted of the following stages:

1. Install bored piles. Type A piles are extended from ground level to shale and are provided with bells. Type B piles are embedded about 10 ft below excavation level and have their tops terminated at mezzanine level. A wall omission is thus created between type A piles above that level.

2. Install steel sheet piles to retain earth temporarily above mezzanine floor.

3. Construct L beam on top of A piles and place precast girders. This is the roof of the station and also supports street loads, including traffic.

4. Excavate to underside of mezzanine floor and cast beams and slab. Beams are overlapped with type B piles and form a rigid connection with the pile wall.

5. Excavate to base-slab level and construct rail floor and platforms.

6. Complete exterior wall between A piles above mezannine floor and remove steel sheet piles.

Evidently type A piles carry all the loads (both dead and live) from street level and mezzanine floor, and they are therefore the main load-bearing elements of the structure. Besides its own weight the base slab carries the loads from moving trains and must also resist some hydrostatic uplift should the water table rise outside the excavation. This slab merely touches the pile wall, and thus it can move freely downward to transfer the loads to the soil underneath.

In lieu of tangent piles as ground support the same construction could be phased with diaphragm walls in a sequence essentially similar to the downward process of Fig. 12-24. The section could be supplemented with one row of prefounded columns or strip panels installed along the center of platform to reduce the span length of the floor slabs and distribute the loads more evenly. Thus it would be possible to avoid the extension of load-bearing elements to shale depth; eliminate the use of sheet piles for any part of the excavation; connect the base slab to the side walls and thus attain a total load transfer at this level; provide more structural continuity, which is desirable in the final equilibrium of loads, stresses, and deformations; and make the construction economically attractive to more bidders.

Subway for Akasaka Block, Tokyo This is a three-story structure located in the central business and commercial district of Akasaka, Tokyo. The wall systems and their sequence of installation exemplify the techniques and principles described in the foregoing sections. This sequence is grouped into four main stages, shown in Fig. 12-30.

1. First stage.
   a. Install exterior sheet-pile walls and interior prefounded columns. These are temporary structures only.
   b. Place temporary decking.
   c. Excavate to level 1 and install first row of tiebacks and lateral strutting beams.
   d. Excavate to level 2 and install second row of tiebacks.
2. Second stage.
   a. Build guide walls at second construction level, which is excavation level 2.
   b. Construct diaphragm wall from this level.
   c. Remove guide walls and prepare ground for top concrete slab.
   d. Place concrete slab using ground as formwork.
3. Third stage.
   a. Excavate to level 3 and install third row of tie backs and bracing beams.
   b. Excavate to level 4 and install lateral strutting beams.
   c. Excavate to level 5, which is final excavation level.
4. Fourth stage
   a. Place base concrete slab using ground as formwork.
   b. Construct walls and columns for lower floor.
   c. Place concrete slab for intermediate level.
   d. Construct walls and columns for intermediate level.
   e. Place concrete slab for the upper level and complete the wall and column construction.
   f. Cut off steel columns and bracing inside the permanent structure, transferring all loads from above to permanent columns through the roof slab.

It is interesting to note the following points: the cast-in-place exterior walls are connected to the diaphragm wall to provide composite action, and in this respect their thickness decreases from the bottom up; the bottom slab acts as a foundation mat for the final structure, so that the transfer of load is accomplished at this level; the roof slab is relatively thick and heavily reinforced since it must carry and transfer all loads above this level; and the steel framing above the roof slab initially supported the temporary decking and street loads but will become part of the permanent framing for a commercial development to be built below grade and above the subway station.

12-11 TRAFFIC UNDERPASSES AND DEPRESSED ROADWAYS

The unique variation of methods and techniques often introduced in subway and underground railway construction is seldom if ever necessary for traffic underpasses unless a covered motorway must be provided under a street to divert or handle more traffic. Motorways have the advantage of regular

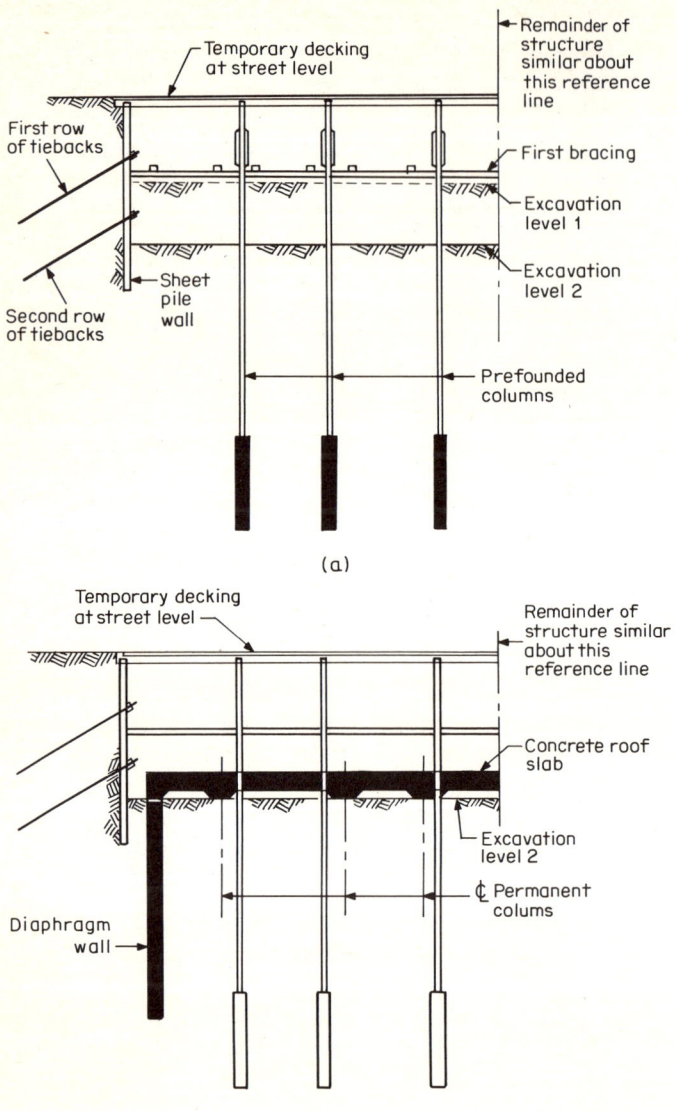

Fig. 12-30 Construction of a subway station in Akasaka, Tokyo. The installation was completed in four stages: (*a*) stage 1: exterior sheet-pile walls, prefounded columns, temporary decking, excavation to levels 1 and 2, and installation of bracing; (*b*) stage 2: diaphragm wall and concrete roof slab; (*c*) stage 3: excavation of levels 3 to 5 and installation of bracing; (*d*) stage 4: final structure and cutting off and removal of temporary bracing.

Uses and Applications 451

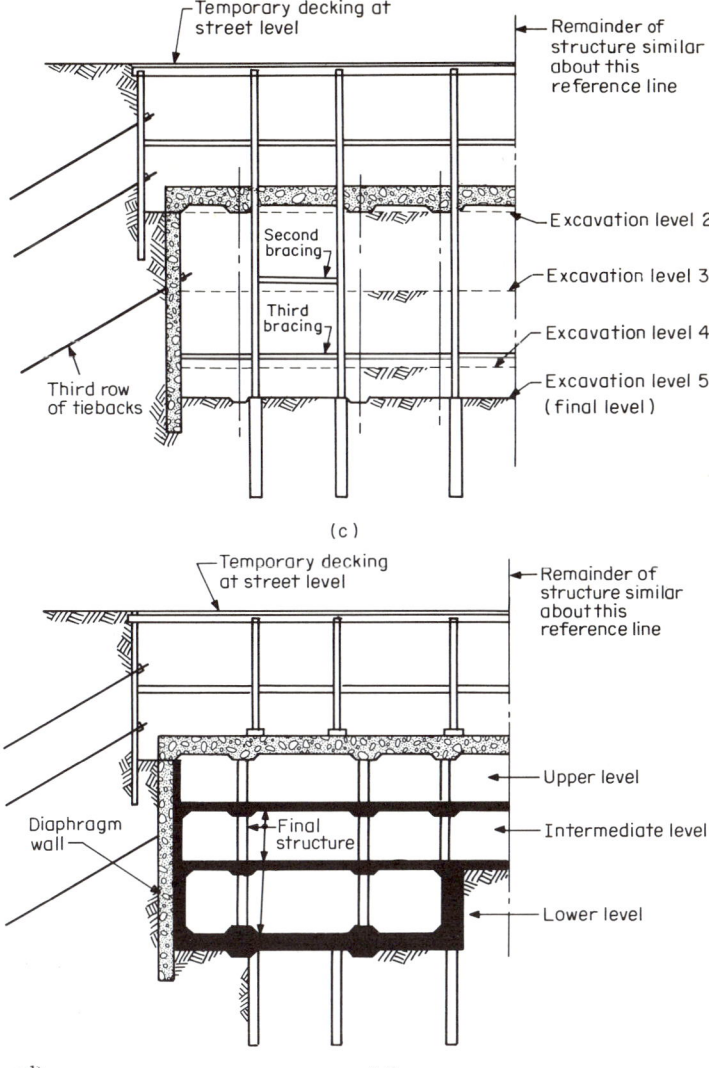

Fig. 12-30 (Continued)

configuration and uniformity of section, and usually they are depressed to provide minimum clearance at intersections and at grade separations. Although the area to be excavated is much wider than for subways, its depth is much less. The selection of ground support is governed generally by the same factors that affect excavations for subways. Policies and standards regarding these considerations are established by local authorities and are beyond the scope of this text.

Construction with Cast-in-Place Walls

The four-lane roadway with safety walks is adopted in most regions in the United States and is also popular abroad. It may require from 50 to 60 ft (15 to 18 m) face to face of walls. For the usual requirements of vertical clearance the depth of excavation from street level is of the order of 19 to 20 ft (about 6 m), which is within the range of vertical cantilever walls, as shown in Fig. 12-31.

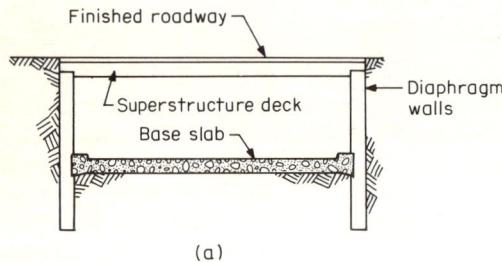

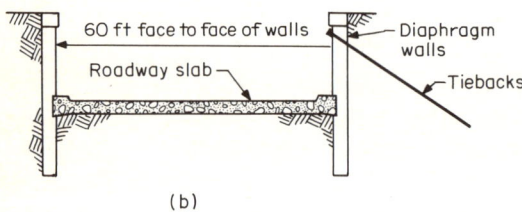

Fig. 12-31 Diaphragm walls for trafffic underpass: (*a*) covered section; (*b*) uncovered section for approach roadway.

In the covered portion the roof slab serves as the uppermost bracing level for the walls; it is therefore cast before excavation so that all earth moving is done under cover. In the uncovered portion the walls are almost certain to undergo excessive movement, regardless of embedment and soil stiffness below excavation level, and therefore it is necessary to provide bracing, usually using tiebacks near the top and the base slab near the bottom. This permanent bracing combined with temporary strutting until the base slab is cast can reduce the wall embedment considerably. Cross walls below excavation level can be effective and also support the base slab against uplift, but their cost is higher than in narrow cuts because of the extra length involved.

Because the walls are exposed to public view, face treatment is usually mandatory. This includes precast panels, brick facing, or a separate concrete facing. For the same reason specifications require fairly waterproof joints, usually the round-tube joint with water stop modified at suitable intervals to function as expansion joint.

Construction with a Center Wall If the number of traffic lanes plus any median results in unusually wide excavation, it is advantageous to use a center wall. Whether this wall is installed as diaphragm or is constructed after the center portion is excavated depends mainly on cost differences, job restrictions, and bracing procedures.

Since the center wall receives more loads than the outside walls, it may have to be deeper for adequate bearing, which is provided mainly by base resistance. A suitable wall is the strip-panel type capped with a continuous beam to distribute the loads.

Construction with Prefabricated Panels

The advantages inherent in precasting are, by contrast to the complexity of subways, more applicable in underpasses. Usually the size of the job will offset the fixed costs of precasting; the uniformity of the section allows the use of standard panel dimensions; sufficient space is available as casting yard; the precast panels are utilized for most of their height; and the smooth wall finish generally satisfies the requirements of face treatment.

Figure 12-32 shows a covered motorway. The exterior walls consist of continuous precast panels, whereas the center wall has strip sections. When all walls were in place the posttensioned concrete roof slab was installed, and earth moving was carried out under cover. A cast-in-place slab provides the roadway bed.

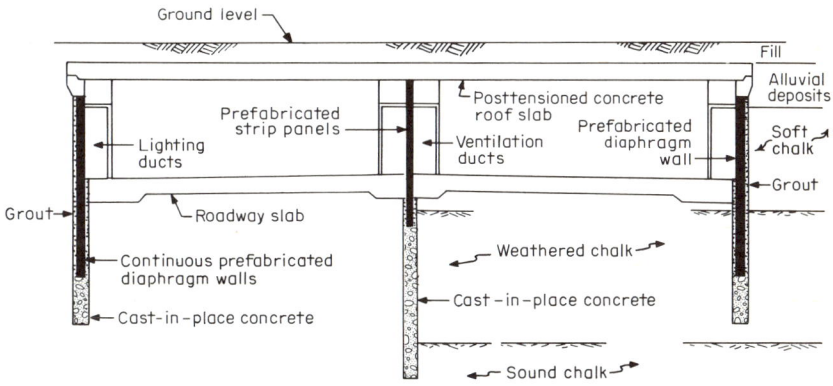

Fig. 12-32 Cross section of a covered motorway built with prefabricated diaphragm walls. (*From Leonard, 1974.*)

As mentioned, heavier loads on the center wall may require a deeper foundation. According to the principles discussed in previous sections, it may be practicable and more economical to fill the bottom part of the trench (usually below the underside of the roadway) with cast-in-place concrete. For the example of Fig. 12-32 plain concrete was chosen instead of a coulis since the latter would not provide sufficient strength and would undergo too much elastic shortening due to the considerable depth. With the trench filled to this level, the prefabricated panels must cover only the exposed portion of the wall and thus considerable reduction in the panel size is possible. For instance, the average size of the panels for the center wall is 9 by 1.80 by 0.40 m (about 30 by 6 by 1.35 ft). Thus the panels were easy to lift, handle, position, and suspend at the top.

The panels for the outside walls could have been set either in single grout (or a coulis) or in displacement grout. However, cast-in-place concrete was used in the lower part of the wall in order to provide a homogeneous foundation. It should also be noted that these panels are set deeper because of lateral stability requirements and are 12 m (40 ft) long.

Construction Details The excavation for the example of Fig. 12-32 was carried out with a kelly rig equipped with a grab 0.60 m wide (about 24 in), so that sufficient grout enveloped the precast panels for watertightness. Guide walls built along each trench also served to hold the prefabricated sections in place. The method of suspension is shown in Fig. 12-33.

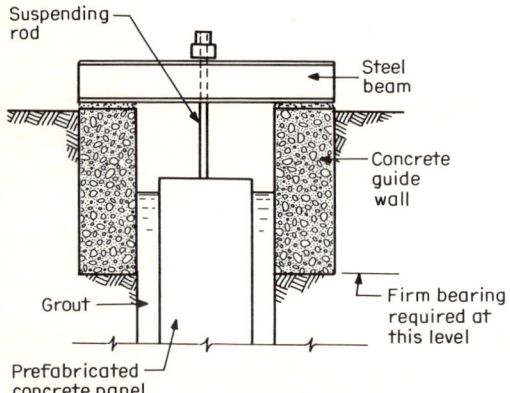

Fig. 12-33 Method of suspending prefabricated panels.

Trench stability was maintained by a single grout consisting of water, bentonite, cement, and a retarding agent to regulate the setting time. The strength-time characteristics are shown in Fig. 12-34. This behavior enabled the grout to remain essentially fluid during the first few days, in order to allow for the excavation, concrete placement, and insertion of precast sections. With time the grout appears to have gained sufficient strength without becoming too hard or reducing the productivity of the kelly rig.

Plain concrete was tremied to about 1 m (3 ft) above the base of precast panels, and while it was still fresh, the precast sections were inserted and seated so that their tips were submerged in the mix. Positioning the panels in this manner can be quite difficult due to their considerable weight and the limited time available for the operation.

Figure 12-35 shows construction-joint details. When the wall was exposed,

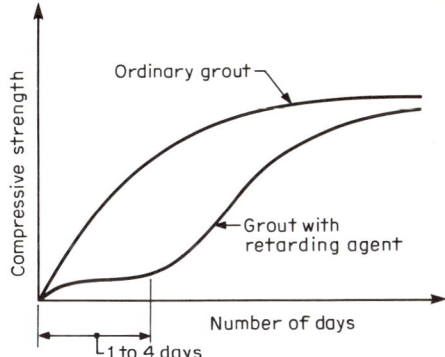

Fig. 12-34 Grout characteristics for installing prefabricated panels.

the wider part of the joint was cleaned to a depth of 2 cm (0.8 in). In order to improve the watertightness of the joint a special resin compound was placed, as shown, and the rest of the joint was filled with mortar. A distinction is appropriate between initial (short-term) and permanent (long-term) watertightness. Initially the grout in the joint and along the back of the wall must enable the excavation to be carried out under essentially dry conditions. Later the grout may shrink due to drying out, and additionally small differential panel movement can cause cracks to develop at the joints. With time, damp patches are likely to be noticed unless the joints are properly treated. Treatment is easier and more effective if the joint width is great enough to accept a watertight resin, and the joint is easier to clean.

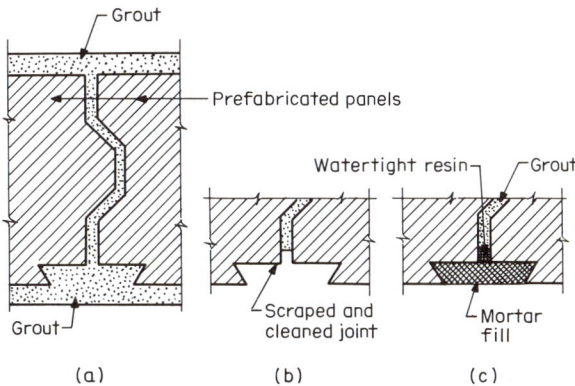

Fig. 12-35 Construction-joint details for prefabricated panels: (a) joint before treatment; (b) joint scraped and cleaned; (c) joint waterproofed with resin and filled with mortar.

Roadways under Parks or Public Land

Covered motorways through parks and recreational land offer a solution to the problem of preserving public space for this purpose, but they are more expensive than any other type of roadway both at construction and during operation. Where land is at a premium the construction of covered motorways can be justified. In this case the adaptation of the technique implies little or no interference with the natural water table, so that the landscape is not affected. The roof of such motorways is cast several feet below finished grade so that a fill can be placed for replanting vegetation and trees.

12-12 UNDERGROUND CONSTRUCTION FOR BUILDINGS

In general this consists in building deep basements in commercial or residential areas. Deep basements can be defined as underground enclosures 6 m deep or more (20 ft), although the definition is relative to the particular conditions and uses of underground space. Factors to be considered for any basement construction, given the client's general requirements, are the influence of subsurface conditions on the choice of method, the effect of plan area and shape of the basement, and the effect of completion time on the proposed use.

Assessment of Deep Basements

To the client the viability of excavating deep depends on the relationship between the proposed expenditure and expected income. The expenditure is represented by the total cost of construction and the income by rental from the use of space. Obviously the finished form of a deep basement, hence the method of construction, depends on the intended use.

From the client's point of view no advance technological solution to the basement problem can be acceptable if it is to prolong the total construction period and increase the final cost. When we accept simplicity, economy, and speed as basic requirements, the following criteria have by large the greatest effect on basement construction: the ability of the permanent wall to supplant the temporary sheeting; the ability to brace the ground support temporarily and permanently at every floor level; the ability to take advantage of the ground as means of supporting the floor slabs by casting directly on the ground and thus avoiding formwork; and the ability to have earth moving within the excavated area carried out rapidly and continuously.

Use of Deep Basement as Support As a basement excavation becomes deeper, the net intensity of load on the ground below is reduced by the weight of ground excavated. For a given imposed load the net load per unit area is decreased with depth, and eventually there will be no increase in the resultant load. Theoretically, therefore, the deeper the basement the cheaper the foundation, but in reality the deeper the excavation the more expensive the ground support and the higher the cost of excavation.

Building a deep basement solely for the purpose of supporting superstructure loads is at best rare and unorthodox. On the other hand, a deep basement allows the transfer of load at a level where more strength usually is available in the soil and also allows reduction of the net bearing pressure by the displacement of overburden. Loads of the order of 150 t/m have been transferred to dense fine sand from 80-cm-thick walls in Croydon, south of London. Diaphragm walls 30 in thick (75 cm) have been used in Chicago as load-bearing elements to carry a vertical load of 25,000 lb/ft (linear) or a bearing pressure of 10,000 lb/ft^2 (Cunningham and Fernandez, 1972). At working loads this transfer occurs as side resistance along the back of the wall, and in deep basements this offers a serious advantage considering the contact area of the outside perimeter.

Current architectural and structural trends dictate heavier column loads due to larger panels used to provide greater unobstructed floor space. In other instances the superstructure is used only above part of the usable underground space. The structural form of the superstructure predetermines the distribution of loads through the basement substructure. In some cases conventional column-and-beam construction allows the loads from the periphery of the superstructure to be transferred to load-bearing walls in the basement, but a high-rise block supported on a central core cannot derive these benefits unless the basement is locally deepened at the core.

Requirements during Construction

It is unlikely that the bottom of ground support for a deep basement can be braced below excavation level before general excavation, as with the narrow cuts described in the foregoing section, since the walls usually are too far apart; therefore the sequence in which the ground support is built and braced has significant influence on stresses and deformations. Although it is erroneous to anticipate that ground movement can be eliminated completely, in most cases it can be significantly reduced by a construction sequence and bracing appropriate to the conditions.

If the completed structure below ground is of such a form that permanent members and parts of them can effectively be used to brace the walls, how this bracing is introduced and its ability to provide the necessary restraint should be determined before construction. Such permanent members are often provided with a jacking system at the boundary-wall face to compensate for strains and deformations due to construction tolerance or because of the shortening of structural members in elastic compression, shrinkage, creep, and temperature effects.

The many uncertainties of site control and the substantial extra costs incurred if the initial methods and sequence must be changed after construction has been started fully justify the choice of an excavation sequence and procedures which are simple and practicable to follow.

Structural Integrity and Reliability Almost all deep basements built to date have utilized some, many, or all the temporary work members as part of the permanent structure. This dual role and function implies a wide variety of loading and support conditions which the wall must withstand, and this in turn results in changes in the type and magnitude of stresses and deformations induced in the system. Thus damage can occur in many ways, e.g., by overstressing in the temporary stage if the unsupported lengths are much greater than are in the permanent condition and if the assumed lateral earth stresses are much smaller than the actual; by excessive cantilever lengths allowed either at the top or at the bottom until the completed structure rectifies this situation; and by inadequate performance due to construction limitations and shortcomings or by failure of the soil-structure interaction such as excessive kicking of the wall inward during excavation to final level and before the placement of the base slab.

The design and construction of a wall to withstand all the stresses and deformations induced in the temporary and permanent condition becomes more difficult because the scope of the structural requirements is limited and because these requirements sometimes conflict from one stage to another. In almost every excavation, especially where the wall is relatively deep and supports soft plastic soil, stresses caused by soil pressures, temperature-distortion effects, and poor bracing can result in some form of damage to the wall or to the construction as a whole. This is especially true if the wall is allowed to be overstressed and if the assumption is made that the effects of temporary overstressing are temporary only. Examples are (1) diaphragm walls which developed cracks because they were built to withstand shear but were not provided with suitable shear-resisting devices; (2) rotation of bearings supporting struts because of poor connection with a diaphragm wall or rotation of a waling beam due to the reaction of inclined strut; (3) temperature changes in a long strut causing strut stress changes exceeding those of the estimated reaction; and (4) accidental movement or displacement of parts of the temporary bracing, e.g., a column supporting a strut, due to construction operations or unforeseen conditions.

Wall Embedment below Excavation Level

Initially the wall embedment is checked for toe stability. In soft or loose soil the toe penetration may be significant. If a much stiffer layer exists at relatively shallow depth beneath the excavation, it will have a beneficial effect on reducing toe movement until the lowest bracing is in place, and therefore it is advantageous to let the wall penetrate into this layer. More often, however, the soil permeability below excavation is likely to determine the embedment (see also Sec. 12-6). In pervious ground the penetration depth of the toe can materially change the rate of seepage and uplift pressure at the base of the excavation, but this problem is not simple to analyze. If an excavation is underlain by impervious soil but pervious layers exist a little below and the wall is terminated as shown in Fig. 12-36, it will not cut off the water completely from the permeable layer. Seepage pressure in this case can lead to rapid softening or actual uplift of the overlying layers.

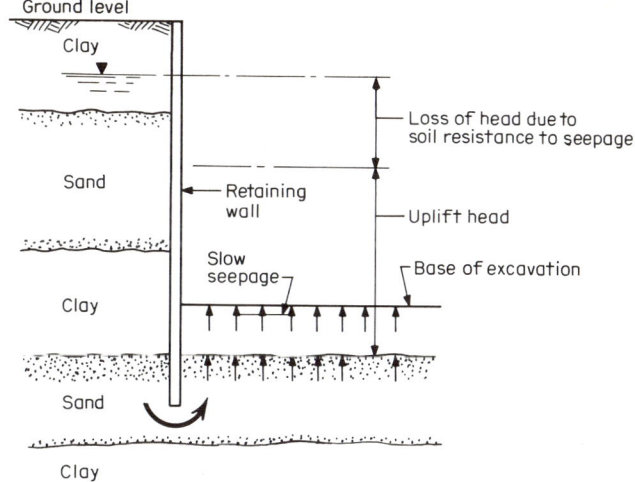

Fig. 12-36 Seepage flow and uplift beneath an excavation. The uplift in this case can be considerable since the wall does not penetrate into the impervious formation.

Support Systems

Bored Piles When the boundary wall is virtually adjacent to an existing basement or foundation, the underpinning problem can be solved by a contiguous or secant-pile wall cast in line around the basement periphery to serve both as temporary and permanent ground support. Stiff homogeneous clays with little groundwater offer suitable soil conditions for construction with contiguous bored piles. As mentioned earlier, gaps of 5 cm (about 2 in) are not uncommon between adjacent piles, and probably more can be expected in deeper basements under the permitted vertical tolerance. If gunite is used to seal the joints, it represents an additional cost.

Secant piles have all the advantages of bored piles without their undesirable characteristics, but they cost considerably more. They make fairly watertight walls even in poor soils. A secant-pile wall can be cantilevered or braced with rakers, struts, or tiebacks.

Diaphragm Walls Deep-basement construction on private property encounters few underground obstacles and utilities. At present more than 80 percent of all diaphragm walls built as boundary support are permanent. Diaphragm walls become more attractive if they carry vertical loads, since this usually requires little or no extension of the normal embedment below excavation level. If the applied loads vary in magnitude, some differential displacement can occur and cause seepage at the joints. Hence, if a high degree of watertightness is desired, the joints must be waterproofed according to the methods described in previous sections or by the use of epoxy-resin strips from the inside applied to the surface of the joints.

The job restrictions on a building site are less serious and seldom govern.

There is usually sufficient space for plant facilities, reinforcement assembly, and storage, and the work does not suffer from lack of time. The contractor has ample opportunity to select the most advantageous panel sequence and size, and vehicular movements in and out of the site are scheduled under more favorable conditions. A common complaint is associated with delays in the delivery of fresh concrete because of traffic congestion in public streets.

Occasionally it is necessary to deal with existing subbasements, usually one story. Whereas the cost of demolition in normal excavation is nominal, the removal of such obstructions in slurry trenches is costly, so that in many instances it is better to excavate to subbasement level first and use this level as construction platform for the diaphragm wall.

Prefabricated Panels From the technical viewpoint prefabricated panels are feasible and advantageous. Exceptions must be noted with respect to certain requirements, e.g., structural continuity between joints under the application of vertical loads. For deep basements precast walls are economically attractive if the size of the jobs is big enough to absorb the fixed costs of precasting; if most of the panel height is exposed and therefore utilized; and if the function of the basement requires a wall finish and face treatment which cannot be provided with cast-in-place diaphragm walls.

Plan Area and Shape

Given the area, the plan shape of a basement can affect the construction method and the cost. For the same area a square basement gives the minimum wall perimeter. Basements of circular or elliptical shape have a certain fascination and are practical for certain uses, e.g., underground parking or storage tanks. Like squares, circular enclosures provide high utilization areas with minimum perimeter.

Where there is prime need for full utilization of space, the result is minimum clearance between new construction and the boundary of adjoining building. A summary of the requirements of various ground-support systems and clearance necessary for the installation is shown in Table 12-1. If the fullest utilization is needed, it is necessary to reduce the wall thickness to a minimum, and in this respect bored-pile walls may be unsuitable compared with linear panels of uniform thickness.

Bracing

Some of the bracing possibilities are shown in Fig. 12-37 and are summarized as follows.

Cantilever Walls These are self-supported, and this is feasible with shallow excavations and firm soil beneath or where ground movement can be tolerated. Unbraced diaphragm walls must be used with caution, in spite of the quick installation and the unobstructed area thus provided.

Ground Anchors These are becoming very common in deep basements for the many reasons mentioned in previous sections. They are especially attrac-

TABLE 12-1 Minimum Distance between Soil Support System and Site Boundary (from Puller, 1974)

Support system	Installation plant	Distance* in	cm
Underpinning	Conventional bulk excavation plant, e.g., hydraulic excavator with hydraulic grab	0	0
Steel sheet piling	Crane and piling hammer	20	50
Contiguous bored piles	Rotary piling equipment 60 cm pile	24	60
	Tripod piling equipment 50 cm pile	6	15
Secant piles 88 cm thick	Benoto rig	42	106
Diaphragm walls	Rope grab, kelly-mounted grab, or Tone reverse-circulation drill	6	15
Soldier piles and horizontal lagging	Rotary piling equipment 60-cm-diameter bore	24	60

*Minimum distance between outer face of support system and site boundary. Distances quoted are those at ground level; considerations must be given to verticality tolerances of support system.

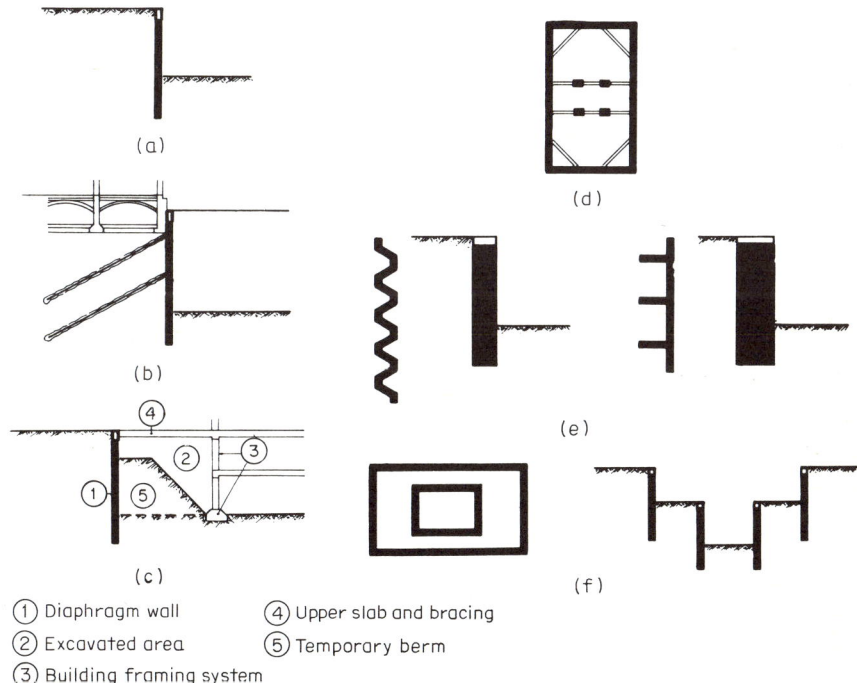

① Diaphragm wall
② Excavated area
③ Building framing system
④ Upper slab and bracing
⑤ Temporary berm

Fig. 12-37 Bracing combinations of excavation for building basements.

tive when anchored into rock. In clay and where anchors are placed under existing buildings care is necessary to prevent their movement.

Temporary Berms Earth berms can be used either in conjunction with the sequence shown in Fig. 12-37c or combined with tiebacks in the upper part of the wall and rakers (inclined struts) in the lower. They can be subjected to long-term effects and creep or move excessively if the installation of permanent bracing is delayed.

Long Flying Struts These are suitable for narrower excavations. Besides impeding construction, the monitoring of jacks is more difficult than with other methods. The corner bracing is economical, but it often results in awkward redistribution of earth stresses since the reaction in the strut from one side must equalize the reaction from the other side.

Walls with Counterforts or Keys These have a far greater moment capacity and therefore need less bracing, but they should not be expected to stop or reduce ground movement. Although they are suitable for waterfront structures, they should be avoided in building construction.

Center-Core Bracing Where the depth of a basement is varied to accommodate a center core, extending the outer perimeter walls below the lowest excavation level is not necessary. Instead, the construction can be carried out as shown in Fig. 12-37f. For this situation two analyses are necessary: (1) determine whether it is better to build the interior walls first from ground level, continue with the construction of the exterior walls, excavate for the central core, and brace the interior walls, since this will reduce materially exterior wall movement or (2) build the exterior walls first, excavate to top of interior wall level, and install interior walls from this level.

At best bracing is provided in a downward construction (see also Sec. 12-7), particularly in deep excavations, with each permanent floor cast on ground as excavation is carried down in stages.

Walls with reentrant corners in plan require frequent resetting of the slurry-plant facilities. Additionally, wall bracing with struts, rakers, or tiebacks is more complicated than in regular shapes. Adjacent walls at a reentrant corner must be braced in a way that will prevent tiebacks from clashing with each other or having a common potential failure wedge. It is sometimes better at a reentrant corner to have the walls tied horizontally and diagonally within the corner across the ground.

12-13 SPECIAL PROBLEMS OF DEEP VERTICAL CUTS

Figure 12-38 illustrates a common problem with deep basement construction. A wall is built and tied back with ground anchors, and then an adjacent excavation is made into rock below the tip of the wall. The vertical component of the force in the ground anchors creates a considerable additional load im-

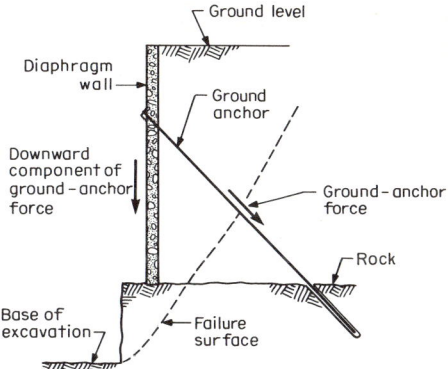

Fig. 12-38 Vertical loads from tie tension and chiseling into rock, causing shear failure in the rock mass below the base of the wall. (*From White, 1974.*)

posed on top of the rock within a short distance of the vertical face. Normal loads are the weight of overburden, the weight of the wall, and surcharge loads.

The stability of bedrock against shear failure is not only a matter of strength but also depends on the presence or absence of fissures, clay-filled seams, weak joints, and cracks. Such defects can deprive the rock mass of strength if they have unfavorable orientation and cause collapse of the rock and the wall, including the overburden. The same type of failure has occurred with conventional struts in subway construction and trench work and is typically recorded as rock slides (White, 1974).

Besides the weakening of rock due to its structural orientation this condition is aggravated by the construction operations. Chiseling or drilling to obtain a rock socket for the wall and blasting for rock excavation can cause splitting of the rock mass and shearing along weak joints and seams. Mishandling of equipment can also damage rock supporting the wall. White (1974) reports a case of rock consisting of various strata of horizontally bedded soft shales and sandstones where a heavy power shovel bucket actually ripped out the supporting rock back of the line and caused settlement of 1 to 2 ft.

The problem is sometimes compounded by an overconservative estimation of the lateral earth stresses. This overestimation results in more prestressing force in the tiebacks than necessary and therefore extra and undue downward thrust applied on the wall.

White (1974) gives a summary of preventive measures for the most common wall types and usual conditions. For diaphragm walls one measure is to construct concrete buttresses to shore the vertical rock face, as shown in Fig. 12-39*a*. A second measure is to use a flatter inclination for the anchors as shown in Fig. 12-39*b*, but this often implies conversion from rock to earth anchors if the quality of the overburden permits. A third measure involves the use of vertical supports, as shown in Fig. 12-39*c*, which function to transfer the vertical component of the anchor load directly to the rock at or below excavation level, thus relieving the rock above of the need to carry this load.

For the diaphragm wall of the World Trade Center in New York City the

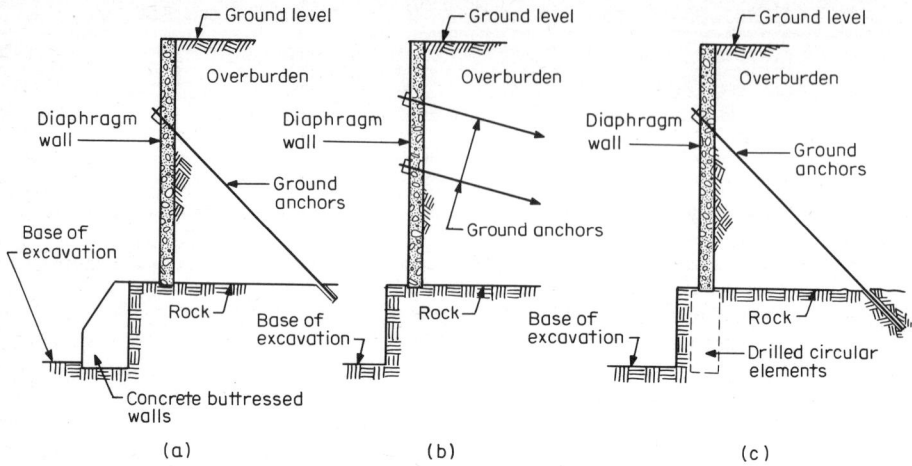

Fig. 12-39 Methods of supporting a vertical rock face below the tip of wall.

entire structure (3000-ft-long perimeter) was built into a 5-ft-deep slot laboriously chiseled into a hard formation of mica schist. At 10-ft intervals, corresponding to the horizontal tieback spacing, concrete buttresses were placed to support the toe of the wall laterally and prevent its movement toward the excavation. In other locations of the structure base stability was ensured by pouring continuous reinforced-concrete slabs inclined with the horizontal and butting against the wall at its base.

12-14 UTILITY TUNNELS

Utility tunnels are structures containing more than one public-utility system. They have been used abroad, whereas in the United States response to this type of work has been favorable, unfavorable, or mixed.

The concept is particularly adaptable for use with cut-and-cover transportation tunnels. The resulting benefits may be a decrease in the disruption time of surface activities and technical and economic advantages to the organizations and entities operating the services. Utility tunnels are more useful where new utility services are contemplated, either concurrently with the construction of the transit line or at some future date. Nevertheless, utility tunnels are not always applicable. Examples of where their applicability may be questionable are a restricted right-of-way or very wide right-of-way making complete relocation of utilities possible, both of which are against the economic justification of the project.

The reason for utility-tunnel installations abroad has varied. In Japan and West Germany, where utility tunnels have been used more widely, three main factors have contributed to the concept: both countries suffered extensive damage during World War II, necessitating complete reconstruction of utilities, and their placement in utility tunnels was found more economical than in direct burial; in both countries rapid urban growth is occurring and implies rapid increase in the demand for utility services; and both countries have adopted the policy of reducing the incidence of street cuts due to modifications and addition of utilities.

The feasibility of utility tunnels usually is determined from technical, economic, and operational considerations. A positive aspect is a concurrent need for a new transit line and expansion of utility services. On the negative side, the possibility of gas explosion or the potential of interference in communication lines may prompt authorities to reject the idea. Considerable effort has been made both in the United States and abroad to establish and classify the merits and the objectionable features of the system, but at present no conclusions can be drawn. With the exception of sewer lines, almost every utility type can be included in the same tunnel, usually in the upper chamber. The gravity-flow sewer system is excluded mainly because its grade requirements differ from the subway profile, although successful applications have occured with this utility located below the transit-line section.

Utility Safety Functional or safety-systems failure has been a main concern in utility tunnels, but problems can also arise in connection with a structure which is not structurally sound.

A problem that might be serious is differential settlement. Even if this is less than the settlement causing structural damage, it can still affect utility function and increase maintenance costs. A second problem, probably less frequent but more serious, is explosion at street level, introducing vibratory excitations of a magnitude that will damage the structure and the services.

If there is insufficient space in the tunnel, the result will be operational inefficiencies, e.g., difficulty in making simultaneous repairs of utilities, problems in transporting equipment, utility damage, and at worst personnel casualties.

A further problem can arise from periodic vibrations and impact caused by railroad or street traffic moving along the roof of the tunnel. Unless the supporting frame has a damping effect, impact will eventually loosen the pipe joints and service connections and result in utility damage. Some concern has also been expressed about the watertightness of a utility tunnel. If the seepage is excessive, it will require frequent pumping and may create unfavorable humidity conditions.

Examples of Utility Tunnels

Utility Tunnel at Bourse Station, Brussels The construction for this tunnel, shown in Fig. 12-40, was executed with diaphragm walls and prefounded columns. The main objective was to relocate the two elliptical sewer sections just below street level used to store storm water, in order to provide space for the mezzanine floor of the station.

The construction was scheduled in several phases, and certain main stages are shown in Fig. 12-40. In stage 1 (Fig. 12-40a) the diaphragm walls are

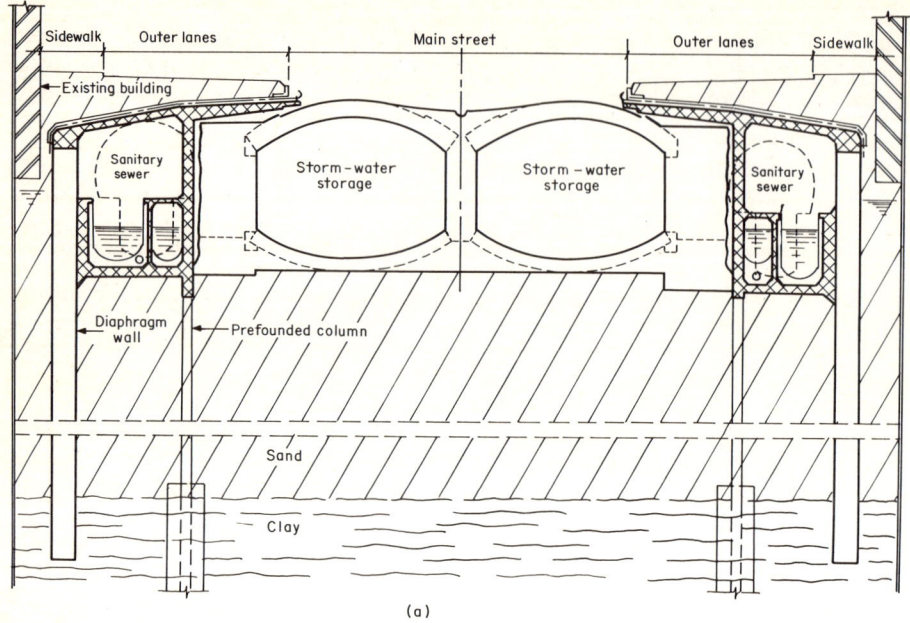

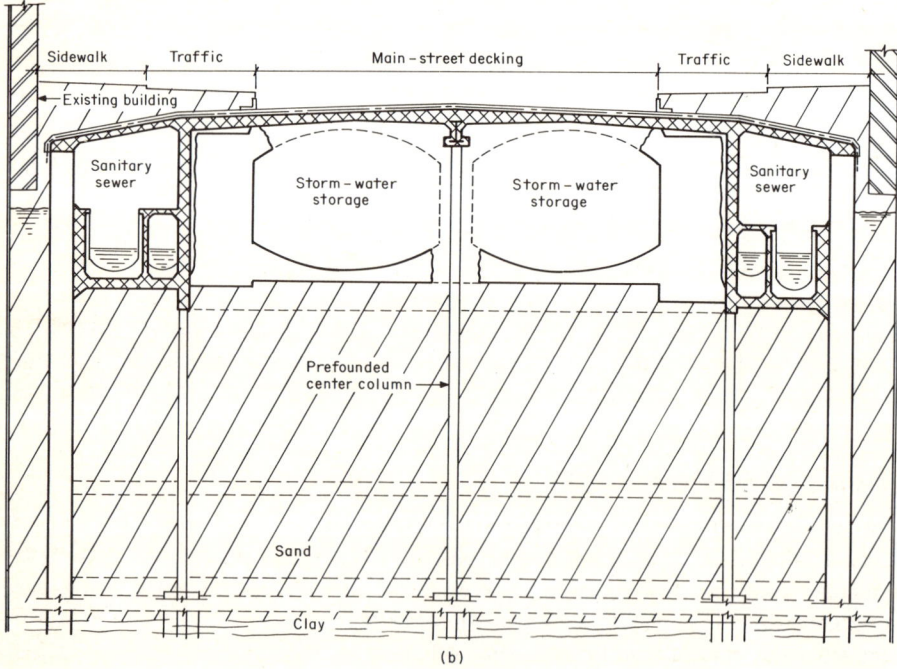

Fig. 12-40 The concept of utility tunnel as applied to the construction of the Bourse Station in Brussels; (a), (b) and (c) construction phases. (*Franki.*)

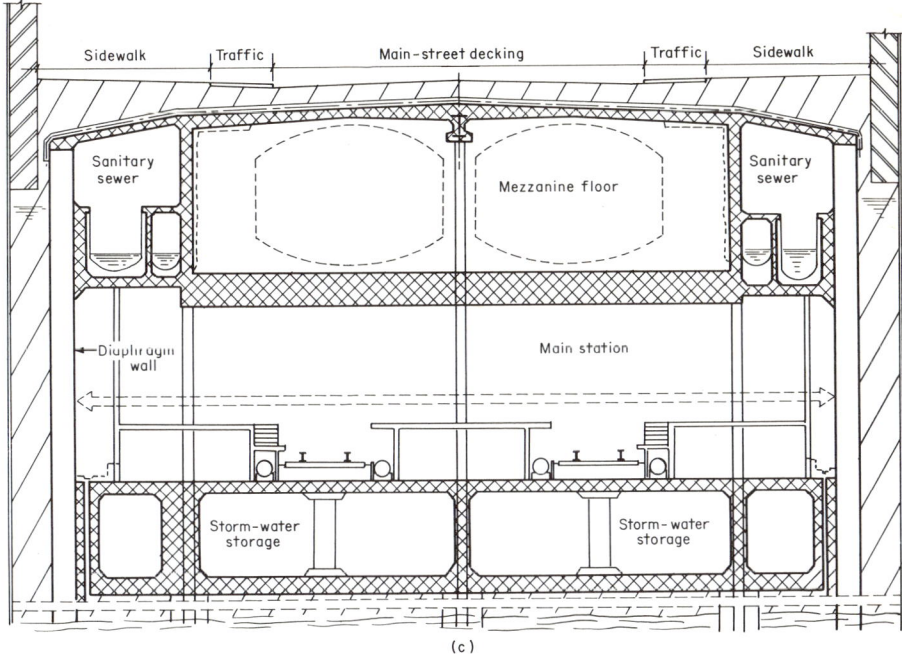

Fig. 12-40 (Continued)

installed, together with the exterior prefounded columns, and the permanent slab is placed along the outer lanes of the street. In stage 2 (Fig. 12-40b) the center prefounded columns are inserted, followed by the completion of the street decking. The excavation was thereafter carried on under the roof, and during this process the elliptical sewers had to be underpinned. The finished subway section is shown in Fig. 12-40c. The new storage space is provided below track level and consists of four separate chambers.

Utility Tunnel in Osaka This tunnel is located under a street in a mixed commercial and residential neighborhood (see Fig. 12-41). It has a depressed roadway immediately below street level, the upper part of which (not shown in Fig. 12-41) is to house cables for communication services, electrical distribution systems, gas lines, and water mains. The roadway section is underlain by a rectangular gallery to handle storm water. The sewer profile is thus independent of the roadway grades and follows the hydraulic requirements.

The construction was mainly in water-bearing ground, consisting essentially of fine loose silt. The ground surface in the immediate vicinity undergoes a normal consolidation settlement of 10 to 20 mm/year (0.4 to 0.8 in), and the contractor was asked to guarantee the structural integrity of surroundings and accept liability for any damage. Diaphragm walls were used as ground support, except at intersections where bored piles were more advantageous for the traffic conditions. The walls have an embedment of 4.8 m (about 16 ft) below excavation level, and in addition to the permanent slabs

Fig. 12-41 Utility tunnel in Osaka. The upper level is a depressed roadway immediately below street level, also housing communications and electric lines, gas lines, and water mains. The lower chamber is for storm-water storage. (*Zenitaka.*)

they were braced at four intermediate tiers. Maximum observed wall movement was 1.7 cm (0.67 in).

12-15 UNDERGROUND PARKING

Besides providing parking areas in the underground portion of buildings, commercial or residential underground parking garages are built as independent facilities. The main problem in this case is the availability of suitable property at reasonable initial cost or rent which can be recovered from the uses of the facility.

For independent structures circular and elliptical shapes provide certain structural and functional advantages. An example is the 11-story underground garage shown in Fig. 12-42, built inside an elliptical enclosure supported by diaphragm walls. The elongated shape of the structure accommo-

Fig. 12-42 Interior of a box-shaped underground elliptical parking garage in Milan, showing the access core for the elevators and the parking levels. (*ICOS.*)

dates a boxlike arrangement for one row of parking on either side of the central access core, and its curved configuration consolidates external lateral stresses in direct compression. This plan offers good possibilities where parking must be provided in congested sites, but the relatively small size cannot accommodate access ramps.

Under appropriate space and plan conditions circular parking provides the highest utilization ratio, particularly if used in connection with double helical ramps and interlocking cross-overs. This design was developed and first applied to Bloomsbury Square parking garage in London. The main structure houses a pair of interleaved spiral access and parking ramps, and a central service core that contains the lifts, stairways, and other facilities (*Ground Eng.*, 1972).

12-16 INDUSTRIAL AND SERVICE INSTALLATIONS

A large number of industrial plants, waterworks, sewage works, and service installations have structures built below grade, e.g., intakes for thermoelectric power stations near the sea, lakes, and rivers; hydroelectric power stations; pumping stations; sewage-treatment plants; water-cooling facilities; underground vaults to house industrial machinery; tanks and reservoirs; and deep shafts. Classification of these structures according to the intended purpose may lead to overlapping. In principle the enclosures for a pumping station or a power station may be much alike. Likewise a circular reservoir, a circular gas tank, and a circular tar tank are very similar. Hence, these structures are classified here according to shape, design, and construction details.

Interlocking Enclosures

If a facility must be wholly placed underground and in plan it is such that it can be embraced in a uniform manner by an enclosure wall, the answer is a circular structure. If machinery and functions are spread irregularly, a solution may be provided with interlocking enclosures, e.g., interesting circular structures which can divide the construction into two or more portions without sacrificing the benefits of circular walls and without making the structure larger than necessary in plan.

Figure 12-43 shows the perimeter enclosure of Beckton surface-water pumping station (England), which consists of two separate circular structures interlocking as shown. At their intersecting points the structures have common panels with counterforts, whereas permanent struts extend across to transfer the compression forces in the open areas from one section to the other.

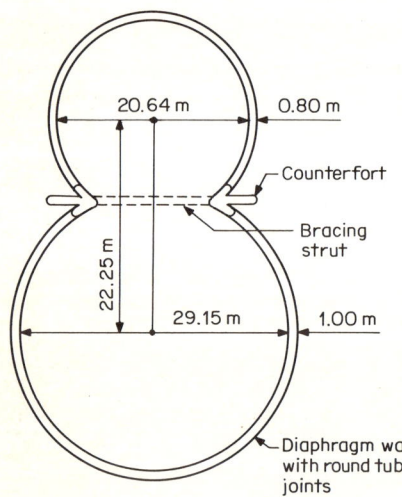

Fig. 12-43 Interlocking circular enclosures for the Beckton (England) surface-water pumping station.

Tanks and Reservoirs

Figure 12-44 shows new forms of tank construction used for storage of industrial liquids, as gasholder tanks, tar tanks, and the like. In Fig. 12-44a the construction has a diaphragm tank below ground, which extends above ground by means of a steel or concrete structure. In Fig. 12-44b the circular diaphragm wall is used as foundation for a steel tank in soft ground. The tank shown in Fig. 12-44c is completely below ground.

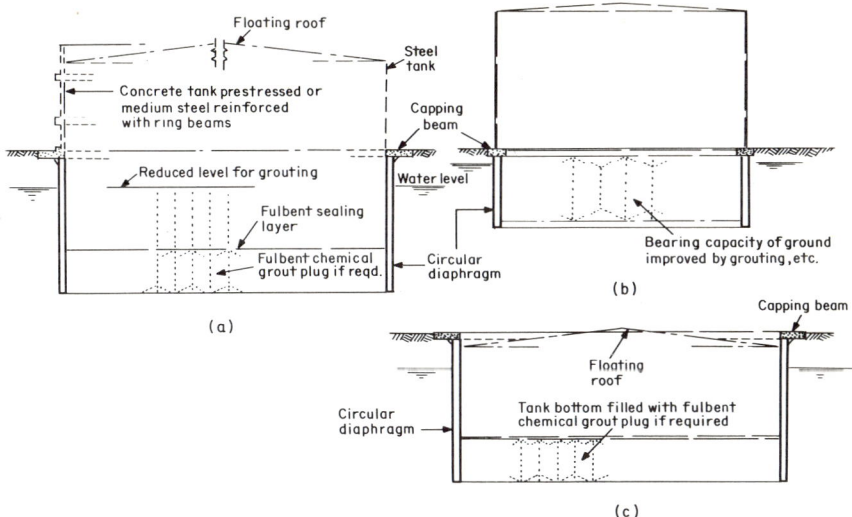

Fig. 12-44 New forms of tank construction: (*a*) diaphragm tank below ground supporting concrete or steel upper structure; (*b*) circular diaphragm supporting steel tank in soft ground; (*c*) diaphragm tank below ground.

Whereas many substances are not injurious to concrete, some are distinctly harmful, e.g., all acids and many salts. In these cases the concrete must be protected, either by special linings and acid-resisting cements or by coating of asphalt, paraffin wax, lead linings, mixtures of pitch and tar, and the like.

Where storage tanks are permanent installations in fixed locations, circular concrete tanks below ground offer great advantages over steel tanks above ground for the storage of crude oil or its products. Cement should not be expected to harm the oil, and a concrete floor at the bottom is useful in protecting the tank from rust caused by water separating from the oil.

However, with relatively large or deep tanks, even when very dense concrete has been used and utmost care exercised during construction, losses have occurred when volatile liquids, e.g., petroleum, percolate through the finished structure, particularly in ground with a low water table. In the past this problem has been remedied by a double wall separated by a void space from the exterior wall. A false bottom is also provided. The cavities in the wall and the floor are filled with water, so that its hydrostatic pressure prevents the liquid from leaking.

Deep Shafts

The advantages of polygonal enclosures are best utilized in the construction of deep and medium to small-diameter shafts for mining, ventilation, access, and hydraulic purposes. Usually these shafts are built in two- or three-sided panels. In order to avoid lack of structural contact, which may cause some movement, the construction joints are grouted from within as the wall is exposed. Grouting is mandatory in water-bearing ground.

When the perimeter wall of the shaft is built and before excavation of the core is started, it is advantageous to install drainage wells to reduce hydrostatic pressure and bottom heave and also allow the core excavation to be a carried out in the dry. Some difficulties will be encountered if it is not possible to lower the water table inside the shaft and the excavation must continue below this level. In some instances the base slab must be placed underwater, and in order to execute the connections with the walls divers must be lowered to the bottom. A further problem can arise in regard to the stability of the main earth core while excavating a multisided panel, particularly if another panel is still open and the rig operates inside the core.

A diaphragm-wall shaft is shown in Fig. 12-45. The shaft is 63 ft deep (19 m) and has a 27-ft diameter (8.2 m). It provides ventilation and exhaust for the second underpass beneath the Mersey River in Liverpool. The wall panels are merely keyed into underlying rock and terminated at this level. Excavation in soft ground was relatively easy until it reached the rock formation. From this level down the excavation was continued by blasting. In spite of the shock of the operation it was not necessary to line or otherwise protect the structure above.

Fig. 12-45 Panoramic view of an access shaft for the Mersey River underpass in Liverpool. (*ICOS.*)

Box Culverts

For open-trench construction of sanitary and storm facilities box culverts and conduits are attractive alternatives to large-diameter sewers. On a regional basis diaphragm walls are widely used in Japan and Italy for the construction of drainage systems in built-up areas.

Certain advantages can be derived where excessive trench loads and large superimposed traffic loads impose extra requirements on large-diameter conduits or storage tunnels or where very deep and very wide elliptical shapes required by horizontal or vertical space limitations result in very deep or wide excavations. Advantages are also evident where it is desired to separate storm and sanitary sewers, where large storm-relief sewers are built, and where special underflow systems are provided for storage of peak flow. In certain instances it is possible to divide the basic structural section into compartments for various uses and types of flow.

Diaphragm walls should not be used in conjunction with sections which are unfinished or left with irregularities and protrusions. The hydraulic performance of such openings is liable to be affected by the final finish. For flat profiles this is important since it can increase hydraulic losses over the limits of available energy and reduce the ability to maintain self-cleansing velocities. In order to avoid these problems the interior of the section should be given the necessary shape and finish by casting plain concrete.

In Japan the high cost of concrete guide walls for long culverts has resulted in the use of special prefabricated portable steel guide walls. They can easily be dismantled and moved to the next location or job. With some modifications the excavating rigs are operated without rails, but they are fixed on, and supported by, the guide walls directly.

12-17 WATERFRONT FACILITIES

Piers, Quays, and Wharves

The buttressed walls, cells, and arched structures described in Sec. 10-6 have wide applications in the construction of waterfront facilities. Docks, piers, and mooring wharves are constructed with linear diaphragms and load-bearing elements combined with cross buttresses or anchor walls to form gravity-type structures or braced cantilevers. Practical solutions and configurations are shown in Fig. 12-46.

The wall shown in Fig. 12-46a can be considered a gravity-type structure and assumed to fail by overturning if the cap slab is fairly stiff and has rigid connections with the wall and if the bottom of the wall is in solid earth. Since with a thin cap slab the wall is likely to perform as a free cantilever, advantage can be taken of the interior fly or rear wall in mobilizing passive resistance. The latter design is probably more realistic and consistent with the actual construction conditions and details and can also be technically advantageous if sufficient passive resistance is developed at low strains.

The wall shown in Fig. 12-46b is partly gravity type and partly free cantilever. It is useful in situations imposing heavy vertical loads (seafront installations, loading cranes, etc.) on the cap slab. For better results the rear diaphragm wall must be built in firm soil. The lateral forces along the back are

474 Slurry Walls

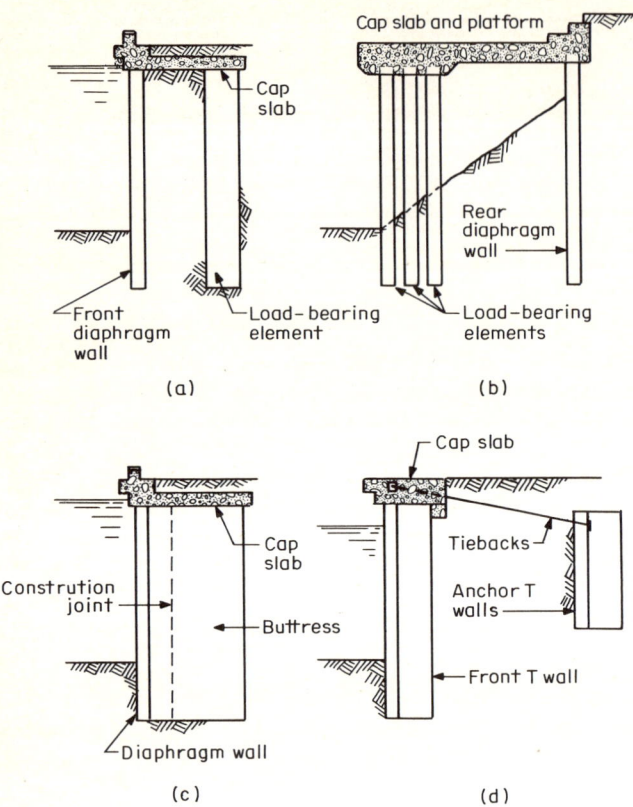

Fig. 12-46 Schematic designs of waterfront facilities by diaphragm walls and related elements.

resisted partly by passive forces on the slope side and partly at the top by the cap slab. The latter reaction is transmitted to the front group of vertical elements, and if this group has high rigidity and moment of inertia, the horizontal reaction is translated into axial forces.

The wall shown in Fig. 12-46c is an open-cell construction and has a front wall and cross walls. If the cross walls are long, the structure can be analyzed as a gravity wall and therefore two requirements must be satisfied; adequate bearing must be provided at the tip, where the pressure is maximum, and gravity loads must balance overturning. The second requirement is satisfied by having the superstructure bear on the walls or by lowering the concrete cap to place an earth fill on top for more weight. Weight also improves the frictional resistance along the base and therefore the factor of safety against sliding. There is usually one cross wall for each front-wall panel. Better resistance against laminar tilting is achieved if part of the cross wall is cast together with the front panel and the construction joint is located as shown.

The T wall shown in Fig. 12-46d relies on the use of an anchor wall to receive and resist the tieback forces. The orientation of the front T wall depends mainly on its function. If the wall can be keyed into rock so that it is laterally supported at the bottom, it is better to reverse the configuration and place the webs outside.

Load-bearing Elements These can be used in the construction of piers and mooring platforms instead of conventional sea piles. The substitution is indicated particularly where heavy loads are expected during service and the sea bottom has soft mud to considerable depth. Construction through seawater requires the use of steel casing, which is also necessary for tremie placement. The casing is driven to penetrate the bottom for some distance and seal the excavation. The seawater is them pumped out and replaced by bentonite slurry. Drilling is started at sea bottom under slurry protection but is likely to encounter unusually large amounts of salt, and therefore the slurry must be prepared for this condition.

Example of Quay Wall Figure 12-47 shows the quay wall for the Peterhead (England) harbor development, based on the concept of building a facility using a land-located plant. The diaphragm wall is 1 m (3.3 ft) thick and was placed through sand fill.

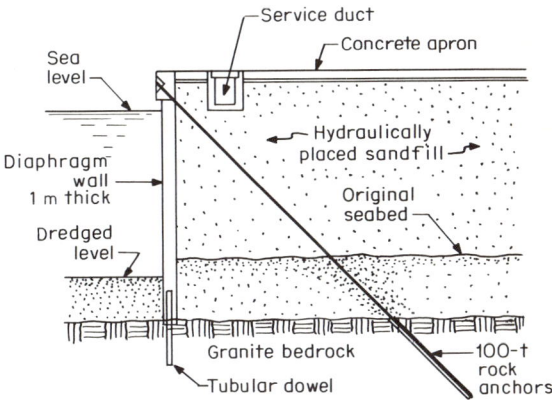

Fig. 12-47 Quay wall for the Peterhead (England) harbor development.

The wall was keyed into bedrock by drilling and placing large-diameter steel dowels through shafts left in the concrete panels. In certain locations softer formations permitted direct excavation into bedrock, and the dowels were omitted. The diaphragm wall is capped by a continuous beam, 2 m deep (6.6 ft), through which sleeves were left to locate the top anchorage position. The top lateral thrust of the wall was resisted by ground anchors installed at 45° and drilled into bedrock. The capacity of the anchors is 100 t. Certain panels were prestressed vertically by inserting 200-t-capacity anchors into rock at the dowel locations.

Behind the wall is a 20-cm-thick (8-in) reinforced concrete pad, which has a large service duct freely resting on the reclaimed land.

Protection of Dykes, Reservoirs, and River Banks

Chapter 1 mentioned some representative examples of protective works along the banks of rivers, such as the Arno River in Florence after the damaging floods of 1966. Besides protection against erosion, diaphragm walls along river banks create new stable river beds, and this is important where channel relocation is difficult. Another application is the protection of sections immediately upstream or downstream from hydraulic projects such as water intakes, basins, and the like. Besides protecting the bank, the walls also serve as retaining structures for earth and surcharge loads.

For protection against erosion the walls are built along the banks. They are keyed into the river bed, and may have a top connecting beam. If they must provide protection to adjoining areas from seepage and water infiltration, they must be made deeper and seated in impervious formations. The construction usually is executed at the toe of the embankment on the river side, and therefore it is necessary to complete the embankment treatment on the earth side of the wall.

Whether the excavation is carried from existing ground level or from a specially prepared working level depends on ground conditions and mainly on the water table. For maximum efficiency several operating rigs must be used at the site serviced by a common slurry plant. The construction material may be reinforced and plain concrete (or plastic concrete if deformability is desired).

Examples The installation of the protective work on the Arno river was initially conceived as a watertight screen for the hydraulic works merely to protect the surroundings from rising water level in the river. Subsequently the design was modified, and the walls were used as foundation elements for the restoration of the quay and in order to provide safe conditions for strengthening adjacent buildings.

A useful application is shown in Fig. 12-48 for the water supply of the industrial area of Nottingham through the artificial creation of an accumulation basin. The top of the basin above existing ground was obtained by a concrete dam protected outside by clay embankment. The stability of the dam is further secured by a system of bored piles and a diaphragm wall built as shown. The diaphragm wall is extended into impervious layers and therefore serves as cutoff below the bottom of the basin. The wall and the piles are rigidly connected to the main concrete core by special joints. Inside the basin similar structures are used to divide the area into smaller sections, and they are built to accommodate varying pond conditions, e.g., one side full and the other empty.

12-18 SPECIAL EXAMPLES

Foundation Anchorage for the Humber Suspension Bridge An unusual foundation solution has been conceived for this project, which at present

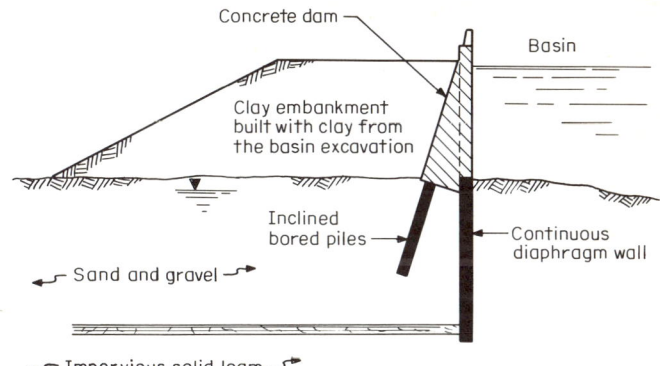

Fig. 12-48 Concrete dam and diaphragm wall for Nottingham (England) water basin.

is the largest suspension bridge in the word. The 36,000-t pull from the suspension cables is eventually transferred into a clay formation at the Barton site by an anchorage of composite construction. Initially the high stress concentration is dissipated by terminating the steel cables in the upper half of the anchorage block. The lower part of this structure consists of five sections 24.5 m deep (80 ft) excavated between diaphragm walls (see Fig. 12-49). The composite construction is over 70 m (240 ft) long and has an average width of 40 m (131 ft).

The main purpose of the composite construction was to penetrate the underlying clay in a controlled manner (*Ground Eng.*, 1974) in order to prevent flooding of the excavation by groundwater which might have softened the clay formation and caused anchorage problems. The possibility of long-term effects was also considered, and the engineers suggested speeding up the construction program to avoid these effects although the simultaneous excavation of two adjoining cells between diaphragm walls was avoided. For the same reason the bottom concrete in the cells was generally placed immediately following the removal of the overburden. By avoiding having the excavation open for too long the contractor was able to control swelling and bottom failure so that no appreciable loss of strength occurred.

The dead weight of the cell structure of the anchorage was increased considerably by pumping sand and water inside. This load application was introduced progressively and represents a new approach to the control of large foundations.

The construction of diaphragm walls for the anchorage block was completed in ground conditions which generally are unfavorable for trenching. Besides a water-bearing ground and a relatively high water table, the problems were aggravated by the need to operate from a level 6 m below the natural ground level of the existing embankment. This induced the risk of trench collapse (see also Chap 2) during excavation should the groundwater level rise above a safe level; in spite of the countermeasures (including groundwater lowering) it was not always possible to counteract the tidal fluctuations in the river, and a few

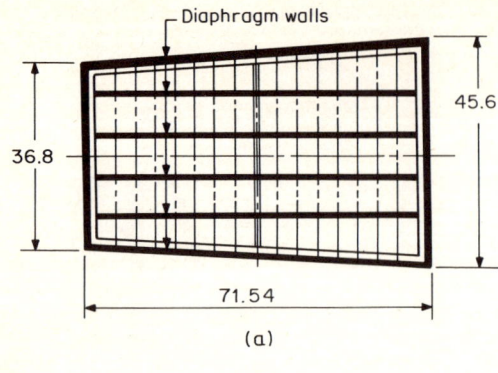

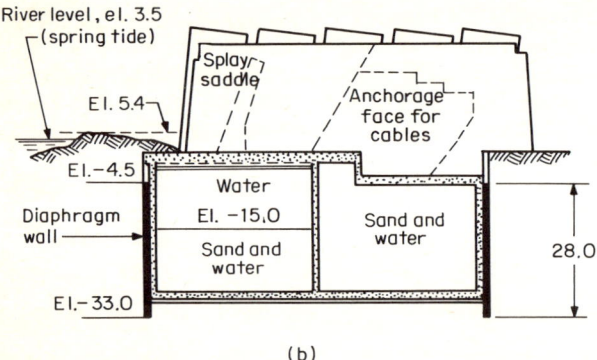

Fig. 12-49 Foundation anchorage for the Humber suspension bridge at the Barton site: (*a*) plan showing diaphragm wall block and anchorage; (*b*) longitudinal section. All dimensions and elevations in meters.

partial collapses of panels were experienced. These collapses required backfilling with lean concrete and the supplemental bracing of the adjacent slopes with steel sheet piling.

The difficulties of trenching were somewhat lessened by allowing the direct discharge of used slurry. The disposal scheme included steel pipeline of a suitable diameter which transported and discharged used slurry directly into the river.

The horizontal bracing of diaphragm walls across the five sections is provided by precast concrete struts placed in sets of two. This type of bracing is reported to be simple to install, and contractors claim it can control wall movement more efficiently than steel props. Its use in this project may be considered the progenitor of the adaptation of precast members in large underground works.

The prefabricated struts (see Fig. 12-50) were placed using special hangers which held the concrete members in position until they were connected to the walls. The hangers were suspended from a row of steel beams used as part of the bracing and set on top of the walls, as shown in Fig. 12-50. The precast units were left as permanent bracing.

Fig. 12-50 Part of the Humber bridge-foundation anchorage showing one of the five cells during excavation. Note the sets of precast-concrete struts. (*ICOS.*)

Ohka River Project, Yokohama The plan for this project is shown in Fig. 12-51a. It consists in the construction of an underground expressway to provide a direct link with the Haneda airfield and an underground railway in twin sections just under the expressway. The plan also calls for the replacement of a vehicular bridge over the Ohka river, which is overlapped by the alignment of the expressway. The new construction barely misses the foundations of an existing railroad truss bridge, whereas the new underground expressway and railway intersect at a severe skew angle.

Since the river is navigable at this point, the new expressway, which comprises the uppermost structure of the new construction, is located so that its roof is well below the dredged level of the river. The expressway approaches to the river are supported by diaphragm walls 1.2 m (about 4 ft) thick. The structural walls for the river portion consist of bored piles, 1.6 m in diameter (64 in). For the installation of the piles steel casings were driven in interlock to final construction level and were left permanently in place. The water was pumped out and excavation below the riverbed was carried out under the protection of slurry. When the construction was completed, the upper part of the casings was cut off. The interlocking of the steel casings served to control the verticality of the installation, and the reduced construction tolerance gave a fairly watertight structure. The watertightness was improved by grouting any leaking joints from within as the excavation proceeded down to final level.

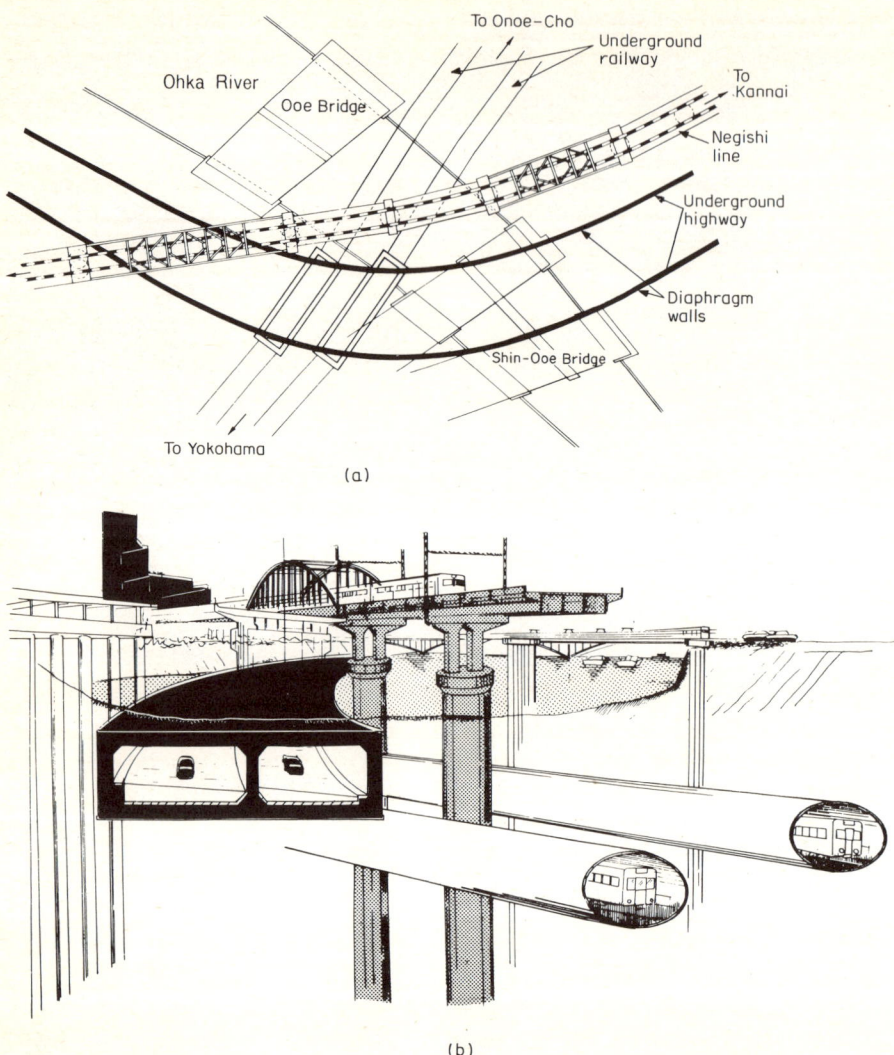

Fig. 12-51 Construction of an underground expressway and railway under the Ohka river in Yokohama: (*a*) general plan showing the existing bridges over the river and the new construction; (*b*) perspective showing the new construction in relation to the existing structures; the twin box section is the configuration of the new expressway, and the twin tubes under it are for the new railway.

The foundations for the existing railroad and vehicular bridges bear on the underlying rock. Although the new construction did not require underpinning of these structures (used in strict terms), the foundations were nonetheless laterally protected by means of heavy bored-pile enclosures which gave them lateral confinement and ability to resist any tendency to deflect horizontally because of the new excavation.

The new twin vehicular bridge is supported on common foundations with the underground expressway. The diaphragm walls are extended to the underlying rock so that their depth at some locations is 150 ft (46 m). The underground railway was constructed using the shield method in twin tubes bored after the construction of the expressway was completed.

Rupel Tunnel, Belgium This project, located between Antwerp and Brussels, carries a major expressway under the Rupel River and the maritime canal. The canal is used by small seagoing cargo ships to reach the Brussels harbor. At the project site the two waterways are nearly side by side, and this is a favorable factor for the feasibility of the project. The tunnel will replace two old drawbridges and will handle an average daily traffic of about 30,000 vehicles.

Figure 12-52 shows an aerial view of the project under construction. The plan is based on the concept of an in situ tunnel in cut-and-cover across the land between the two waterways using diaphragm walls and precast partially prestressed concrete box sections installed as sunken tubes across the two waterways. On the same site diaphragm walls provide protection for the dykes of the canal and along the north bank of the river, where they retain access ramps. Diaphragm walls are also used as foundations for a bridge and for two underground pumping stations built to handle the roadway drainage (Barr, 1977).

The cut-and-cover tunnel is 260 m (850 ft) long. The exterior diaphragm walls are part of the permanent structure. They are 1 m thick (40 in) and 26 m deep (85 ft). During excavation the walls were braced at three levels with cast-in-place heavy concrete ribs, also serving to support the floor slabs at these levels, connected directly to the exterior walls. The ribs are clearly shown at the upper half of Fig. 12-52.

Inside the tunnel the traffic is separated by a center wall, but in case of emergency it is possible to divert all traffic to one tube only through access openings. Although the structure is not strictly a utility tunnel, a special chamber is provided at the center of the section to house pipes, cables, and other public distribution networks.

The plans called for the tunnels under the waterways to be built without significantly interrupting navigation. For the river section it was necessary to prepare a dry dock formed by a steel-sheet-pile cofferdam on the north bank of the river. The dock was used as casting yard to cast and prestress two concrete tube sections, each 99 m long (325 ft). For the canal section the casting yard was created in a dry dock formed by earth banks south of the canal. The crossing at this location consists of a single tube section 138 m long (452 ft).

When the sunken tubes were completed, the contractor removed the steel cofferdam and the earth embankment, flooded the site, and allowed the precast tubes to float out (Fig. 12-52 shows the canal dock already flooded and the tube floating). The tubes were sunk and placed in preexcavated trenches at the bottom of the waterways. The dry docks were resealed and dewatered once again for the casting of tubes for the tunnel-ramp approaches. Figure 12-53 shows typical cross sections for the sunken tube and for the cut-and-cover tunnel.

Fig. 12-52 Aerial view of the Rupel Tunnel in Belgium, (from bottom to top) showing the center portion, consisting of a cut-and-cover tunnel built with diaphragm walls; the river tunnel, consisting of two sunken tubes; and the canal tunnel, consisting of one sunken tube shown floating. (*Franki*.)

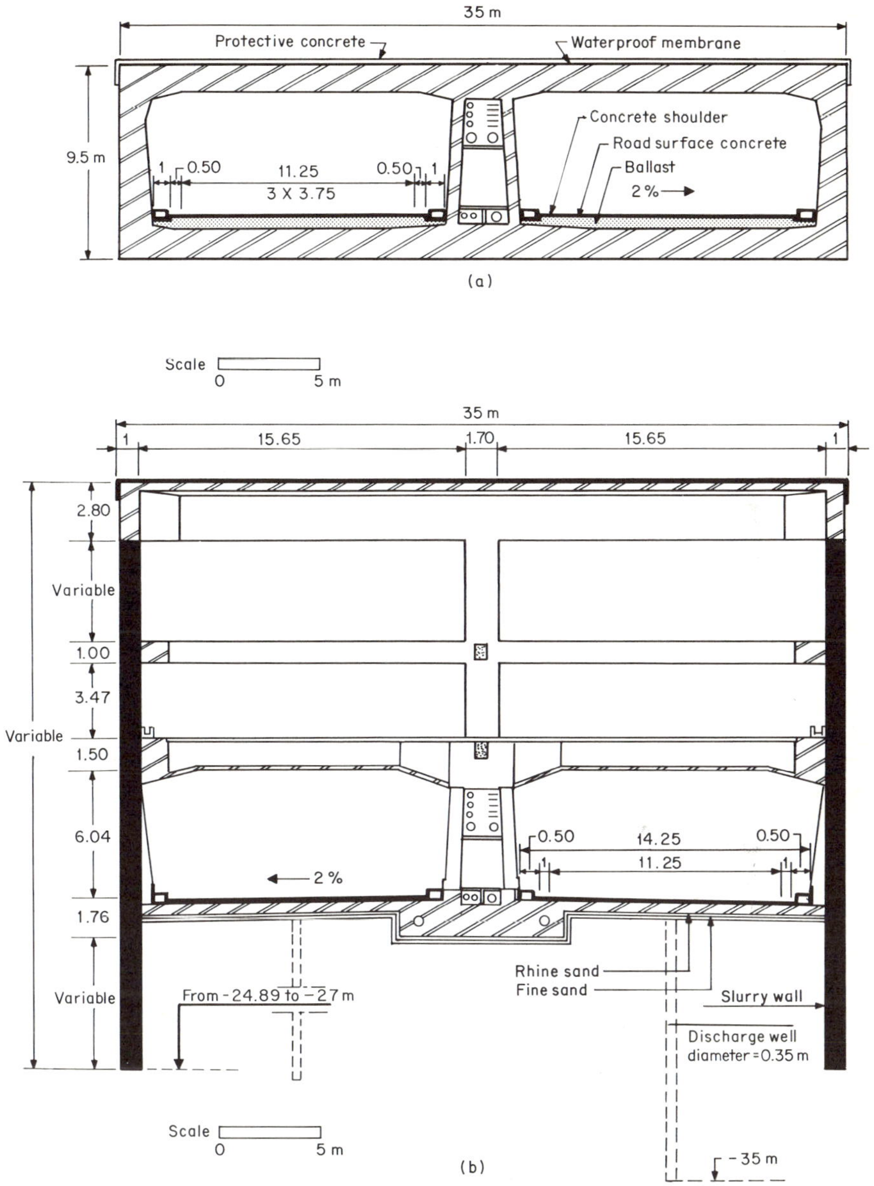

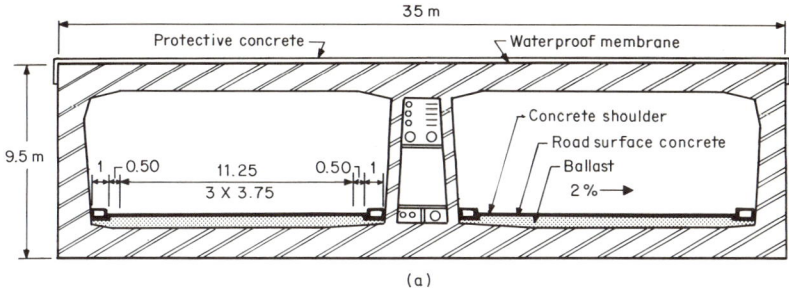

Fig. 12-53 Rupel Tunnel, Belgium: (*a*) typical cross section of the sunken tubes used in the Rupel river and the maritime canal; (*b*) typical section of the cut-and-cover tunnel used in the land portion between the waterways. (*Franki.*)

REFERENCES

Barr, M., 1977: The Rupel tunnel, personal communication.
Cunningham, J., and J. I. Fernandez, 1972: Performance of Two Slurry Wall Systems in Chicago, *Proc. ASCE Conf. Performance Earth Earth-supported Struct., Purdue Univ.*
Eide, O., G. Aas, and T. Josang, 1972: Special Applications of Cast-in-Place Walls for Tunnels in Soft Clay in Oslo, *Proc. 5th Eur. Conf. Soil Mech. Found. Eng., Madrid.*
Fenoux, Y., 1971: "Deep Excavations in Built-up Areas", Soletanche Enterprises, Paris.
Ground Eng., 1972: Bloomsbury Square Underground Car Park, vol. 5, number 4, July.
Ground Eng., 1974: Humber Suspension Bridge, reprint.
Hodgson, T., 1974: Design and Construction of a Diaphragm Wall on Victoria Street, London, *Proc. Diaphragm Walls Anchorages, Inst. Civ. Eng., Lond.*
Jobling, D. J., 1975: Diaphragm Walls and Secant Piles in Subway Construction, *U.S. Dept. Transp., Urban Mass Transp. Adm. Proc. Semin. Underground Constr. Probl., Techniques Solutions.* Chicago.
Leonard, M., 1974: Precast Diaphragm Walls Used for the A13 Motorway, Paris, *Proc. Diaphragm Walls Anchorages, Inst. Civ. Eng., Lond.*
Puller, M. J., 1974: Economics of Basement Construction, *Proc. Diaphragm Walls Anchorages, Inst. Civ. Eng., Lond.*
Schnebel, H., 1971: Sloped Sheeting, *Civ. Eng.*, **41**(2): 48-50.
Thon, J. G., and R. C. Harlan, 1971: Slurry Walls for BART Civic Center Subway Station, *ASCE J. Soil Mech. Found. Div.*, September.
White, R. E., 1974: Anchored Walls Adjacent to Vertical Rock Cuts, *Proc. Diaphragm Walls Anchorages, Inst. Civ. Eng., Lond.*

Chapter Thirteen
TOPICS RELEVANT TO ANALYSIS AND DESIGN

In the analysis and design of underground structures, particularly those involving a deep excavation, the engineer must first ensure the feasibility of the proposed construction method, then predict the probable effects on surroundings, and finally proportion the structure for strength and appropriate stiffness. They key steps of this work usually are carried out according to the following sequence.

1. Predict the pattern, distribution, and magnitude of ground movement due to excavation, and establish the corresponding probable settlement of adjacent buildings and utilities.

2. If limitations must be imposed on the permissible or acceptable movement, determine the constraints on the construction method and sequence and therefore on the design.

3. Estimate the magnitude and distribution of lateral stresses as a function of displacement and strain in various construction stages as well as after construction.

4. Analyze and design the structure and its parts including the bracing. The analysis may be based on semiempirical methods, or it may be checked by numerical techniques.

5. Consider the merits and the feasibility of observing and monitoring the first sections of the project to check the predictions and the accuracy of the analysis and if necessary modify the initial design.

486 Slurry Walls

This chapter deals with these topics, and mainly with how they apply to certain types and classes of semirigid concrete structures built by the slurry-trench method.

13-1 SETTLEMENT DETRIMENTAL TO SURROUNDINGS

The usual approach to the settlement problem is to compile a record of existing buildings, structures, streets, and utilities and supplement it with an up-to-date survey. The next and most important step is to establish a range of permissible differential settlement, but the wide disparity of observed results and views on the subject make general guidelines difficult to lay down.

The settlement caused by open-cut excavation which a building can withstand is less than the settlement which the same structure can undergo without damage under the effect of its own weight. Settlement caused by subway construction may occur rapidly and produce erratic effects. Damage to nearby buildings will depend first on the type, magnitude, and distribution of movement and second on the structure type, foundation, age, and general condition of the building (D'Appolonia, 1971). The first comprehensive summary of settlement that may be damaging to surroundings was provided by Skempton and MacDonald (1956). Useful classifications of allowable and detrimental settlements in various conditions are also given by Sowers (1962) and more recently by Grant et al. (1972).

Differential movement, also expressed as an angular distortion, is of much greater concern than uniform settlement or tilting. The allowable angular distortion or tolerance has been studied theoretically, by tests on large models of structural frames, and by field observations. It appears that a building undergoing an angular distortion greater than $\frac{1}{300}$ will probably suffer some form of damage, although this is not necessarily true in every case. The settlement corresponding to a distortion of this order will generally vary and depend on the soil type, rigidity of the foundation, and distance between exterior columns and walls. Although considerable data have been compiled, the record is still too spotty and limited to allow a simple and reliable correlation of settlement and angular distortion for any type of building and structure.

For a given magnitude of differential settlement a building resting on sand may suffer more damage than one founded on clay. This is generally true for buildings that settle owing to their own weight, the reason for the difference being that settlements in clay occur over a longer period, so that the structure has more time to adjust to the deformation. The same guidelines should not be used for settlements due to excavation. In this case, the settlement may be irregular in pattern, irrespective of the soil type, and may occur over the excavation period. Thus, it is better to assess the associated effects by estimating the actual differential settlement based on a frame analysis, and its distribution. Buildings for special use or with sensitive framing construction should receive an additional study to determine the final relation between differential settlement and forces in the structure.

The tolerance of underground services and utilities to differential movement is even more difficult to quantify. The most susceptible are older sewers and tunnels constructed of brittle materials, i.e., clay or cast-iron pipes, and where joint leakage is caused by very little disturbance. Utilities made of more ductile materials and having flexible and telescopic sealed joints can deform without distress even under the rigorous conditions imposed in areas of mining subsidence. Problems may still be encountered if there is a large relative movement across a structural discontinuity, e.g., the entrance of a pipe to a building or a manhole.

The present trend here and abroad is to allow damage in some instances to occur and repair it afterward. The basis for this approach is that the cost of repairs may be much less than the cost of special construction procedures adopted solely for the purpose of preventing damage, and this is true for buildings of a small value and limited functions. Many large contracts are handled on this basis, and the supervisory authority makes appropriate inclusions in the technical documents for remedial work after the job is completed. Whether this is always justified should not be a question merely of convenience but also a matter of technical and legal competence, since litigations have resulted not only with structural damage but whenever the appearance of a building was altered and showed architectural defects.

13-2 GROUND MOVEMENT AND SETTLEMENT DUE TO EXCAVATION

For almost any type of ground support, the movement and settlement due to excavation depend upon the soil characteristics and the groundwater conditions; the size of the excavation, particularly its depth; the rigidity of the support system; the details, installation sequence, rigidity and preloading or prestressing of the bracing; and the general workmanship of the construction. As a result of the variety, number, and diversity of the factors influencing movement the best practice has been to observe, measure, and analyze ground movement around actual excavations and use this record to supplement theoretical predictions. For all types of support, but particularly with flexible walls, the soil strength appears to have the most decisive effect on movement (Peck, 1969).

Typical patterns of movement are presented in Fig. 13-1. In Fig. 13-1a the movement occurs as lateral yielding of soft soil above stiff soil for a cut supported by a flexible wall. In Fig. 13-1b the lateral yield of soft clay occurs above and below the base of the cut, whereas the underlying soil tends to move upward. Figure 13-1c and d is for the same excavation but for a rigid support. In both cases the movement is associated with (1) lateral displacement of the support above and below the level to which the cut has progressed, as the

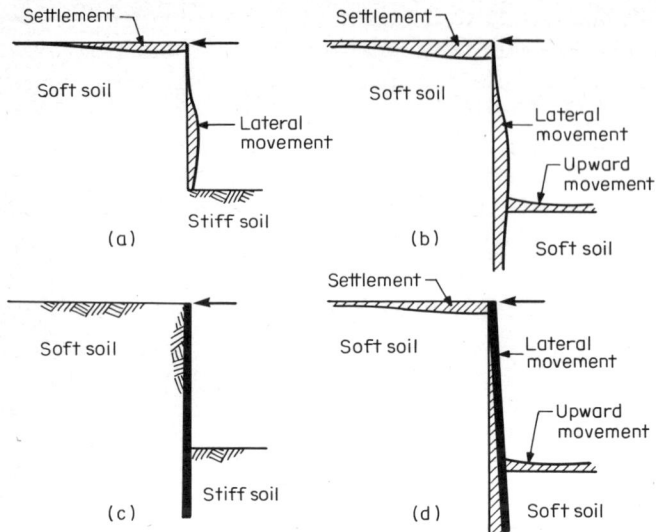

Fig. 13-1 Settlement due to lateral movement: (*a*) and (*b*) flexible wall; (*c*) and (*d*) rigid wall. The top of the wall is restrained against movement and may or may not have intermediate bracing.

excavation proceeds to a lower level and before the next bracing is installed; (2) some yielding of the bracing already installed due to more load; and (3) elastic deformation of the wall between bracing supports.

Movement above Excavation Level The installation of any bracing tier normally is preceded by excavation to an appropriate level. The stress release accompanying this process and the relative freedom of the wall below the last bracing cause it to move. Wall movement above the excavation level reached at any time is controlled mainly by the procedure and execution of bracing, e.g., bracing intervals, time elapsed between earth moving and bracing, degree of preloading, and stiffness of the medium where the bracing forces may eventually be transferred. If these details are ignored, or if they do not receive a comprehensive treatment, lateral wall movement above the excavation can be large and thus contribute significantly to the total movement and settlement outside the excavation.

Movement below Excavation Level Lateral yielding of the support below a given excavation level normally occurs irrespective of the capacity and preloading of the bracing above that level. The lower part of the wall is in an elastic medium and will therefore continue to move as the effect of the adjacent overburden is intensified with the removal of more soil from the cut. Since the wall is not yet braced in the lower part, its tendency to move must be resisted mainly by the strength of the soil in the embedded portion and by the stiffness of the structure. Experience has confirmed that this movement increases in soft or in very compressible soil, with increasing wall flexibility, and with reduced wall embedment. Where the stability number N is greater than 6 for excavations of ordinary dimensions and shape, the conditions of base instability by heave are rapidly approached, leading to more movement.

Further Causes of Movement At best the elastic distortion of ground can occur without volume changes. Besides this idealized situation movement often is the result of changes in the ground volume caused by migration of water in the voids as the ground gradually adjusts to the loading imposed by the completed structures. This problem involves an estimation of the long-term changes in the effective stress.

Vertical movement outside an excavation can also occur irrespective of ground support and bracing. The removal of vertical load from within the excavation indicates the possibility of such movement. Examples are given by Ward (1963) and by Ward and Burland (1972) for excavations in London clay. In these cases the ground outside the excavation settled initially, but long-term net relief of load beneath the excavation caused the ground to swell and give rise to upward movement outside the excavation, which exceeded the initial settlement. The conclusion from these investigations is that the vertical movement outside the excavation caused by changes of vertical load is less than the vertical movement within the excavation.

Control of Settlement Due to Excavation The settlement outside an excavation can be estimated from the wall deflection. The volume of soft plastic clays remains constant, and therefore the volume of settlement is roughly equal to the volume of inward movement adjusted for the volume of heave (Peck, 1969). This leads to the conclusion that, excluding poor construction procedures and workmanship that cause erratic effects and loss of ground, the settlement due to excavation can be controlled only if inward wall movement is controlled. This is true regardless of the soil type and where the settlement volume is not equal to the volume of lateral movement.

The role of water is of prime importance in the installation of ground support. Large erratic settlements have occurred with excavations in water-bearing sands and are associated with a sudden loss of ground due to migration and flow of earth particles into the cut. These settlements depend on accidental features and construction events, and they are neither predictable nor controllable. With continuous diaphragm walls settlements of this type are completely eliminated since the method is insensitive to groundwater. With bored-pile walls this problem may persist if gaps are formed between piles and are not grouted before excavation or if the installation uses lagging between piles.

The foregoing discussion provides a basic understanding of ground movement and settlement. Having explained the causes and the characteristics of movement, we shall discuss the subject again in Sec. 13-11, which presents an analysis of ground movement for excavations supported with diaphragm walls.

For the example of Fig. 13-1 the effect of increased stiffness is significant, particularly in conjunction with a suitable embedment. In Fig. 13-1c the wall stiffness precludes elastic deflection above excavation level. In Fig. 13-1d the presence of soft soil still influences ground movement, but thanks to its greater stiffness the wall will not move as much as does in Fig. 13-1b.

13-3 HEAVE IN NARROW EXCAVATIONS SUPPORTED BY DIAPHRAGM WALLS

In soft plastic clays bottom instability is approached where the weight of soil beside an excavation tends to displace the underlying soil toward the excavation. The clay beneath the base acts primarily as frictionless material under undrained conditions. In the past the problem has been avoided by adapting special construction procedures whereby sheet piles with a suitable embedment are first installed, then the roof is cast, and the excavation is completed under compressed air. Other methods consist in soil freezing or excavation by dredging under water, but very often these methods are impractical or too expensive.

The heave problem is not yet completely understood. It appears that there is a mutual relationship between the transition from elastic to plastic failure within and below the zone of influence of the cut, the influence of the lateral support, including its stiffness and embedment below excavation level, and the presence of a firm stratum, which the wall may or may not reach. Other factors that influence further base instability by heave are the configuration of the cut and the strength of the material acting as surcharge above excavation level.

A rough theoretical model indicating the contribution of ground support to base stability of a cut in cohesive soil is presented in Fig. 13-2. An element outside the wall is considered in the active state, and an element below the

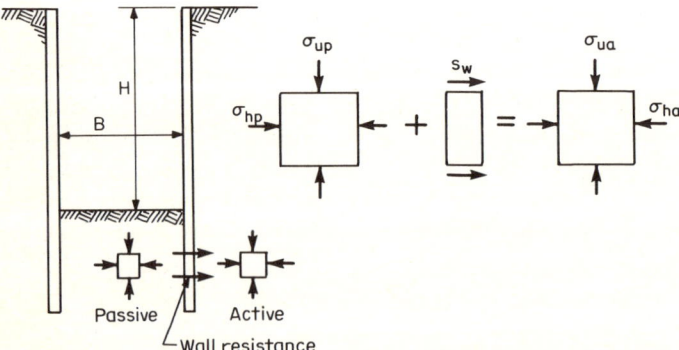

Fig. 13-2 Rough theoretical model for inelastic movement and base instability for cuts in soft clays:

$$\sigma_{ha} - s_w - \sigma_{hp} = 0 \qquad \sigma_{va} - \sigma_{ha} = 2s_u$$

$$\sigma_{hp} - \sigma_{vp} = 2s_u \qquad s_w = \gamma H - 4s_u$$

(From D'Appolonia, 1971.)

base is in the passive state. The difference between the active stress and the passive resistance must be sustained by the strength of the wall in a comparative form; otherwise plastic failure will occur.

At present base instability can be analyzed only approximately. The first step is to consider the deformations of the ground beside and beneath the excavation. During removal of earth, shear stresses are set up beneath the excavation as a result of the surrounding soil pushing the clay inward and up onto the excavation. If the shear strength of the clay in the zone involved is sufficient, the base is stable. If this strength is exceeded, the ground beside the excavation will start sinking vertically, forcing the bottom upward. This stability problem has been discussed by Terzaghi (1943), Terzaghi and Peck (1967), and Tschebotarioff (1973).

If the ground support extends below excavation level, the stability is improved by considering the shear strength at the interface of the embedded portion and assuming that this will be mobilized with the tendency of the clay mass to move upward. If the ground support is a diaphragm wall, assuming that the problem is investigated primarily for soft clay, the shear strength at the interface will be the undrained shear strength of the soil so that the adhesion factor α (see also Chap. 11) is unity. This analysis is reliable for temporary excavations, and no attention is directed to changes in shear strength which may occur with time in permanent cuts. Base stability is further improved if cross-walls like those mentioned in Sec. 12-10 are combined with the longitudinal walls to increase the contact area.

Heave Analysis of Ordinary Strutted Excavations Such an excavation is shown in Fig. 13-3. There is no embedment of the ground support, or where there is embedment its effect is disregarded. The word "strutted" is used to the extent that the supported ground is prevented from moving into the cut. The analysis of stability is based on the consideration of a stability number N_c. Values of N_c for cuts of usual shape and configuration have been established by Bjerrum and Edie (1956). This number depends on the form of the excavation and mainly on the width-to-length and depth-to-width ratio. The values of N_c are in this case the same as those given by Skempton (1951) for bearing-capacity failures.

For the cut of Fig. 13-3 the factor of safety against base failure by heave is expressed by

$$F = N_c \frac{s_u}{\gamma H + q} \qquad (13\text{-}1)$$

where F = factor of safety
H = height (or depth) of excavation
γ = soil density
s_u = undrained shear strength of clay below and around excavation level
q = surcharge load at ground surface
N_c = stability number

492   Slurry Walls

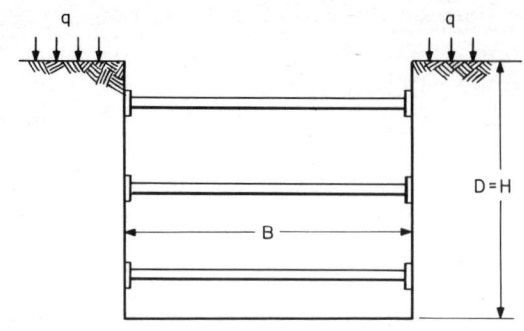

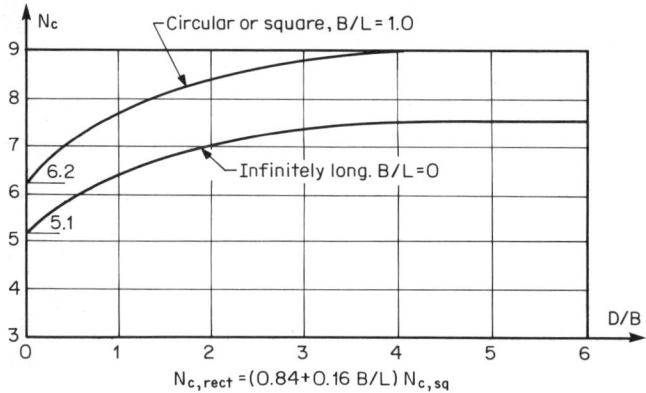

Fig. 13-3 Values of stability number N_c for heave analysis of braced excavations. (*From Bjerrum and Eide, 1956.*)

Values of N_c are estimated with the aid of the diagrams of Fig. 13-3 using $D = H$ and then are entered in Eq. (13-1). The results are not completely accurate but are nonetheless satisfactory for most practical situations and where the cut is temporary.

Heave Analysis with Diaphragm Walls   In this case the analysis includes the shear strength mobilized at the interface of the embedded portion of the walls. It is not known what this effect is in a precise sense, but evidently it becomes more pronounced with increased embedment and where the undrained shear strength increases with depth. For infinitely long excavations ($B/L = 0$), the factor of safety is given by the expression

$$F = \frac{N_c s_u + (2h/B)\alpha s_u}{\gamma H + q} \tag{13-2}$$

where h = wall embedment below excavation level
α = adhesion factor (see Chap 11), taken as 1 for soft clay
B = width of excavation (see also Fig. 13-4)

and the other symbols are as before. N_c is now estimated from Fig. 13-3 using $D = H + h$.

Heave Analysis with Diaphragm Walls and Cross Walls, Single Opening Likewise for excavations supported by diaphragm walls and cross walls, as shown in Fig. 13-4, the stability against bottom heave is expressed as

$$F = \frac{N_c s_u + [2(B+L)h/BL]\alpha s_u}{\gamma H + q} \quad (13\text{-}3)$$

in which D, H, h, B, and L are according to the configurations of Fig. 13-4 and N_c is again computed from Fig. 3-3 considering the ratio D/B.

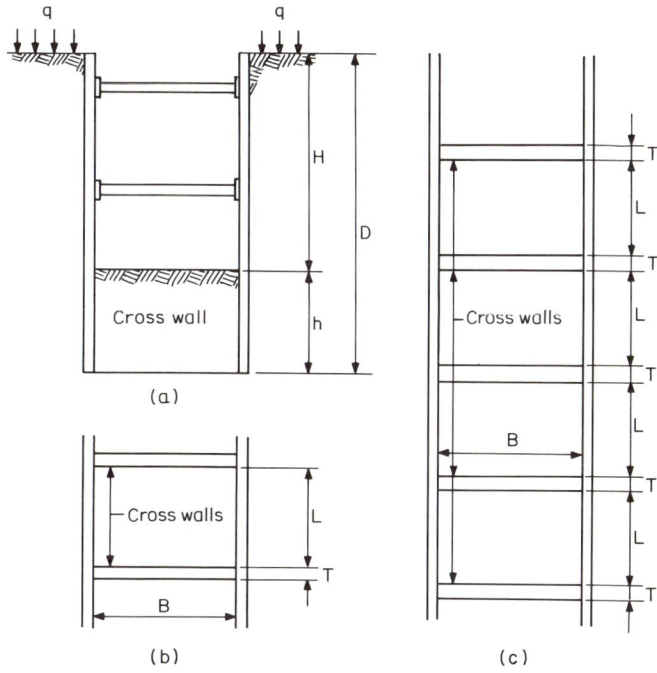

Fig. 13-4 Typical plan and section for analyzing stability against bottom heave: (*a*) section showing longitudinal walls and cross walls; (*b*) plan for a single opening; (*c*) plan for analyzing stability through two or more openings.

Since the bottom heave is analyzed through a single opening, the depth-to-width ratio should be taken as D/L if L is less than B in estimating N_c. If the walls are extended deeper by means of piles, subpiers, or stilts, so that their tip is in incompressible materials, the shear stress at the exterior interface mobilized when the overburden begins to settle as result of the tendency to heave may be included since it will increase the factor of safety. This effect is not included in the present analysis.

Heave through Two or More Openings The possibility of bottom heave must also be considered for two or more openings. For this case a further factor improving stability is the downward resistance on the underside of the transverse walls (Eide et al., 1972), which is assumed evenly distributed over the

494 Slurry Walls

area under consideration. For two openings the stability factor expressing this effect is (Fig. 13-4c)

$$q_1 = \frac{TBNs_u}{B(2L+T)} \tag{13-4}$$

in which N can be taken as 7.5. The factor of safety for two openings is therefore

$$F = \frac{N_c s_u + [4(B+L)h/B(2L+T)]\alpha s_u + q_1}{\gamma H + q} \tag{13-5}$$

Likewise the factor of safety against heave through an infinite number of openings (long excavation) is

$$F = \frac{N_c s_u + [2(B+L)h/B(L+T)]\alpha s_u + q_1}{\gamma H + q} \tag{13-6}$$

in which q_1 is not the same as in Eq. (13-4) but is computed from the expression

$$q_1 = \frac{TBNs_u}{B(L+T)} \tag{13-7}$$

and N in Eq. (13-7) can again be taken as 7.5.

The factor of safety should preferably be 1.5, particularly where the excavation is expected to stay open for a prolonged period. If the undrained shear strength can be estimated with satisfactory accuracy and the excavation is for a short duration, the factor of safety may be reduced to 1.3. The tendency of the base to heave will be aggravated if the cut is not well braced and the walls begin to move excessively. Where the upper clay layers are relatively stiff compared with the clay below excavation level and the cut is relatively shallow, base failure by heave is influenced to a lesser degree by the bracing of the walls.

Example 13-1 The analysis of base failure by heave will be investigated for a narrow excavation to illustrate how the foregoing procedure helps alleviate the serious problem of base instability. Referring to Fig. 13-4, the following are given: $D = 60$ ft, $H = 40$ ft, $B = 30$ ft, $L = 12$ ft, $T = 3$ ft, $q = 100$ lb/ft². The soil is soft clay with $s_u = 600$ lb/ft² and $\gamma = 120$ lb/ft³. The value of the adhesion factor α will be taken as unity.

STABILITY WITHOUT DIAPHRAGM WALL. Evidently $D/B = H/B = 40/30 = 1.33$, and $B/L = 0$, so that $N_c = 6.6$ (from Fig. 13-3). From Eq. (13-1) the factor of safety is

$$F = \frac{6.6(600)}{120(40)+100} = \frac{3960}{4900} = 0.81 \quad \text{not enough}$$

STABILITY WITH DIAPHRAGM WALLS BUT NO CROSS WALLS. Evidently $D/B = 60/30 = 2$, and $B/L = 0$, so that $N_c = 7$ (from Fig. 13-3). When Eq. (13-2) is applied, the factor of safety is

$$F = \frac{7(600) + 2(20)2(600/30)}{4900} = \frac{4200 + 800}{4900} = 1.02 \quad \text{still inadequate}$$

STABILITY WITH CROSS WALLS, SINGLE OPENING. Evidently $L<B$, so that $D/L = 60/12 = 5$ and $N_c = 9$. The factor of safety is now estimated from Eq. (13-3) as

$$F = \frac{9(600) + 2(30 + 12)(20)\,[600/30(12)]}{4900} = \frac{5400 + 2800}{4900} = 1.67 \quad \text{OK}$$

LONG EXCAVATION WITH CROSS WALLS. First the factor q_1 is computed from Eq. (13-7):

$$q_1 = \frac{3(30)(7.5)(600)}{30(12 + 3)} = 900$$

$D/B = 60/30 = 2$ and $B/L = 0$, so that $N_c = 7$ (from Fig. 13-3). Next compute the factor

$$\frac{2(B + L)h}{B(L + T)}\alpha s_u = \frac{2(30 + 12)(20)(600)}{30(12 + 3)} = 2240$$

From Eq. (13-6)

$$F = \frac{4200 + 2240 + 900}{4900} = \frac{7340}{4900} = 1.50 \quad \text{OK}$$

13-4 DIAPHRAGM WALLS IN LIEU OF UNDERPINNING

The settlement effects discussed in Secs. 13-1 and 13-2 are of special concern if an excavation is adjacent to an existing relatively shallow foundation or footing. Whether the diaphragm wall protecting the excavation can also protect the adjacent foundation and therefore serve instead of underpinning will depend on two considerations: (1) the ability of the wall to stop ground movement or reduce it to a magnitude which is not detrimental to the adjacent structures and (2) the feasibility of strengthening the soil beneath the foundation if it should become weaker or if the pressure distribution should be confined to a smaller area because of the adjacent excavation.

In order to understand these changes it is necessary to predict first the lateral wall movement and then estimate the soil-bearing capacity either immediately beneath the foundation or at some other critical depth, taking into account the new geometry of the failure surface. At best these steps can be combined into one investigation, but more often they are carried out as two independent analyses. If, for example, the Terzaghi method is used for the bearing analysis, the problem is as presented in Fig. 13-5. The existing footing is under the action of load P tending to push the wedge abc into the ground underneath, and this will result in the displacement of zones II and III. This tendency is resisted by shear stresses developed along the slip planes ab and ac and by the weight of the soil in these zones.

The adjacent excavation intercepts the right zone III and thus deprives this

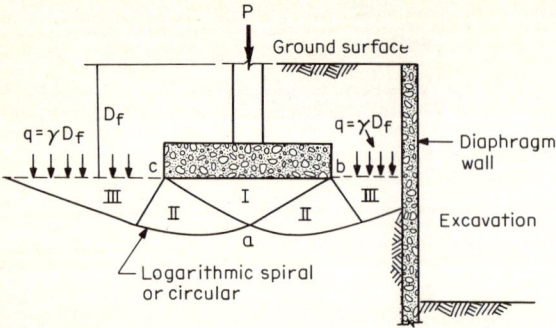

Fig. 13-5 The bearing-capacity solution according to the Terzaghi method, modified to include the effect of excavation.

zone of soil weight. Furthermore the removal of earth may cause long-term changes in the shear strength of the soil with a corresponding effect on the bearing capacity of the foundation. The possibility of these changes must be included in a quantitative form in the bearing-capacity equations. After these adjustments, if the new bearing capacity at the base of the footing is not adequate, ground strengthening will be necessary.

If a weak zone exists at a given distance from the base of the foundation, it may be necessary to investigate the soil pressure at this depth. The simplest method is to use a convenient load distribution, e.g., a 1:2 slope, although this does not completely correspond to reality. If the results must be highly reliable, a *crater* pressure diagram should be introduced (Koegler and Scheidig, 1948). The excavation will intercept the distribution zone as shown in Fig. 13-6, so that the result will be more pressure at depth z unevenly distributed. Thus the soil pressure p is now increased from a value $P/(B + z)(L + z)$ to an average value $P/(B + z)D$. If the latter exceeds the allowable, either for bearing or for settlement, the ground strengthening must be extended to this level.

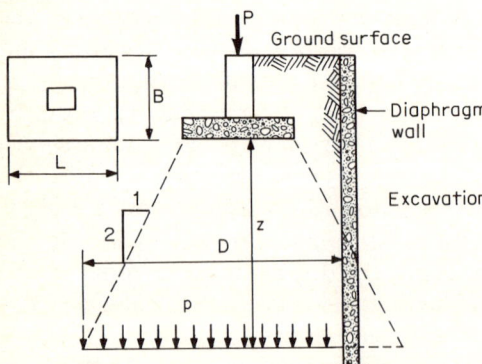

Fig. 13-6 Load distribution and increase in the average soil pressure at depth z because of adjacent excavation.

For a description of bearing-capacity theories and a derivation of the bearing-capacity equations, including the Terzaghi method, see Bowles (1968).

13-5 STABILITY OF CIRCULAR, POLYGONAL, AND ARCHED STRUCTURES

Circular Walls

The analysis and design of circular walls built underground is usually based on the assumption that the finished structure, when acted upon by the exterior loads, will behave essentially as a compression ring. The compression force on a cross-sectional area 1 ft deep is $P = r\sigma_h$, where r is the mean radius and σ_h is the uniform lateral earth stress at the depth under consideration (see also Sec. 10-4). A trial wall thickness is thus selected and checked for its adequacy to resist the maximum compressive stress. For relatively small ratios of wall thickness to wall diameter, wall depth, or panel length, the enclosure is more likely to perform like a thin shell, so that it must be investigated for buckling response and buckling loads. Since the external forces commonly consist of vertical as well as lateral loads, a thin wall must be analyzed for vertical compressive buckling causing deformations in the plastic region, for torsional buckling causing twisting and buckles of a spiral shape, and for flexural buckling in both the horizontal and the vertical directions.

Since the minimum practical wall thickness, regardless of wall diameter, is seldom less than 18 in (45 cm), it follows that buckling might be a problem primarily with large-diameter walls. For the purpose of analysis circular walls are grouped as follows: small diameter, up to 35 ft (10 m); medium diameter, 35 to 130 ft (10 to 40 m); and large diameter, over 130 ft (40 m). The analysis must further take into consideration the actual construction details, especially the compressibility and take-up of the joints.

For well-constructed enclosures and in uniform conditions of loading the above procedure usually leads to a conservative design. In spite of a careful analysis, however, large-diameter walls are often overstressed or become prone to local buckling because of (1) misalignment or deviation from the true circular configuration; (2) base restraint and fixity by rigid connection to a base slab or by embedment in dense material below excavation level; and (3) nonuniform or asymmetrical loading acting laterally upon the exterior face.

Useful information regarding these effects has been obtained from representative model tests. Rigden and Rowe (1974), for example, have investigated the stability of circular diaphragm walls, designed as thin shells in unreinforced concrete, by building 1:80 scale models and carrying out a series of centrifuge tests. The plan of the original structure, shown in Fig. 13-7a, consists of a circular enclosure for an underground car park in Amsterdam. The intermediate floors comprising six parking levels are supported on columns and a central core built independently of the exterior wall, which is thus

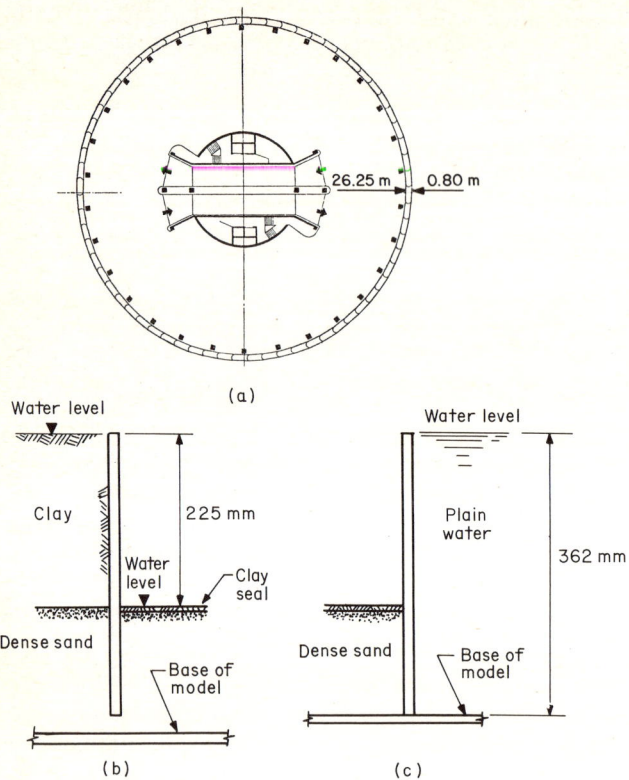

Fig. 13-7 Underground car park in Amsterdam, built with a diaphragm-wall circular enclosure: (a) sectional plan of the actual structure; (b) and (c) sections through model walls used to check the effect of base restraint.

unbraced. The inside diameter is 52.5 m (172 ft), but the diaphragm wall is only 80 cm (32 in) thick, giving a wall-thickness-to-wall-diameter ratio of 0.015. The depth of the excavation from ground surface is about 18 m (59 ft), and the wall is embedded about 11 m (36 ft) below excavation level in gray medium-fine sand.

Five models were constructed and tested to investigate the influence of noncircular construction, base fixity, and nonuniform loading. The conclusions from these tests are discussed in the following paragraphs.

Effect of Base Restraint In one of the two models used to study this effect the significant stress was the one caused by vertical bending; it was noticed at or just above excavation level. This stress exceeded the tensile strength of concrete (6 N/mm²), and as a consequence horizontal cracks appeared at the same level. The model retained an undisturbed sample of organic silty clay above its base and dense sand (corresponding to $\phi' = 40°$) in the embedded part (see part Fig. 13-7b). During the test excess pore pressure occurred at the center of the clay layer and resulted in an increase in the total lateral stress in the same zone. Without this excess pressure the model would have remained undamaged and free of cracking at an overload factor of 1.35.

The second model, shown in Fig. 13-7c, was tested under a full water head of 362 mm on the exterior face and with no support on the inside of the model except the dense sand below excavation level. The centrifuge was accelerated to 135g before the test was terminated. The highest vertical bending stress occurred at maximum acceleration level and was noticed 250 mm from the top, i.e., just below the base of the excavation.

Effect of Noncircular Construction   The model for this test was similar to the one shown in Fig. 13-7b except that the clay was remolded and consolidated under a pressure of 140 kN/m². The noncircular condition was simulated by increasing and decreasing two perpendicular diameters by 5 mm each, so that the new shape resembled a relatively flattened circle. During the test the model was sustaining an external lateral load due to the earth stress and a static groundwater pressure corresponding to the construction sequence of the actual structure. The centrifuge was first accelerated to 80 and 102g and then stopped to allow examination of the wall. No cracks or signs of distress were noticed. The vertical bending stress measured at 80g was again near the dredge level (indicating the effect of wall embedment) and varied between 4 and 5 N/mm². At an overload factor of 1.3 for 102g, the maximum vertical bending stress just exceeded 5 N/mm².

Following the first cycle the test was repeated to 147g without flooding the interior of the model, and during this acceleration some swelling of the clay occurred, indicated by the excess pore pressure recorded. The model was undamaged at 147g (or an overload factor of 1.84), while the maximum vertical bending stress at or near dredge level did not exceed 5.9 N/mm². Evidently this stress was not as high as might have been expected from the results of the first cycle.

Effect of Nonuniform Loading   The model for this case was similar to the previous one but modified with the addition of two surface loads diagramatically opposite and each acting over one-eighth of the model perimeter. When an acceleration of 80g was reached, a load of 2 t/m² was first applied at the surface, followed by 4 t/m² applied directly opposite. The model was finally accelerated to 135g.

The maximum vertical bending stress at 80g before any surface load was applied ranged from 3.8 to 4.5 N/mm² at a depth of 210 mm, or just above dredge level. The first surface load of 2 t/m² had little effect on the stress at or below the dredge level but caused a marked increase of stress in the zone 130 to 170 mm. At 80g and with two surface loads of 2 t/m² acting diagramatically opposite, the maximum stress at 210 mm was increased to 4.75 N/mm². For a nominal tensile strength of 6 N/mm², the factor of safety was thus decreased from 1.33 to 1.26. At 80g and with two loads, one 2 t/m² and the other 4 t/m², the maximum stress was 5.5 N/mm² for a factor of safety of 1.09.

As the model was accelerated over $80g$ and with both surface loads a stress between 5 and 6 N/mm^2 was measured between 85 and $110g$. The model stood at $130g$, whereas at $135g$ it was extensively cracked but still structurally sound.

Conclusions Although these tests do not allow field comparisons, they indicate the influence of certain factors on the stability and performance of the actual wall. If the wall is restrained at excavation level, vertical tensile stresses will develop there and may account for the overstressing of the section. These stresses can practically be eliminated by reducing the depth of penetration and therefore the degree of fixity. Any portion of the wall below the base of the excavation required to provide a cutoff should be formed with plastic and deformable materials.

Deviations from the true circular construction do not appear to have as critical an effect as might be expected, and for the construction tolerance normally allowed this effect may be disregarded. The effect of asymmetrical and nonuniform load can be serious in the midwall area and also in the zone near the base if the wall is restrained there.

Polygonal Walls

Circular structures are actually polygons since they are generally constructed along a series of chords. Those of relatively small diameter appear as true polygons in plan. The division between the two types for analysis and construction is usually arbitrary. A convenient criterion is the deflection angle of two chords meeting at a corner. If this angle is less than 15°, the structure approaches a circle. If it is greater than 15°, the structure should be looked upon as a polygon.

Figure 13-8a shows partial plan for a polygon wall. Each panel consists of two sides cast together and has simple construction joints formed with round tubes. The assumption that the exterior uniform lateral load induces direct (compression) hoop stresses has the merit of simplicity and is usually justified, but often it does not correspond to reality. Although a large majority of polygonal walls have been built without reinforcement and stood well, it is not uncommon for structures of this type to undergo, besides compression, considerable bending in a horizontal plane. This is conceivable with very deep walls, with nonuniform construction loading and surface surcharge, and where the permanence of the installation increases the variety in the type and magnitude of service loads. The analysis in this case can be based on frame action, assuming rigid joints at the corners, or it can consider the principles of a three-hinged arch, depending on the details of panel connections. Each method will introduce transverse shears and moments. The design is thus approximated in a manner which is largely arbitrary and often overconservative but nonetheless acceptable in the absence of a more satisfactory and precise solution. For a usual wall thickness 2 ft (60 cm) or greater the amount of reinforcement is nominal and in the range of 50 to 70 lb per cubic yard of concrete.

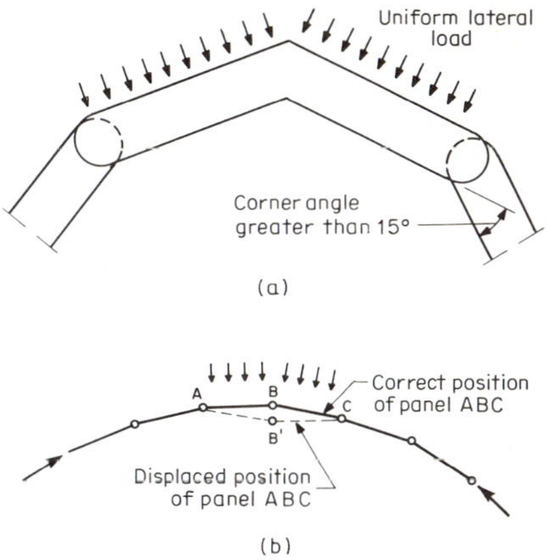

Fig. 13-8 Examples of plan configurations that can influence the distribution of load and the performance of the wall: (a) partial plan of polygonal wall; (b) stability of a circular wall with relatively large diameter.

For enclosures of relatively large diameter the in situ performance and stability may be influenced to a great extent by the manner in which the individual panels are fitted against each other and stay clamped together. Although this is not a theoretical problem, it indicates the possibility of serious distortions and even failure of the wall if one or more panels become loose at the joints and move out of position. In order to understand how this can happen, it is helpful to consider the true geometry of the structure in plan. For example, the wall shown in Fig. 13-8b approaches a true circle very closely, and as such it might be expected to act like a closed ring. Now consider panel ABC isolated from the remaining of the wall, and let the structure to the left of corner A and to the right of corner C be assumed stationary and fixed. A new action will be possible under these conditions, very different from the assumed closed ring. The lateral loads will tend to push the panel inward, and if the take-up of the joints is broken, all that is needed for this panel to become loose and start moving toward the excavation is a minor distortion, as shown by the displaced position $AB'C$. Once this happens, the structure as a whole is on the verge of complete collapse.

The situation just described can be avoided if (1) the panel is adequately supported at the ends A and C; (2) it can resist bending and shear in span AC; and (3) the structure is supported in a manner which precludes a progressive collapse, e.g., by bracing alternate panels.

Arches, Buttressed Walls, and Cells

These structures have been selected for most waterfront facilities, where it is necessary to have a relatively high support and retain a landfill or poor ground. This condition usually precludes the use of tiebacks and ground anchors to brace the walls, and therefore consideration must be given to stiffening the support with tees or by the formation of arches and cells. For such walls the stability concept ranges from stiffened cantilevers to gravity structures (see also Secs. 10-6 and 12-17).

The shape of the section offers the advantage of arranging the wall to span vertically rather than horizontally. The analysis normally must consider the composite moment of inertia of the horizontal cross section since this considerably improves the stability against overturning; consequently the vertical shear along potential failure planes, e.g., construction joints, in the fin section must be resisted to provide restraint against laminar tilting (see also Fig. 10-31). As explained in Sec. 10-6, this requirement is satisfied either by providing shear connectors across the joints or by having the toe of the wall securely keyed into firm soil while the top is framed into a rigid superstructure.

The inclusion of shear resistance (friction or adhesion) at the wall-soil interface or along some other critical plane influences the stability significantly, and so does the inclusion of the weight of soil in zones where this is justified. The principles of load transfer discussed in Chap. 11 are useful as general guidelines as long as the pattern of ground movement and deformation can be predicted, particularly with respect to the movement of the wall, although a great deal of reliance still must be placed on experience and retrospective analysis.

For gravity-type walls consisting of cells and arches the assumption of a rigid structure which cannot yield to mobilize fully the shear resistance of the soil (active state) is in most cases invalid. This has been shown in prototype tests and confirmed by direct field measurements (Fisher, 1974). After the removal of earth from the front of the structure (dredging) the wall will yield probably by as much as $\frac{1}{500}$ times the excavation depth or even more, indicating rotation about the front toe. This forward movement normally will be sufficient to reduce the earth pressure behind the wall to the active state.

13-6 FAILURE CRITERIA FOR UNREINFORCED STEEL-AND-CONCRETE PANELS

The failure of the unreinforced steel-and-concrete panels described in Sec. 10-3 must not be associated with the cracking of plain concrete in bending. The wall continues to resist the stresses developed in beam action after initial cracking is observed on the tension side, and failure by fracture actually occurs when these stresses exceed many times the tensile strength of concrete. Ordinarily the flexural strength is known to be much greater than the strength of concrete in pure tension. Additionally, the relative short span (compared with the depth of the section) over which bending acts appears to have a significant influence on the behavior of concrete beyond the elastic range and thus contributes markedly to increasing the capacity of the wall.

This action has been demonstrated by tests on steel-and-concrete panels conducted by Miyoshi et al. (1976). Such a panel, shown in Fig. 13-9, has a close beam spacing, 1 m or 3.3 ft, giving an effective span-to-thickness ratio of 2 (Fig. 13-9b). The plain concrete has a compressive strength of 250 kg/cm^2

(3600 lb/in²) and a tensile strength of 22 kg/cm² (315 lb/in²). The composite wall was used in the construction of a subway section for the Tokyo Metropolitan Transit Authority.

The panel was subjected to beam action until tension cracks appeared at the center of the span under the loading shown in Fig. 13-9c. When the load intensity was increased further, the wall withstood the bending action satisfactorily. As q approached 400 t/m² (about 82,500 lb/ft²), diagonal cracks developed at the edge of the steel flange on the tension side and then spread across the section to the compression side, causing fracture, as shown in Fig. 12-9d.

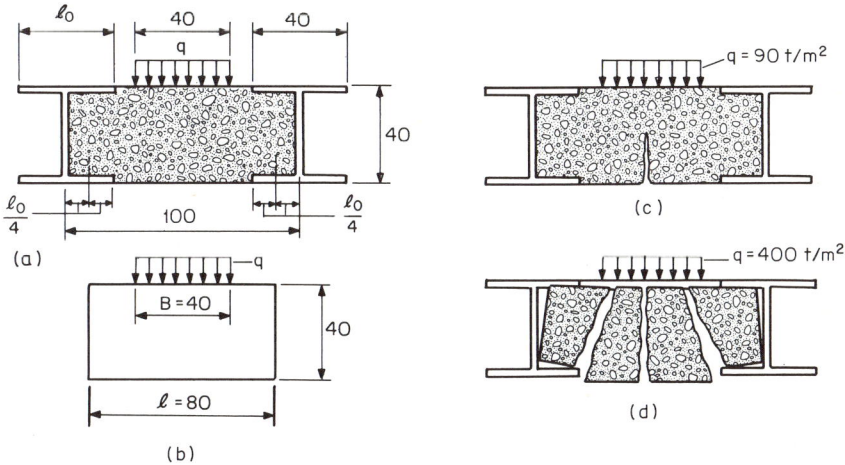

Fig. 13-9 Failure of composite-steel and unreinforced-concrete panels: (a) panel configuration and loading; (b) span length; (c) initial cracking on tension side; (d) fracture by transverse and diagonal cracks. All dimensions in centimeters. (*From Miyoshi et al., 1976.*)

For a beam loaded as shown in Fig. 13-9a and taking the effective span as shown in Fig. 13-10b the maximum bending moment M and shear V are, respectively,

$$M = \left(\frac{Bl}{4} - \frac{B^2}{8}\right)q \quad \text{and} \quad V = \frac{Bq}{2}$$

For $B = 0.4$ m and $l = 0.8$ m, expressing q in tons per square meter, we have

$$M = 0.06q \text{ m} \cdot \text{t} \quad \text{and} \quad V = 0.2q \quad \text{t}$$

so that the maximum stresses in the concrete are

$$\text{Flexural stress} = f = 2.25q \quad \text{t/m}^2 \quad (13\text{-}8)$$
$$\text{Shear stress} = v = 0.75q \quad \text{t/m}^2 \quad (13\text{-}9)$$

The graphs in Fig. 13-10a to c show the flexural stress plotted as a function of the distance from the neutral axis of the beam, for values of q 25, 50, and 75 t/m², respectively. The graph of Fig. 13-10d shows the maximum shear

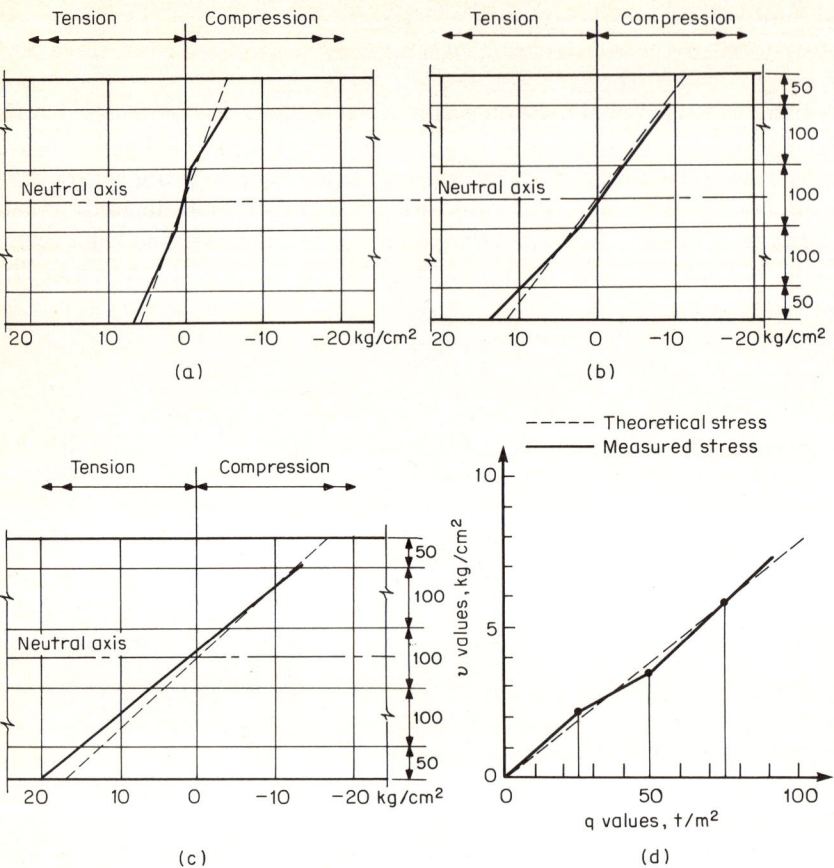

Fig. 13-10 Theoretical and measured stresses for the panel of Fig. 13-9. Flexural stresses for q values of (a) 25, (b) 50, and (c) 75 t/m²; (d) maximum shear stress from Eq. (13-9), (*From Miyoshi et al., 1976.*)

stress v plotted vs. q. The dashed lines represent theoretical stress from Eqs. (13-8) and (13-9), i.e., derived according to the flexural theory, and the solid lines represent actual stresses measured from the test. Within this range of loading the concrete behaves like an elastic material, and the measured stresses are very close to the calculated.

It can be concluded from these tests that the concrete section actually does not fail until the load intensity is many times the load that causes initial cracking, in this case 4.4 times. This has already influenced current design procedures. Several countries abroad have introduced more realistic methods based on the foregoing results, and a similar trend is gaining favor in the United States.

It is interesting to apply these data to an actual excavation. Consider, for example, a cut 20 m deep (66 ft) in soil with density 1.6 t/m³ (100 lb/ft³) and an earth-stress coefficient 0.7. The maximum lateral earth stress at excavation level is 20(1.6)(0.7) = 22.4 t/m² (or 4500 lb/ft²). If this is applied as a continuous uniform load upon the wall, and if the bending action is correlated with Fig. 13-9c, the factor safety is 3.2, which is more than ample. If the strength of concrete is considered beyond the elastic range, so that failure is expressed by the criterion shown in Fig. 13-9d, the factor of safety increases to 14.3. In the author's opinion the former is too conservative, whereas taking the fracture load shown in Fig. 13-9d as the failure criterion and applying a factor of safety to this load will generally be very unsafe. In many instances it will be sufficient to use a working stress somewhat less than the tensile strength of concrete, say 80 to 90 percent, in order to avoid the possibility of tension cracks, which may cause appearance problems or make the wall prone to seepage.

13-7 MISCELLANEOUS PROBLEMS RELATED TO CONSTRUCTION CONDITIONS

Theoretical predictions are valid if they are based on a complete and concise understanding of the actual construction conditions. Unless this criterion is fulfilled, the results of the analysis are likely to be misleading. This can be demonstrated by the three examples cited in this section dealing with (1) a case of unanticipated ground movement; (2) the presence and action of loads differing from the design loads; and (3) the possibility of structural response at variance with the original analysis.

Bored Piles with Lagging Walls of this type are shown in Fig. 10-11, and ordinarily they are used where the ground is competent to stand temporarily. The arrangement shown in Fig. 10-11a requires less soil trimming at the general excavation face, but fitting the lagging into the preformed grooves can be a serious difficulty where the piles are out of plumb. The arrangement shown in Fig. 10-11b is more consistent with a cast-in-place concrete panel between piles and along their centerline but requires more soil trimming to insert the temporary lagging. Bored piles are often chosen because of their greater rigidity in situations requiring close control of ground movement.

The problem, nevertheless, is very similar to the problem of conventional soldier piles with lagging (Peck, 1969). In both arrangements of Fig. 10-11a and b the soil outside the initial excavation line tends to yield toward the open space created between the piles as the soil there is trimmed off to make room for the lagging. While this is happening, the lateral earth stress is transferred directly by the soil mass to the piles at the contact zone, precisely as for a stiff face supported in a similar manner, so that very little pressure is left to be taken by the lagging. However, in Fig. 10-11b the heavily loaded soil around and behind the piles is cut away to make room for the lagging, the spacers, and the like. Upon removal of the highly stressed soil from the support region, a new mass of soil moves vigorously toward the piles with a corresponding loss of ground.

Permanent Walls in Deep Basements Ahead of Main Construction The fallacy of inserting the ground support before completing the details of the main structure, sometimes done to utilize seasonal advantages, will be illustrated with the aid of Fig. 13-11. In this case the exterior diaphragm wall is inserted before the main construction and irrespective of the interior foundation and base-slab details. The construction is assumed to be in soft plastic clay, and therefore the basement must be protected against heave and hydrostatic uplift.

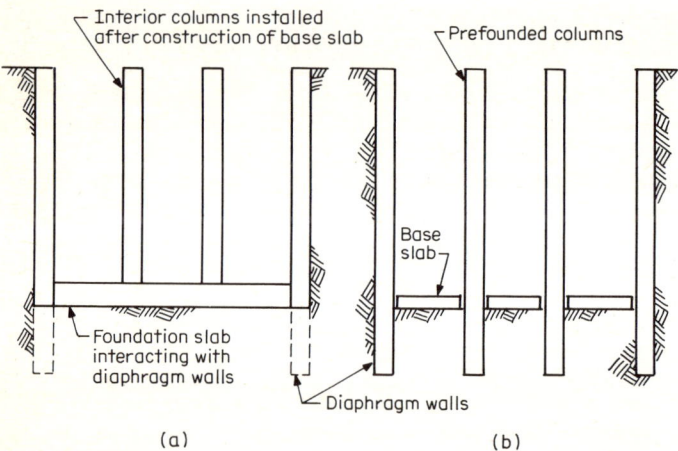

Fig. 13-11 Alternate designs for a deep basement: (*a*) continuous ground-bearing slab supporting interior columns; (*b*) prefounded columns and separate base slabs.

The solution shown in Fig. 13-11*a* consists in the use of a foundation-bearing slab, or mat, which also receives loads from the interior columns; it is essentially a flotation scheme. The basement slab is treated as one large footing, so that the maximum bearing capacity of the soil beneath the slab is influenced by the amount of the overburden removed. The slab must resist full heave and uplift pressure, and therefore it will develop vertical upward forces against the walls. The latter will induce loads in the slab on account of the lateral earth stresses.

In Fig. 13-11*b* since the prefounded columns lie within a large excavation area, no allowance is made for relief of overburden in estimating the bearing capacity of the soil. The use of a slab which is separated from the columns reduces the relative heave movement considerably, so that heave pressures can now be taken up by thin slab sections. However, because this design cannot deal directly with water underseepage, it will be necessary to use filter beds or rely on a deeper exterior wall to reduce the amount of seepage.

The complete design of each scheme is therefore seen to depend on a number of factors, including (1) the wall embedment irrespective of the conditions of the initial excavation; (2) the type, magnitude, and source of loads to be resisted by the final structure; and (3) the interaction of the various members with the wall.

Composite Sections The composite box section shown in Fig. 12-22b becomes an integral structure by attaching the interior box, usually cast monolithically, to the exterior diaphragm walls by means of suitable shear devices. At present there are no standard types of shear connectors, and these details are often left to the specialty contractors together with the method of installation.

The principal advantage of composite construction is the ability to combine the requirements of the temporary and permanent conditions. Initially the walls serve to handle stresses and conditions during excavation, taking into account an appropriate (temporary) bracing sequence, and then they are thickened as needed at the box section for the actual (permanent) unbraced lengths and the actual postexcavation loads. The concrete in the wall is thus stressed twice, first during excavation and then as an integral part of the composite section. The loads acting on the walls alone are the ones applied before insertion of the box, and the loads acting on the composite section are the ones applied thereafter, e.g., the weight of the fill above plus any loads representing long-term changes and adjustments in the soil stress.

Figure 13-12 demonstrates the effect of construction details on the distribution of loads within the finished structural frame. In Fig. 13-12a only the slabs are rigidly connected to the exterior diaphragm walls, but in Fig. 13-12b the walls of the composite sections are tied together with shear connectors over the entire height. In spite of the unique effects of each scheme, there have been instances where, surprisingly, the difference in the two details was not identified, and neither method was shown on the plans. Yet the two walls in Fig. 13-12b can in reality be anticipated to act as one unit having the stiffness of the whole section. In Fig. 13-12a the distribution of loads is more likely to depend on the two walls acting as separate members meeting at a rigid joint, so that the analysis must take into account the individual stiffness of members AB and CD. If this is true, both the magnitude of stress and the stress pattern in Fig. 13-12a will differ materially from the ones in Fig. 13-12b.

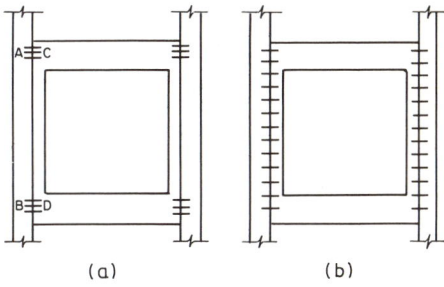

Fig. 13-12 Composite box section; in (a) the section becomes integral by connecting the top and bottom slabs to the wall; in (b) the shear connectors extend for the entire height of the walls.

13-8 POSTTENSIONED DIAPHRAGM WALLS

The construction fundamentals of posttensioned cast-in-place diaphragm walls were discussed in Sec. 10-5. This section deals with certain analytical and design concepts.

The basic principle of posttensioning a wall which is embedded in the ground is to prestress the concrete before the application of the exterior lateral loads, so that any tension induced by them will be wholly or partly offset by the compression of the prestressing force. This principle was first applied to diaphragm walls in 1969, and since then more than 60,000 m² (650,000 ft²) of posttensioned walls have been built, mainly in Italy, Switzerland, and England. The United States had its first posttensioned wall in Chicago (Xanthakos, 1977).

Prestressed precast diaphragm walls have been introduced abroad on a trial basis, and so far they have not received general acceptability. It appears that the posttensioning of precast walls in situ may offer certain advantages over prestressing, and mainly the ability to utilize the stiffness of the surrounding soil to counteract the initial effect of the prestressing force. Posttensioning tendons could be placed in the form before the section is cast, but some degree of prestressing or conventional reinforcement should also be provided to resist handling stresses. A number of good references for prestressed precast wall units, e.g., Lin (1955), PCI (1972), and Martin et al. (1977), provide a detailed treatment of the subject. This section considers only posttensioned walls.

Posttensioning Assembly

An arrangement of prestressing strands in posttensioning units was shown in Fig. 10-24. The reinforcing cage serves primarily as an attachment for holding the casing ducts that contain the strands and resists handling stresses. Once in place, the cage no longer has a static function except that of preventing the distortion and displacement of the ducts under the effect of differential pressure from the rising fresh concrete.

Although the cage serves temporarily, it must be constructed with a fine tolerance. Gysi et al. (1975) have proposed the standard assembly details shown in Fig. 13-13. The cage is spot-welded for increased stiffness, but it may still be deformable because of its small weight and (usually) large size. The position of the ducts should therefore be checked while the cage is held vertically and before it is inserted into the slurry trench (see Fig. 13-14).

The most commonly used method for posttensioning is jacking. In Great Britain, for example, contractors have been able to fix and firmly hold the BBRV units within the cage. Jacks are then used to pull the steel against the hardened concrete. Problems, however, can arise with the anchoring of the heavily stressed strands in the base of the wall. Although anchor blocks and similar devices operating on the principle of direct bearing are dependable

Fig. 13-13 Method of detailing and assembling a reinforcing cage for posttensioned cast-in-place diaphragm walls: (*a*) upper perimeter bar; (*b*) cable ducts; (*c*) cable brackets with vertical reinforcement; (*d*) lower perimeter bar with vertical reinforcement; and (*e*) cage in the assembled position. (*From Gysi et al., 1975.*)

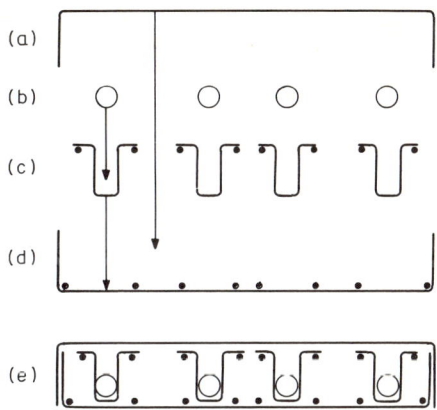

Fig. 13-14 Reinforcing cage held in vertical position to be checked for exact location of cable ducts and possible distortion. (*ICOS, Great Britain.*)

systems in conventional posttensioning, their use in this case has a serious shortcoming because of the possibility of yielding if contaminated concrete and bentonite mud accumulate on the upper side of the block and remain there as residual materials. These devices have thus been avoided, and instead the method of looping the bars in a U shape around the concrete has been found more satisfactory. The stress concentration resulting from the small loop radius is resisted by special slot-tension bars placed in the loop area.

Special Concepts of Prestressing

Elastic Shortening of the Wall Besides the dependability of the prestressing method and the details of the end anchorage, the effectiveness of the application requires that the prestress will not dissipate and will not otherwise be absorbed. The prestressing force normally will cause the wall to shorten. Since the wall is confined in the ground, one might suspect that the soil in contact will offer an impediment to the elastic shortening and retard or even stop the process altogether. Comparison of the modulus of an average soil with the usual modulus of concrete, however, shows that the latter is from 150 to 300 times as stiff; hence the only impediment will be provided by shear resistance at the face (friction or adhesion) acting opposite to the direction of the prestress.

The existence of frictional forces along the face of a wall being prestressed is very uncertain and at best a highly indeterminate problem, so that any attempt to introduce a precise analysis may fail even with the best numerical techniques. For routine jobs in average conditions the response of the wall to prestressing can be inferred merely by estimating the probable elastic shortening. For example, a wall 60 ft deep and 24 in wide acted upon by a prestressing force of 200,000 lb per foot of length will probably undergo an elastic shortening of 0.15 in (0.4 cm), based on a concrete modulus $E_c = 3 \times 10^6$ lb/in². The corresponding relative displacement anywhere in the wall will be insufficient to mobilize shear stresses in any type of soil, and therefore the entire prestress will be transferred to the concrete.

With the excavation in progress on one side, the relative wall movement with respect to the soil will probably cause friction along the face of contact. If the ground moves toward the active state, it will result in a downward movement of the sliding wedge, as shown in Fig. 13-15, producing a similar downward friction along the back face to supplement the initial prestress. Posttensioned walls usually support ground that stretches in the foregoing manner. In certain cases, however, the outward stretching will be combined with inward compression, causing friction which is either positive or negative, so that its sign will now have to be determined from a consideration of the motion expected in each case. The analysis is again an indeterminate problem and can only be approximated by applying an appropriate reduction factor to the initial prestress to compensate for these effects.

Effect of Soil Stiffness The arrangement and position of strands and the amount of prestress in the concrete must counteract the stresses resulting from the external loads acting upon the wall at the end of excavation. Little or no attention is paid to the boundary conditions of partial excavation or with the wall fully embedded in the ground. If the tendons are laid out to supply the most desirable and most advantageous system of prestress to the concrete for the final loads, the result can be a considerable initial tensile stress, which is normally true for unrestrained posttensioned elements.

A posttensioned wall embedded in the ground, however, has the serious

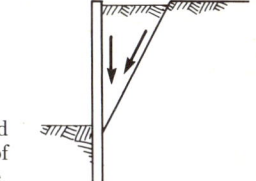

Fig. 13-15 Downward friction along the back of wall caused by active state.

advantage of confinement in a restraining medium (Xanthakos, 1974). The consensus of opinion based on field observations suggests a very different and more favorable response of the wall and the surrounding soil. Tests on the posttensioned walls of the Viale Monte Grappa project in Milan (Braun, 1972) have shown that the introduction of prestress could not cause tension in the concrete, even if the force was applied outside the middle third and the strands were overstressed. These tests lasted for 4 months, allowing more than 100 readings of stress measurements to be taken. The conclusion was that the surrounding soil acted as a stressing bed of infinite strength.

The absence of full flexural deformation in the embedded wall during the eccentric application of the jacking force indicates a considerable intensity of soil response, so that passive resistance is partly developed without the movement predicted in soil mechanics taking place. An important point to be considered is the fluidity and flowability of the high-slump tremie concrete, causing a net gain in the initial stress at rest in most soils, and this can partly account for the observed high-elastic soil response. This clamping effect must be expected in most soils, except for very soft or very loose materials such as turf and organic normally consolidated sea sediments. Better results, however, will be achieved if the posttensioned wall is in relatively stiff or dense formations.

Using a theoretical approach which was later confirmed by large-scale tests and in situ measurements, Gysi et al. (1975) derived the solution shown in a graphical form in Fig. 13-16. The diaphragm wall is embedded in silty moraine having an elastic modulus $E_s = 500$ kg/cm² (about 6900 lb/in²). The concrete modulus is $E_c = 200,000$ kg/cm² (2.75×10^6 lb/in²); that is, the concrete is 400 times as stiff as the soil. The wall is 20 m (66 ft) deep and 80 cm (32 in) thick.

Figure 13-16a shows the arrangement of prestress. Evidently the axis of the tendons is a parabolic curve below 6 m and reaches a maximum eccentricity of 30 cm, well outside the middle third. Figure 13-16b shows the soil pressure against the wall mobilized upon the application of the jacking force, estimated from a mathematical consideration of the soil response. Figure 13-16c shows the vertical bending moments due to posttensioning, the moments due to the earth pressure, and the resultant bending moments. Figure 13-16d shows the stress in the concrete section due to the resultant bending moments at various depths. In spite of the large eccentricity the concrete is always in compression.

Without the restraining effect of the soil the wall would have been stressed in tension in the zone of large eccentricities. For a maximum prestress mo-

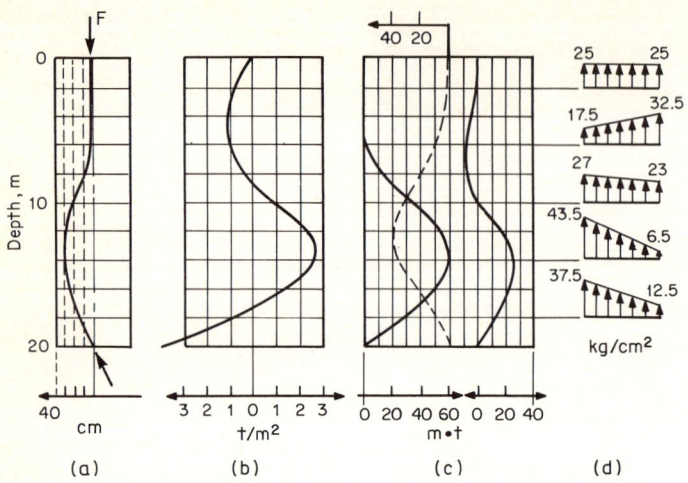

Fig. 13-16 Prestress, earth pressure, bending moments, and stress in a posttensioned diaphragm wall embedded in silty moraine: (*a*) diagram showing the eccentricity of prestress; (*b*) earth pressure caused by the prestress; (*c*) bending moments caused by the prestress, the earth pressure, and resultant bending moments; (*d*) compressive stresses in the wall at various depths (*From Gysi et al., 1975.*)

ment of 60 mt corresponding to the largest eccentricity, the tensile stress would have been 11 kg/cm² (155 lb/in²).

Loss of Prestress Owing to the scarcity of observed or published data, no comprehensive procedure can be established for estimating the loss of prestress. The author accordingly suggests the following evaluation.

Since the force in the posttensioning system is usually measured after the elastic shortening of the concrete has taken place, it is not necessary to make an allowance for a loss of prestress due to this factor. Likewise the loss of prestress due to the local yielding of weak or contaminated concrete in the loop zone will occur as soon as the jacking force is introduced, and therefore it does not represent a final loss. Although the effect of elastic shortening and local yielding can be counterbalanced, the loss due to creep cannot be easily compensated for. Creep is a time-dependent deformation resulting from the presence of stress, so that both the steel and the concrete will continue to flow under no increase in stress. Creep will be more serious if the prestress in the steel is low and the compression in the concrete is high, but fortunately the latter condition can be avoided with a relatively large cross-sectional area of the wall. On the other hand, it is not desirable to overstress the wire strands because this will enhance the possibility of further local yielding of the concrete in weak zones. Assuming an average prestress in the concrete of 1000 lb/in² (about 70 kg/cm²), the loss of prestress due to creep may be from 6 to 7 percent.

The amount of prestress loss due to shrinkage varies first with the proximity of the concrete to water during hardening and then with the time of application of prestress. The favorable curing conditions and the moist environment keep the initial shrinkage low, but this is not of great importance since the posttensioning is applied much later. The concrete, however, is prestressed and then subjected to a dry atmosphere as excavation proceeds; hence further shrinkage should be expected with a corresponding loss of prestress.

Although it is difficult to generalize the amount of loss of prestress, it is conceivable that for the usual curing conditions, type of concrete and steel, amount and time of application of prestress and methods of posttensioning, the loss will be less than 15 percent. Hence the recommended effective prestress is 85 percent of the initial prestress.

Walls Suitable for Posttensioning

The following categories of walls are suitable for prestress: (1) vertical cantilevers fully restrained at the base (by sufficient embedment) and (2) walls restrained at the bottom (either by sufficient embedment or by unyielding support) and braced at or near the top. These types are shown in Fig. 13-17 with the corresponding lateral-earth-stress diagram.

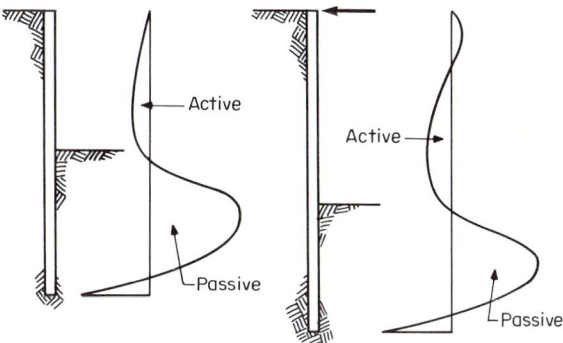

Fig. 13-17 Walls suitable for posttensioning and patterns of lateral-earth-stress distribution.

The walls in the first group depend on an adequate embedment into the ground below the excavation line, not only because their stability is provided by the developed passive resistance in front of the support but also because the lateral deflection will be relatively large. These walls must therefore be used primarily in stiff or dense soil. The bracing at the top for the walls of the second group helps not only to reduce the bending moments but also to decrease the lateral wall deflection and embedment.

The pattern of the lateral-earth-stress distribution developed by the application of prestress may be assumed to be essentially similar to the same pattern with the excavation completed. This pattern is presented in Fig. 13-17 next to the walls, and with the bracing at the top it takes into account the preloading of the struts. Only the magnitude and at some point the sign of the stress in the stress diagram will change as the excavation progresses, gradually leading to the final active-passive earth-stress distribution. The two conditions for which the wall must be analyzed are the initial phase of prestress with the wall fully embedded and the final stage with the full excavation completed. Intermediate phases with partial excavation will seldom govern and need not be considered.

This important conformity at the various stages of loading is no longer possible with the wall braced at several levels. At each excavation stage the newly created combination of spans, span lengths, and number of supports will be in some conflict with the preceding or with the following stage insofar as the eccentricity of the tendons and the amount of prestress are concerned, and this will result in intricate complications. The prestress and its composite action with the concrete might have to be applied in stages or even delayed until the final state. In other words, diaphragm walls are suitable for eccentric prestressing only where the pattern of elastic deformation stays uniform from the initial (no-excavation) to the final (full-excavation) condition.

The concept of concentric tendons for multibraced or multianchored walls, mentioned briefly in Sec. 10-5, deserves attention, but there are some doubts regarding the economic justification of this solution compared with conventionally reinforced walls. The concentric application of the prestress is intended to alleviate the complications mentioned in the preceding paragraph by introducing the simple action shown diagramatically in Fig. 13-18. The prestress force is chosen to counteract the tensile stress due to a maximum moment acting any time and anywhere in the wall. The possibilities and the limitations of this concept are demonstrated by the following example.

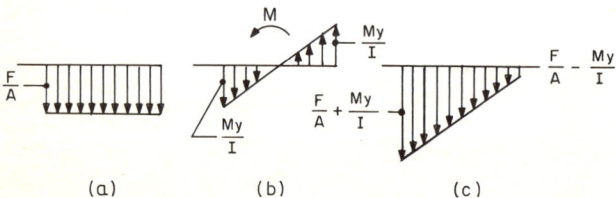

(a)   (b)   (c)

Fig. 13-18 Stress distribution across a concentrically posttensioned wall: (a) due to prestress F; (b) due to external moment M; (c) due to F and M.

Consider the wall shown in Fig. 13-19 for an excavation braced at five levels. Assuming that the wall rotates around its top as the excavation is taken down, the apparent pressures may be as shown in Fig. 13-19b. The maximum moment occurs at full excavation for an effective earth stress and full hydrostatic pressure as shown. According to these diagrams the maximum bending moment is about 40 ft·kips (about 5.5 m·t). The cross-sectional area and the moment of inertia per foot of length are 216 in² and 5832 in⁴, respectively, so that the tensile stress due to bending is $My/I = 740$ lb/in² (52 kg/cm²). The

Topics Relevant to Analysis and Design 515

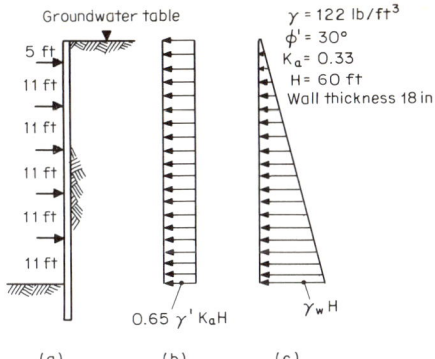

Fig. 13-19 A concentrically posttensioned diaphragm wall braced at five levels: (a) wall section showing the bracing intervals; (b) lateral-earth-stress diagram; and (c) water-pressure diagram.

required prestress force per foot of wall is F so that $0.85F = 0.74(216)$, or $F = 190{,}000$ lb (86 t). The total compressive stress in the concrete will therefore be $f_c = 740 + 740 = 1480$ lb/in² (104 kg/cm²). For $f'_c = 4000$ lb/in² (about 280 kg/cm²) the allowable concrete stress (see Sec. 8-5) is $0.38(4000) = 1500$ lb/in² (105 kg/cm²); hence the analysis is just right. The concrete will be overstressed in tension or in compression, however, if the span lengths become larger or if the excavation becomes deeper in the same soil unless a thicker wall is chosen.

If the same wall of Fig. 13-19 is a load-bearing wall, the analysis should not be materially different. Since the maximum prestress will probably not be needed until the excavation is completed, permanent loads, e.g., the weight of the wall or external dead loads, can be considered part of the total prestress, whereas temporary loads can be accounted for by calculating the additional compressive stress.

13-9 LOADS ACTING ON GROUND SUPPORT

Lateral Loads

A ground support must resist (1) lateral earth stresses, which are generally dependent on the magnitude of strains developed in the ground; (2) lateral pressures caused by surcharge loads acting at the ground surface; (3) lateral stresses induced by concentrated loads, e.g., footings, acting within a mass of soil; and (4) pore-water pressure.

Lateral Earth Stresses These are of great importance because they persist throughout the service of the ground support and because they may undergo changes in magnitude and distribution with time. Lateral earth stresses for rigid supports are discussed in some length in the last sections of this chapter.

The lower limit, or active state, is reached when the wall moves far enough away from the retained ground, mobilizing the full shear strength on the failure surface. The soil thus supports itself to the limit of its strength. The coefficient K_a is the ratio between lateral and vertical (total or effective) stress at this stage. Further wall movement may decrease the soil strength as result of larger strains and thus increase the lateral earth stress. This phenomenon is known as *progressive failure*.

The upper limit, or passive state, is reached when the wall is forced to move against the retained ground, mobilizing the shear strength on the failure surface but in the opposite direction. The coefficient K_p is the ratio of the lateral to the vertical stress at this stage.

Of special interest is the earth stress at rest existing at the natural state and the coefficient K_0, which defines the ratio of the lateral to the vertical stress at rest. K_0 may vary with depth and probably varies with direction. Appreciable changes in the at-rest stress can occur on account of processes like deposition and overconsolidation which change the vertical stress and changes in the stress environment due to excavation or the replacement of earth by other materials.

Water Pressure Diaphragm walls are commonly subjected to full hydrostatic pressure. The groundwater table is not altered, and the analysis must provide for any fluctuations. Accurately estimating the high water table and the source of water is therefore very important since the analysis is based on effective stresses. This is recommended even when dealing with the stability of temporary stages of the excavation since it is possible that in spite of an anticipated short duration of loading delays can occur beyond that period. The condition of effective stress and full pore-water pressure is commonly more critical than an analysis based on undrained strength.

Effect of Surcharge Loads at Ground Surface For uniform surcharge loads acting at the ground surface the resulting lateral stress usually is obtained by applying an appropriate coefficient K_a or K_0, so that the surcharge is considered part of the effective lateral earth stress at a particular state. In the absence of more precise methods the effect of concentrated loads acting at the surface is analyzed according to elastic theory (see also Sec. 2-9), which is combined with empirical data when they are available and modified as necessary to account for the increased stiffness of rigid walls (Terzaghi, 1954; Spangler, 1951). Other investigators have considered the soil elastic but introduced a modulus of elasticity increasing linearly with depth (Turabi and Balla, 1968). Although comparison of the measured stresses from the few field cases available with stresses calculated from elastic theory indicates a surprisingly good agreement, engineers are cautioned to expect possible errors of the order of ±30 percent or higher.

An example of lateral load distribution due to an external load at the ground surface, presented in Fig. 13-20, presumably represents the effect of traffic and construction equipment. It is based on an assumed surcharge of 600 lb/ft² acting alongside the trench and represents mainly a strip load. The ground where this load is distributed consists of an upper layer of fill and loose silty fine sand underlain by firm sand. Evidently, the lateral distribution is

Topics Relevant to Analysis and Design 517

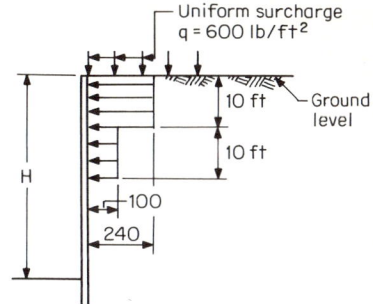

Fig. 13-20 Lateral load distribution due to external load at ground level. Gradients shown are pounds per square foot per foot of depth. The diagram is applied to single- and multiple-braced excavation.

independent of the excavation height and affects primarily the upper portion of the wall and its bracing.

Loads within a Soil Mass A simplified distribution for such a load is shown in Fig. 13-21 and represents the effect of a foundation mat or a continuous basement slab in close proximity to the face of the excavation and in ground conditions similar to those in Fig. 13-20. The essential points are that (1) the total load of the foundation is reduced by the weight of the overburden; (2) the lateral load is ignored above the level of load application; and (3) the effect disappears when the edge of the loaded area exceeds a certain distance from the wall.

For a load within a soil mass the elastic analysis is complicated by the extension of the elastic soil medium above the plane of load application, so that it is more convenient to use a simple load distribution. Figure 13-22 shows two such methods for converting a footing (strip) load into a lateral stress. In Fig. 13-22a beneath the foundation the load is distributed first vertically according to an angle of inclination α and then horizontally, as shown. The coefficient K is the ratio of the horizontal to the vertical earth stress. In Fig. 13-22b the angle of inclination is considered together with the failure wedge, and only the portion of the load within this wedge is assumed to cause lateral stresses. The resultant of the lateral stress is then estimated from this load by applying the coefficient K. The two methods have similarities as well as differences.

For external loads acting either at or below ground surface, the lateral distribution is thus based on both elastic and limit theory, and very often there is an awkward crossover and shifting from one method to the other as the analysis considers more and different types of load. An approximate method which comes closer to reality than either of these two procedures is to establish a conformity between soil deformation and wall movement by iteration. This has been attempted by Haliburton (1968), assuming a discontinuous elastic-plastic deformation as an approximation for the soil, for a multiple-strutted wall. This work is in the right direction, but it still does not provide a satisfactory conformity to reality because (1) the total wall movement depends not only on the position of struts but also on the excavation

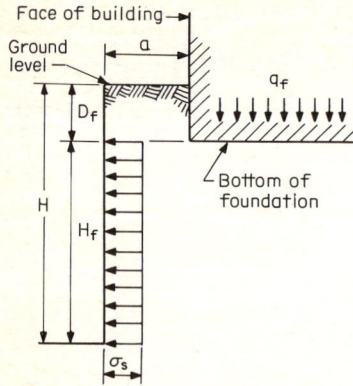

Fig. 13-21 Lateral load distribution due to building foundations;

q_f = total live- and dead-load foundation pressure, lb/ft$_2$

$ = \dfrac{\text{building load in width } H_f}{\text{area of width } H_f}$

$q_n = q_f -$ weight of overburden $= q_f - \gamma D_f$

$$\sigma_s = \begin{cases} 0.5 q_n \left(1 - \dfrac{a}{1.5 H_f}\right) & \text{for } 0 < \dfrac{a}{H_f} < 1.5 \\ 0 & \text{for } \dfrac{a}{H_f} > 1.5 \end{cases}$$

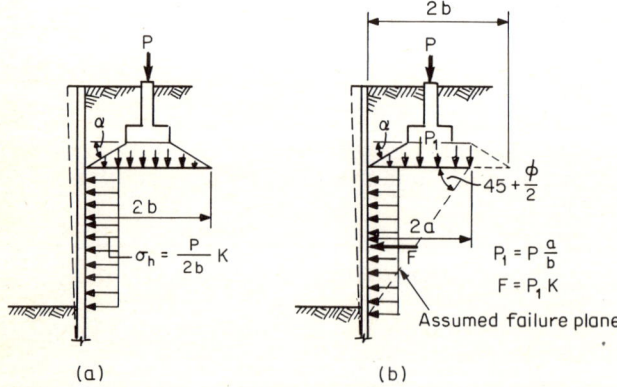

Fig. 13-22 Lateral stress distribution due to external load within a soil: (a) vertical and horizontal distribution according to an inclination angle; (b) distribution taking into account the zone of influence within the failure wedge.

sequence, time, etc.; (2) the actual deformation modulus of the soil varies with the previous deformations; (3) the wall sometimes moves toward the retained soil and sometimes away from it, and this reversal of the displacement continues until the excavation is completed; and (4) there is a repeated change from loading to unloading as some of the external loads are applied and removed.

Tests on small-scale models simulating a strutted excavation and carried out to study the lateral effect of foundations have shown that the distribution of external load depends on the bracing position and excavation sequence much as the lateral earth stress does (Breth and Wanoschek, 1972). Whereas the magnitude of the lateral stress is influenced by the position of the external load and its proximity to the wall, its distribution hardly follows an elastic or limit theory and is decisively governed by the excavation and bracing process. The same conclusion was reached following field measurements to check the influence of building loads on walls supporting open-cut excavations in Frankfurt.

Vertical Loads

Besides the weight of the structure the vertical loads acting directly on the wall are the reactions from floor slabs and beams. Diaphragm-wall connections result in a certain eccentricity of these loads with respect to the centroid of the section and generate unbalanced moments.

Other significant loads will act on the completed structure and will reach the walls indirectly. Besides the forces and loads transmitted from above, the entire structure may be subjected to the action of upward forces caused by the ground reaction, heave pressure, and hydrostatic uplift, but usually these do not take full effect until some time after the excavation. The magnitude of these forces can be estimated with accuracy, but the conditions giving rise to this action are time-dependent and may vary from point to point within the excavation, so that a representative analysis is sometimes difficult.

Another important vertical load is friction along the back face, caused by the relative movement of the wall with respect to the soil. Situations where this force can exist are discussed in subsequent sections.

Construction Loads

For a braced or anchored excavation the construction sequence and method of bracing influence the lateral loads in two ways: (1) by changing the manner in which the soil supports itself so that the actual earth stresses are markedly different from the stresses existing at a limiting state and (2) by inducing loads associated with the construction operations. The latter comprise a group of forces in the bracing system intentionally introduced by jacking struts or by prestressing tiebacks and loads which are due to the structural response of the entire support system, e.g., shortening or extension from shrinkage or elastic compression and thermal expansion or contraction.

Bracing or anchoring loads are introduced where it is necessary to limit ground movement by limiting wall displacement. For the preloading the struts or ties are provided with a jacking system or with a prestressing device. The condition of excavating in stages tends to force the lower (unbraced) portion of the wall to move away from the supported ground, whereas the introduction of bracing load tends to force the wall to move toward the supported ground mainly in the upper (braced) part of the structure. This combination gives rise to lateral stresses which deviate considerably in magnitude and distribution from the stresses of limit theory and are much larger than the active stresses. These facts have been known since Terzaghi's initial work (1934) and tests on retaining walls. For strutted excavations which are not provided with a jacking system a reduction in the initial stress may be accepted assuming that favorable ground strains will occur because of a temporary reduction of support load during excavation followed by structural shortening due to shrinkage and compression. This has been applied in certain projects (Collingridge and Tuckwell, 1960; Heydenrych and Isaacs, 1967), but how much reduction is justified is a matter of judgment, even with the most advanced analytical methods.

The effect of thermal expansion and contraction of the bracing of two opposite walls is a reversal of ground strain. For relatively long or wide strutted excavations too much contraction of the bracing may cause active stresses but ground movement exceeding the allowable, and too much expansion can induce lateral stresses much higher than the walls were designed for. In dissimilar ground and under varying initial loading conditions the expansion of bracing supporting two opposite walls can result in an awkward redistribution of lateral stresses which is very difficult to understand and express analytically.

Accidental loads of the type mentioned in Sec. 12-12 are normally unpredictable, and therefore they can be compensated for only by the factor of safety.

13-10 EXPERIMENTAL COMPARISON OF THE STABILITY OF RIGID AND FLEXIBLE WALLS

The suggestion that rigid* walls do not correspond to the behavior of flexible supports, e.g., sheet-pile walls or soldier piles, has received increasing attention. In one of the first investigations Verdeyen and Gillet (1969) used two-dimensional models and a special roller-type apparatus to simulate the compressibility of a soil and studied the mechanical process induced in rigid and flexible walls supporting an excavation in cohesionless and with $\phi = 22°$, $\gamma = 2$ t/m^3, and a void ratio $n = 0.28$.

---

*The word "rigid" as used herein does not mean absolutely rigid but indicates the stiffness of a concrete diaphragm wall 18 in (45 cm) thick or more.

Free-Standing Vertical Cantilever Walls Figure 13-23a shows the configuration and notation of such a wall. Under a progressive and continuing excavation on one side the wall yields laterally, and the displacement y at the top is measured at various excavation levels for a smooth- and a rough-wall interface. The ratio y/H of displacement to the total wall height is plotted vs. the ratio f/h of the wall embedment to the exposed wall height and is shown in Fig. 13-24 for both the smooth and the rough surface.

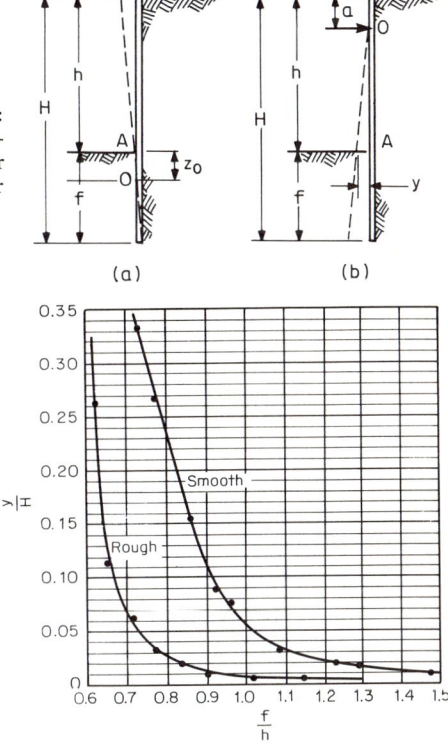

Fig. 13-23 Two types of walls: (a) free-standing vertical cantilever wall; (b) wall braced at or near the top. Point O is the center of rotation.

Fig. 13-24 Relationship between lateral displacement, embedment, and exposed wall height for a free-standing vertical cantilever wall. Roughness of wall simulated in two-dimensional models. Compressibility of soil corresponding to $\phi = 22°$. (From Verdeyen and Gillet, 1969.)

If a wall rotates around point O located at distance z_0 from excavation level, and if the soil in front of the wall and behind it is stressed to its limit, it is possible to correlate the ratio K_p/K_a with the ratio f/h. This is shown graphically in Fig. 13-25, which is a plot of a fourth-degree equation for f/h (Verdeyen and Gillet, 1968); the ratio K_p/K_a is a variable. The graph is general and shows that if the density of sand (or stiffness) is increased (meaning an increase in the K_p/K_a ratio because of an increase in the angle of friction), for the same exposed wall height the embedment is decreased. Combining the data from Fig. 13-24 and 13-25 makes it possible to relate K_p/K_a and y/H, which is shown graphically as curve L_1 and L_2 in Fig. 13-26 for the smooth

522 Slurry Walls

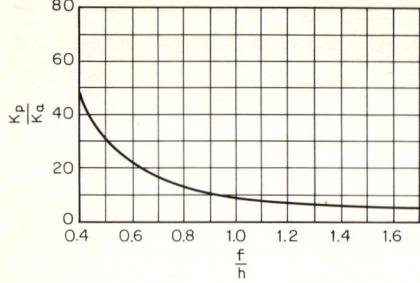

Fig. 13-25 Relationship between the K_p/K_a ratio of lateral-earth-stress coefficients and the ratio of embedment to exposed wall height for free-standing vertical cantilever walls. Results obtained from a theoretical analysis.

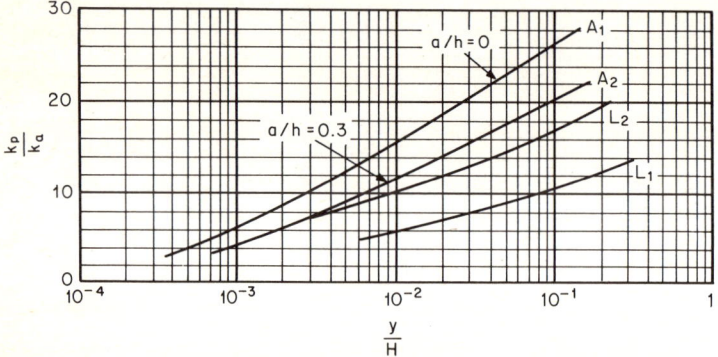

Fig. 13-26 Relationship between the ratio of lateral-earth-stress coefficient and the ratio of displacement to total wall height. (*From Verdeyen and Gillet, 1969.*)

and the rough face of the model wall; note the assumption that the graphs of Fig. 13-24 apply to a soil regardless of its density.

Walls Braced at or near the Top Figure 13-23b shows the configuration and notation for this wall. The bracing near the top precludes displacement of the support at this level, so that as the excavation begins and proceeds to final level, it induces a wall movement which is rotation around the bracing point. The resulting displacement at excavation level is then measured for a varying excavation height and wall embedment and for five values of a, namely 0, 5, 10, 15, and 20 cm on the model scale. The ratio y/H is likewise plotted vs. the ratio f/h in Fig. 13-27, where the thick solid line represents the average curve for the five cases. Assuming again that the soil on either side of the wall is stressed to its limit and that a parabolic lateral stress distribution is appropriate to a braced cut, we obtain a general relationship between the ratios K_p/K_a and f/h shown graphically in Fig. 13-28 for two values of a/h, 0 and 0.3.

Combining the results from Fig. 13-27 and 13-28, we get a relationship for he ratios K_p/K_a and y/H, plotted in Fig. 13-26 as curves A_1 and A_2 for $a/h = 0$ and 0.3, respectively. Likewise, these graphs relate the results of model tests to a soil of variable density and stiffness.

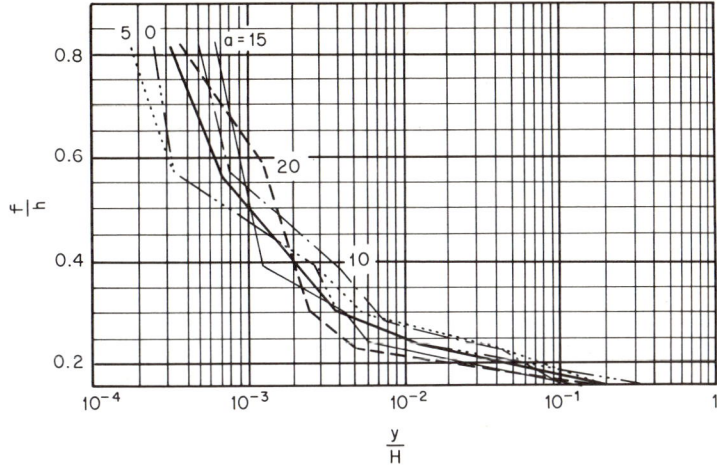

Fig. 13-27 Relationship between displacement, embedment, and exposed wall height for a braced wall. Results from model tests. (*From Verdeyen and Gillet, 1969.*)

Fig. 13-28 Relationship between the K_p/K_a ratio of lateral-earth-stress coefficients and the ratio of embedment to exposed wall height. Walls braced at or near the top. Results obtained from a theoretical analysis.

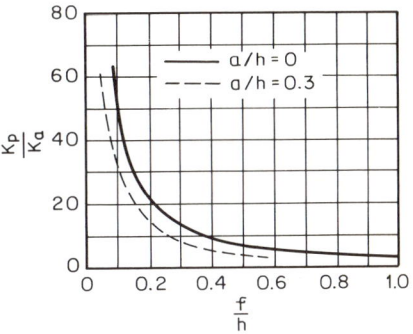

Interpretation of Results For the cantilever wall the assumption of a triangular active-stress distribution is in good agreement with the actual wall movement. The wall roughness and shear strength at the interface have a significant effect on stability provided the embedment is adequate. This is apparent by reference to Fig. 13-29, where the ratio $y_{\text{rough}}/y_{\text{smooth}}$ is plotted vs. the ratio f/h. On the model scale the displacement of the rough wall is only a fraction of the displacement of the smooth wall, and the ratio of the two displacements is reduced to 0.10 for f/h equal to 0.9.

If $f/h > 1.1$ for the smooth wall and $f/h > 0.8$ for the rough wall, the lateral displacement y increases very gradually and almost linearly with decreasing f/h ratio, as can be seen from Fig. 13-24. This means that in this region the soil, already in a limiting state, has developed its passive resistance to balance the active stress. If the embedment is decreased below the above values, the strength of the soil is exceeded and excessive displacement follows.

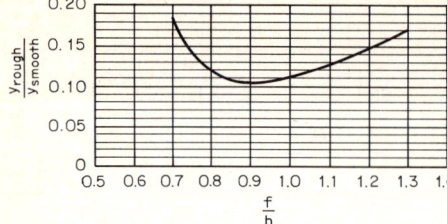

Fig. 13-29 Effect of surface finish and roughness on wall stability. (*From Verdeyen and Gillet, 1969.*)

For the braced wall the assumption of a parabolic stress distribution is in good agreement with the results of the test. Likewise the displacement ratio y/H is dependent on the wall embedment f. When f exceeds about 0.3 or 0.4 times h, the displacement y is very small. When the embedment decreases ($f/h < 0.3$), the wall displacement increases very sharply, as can be seen from Fig. 13-27, indicating that the soil strength is first mobilized and then exceeded.

The effect of bracing on wall movement is more significant if the bracing point is at or near the top ($a/h < 0.3$). Although this may first seem paradoxical, it shows that the wall moves as the excavation proceeds to the bracing level and before the bracing is installed. Bracing at the top precludes this movement and allows only movement below excavation level. Examination of curves A_1 and A_2 in Fig. 13-26 shows an increase in the ratio K_p/K_a by as much as 40 to 50 percent for the same displacement ratio y/H as the bracing moves from the $0.3h$ point to the top.

Figure 13-30 shows the relationship between the ratios y/H and H_f/H_b, where H_f and H_b are the total heights for the free-standing and the braced wall, respectively. For the same excavation depth h and the same relative displacement y/H the use of bracing reduces the total wall height by as much as 30 to 40 percent merely by reducing the wall embedment. More significant in this case is the reduction of wall movement which is possible with the braced structure.

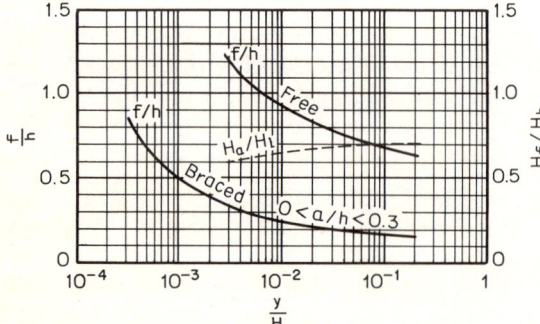

Fig. 13-30 Relationship between the ratios y/H and H_f/H_b for the free-standing and the braced wall, respectively. (*From Verdeyen and Gillet, 1969.*)

Certain results from the foregoing model tests might be useful for average excavations but only as general guidelines since it is not possible to correlate the model scale with the actual wall. The benefit of introducing bracing at the top can be utilized in a practical manner, first to control ground movement and then to reduce wall embedment. For rigid walls the stability should be improved by the actual weight of the structure, the considerable friction at the wall-soil interface, the friction under the tip of the wall opposite to movement, the wall thickness which increases the lever arm resisting rotation, and to some extent by the considerable watertightness of the support.

These conclusions can be considered for walls built as ground support for depressed roadways and traffic underpasses or posttensioned walls. The main similarity is that these structures are either free cantilevers or have a single bracing at or near the top.

13-11 GROUND MOVEMENT AND SETTLEMENT IN EXCAVATIONS SUPPORTED BY DIAPHRAGM WALLS

In the last several years a considerable record has been accumulated regarding the observed performance of diaphragm walls. Although this record is not necessarily plentiful, it represents most soil types and ground conditions as well as different bracing systems and excavation procedures. The evaluation of the actual wall performance and the analysis of instrumented field cases have thus become a valuable tool in making predictions of ground movement and settlement, and in many instances this may be more reliable than the theoretical procedures and parametric solutions which are presently under development.

An important factor influencing the magnitude and pattern of movement is the type and strength of the soil, including the groundwater conditions, and this is true with diaphragm walls. Nonetheless it has been found advantageous and convenient here to classify the field data according to the construction sequence and type of bracing. There are two reasons for this preference: (1) ground movement is generally critical in soft or in very loose soil and seldom serious in stiff or dense ground, so that the cases selected for study deal to some extent with the former type, and (2) the adaptation of diaphragm walls has resulted in the improvement of bracing methods and also made the development of new ones possible, so that with an intelligent utilization of suitable bracing techniques ground movement can be significantly reduced even in soft ground.

Thus, in this section ground movement and settlement are analyzed for (1) walls braced before excavation is begun; (b) walls braced with struts; (3) walls braced with permanent floors in a downward process; (4) walls braced with tiebacks and ground achors; (5) free cantilevers, and (6) walls supported by a combination of bracing, such as tiebacks, struts, rakers, and berms.

Walls Braced before Excavation

Wall movement can be reduced to an inconsequential amount and even eliminated if the lateral support and the bracing can be inserted and constructed in its final position before excavation. With the structure and its bracing in place, the lateral displacement will depend only on the elastic deflection between bracing points and on some yielding of the bracing already installed. In this manner ground movement can become independent of the soil type and characteristics and will be governed primarily by the stiffness of the supporting structure as a whole.

In the past this idealized procedure existed only in imagination. With diaphragm walls it can be approached very closely by providing the permanent bracing at the top and at the bottom before any excavation. Although for relatively deep cuts intermediate bracing is needed and is usually installed as the excavation proceeds, the top and bottom bracing will reduce wall movement significantly. The procedure is not effective in reducing ground settlement, however, unless the arrangement in the lower part of the wall is such that it will restrict the rise of the base that may occur in soft plastic clays as a result of heave. The application is further irrespective of changes in the pore pressure leading to consolidation settlement.

Diaphragm Walls Braced with Cross Walls This procedure is feasible primarily in narrow cuts like that shown in Fig. 13-31. This example is taken from the construction of a subway extension in Oslo, which was carried out using the techniques described in Sec. 12-10 and exemplified in Fig. 12-27. The transverse cross walls braced the longitudinal diaphragm walls below excavation level and were installed in slurry trenches from the ground surface.

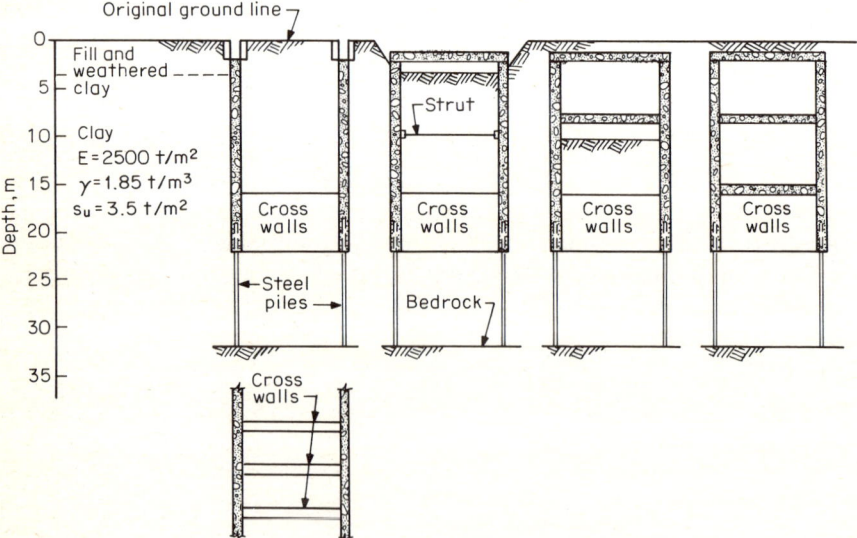

Fig. 13-31 Construction of a subway section in Oslo. Diaphragm walls braced at th bottom with cross walls. (*From Eide et al., 1972.*)

The excavation depth of 16 m (52 ft) is almost twice the depth to which an excavation in soft Oslo clay can be carried out safely without causing base heave. The cross walls were therefore dimensioned and spaced according to Sec. 13-3, so that the structure as a whole became stable against base failure by heave. The top slab was cast before excavation, and in this manner the walls were actually braced at the top and at the bottom before any earth moving was begun. The excavation was carried out under the roof, and the intermediate floor slab was used to brace the walls at midheight.

During the initial stages of construction all the imposed loads were carried by the walls. The absence of an appreciable base bearing at the tip and the very low shear strength at the wall-soil interface limited the load-carrying capacity of the walls significantly, and thus the entire structure was supported on steel piles driven to rock through special casings left in the walls. An alternative would have been to extend and key the walls into bedrock, which exists at the site at a depth of 30 to 40 m (100 to 130 ft), but engineering estimates indicated that this probably would have cost more.

The actual wall movement at midheight was somewhat less than 5 mm (0.2 in) and occurred as elastic deformation between lateral supports. The general excavation inside the structure resulted in considerable drainage through rock where the tip of the wall is close to the rock surface, with a corresponding reduction in the pore pressure. Under these conditions a prolonged normal consolidation could produce large settlements outside the excavation in spite of the negligible lateral wall movement. In anticipation of this problem tubes were initially set in the walls for grouting the leaking rock zones.

Walls Braced by Embedment into Rock If the transverse distance between the walls is considerable and precludes the economic use of cross walls as bottom bracing or as a solution to the heave problem, the walls can be keyed into underlying bedrock if it exists at a shallow depth below the excavation level. An example of bottom bracing provided in this manner is shown in Fig. 13-32.

This structure is for an office building in Oslo (DiBiagio and Roti, 1972). The soil conditions in the vicinity are typical in central Oslo. Bedrock exists at a depth from 10 to 21 m (33 to 69 ft) and is overlain by a thick deposit of normally consolidated marine clay of low sensitivity. As shown in Fig. 13-32a the undrained shear strength of the clay varies from 2.5 to 4.0 t/m² (500 to 800 lb/ft²). The lower crust of the clay layer is occasionally sandy and contains some gravel.

The diaphragm walls are 1 m thick (40 in), and are keyed into the rock for bottom lateral support. The maximum wall depth is 20 m (66 ft). The walls are braced at three levels by the permanent concrete floor slabs, which were cast against the ground at each excavation level. The slabs are 40 cm thick (16 in). During excavation the decrease of pore pressure was small and re-

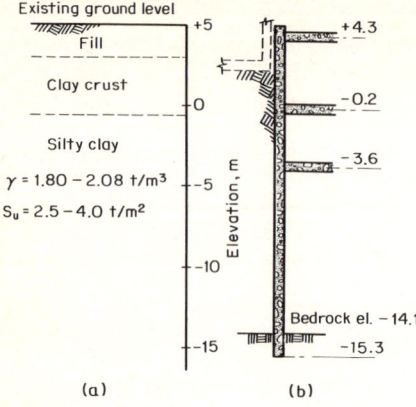

Fig. 13-32 Construction of a deep basement in Oslo. Diaphragm walls braced at the bottom by embedment into rock: (a) soil data; (b) section through wall. (*From DiBiagio and Roti, 1972.*)

sulted in a loss of 1.5 m (5 ft) in the hydraulic head. However, when bedrock was reached, a marked decrease in pressure occurred as a result of dewatering through the rock.

Wall movement and ground settlement are shown in Fig. 13-33. The settlement is given relative to day 46, which is the last set of observations taken before starting excavation. The curve for day 238 shows a maximum settlement of 23 mm (0.92 in). The rate of settlement is relatively constant for the entire period of excavation, and since there was no significant change in pore pressure until the excavation reached the bedrock level, this settlement must have been the result of the lateral wall movement. The bottom of the excavation remained open and unsealed for a considerable time, however, to allow the construction of the basement, and during this period the draingae through the bedrock continued unrestricted. The result was a continuation of settlement beyond day 238, as is evident from Fig. 13-33a, which increased to 33 mm (1.3 in) until the basement was sealed in the rock socket and the pore pressure in the bedrock and the clay layer began to build up again. The additional settlement is therefore due to consolidation effects.

The lateral wall movement shown in Fig. 13-33b refers to day 142, nearly $2\frac{1}{2}$ months after starting excavation, at which time more than 3 m (10 ft) of earth had been removed. The data have been adjusted slightly from the measured values so that all curves pass through a point corresponding to zero movement at the uppermost bracing level. The movement is extremely small for an excavation of this depth in spite of the soft soil. It reaches a maximum value of 15 mm (0.6 in) at a depth of 12 m, but it is conceivable that the total wall movement from the beginning of excavation might have been greater and close to 25 mm (1 in).

It is quite interesting to analyze this case with reference to the data of Fig. 13-34, showing a similar excavation also in soft Oslo clay with almost identical properties and strength but supported by a flexible wall. The wall consists of steel sheeting driven to rock but not penetrating the hard material, so that it is free to move laterally as the ground moves. The data are taken from a set

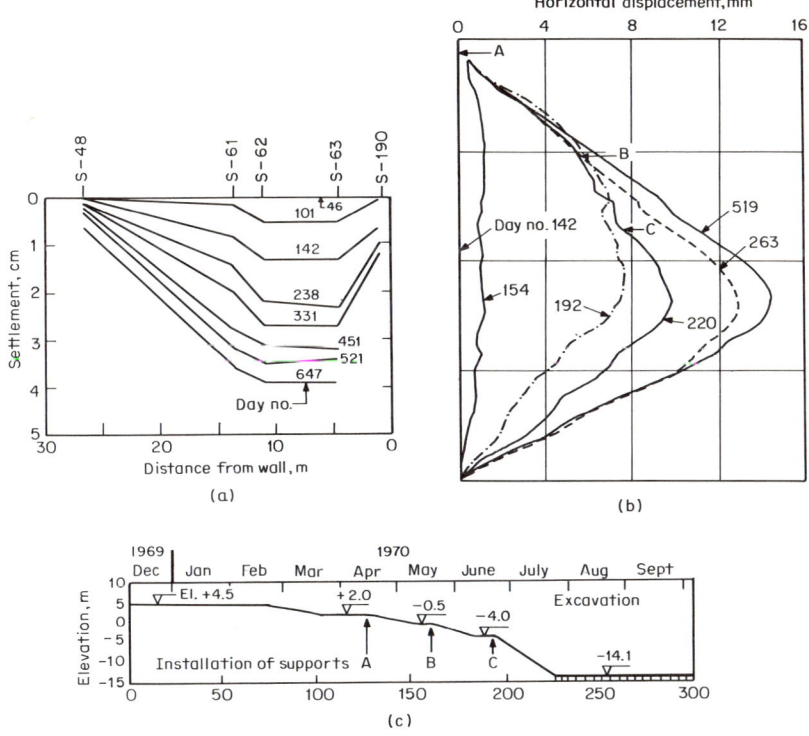

Fig. 13-33 Observed excavation for the office building in Oslo, shown in Fig. 13-32: (a) ground settlement as function of time; (b) lateral wall movement at various excavation stages; (c) construction sequence as function of time (*From DiBiagio and Roti, 1972.*)

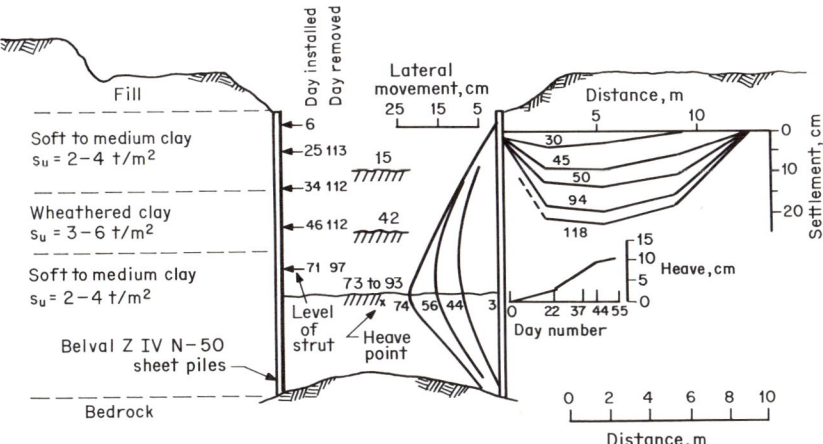

Fig. 13-34 Excavation, lateral movement of sheeting, and settlement of adjacent ground. Open cut in soft Oslo clay. (*From NGI 1962-1966.*)

529

of comprehensive observations (NGI, 1962-1966), and show wall movement and ground settlement in relation to the insertion of the struts as the excavation deepened. They do not represent long-term changes. Both the inward wall movement and the ground settlement are nearly 10 times those of the previous case, and they have been caused by movement below excavation level, base heave, elastic deflection of the flexible walls, and yielding of the bracing.

Walls Braced with Struts

Embarcadero Station, BART Figure 13-35 shows the excavation, soil profile, and the ground support for one of the three stations of the BART project where diaphragm walls were used (Kuesel, 1969). The ground at the site consists of an upper layer of sandy fill underlain by soft bay clay. The excavation is about 60 ft (18 m) deep, and would thus aggravate the heave problem with ordinary ground supports. The stability against base heave is first analyzed from Eq. (13-1). For $B = 55$ ft and $D = H = 60$ ft, $N_c = 6.4$ (from Fig. 13-3). For an undrained shear strength 1000 lb/ft² below and around the excavation level the factor of safety is $6.4(1000)/120(60) = 0.88$, which is inadequate. With a sheet-pile support the probable maximum settlement corresponding to this factor of safety might be of the order of 8 to 10 in (20 to 25 cm).

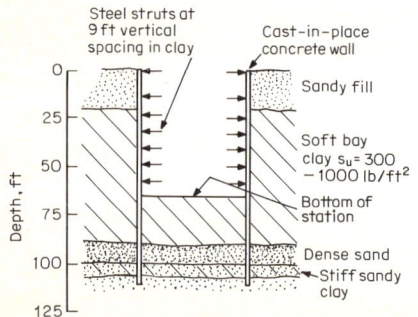

Fig. 13-35 Excavation, soil profile, diaphragm walls, and bracing for the Embarcadero Station, BART. (*From Kuesel, 1969.*)

The diaphragm wall has an embedment of 50 ft (15 m) below excavation level. For $D = H + h = 60 + 50 = 110$ ft, the stability factor N_c is 7.0 when $D/B = 110/55 = 2.0$ is used. From Eq. (13-2) the factor of safety is now $[7(1000) + 2(1000)]/7200 = 1.25$, which is adequate. Protection against base heave is further improved by the presence of the firm layer of dense sand and stiff sandy clay into which the wall is extended.

The cast-in-place wall is 4.5 ft thick (1.35 m) and consists of steel I beam and concrete panels. At the end of excavation and with all the struts in place the maximum wall deflection was from 1 to 1.5 in (2.5 to 3.75 cm). Admittedly the insertion of a top strut before earth moving and the close strut spacing maintained thereafter must have prevented excessive movement as the excavation proceeded to lower levels before the next bracing was installed. Nonetheless, the small inward displacement shows the influence of bottom bracing, which in this case was possible by extending a rigid wall into firm ground.

Excavation in Medium Boston Clay Figure 13-36 shows miscellaneous data for an excavation in section D of the subway extension of the Massachusetts Bay Transportation Authority. The medium blue clay is slightly overconsolidated in the upper crust, and the undrained shear strength actually decreases somewhat with depth from 1000 to 600 lb/ft², where the clay becomes normally consolidated. The excavation depth is 50 ft, giving a factor of safety against base heave of about 1.0 or slightly greater (D'Appolonia, 1973).

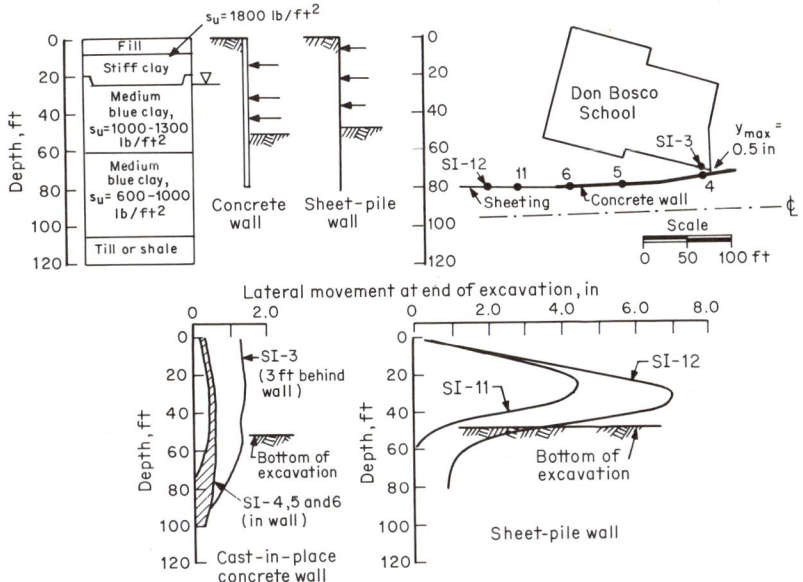

Fig. 13-36 Miscellaneous data and comparison of wall movement for cast-in-place diaphragm wall and sheeting. Excavation in Boston medium clay. (*From D'Appolonia, 1973.*)

A sheet-pile wall served as temporary ground support for most of the length of the section. Because of the close proximity of the seven-story Don Bosco School, which rests on shallow foundation, about 200 ft of the excavation was supported by 3-ft. thick diaphragm walls used also as underpinning at this location. The soil profile, wall depth, number, and location of bracing levels are shown in Fig. 13-36. These features are clearly almost indentical for both the sheeting and the diaphragm-wall section, the only significant difference being the stiffness of the support. The measured pore pressure behind the diaphragm wall was generally 10 to 15 percent less than the static (Lambe, 1970).

532   Slurry Walls

The lateral wall movement at the end of excavation is shown for both supports. Excavation of the slurry trench caused an inward movement of the earth face of almost 1 in (2.5 cm). Actual movement of the concrete wall was less than 1 in. The placement of the diaphragm wall caused an initial increase in pore pressure of about 4 ft of head and a very slight heave of the adjacent school building. The horizontal movement of the sheet pile wall was from 4.5 to 7 in (11.2 to 17.5 cm), or about 8 times the movement of the diaphragm wall.

Utility Tunnel in Osaka This structure has already been described in Sec. 12-14 and shown in Fig. 12-41. Figure 13-37 shows a typical section of the diaphragm wall and the location of bracing. The first 23 m (75 ft) below ground surface consists of very loose to loose sandy silt (standard-penetration-test blow count less than 5). Below this depth are layers of hard clay and dense to very dense sand and gravel showing a blow count of more than 50. The seasonal groundwater table may come to within a few feet of the ground surface. The frequent fluctuations in the water table have caused a continuous consolidation of the loose ground, resulting in a surface subsidence of 10 to 20 mm/year.

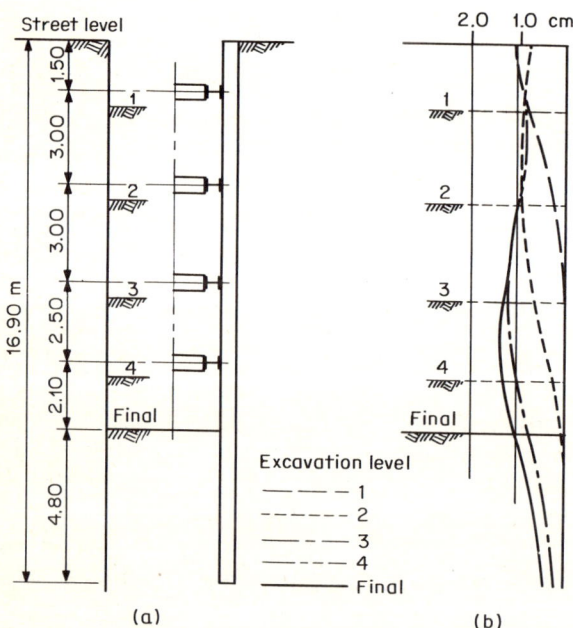

Fig. 13-37 Diaphragm walls for a utility tunnel, Osaka: (a) excavation and bracing sequence; (b) predicted lateral wall movement at various stages of excavation.

The walls were designed for an effective stress computed for a K_0 value of 0.8, full pore-water pressure, and a live-load surcharge of 2.5 t/m² (about 500 lb/ft²). The actual base bearing at the tip of the walls is practically none, and the face shear resistance is negligible. Hence, no load transfer was assigned to the walls either during excavation or in the final service, and all the loads, including the weight of the structure, are carried by the base slab, which serves as foundation mat. The excavation proceeded to the final level under temporary bracing, and the structure was completed from the bottom up and the slabs were connected rigidly to the walls.

The predicted wall movement is shown in Fig. 13-37b for each excavation level. The struts are clearly assumed to restrain further movement in the braced portion of the walls except for some yielding, so that movement occurs in the unbraced section as well as below excavation level. The movement is predicted for the actual wall stiffners (24 in or 60 cm concrete thickness) and wall embedment and reaches a maximum value of about 1.5 cm (0.6 in) between the third and the fourth excavation levels.

The pattern of the actual wall deformation was very close to the one shown, and the maximum observed wall movement was 1.7 cm (0.7 in). The maximum observed settlement at ground surface was 4 cm, or about 1.6 in (Kitagushi, 1976) and included the settlement due to excavation as well as the normal consolidation settlement.

Powell Street Station, BART Figure 13-38a shows a typical section for this structure, a three-level station which consists of the main train platform at the lower level, an intermediate platform level, and the mezzanine level. The exterior support is provided by a composite steel-and-concrete diaphragm wall with a nominal thickness of 30 in (75 cm) and a stiffness EI of about 1×10^6 kip·ft². The walls are 80 ft deep (24 m). The surrounding soil consists of compact fine sand with lenses of silty clay. This material is substantially incompresible, so that exterior dewatering was allowed by deep wells.

The walls were laterally supported during excavation at six levels by cross struts preloaded to 25 percent of their estimated design load. The third and fifth struts at the mezannine and intermediate platform levels have been incorporated into the permanent floors.

The actual wall movement determined from slope indicators placed on opposite walls is shown in Fig. 13-38b. When the 67-ft excavation level (22 m) was reached in March 1969, the inward movement was only 0.2 in (0.5 cm). The movement of the two opposite walls was not consistent. The north wall is seen to have moved toward the excavation, whereas the top of the south wall moved away from it. Additional movement occurred from March to November 1969, when the roof slab was cast, bringing the total maximum wall displacement close to 0.4 in (about 1 cm). Since the excavation was accompanied by the dewatering of the outside area, the very small actual movement must have been the result of the reduced lateral stress. Although the measurements are not particularly consistent, they show the effect of wall stiffness and embedment on reducing movement below excavation level.

534 Slurry Walls

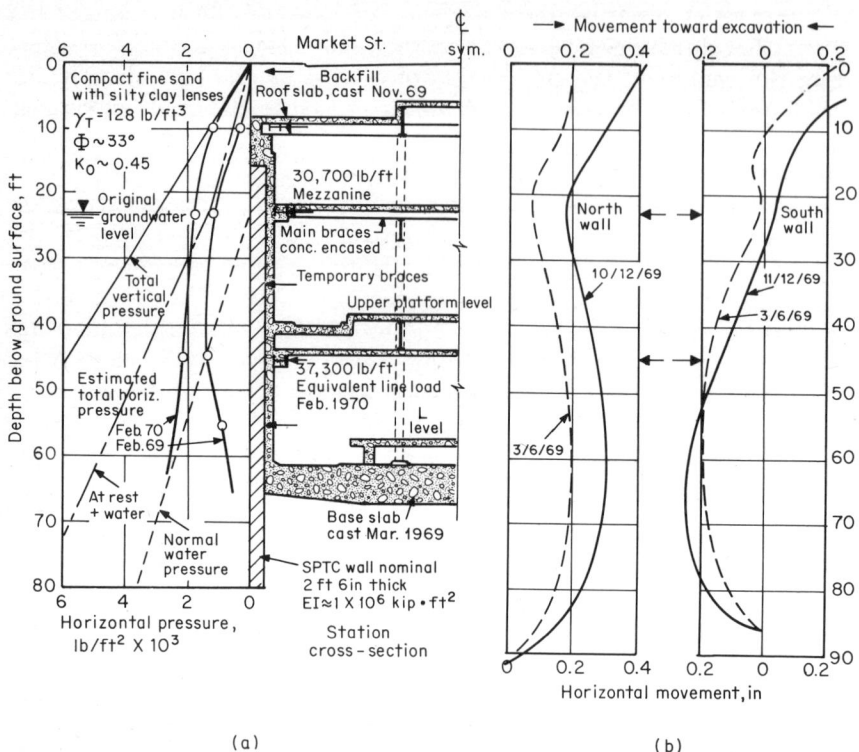

Fig. 13-38 Powell Street Station, BART: (a) typical section through the structure and pertinent soil data; (b) observed lateral wall movement. (*From Gould, 1970.*)

Walls Braced by Permanent Floors

New Palace Yard Car Park, London This structure is 16 m (52.5 ft) deep and was constructed in fissured London clay. The exterior diaphragm walls are within a few meters of the foundations of both the clock tower (Big Ben) and Westminster Hall. The first stage of construction was the installation of the diaphragm walls and the foundation part of the interior columns. The steel columns were lowered into cased boreholes and grouted into position. The construction was then completed in a downward process, in which each slab was cast directly on the ground and the concrete floors were used as bracing.

Although exact soil data are not available, it can be assumed that the ground consists of the typical London clay overlain by a layer of water-bearing gravel and fill. The clay has an average undrained shear strength of 100 kN/m² (about 2400 lb/ft²) and a K_0 value varying from 2 to 3.

Wall and ground movements are shown in Fig. 13-39 and refer to the time when the excavation was completed. Figure 13-39b indicates the profile of the

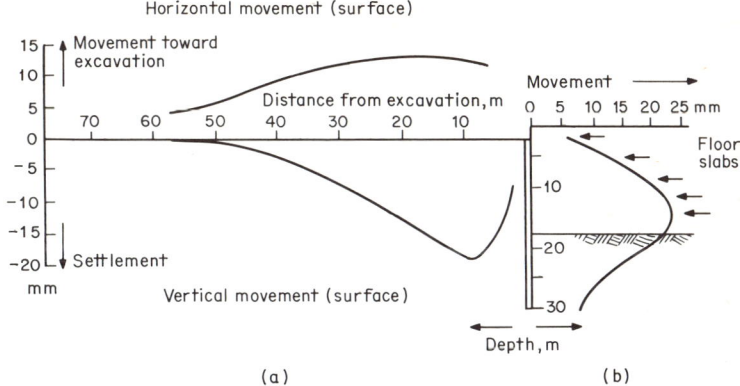

Fig. 13-39 Observed wall and ground movement, New Palace Yard car park, London: (a) vertical and horizontal surface movement and (b) movement of wall toward excavation.

lateral wall displacement with all five floor slabs in place but just before the base slab was cast. The diagram suggests two important conclusions: (1) part of the wall movement, in this case slightly more than 5 mm, was due to initial shrinkage of the concrete floors, which depends mainly on the temperature conditions and the distance between walls; and (2) the stiffness of the wall and its sufficient embedment were effective in limiting movement below excavation level. Two further factors must be considered: (1) the permanent floors do not become effective as bracing until they harden sufficiently, and as time is allowed for this process, the walls are free to move, indicating the possibility of a time-dependent movement due to the continuing softening of the clay outside the excavation; (2) preloading is not possible, as it is with struts, and this precludes the possibility of partial compensation of movement. In spite of these unfavorable factors the wall movement in this case was very small (about 2.5 cm, or 1 in).

The ground-movement diagram presented in Fig. 13-39a shows the vertical settlement at the surface and the horizontal movement toward the excavation as a function of the distance from the wall. The settlement curve shows the distinct effect on settlement of the shear strength mobilized at the wall-soil interface. Although the ground movement is inconsequential, it extends into a zone which is almost 4 times the depth of excavation.

Excavation under a High Water Table Figure 13-40a shows a cross section for a four-level basement built at a site where the water table is normally 4 ft (1.2 m) below the surface (Huder, 1969). The ground consists of glacial till, varved decomposed moraine, and lacustrine deposits. The materials of the moraine are mainly of a clayey nature, somewhat sandy to gravelly, and in some places with lenses of silt. The undrained shear strength is given as 0.2 kg/cm^2 (400 lb/ft^2) for the lacustrine deposit and 1.4 kg/cm^2 (2800 lb/ft^2) for the undisturbed moraine. The depth of the excavation was 56 ft (17 m).

536 Slurry Walls

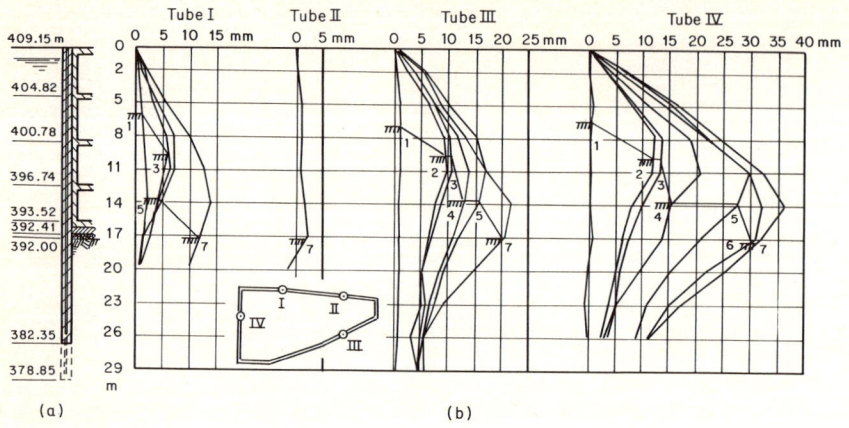

Fig. 13-40 (a) Section through a four-level basement in Switzerland; (b) observed wall movement. Bottom of wall elevation: tube one; = 382.35 m; tube two: 380.00 m; tube three; 380.00 m; tube four; 382.35 m. (*From Huder, 1969.*)

The wall was carried down to unusually great depths, normally between 88 and 114 ft (27 to 35 m) below grade, to provide a cutoff barrier for the groundwater. Outside the excavation the water table was left undistrubed, whereas the interior of the excavation was dewatered to a level 50 ft below surface. It is conceivable that a water head of 40 ft (12 m) existed at the end of excavation.

Lateral wall movement is shown for four locations marked I, II, III, and IV on the wall plan. Evidently the wall depth is about 27 m (88 ft) for tubes I and IV and 29 m (95 ft) for tubes II and III. Initially the ground floor was cast at grade and braced the walls at the top. Bracing was provided thereafter by the intermediate floor slabs in a downward process, whereby the concrete was placed against the ground at each excavation level. The concrete floors were supported in the interior of the basement by steel piles.

Excavation to the first level, 6.5 m (or 21 ft) below surface, caused a negligible wall movement in all four locations. The walls deflected inward more as the excavation was taken down to lower levels. Although zero deflection is indicated at the surface, it is conceivable that some movement must have occurred there. The profiles show that movement continued at the upper floors as the next lower floor was placed, indicating again the effect of shrinkage and elastic shortening of the slabs upon the application of more load. Although tubes I and IV are located in panels of the same depth, tube IV registered more than twice the deflection of tube I. The wall at tube IV is surrounded by varved silt. The greater wall movement at this location is due to more lateral load owing to loose soil; more movement below excavation level resulting from a less stiff reaction below subgrade; and more shrinkage, creep, and elastic shortening of the floor slab because of its greater length.

The walls at tubes II and III are adjacent to the moraine, and both have the same depth. However, the natural ground profile rises some 20 ft (6 m) above the top of the wall behind tube III, so that there is more lateral load against the wall at this location and a subsequent tendency for a net movement of the entire structure from that side to the side containing tube II. The wall movement for tube II is therefore an adjusted deformation profile and includes the interaction with the wall across the excavation. Huder (1969) reported that the exact correlation of movement at various points along the wall was further complicated by variations in wall stiffness. The theoretical wall thickness was 2.6 ft, giving an uncracked EI of about 0.4×10^6 kip·ft^2, but it is conceivable that construction imperfections changed the thickness and resulted in deviations of ±30 percent in the wall stiffness.

CNA Building, Chicago Figure 13-41 shows a typical section through the basement of the CNA building in Chicago (Cunningham and Fernandez, 1972). For the construction of the diaphragm walls the site was cut approximately to elevation +7.00, and after the walls were completed, a capping beam was placed on top and formed to elevation +14.50. The caissons shown adjacent to the west wall support high column loads and are extended to bedrock at elevation −92.00. The west wall is thus only a retaining wall and is extended to elevation −45.00. The east wall is designed to carry a vertical load of 25,000 lb per linear foot, corresponding to a bearing pressure of 10,000 lb/ft^2 at the base. This wall is extended to elevation −50.00. The east and west walls were cross-lot-braced during construction in the east-west direction using the permanent steel frames at floor levels. The south wall (not shown in Fig. 13-41) was allowed to cantilever to the first basement level and braced at this level (elevation +1.00) by diagonal steel bracing transferring the lateral loads to a subbasement level.

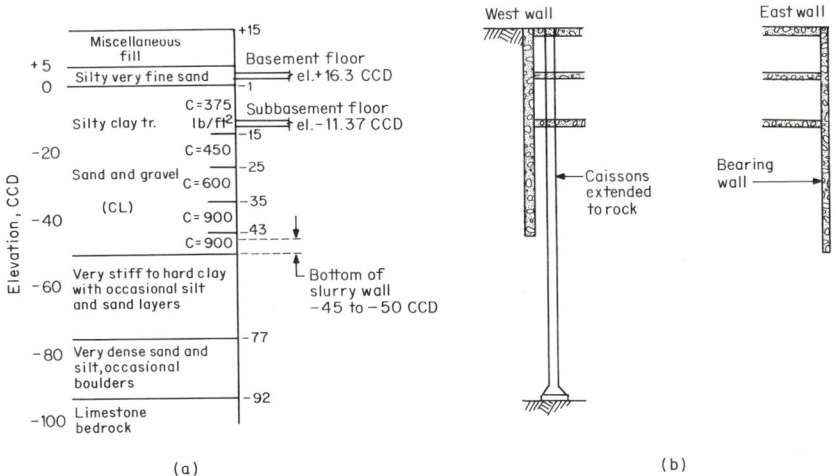

Fig. 13-41 Basement perimeter walls for the CNA building in Chicago: (*a*) soil profile; (*b*) section through basement in an east-west direction; CCD = Chicago city datum.

538 Slurry Walls

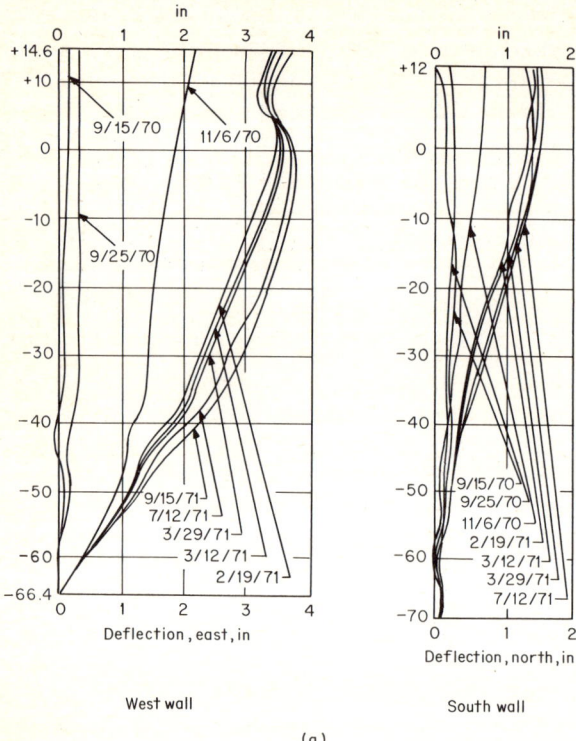

Fig. 13-42 Observed performance of diaphragm walls for CNA building, Chicago: (a) movement of the west and south wall from inclinometer data; (b) top of wall movement from survey measurements; and (c) ground settlement as function of excavation depth. (*From Cunningham and Fernandez, 1972.*)

Movements of the west and south wall obtained from inclinometer measurements are shown in Fig. 13-42a. Lateral and vertical top-of-wall movements for all three sides obtained from surface survey data are shown in Fig. 13-42b. The total surface settlement along lines perpendicular to the walls is shown for the west and south wall at the inclinometer locations in Fig. 13-42c. This settlement is expressed as a percentage of the excavation depth (about 27.5 ft), and therefore each 0.29 percent is approximately 1 in (2.5 cm). It is important to note that the caissons adjacent to the west wall were installed from August 27 to November 6, 1970.

From the inclinometer data it appears that the west wall moved more than 2 in (5 cm) between September and November 6, 1970, i.e., during the construction of the caissons and before any major excavation. Most of this movement was extended for the full depth of the wall, but some movement occurred also below the tip and stopped at the hard clay layer. This wall displacement is thus irrespective of the main excavation and must be solely

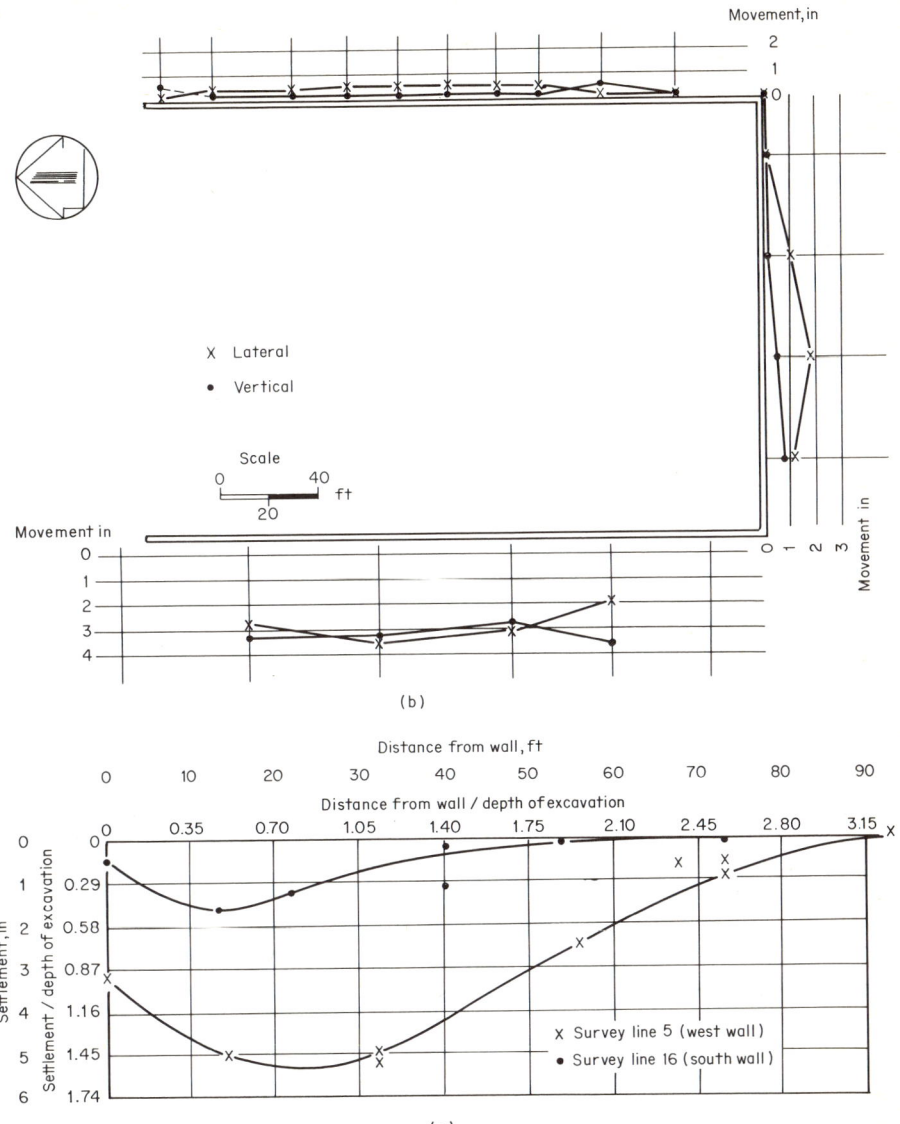

Fig. 13-42 (continued)

attributed to the construction procedures for the caissons. The movement of the west wall caused by the excavation is the difference between the total movement and the movement recorded at the end of caisson construction. Although this wall was braced by the steel framing at the first floor and at the first basement level in a downward process, some movement continued below excavation level after the insertion of the bracing, indicating long-term effects of a rigid structure pushing against soil that had previously been disturbed.

The lateral movement of the east wall is probably a more reliable indicator of the effects of excavation. From Fig. 13-42b it appears that the top of this wall moved laterally an average of $\frac{1}{2}$ in (1.25 cm), which is very small compared with the displacement of the west wall. Although the exact profile of movement is not available, it is conceivable that most of the movement occurred above excavation level.

The movement of the south wall from the inclinometer data is $1\frac{1}{2}$ in (3.75 cm), and one-half of that occurred during construction of the caissons. Most of the remaining movement took place during excavation to the first bracing level and while the wall was allowed to cantilever to this level. Movement below excavation level was negligible.

The ground settlement adjacent to the wall shown in Fig. 13-42c was the result of the lateral wall movement. From the settlement curve it is evident that the weight of the sinking soil mass developed negative (downward) skin friction along the back of the wall. This is particularly true for the west wall, and the data confirm that both the structure and the ground immediately adjacent to it setttled the same amount (about 3 in).

Civic Center Station, BART Figure 13-43a shows a cross section without mezzanine extension for this station (see also Sec. 12-10). The walls were braced by the permanent framing at each floor level and by a temporary tier as the excavation advanced to the final level. The excavation depth is about 74 ft (23 m). The diaphragm walls (steel and concrete) were carried down to 94 ft (29 m) below surface and have an EI value of 1.4×10^6 kip·ft². The soil profile

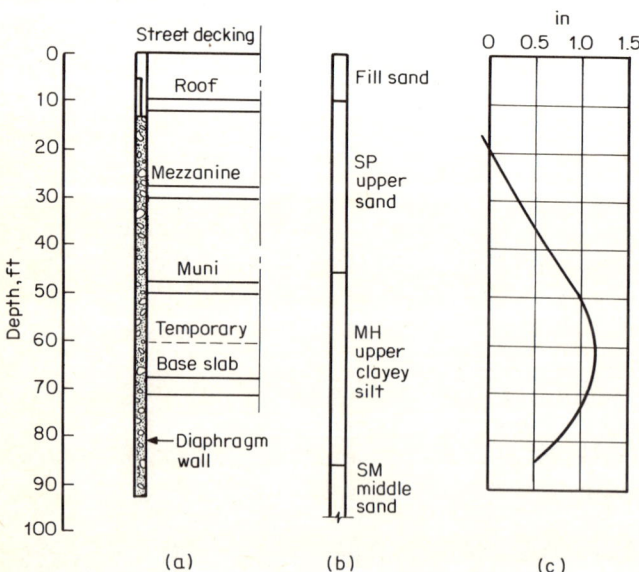

Fig. 13-43 Civic Center Station, BART: (a) section through station; (b) soil profile; (c) lateral movement of the diaphragm wall at the end of excavation. (*From Thon and Harlan, 1971.*)

is shown in Fig. 13-43b and consists of a 50-ft layer of relatively compact sand with peat lenses in its lower portion underlain by clayey silt occasionally organic. The water table can rise to within 10 ft of the surface, and since the soil is at least moderately compressible, the groundwater level outside the excavation was maintained with a recharge system. It is conceivable that a differential head of 30 to 35 ft existed at the end of excavation.

The lateral wall movement at the end of excavation is shown in Fig. 13-43c. The maximum inward displacement, $1\frac{1}{4}$ in (3.1 cm), was registered about 12 ft from the base. At least $\frac{1}{2}$ in (1.25 cm) or more must have occurred below excavation level, so that the remaining is mainly elastic deflection and yielding of the bracing. It is interesting to note the relatively large unbraced spans in the upper bays (about 20 ft).

Walls Braced with Tiebacks and Ground Anchors.

Neasden Underpass, London A cross section for this excavation, including also the instrumentation, is shown in Fig. 13-44a. The excavation is 8 m (26 ft) deep, and the diaphragm walls are extended 13 m (42 ft) below the surface. At the site London clay exists to a depth of 28 m (92 ft) and is underlain by the much stronger Woolwich and Reading beds, which have an undrained shear strength of 300 kN/m² (about 7200 lb/ft²).

The walls were braced by four rows of tiebacks extending in back to at least twice the depth of excavation. The tiebacks have a moderate inclination relative to the horizontal, so that the anchorage zone is in the typical London clay. As the earth moving was begun, the walls began to move horizontally toward the excavation. This movement extended to a considerable depth and well below the bottom of the inclinometer tubes. As earth moving continued, the ground behind the walls continued to move. Toward the end of excavation and subsequently the walls have undergone significant settlement as well as inward movement (St. John, 1975). The displacement trajectories of various points showed that outside the anchorage zone the movement of the ground was essentially horizontal. The vertical displacement of the wall initiated toward the end of excavation must have been caused largely by the vertical component of the ground anchor load, in spite of the flat profile of the anchorage area.

Figure 13-44b shows the lateral movement of the tops of the three inclinometer tubes as a function of time. Time-dependent movement has clearly taken place following the excavation period and in spite of the relatively shallow depth. A considerable movement extends back beyond the anchorage region. The ground anchors therefore appear to be the cause of ground movement and its extension to a great distance behind the walls.

Guildhall Precincts Development, London Figure 13-45a shows a section through the diaphragm wall supporting the excavation. On the same section are also shown pertinent soil data and the tieback bracing. The excavation depth is 10.4 m (34 ft), and the wall thickness is 20 in (50 cm).

542 Slurry Walls

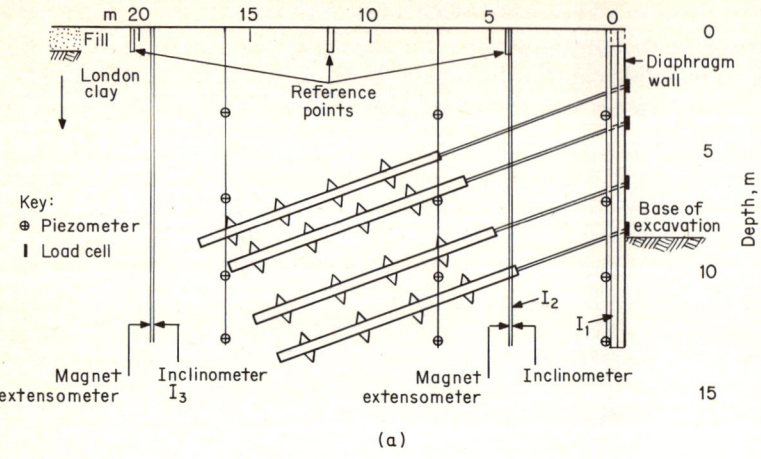

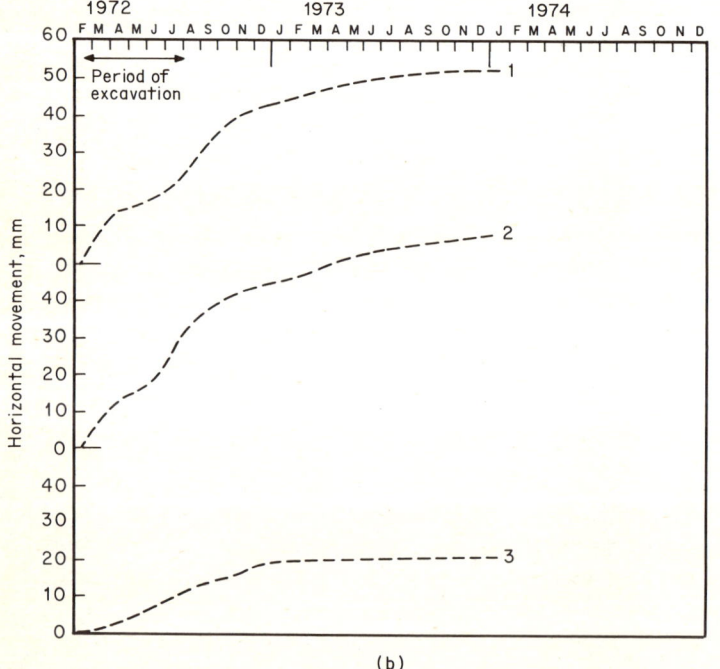

Fig. 13-44 Neasden Underpass, London: (*a*) section through diaphragm wall showing tiebacks and instrumentation; (*b*) horizontal movement of the top of the three inclinometer tubes: 1 = in the wall; 2 = 4.3 m behind the wall; 3 = 19.3 m behind the wall.

Topics Relevant to Analysis and Design   543

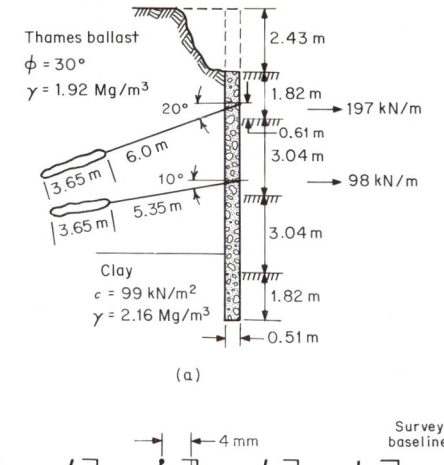

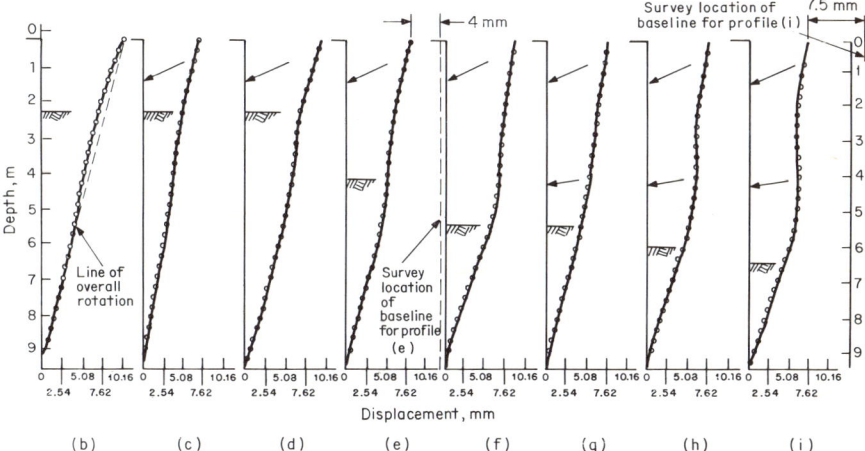

Fig. 13-45 Guildhall Development, London: (a) section through diaphragm wall; (b) to (i) observed wall movement during excavation. (*From Littlejohn and MacFarlane, 1974.*)

The lateral wall movement at various stages of the excavation is shown in Fig. 12-45b to i (Littlejohn and MacFarlane, 1974). The displacement profiles are plotted relative to the toe of the wall, so that no account is given of the overall wall displacement except where the movement was measured during a general survey, as in profiles (e) and (i). It is conceivable that a certain percentage might be added to the movements shown to indicate the total wall displacement.

Profile (b) shows a rotation of the wall toward the excavation which is superimposed on the cantilever action above the first excavation level (4.25 m or 14 ft). The maximum differential displacement between the toe and the top occurred at this stage and is 10 mm, corresponding to a rotation or slope of $\frac{1}{970}$. The effect of prestressing the top anchor is evident in profile (c), showing the wall drawn back toward its initial profile. The difference between profiles (c) and (d) reflects time effects, in this case 4 days, and indicates the displacement along the tieback axis necessary to mobilize full shear in the anchorage zone. Likewise the bulging that occurred below excavation level in profile (*f*) as earth moving continued to the second level is seen to have disappeared when the second tieback was prestressed in profile (*g*). The bulging of the wall continued as the excavation was taken to the final level, as shown in profiles (*h*) and (*i*).

The measured vertical settlement indicated negligible scattered changes around the wall and at the ground surface. The upper anchors were monitored for 103 days, and during this period the prestress force remained essentially constant, indicating that there was no significant relative movement between the fixed-anchor zone and the wall.

Keybridge House, Vauxhall, London The diaphragm wall for this project is shown in Fig. 13-46*a* together with the location of ground anchors and pertinent soil data. The wall is 24 in (60 cm) thick and supports an excavation 14.5 m (48 ft). The wall embedment below excavation level is 2.2 m (7.2 ft). The soil profile consists of an upper gravel layer underlain by London clay, stiffening progressively from an undrained shear strength of 45 kN/m^2 (1080 lb/ft^2) to 154 kN/m^2 (3700 lb/ft^2) at and below the base of the cut. During excavation certain cables had to be made longer in order to reach suitable strata for the grouted fixed-anchor zone.

The lateral wall movement at various stages of the excavation is shown in Fig. 13-46*b* to *i*. From profile (*b*) it appears that the wall underwent a rotation about its toe, indicating the response of a soil which becomes progressively stiffer. Stressing of the ground anchors to 450 kN drew the wall back with an apparent toe rotation [profile (*c*)], but considerable resistance to this displacement was provided in the lower portion of the wall by the stiffer clay. Some 29 days after the prestressing of the anchors the wall had reverted to the shape of profile (*b*), as shown in profile (*d*), although no major change in prestress load was detected.

Following excavation to the second level (6.8 m) the wall deformation below a depth of 6.5 m shown in sketch (*e*) remained identical to the wall deflection of profile (*d*), but an inward toe displacement of about 2.5 mm occurred toward the excavation owing to the consolidation of the clay on the cut side. The development of beam action below the anchor level offset the initial cantilever effect, and this, combined with the considerable wall stiffness, resulted in a very low degree of bending.

With two levels stressed as shown in profile (*f*) the relative displacement between the top and the toe became even smaller. However, following excavation to third level (10.4 m) the differential displacement between top and toe increased to about 13 mm (0.5 in) in spite of the stressing of all anchor levels. At the final excavation level the toe was displaced about 1.5 mm toward the cut. It might appear from profile (*h*) that the wall underwent some rotation about a point between the two upper anchors, which reduced the differential displacement between the top and the toe to nearly 10 mm (0.4 in). The

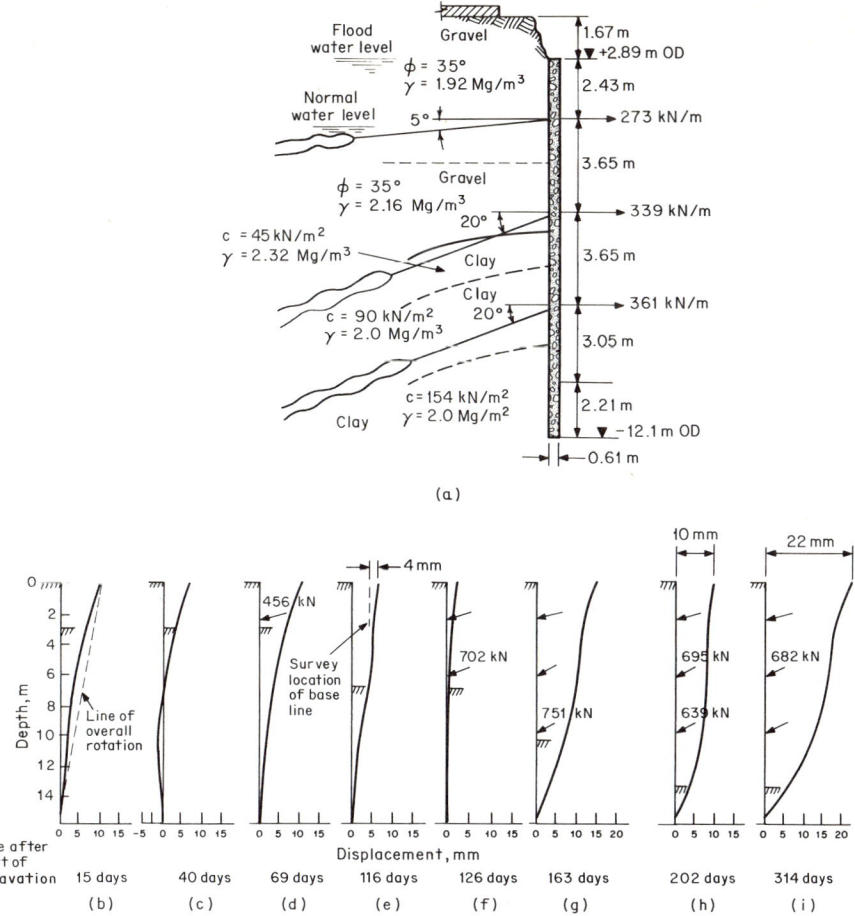

Fig. 13-46 Excavation for the basement of the Keybridge House, Vauxhall, London: (a) section through diaphragm wall; (b) to (i) observed wall movement at various construction stages. (*From Littlejohn and MacFarlane, 1974.*)

vertical settlement at the final stage measured at the top of the wall was 12 mm (0.5 in) and was probably caused by the vertical component of the tieback load until side shear and base bearing were fully mobilized.

The time effects on wall movement become evident from profiles (h) to (i), corresponding to a period of 3 months. During this period the differential displacement between the top and the toe more than doubled, although the load for the center anchor exhibited only a slight loss of prestress. This might indicate the possibility of consolidation of the highly stressed soil surrounding the fixed anchorage zone, but it is more likely that an overall movement of the entire ground mass behind the wall occurred and within the zone of influence of the tiebacks, as was the case with the Neasden underpass.

546   Slurry Walls

World Trade Center, New York   This example is different from the previous cases discussed in that the tiebacks were anchored into rock. The tiebacks served as temporary bracing during excavation. The final permanent support was provided by the floor system, and the tiebacks were eventually distressed. Typical wall sections are shown in Fig. 13-47 for the three monitored panels. The total depth of the excavation is 70 ft. The diaphragm walls were keyed into bedrock, mainly for the transfer of the considerable vertical load imposed during the service of the temporary bracing. Each panel was supported by four to six levels of tiebacks installed at 45° and penetrating 30 to 35 ft (9 to 10.5 m) into rock. The tiebacks were prestressed to their design loads of as much as 600,000 lb. The wall thickness is 3 ft, giving a stiffness EI in the uncracked stage of 1.5×10^6 kip·ft².

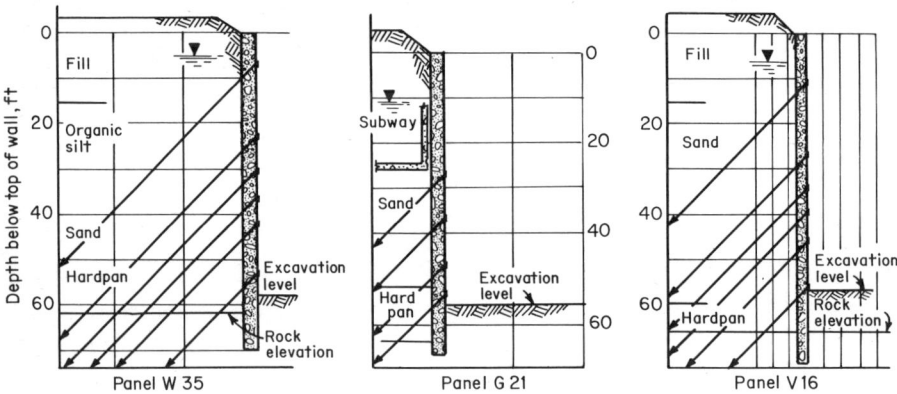

Fig. 13-47   Diaphragm walls for the World Trade Center, New York. Sections through panels W35, G21, and V16 (*From Saxena, 1974.*)

Figure 13-48a shows the lateral wall movement for panel W35 (Saxena, 1974). The tiebacks were stressed as shown to 100 percent of their design loads. The deformation profile shows that the wall continued to move toward the soil as the excavation proceeded to lower levels. The maximum movement measured at the top is 0.2 ft (6.6 cm), and most of it occurred from the beginning to the end of excavation. The wall moved very little in the same direction during the 10 months that followed completion of the excavation. As result of this movement the tieback loads decreased continuously, as indicated by the magnitude shown in the final wall position. Creep in the grouted-rock anchorage zone contributed very little to the tie-load decrease (Saxena, 1974), and the loss of prestress was attributed solely to the reduction in the tieback elongation accompnying the wall movement.

The horizontal movement of panel G21 is shown in Fig. 13-48b. The presence of the subway section precluded the use of tiebacks in this part of the excavation, and the first bracing was installed some 25 ft (8 m) below the top of the support, allowing the wall to cantilever above this level. Additionally, the ties at this location were stressed only to 40 percent of their design load in

Topics Relevant to Analysis and Design 547

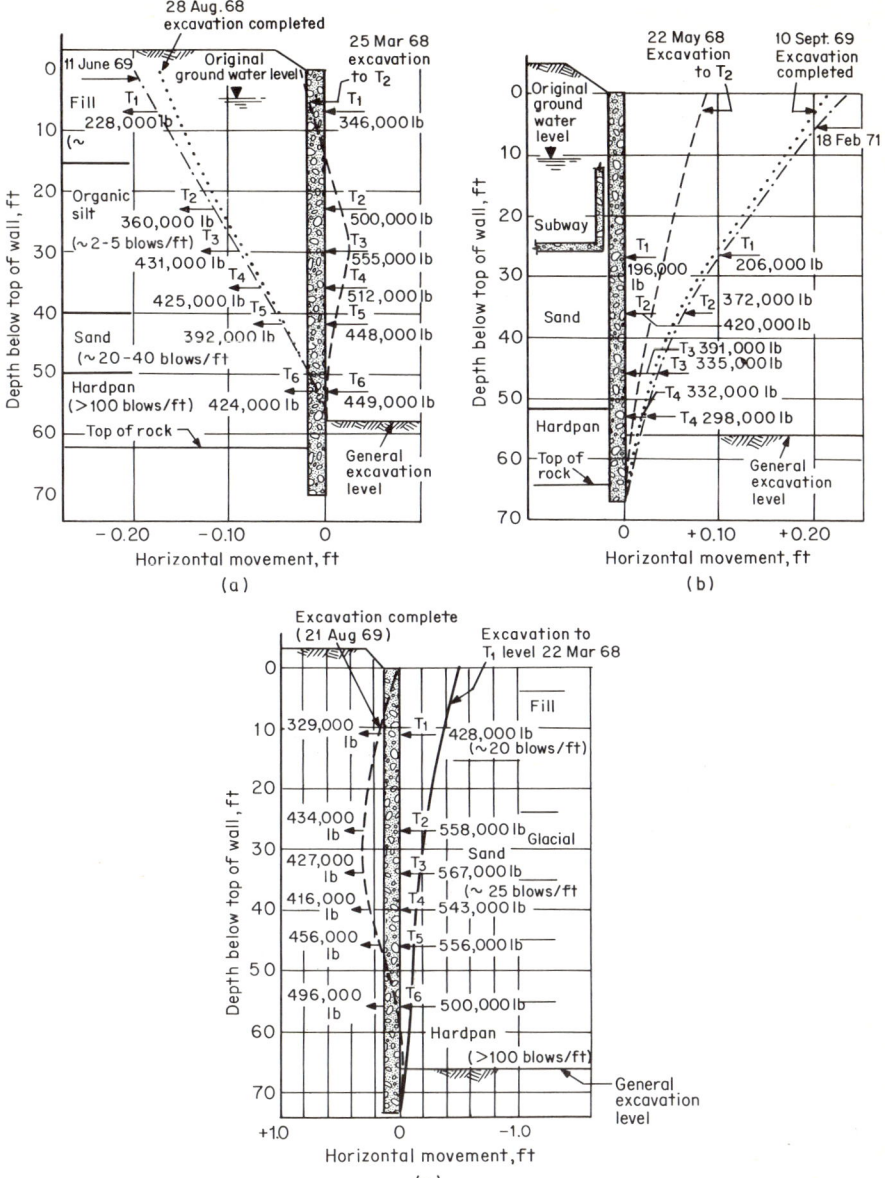

Fig. 13-48 Lateral movement of the diaphragm walls for the World Trade Center, New York: (a) panel W35, (b) panel G21, and (c) panel V16. (*From Saxena, 1974.*)

order to avoid overstressing the subway section. The wall moved continuously toward the excavation in spite of prestressing the tiebacks at the lower bracing levels. Maximum inward wall movement slightly exceeded 0.2 ft (6 cm), and almost one-half occurred during excavation to the second tieback level. The

548　Slurry Walls

apparent small increase of load for tieback T_1 reflects a small increase in lateral load there. The loss of prestress experienced by the remaining three tieback rows in spite of a small elongation may have been caused by slippage between ties and grout, creep, or a combination thereof (Saxena, 1974).

Figure 13-48c shows the wall movement for panel V16. Although this panel was braced at six levels, like panel W35, the first tieback was placed some 12 ft (3.7 m) from the top. The wall moved toward the excavation like a cantilever beam as the earth moving was taken down to the first tieback level, indicating rotation about its toe similar to that in the Keybridge House wall. However, as the remaining ties were installed and prestressed, the wall moved

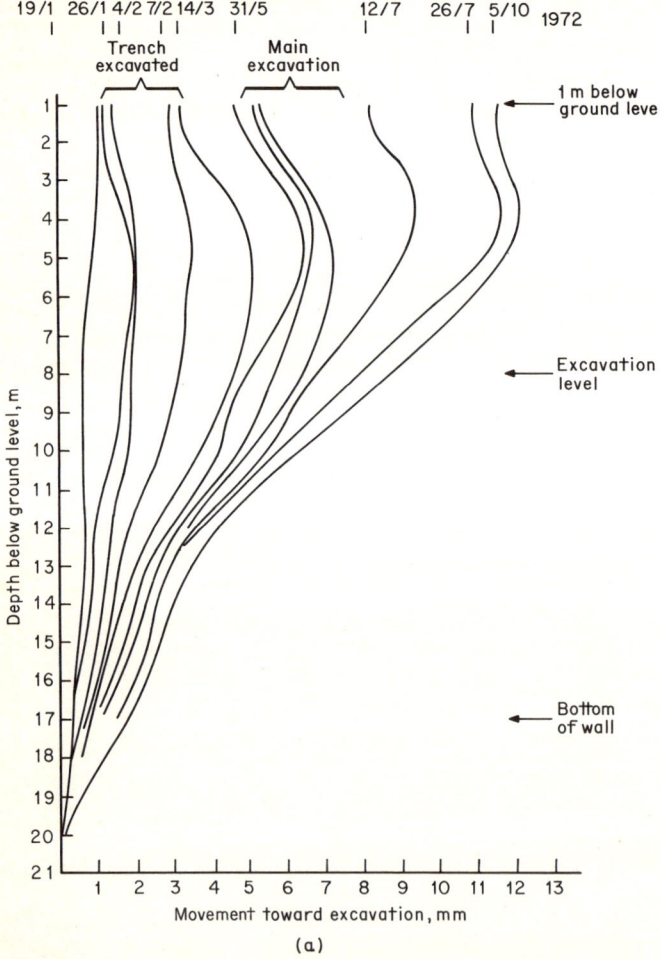

Fig. 13-49 (a) Lateral movement of the cantilever diaphragm wall for the Reading bypass in London; (b) horizontal and vertical ground movement. (*From St. John, 1975.*)

away from the cut. Upon completion of the excavation the tip and the top of the wall were almost at their initial position. At the final stage the wall resembled a beam deflected between its uppermost support and the bedrock, although it retained the initial cantilever effect above the top bracing level.

Free Cantilevers

Examples of diaphragm walls which acted initially like free cantilevers are the structures shown in Fig. 13-42a south wall, 13-46b, and 13-48c. In all these cases the excavation was carried down to a certain level for the installation of the first bracing, and during this phase the wall was allowed to cantilever above excavation level.

Cantilevered walls have seldom been used in urban construction or in deep cuts, but they offer structural possibilities in relatively shallow excavations, e.g., the uncovered portion of a depressed roadway, provided the resulting lateral movement is small or can be tolerated.

Figure 13-49a shows the lateral movement of a cantilever wall at various times and stages of construction. This excavation is for the Reading bypass in London. The depth of this cut is identical to that for the Neasden underpass, shown in Fig. 13-44, but in this case instead of ground anchors the cantilever principle was adopted, the wall extending to 17 m (56 ft) from ground surface. The ground consists of London clay in the upper crust, but the toe of the wall was keyed into the much stronger Woolwich and Reading beds. Figure 13-49b shows the distribution of surface movement.

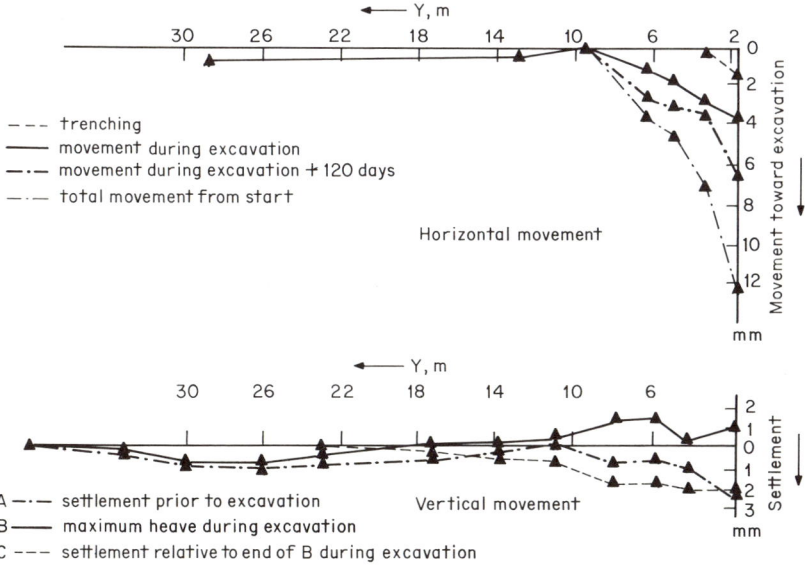

(b)

Fig. 13-49 (continued)

Walls Supported Laterally by a Combination of Bracing

In relatively wide excavations it is customary to leave a sloping berm of earth against the wall while the central portion is excavated. Inclined struts or rakers are then installed through a cut in the berm and are wedged against the wall at the upper end and against the edge of the foundation slab at the base. If the total depth of the cut is considerable, a row of tiebacks can be used to brace the upper part of the wall. The berm is cut away in sections, and if necessary, additional bracing is inserted to provide lateral support for the lower part of the wall. Alternatively, the wall can be provided with sufficient embedment below excavation level.

Although the stability of the temporary berm and the wall itself may be quite adequate, the movement of the uppermost portion of the support may be excessive, particularly if the excavation is in cohesive soil and the insertion of the permanent bracing is delayed for too long.

Central YMCA, London Figure 13-50a shows a section through the underground portion of this building. The basement is 16 m deep (52.5 ft) and is supported by a diaphragm wall extending 18 to 19 m (59 to 62 ft) from ground surface, so that the embedment is only 2 to 3 m (6 to 10 ft). The ground is typical and consists of 7 m of gravel and fill overlying London clay.

The excavation was carried out in three basic stages, presented diagrammatically in Fig. 13-50a. The first stage was to a level 10 m (33 ft) while the wall was temporarily braced in the upper part by a row of tiebacks installed some 3.5 m from the top. A 30-cm-thick (12-in) concrete slab was cast against the ground at the first excavation level and braced the wall at this point. The excavation then continued below this level under the floor. Because of the small wall embedment large movement was anticipated below final excavation level if the wall were left unbraced for any length of time after earth moving was complete.

The portion of the basement below the intermediate slab was therefore excavated in two stages, first excavating the central portion and casting the lower basement slab and then removing the perimeter berm in short lengths, the basement slab being completed each time a portion of the berm was removed.

Figure 13-50b and c shows the wall movement and ground deformation, respectively, measured during construction. Excavation to the 10-m level resulted in an inward bulging of the wall. The removal of the perimeter berm caused a movement which increased rapidly with time. This movement occurred in the form of rotation about the 10-m floor slab. It is conceivable that if the 16-m slab had been delayed, a plastic hinge might have been developed at the 10-m level followed by a rotational slip below this level. An immediate effect of the last stage of the excavation was a corresponding increase in surface settlement behind the diaphragm wall.

Topics Relevant to Analysis and Design 551

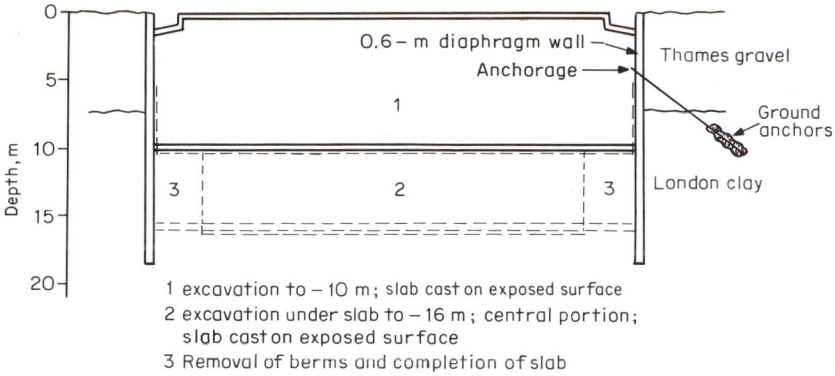

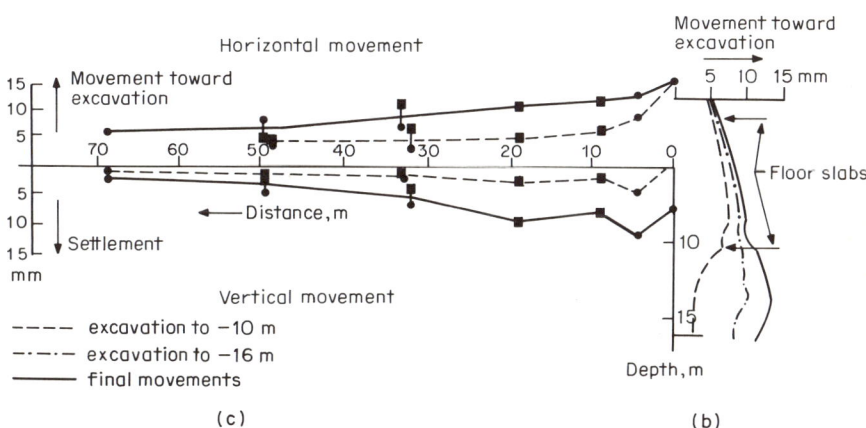

Fig. 13-50 YMCA building, London: (*a*) diagramatic presentation of the excavation sequence; (*b*) lateral movement of the diaphragm wall; (*c*) ground movement and settlement. (*From St. John, 1975.*)

Brittanica House, London The method of construction of the 20-m-deep (66-ft) basement for this building is shown in Fig. 13-51. The tower raft is 27 m wide and 70 m long (88.5 by 230 ft) and is located almost centrally in the east-west direction and close to the north site boundary in the north-south direction. Figure 13-51*a* shows a section though the east and west wall, and Figure 13-51*b* shows a section through the north wall.

The site level was first reduced from +11.0 m (ODN) to about +7.00 m to provide a better construction platform for the diaphragm walls. The walls are 80 cm thick (32 in) and extend to a depth of about 22.5 m (74 ft) below the original grade. Bulk excavation was carried out to the level shown in Fig. 13-51 working southward away from the north wall. A continuous berm was left along the perimeter to brace the walls temporarily, and each excavation slope was covered with a thin protective reinforced-concrete layer to delay the

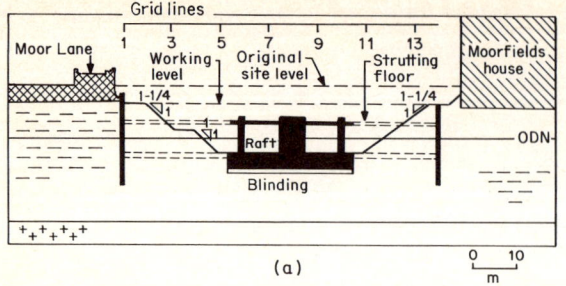

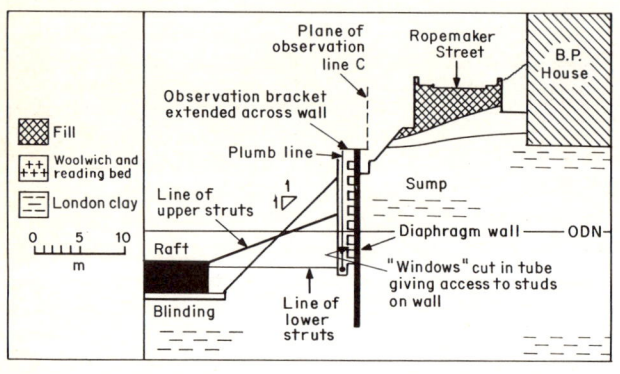

Fig. 13-51 Construction of the 20-m-deep basement for the Brittania House, London: (a) section through the east-west walls; (b) section through the north wall; (c) pertinent soil data; ODN = ordinance datum Newlyn. (*From Cole and Burland, 1972.*)

deterioration of the clay soil. As soon as the excavation reached the formation level for the raft, a 0.9-m-thick layer of unreinforced concrete was placed to prevent swelling and subsequent reconsolidation of the underlying clay. The 3.3-m-thick raft was then constructed in alternate bays to allow for shrinkage effects (Cole and Burland, 1972).

The progress of excavation along the north wall is shown in Fig. 13-52a along grid line 8, which is approximately at the center of the wall between the east and west boundaries. The lateral movement of the top of the north wall at various times is shown in Fig. 13-52b. Evidently considerable support was given to the extremities of the wall by the east and west slopes excavated as shown in Fig. 13-51a. The movement of the north wall at three levels as a function of time is also given in Fig. 13-52c, together with the settlement readings of a leveling stud in the street adjoining the north wall.

The first of the top struts shown in Fig. 13-51b was placed on July 21, 1963, or about 10 days after the concrete raft was cast. Excavation of the top of the north berm continued, so that by August 19, 1963 the entire row of upper rakers was in place and wedged against the raft. The removal of berm beneath this bracing took place after August 19, 1963, and placing of the bottom rakers started at the end of that month.

The excavation procedure for the west and east walls was slightly more complex. Earth moving initially progressed from north to south, as shown in Fig. 13-52a. The movement of the top of the west wall at various times is shown in Fig. 13-53 along the same section of Fig. 13-51a. Excavation from the working level at grid line N (which is the line where the movement is shown) started after June 21, 1963 and was substantially completed by the end of July for the construction of the raft, which was cast on July 30. Partial removal of the berm took place on September 20, and on November 20 the construction of the propping floor braced the wall, as shown in Fig. 13-53.

From Fig. 12-52b and c it appears that the ground response to the excavation was very slow. With the excavation taken almost half-way down (level ODN on November 6, 1963) the movement of the top of the north wall was only 3.5 mm (0.14 in), but it accelerated rapidly as the excavation approached full depth and amounted to 13 mm (0.5 in) on June 21. The movement continued thereafter steadily until July 2, when the excavation reached the profile shown in Fig. 13-52a, at a rate of 1.8 mm/day. After that date the movement continued at a slower rate, and when the struts were placed, this rate was about 0.4 mm/day. Likewise, after an initial lag the movement of the west wall accelerated, reaching a maximum rate of 0.8 mm/day, but at the time the raft was cast, the rate of movement had been reduced to 0.25 mm/day. Thus, this wall moved some additional 16 to 18 mm during the 2 months following the excavation when it was braced only with the earth berm. The total movement of the west wall had increased to 62 mm (2.5 in) by the time the propping floor was in place.

It appears that the movement associated with this excavation was largely time-dependent, and for both the north and west wall the movement which

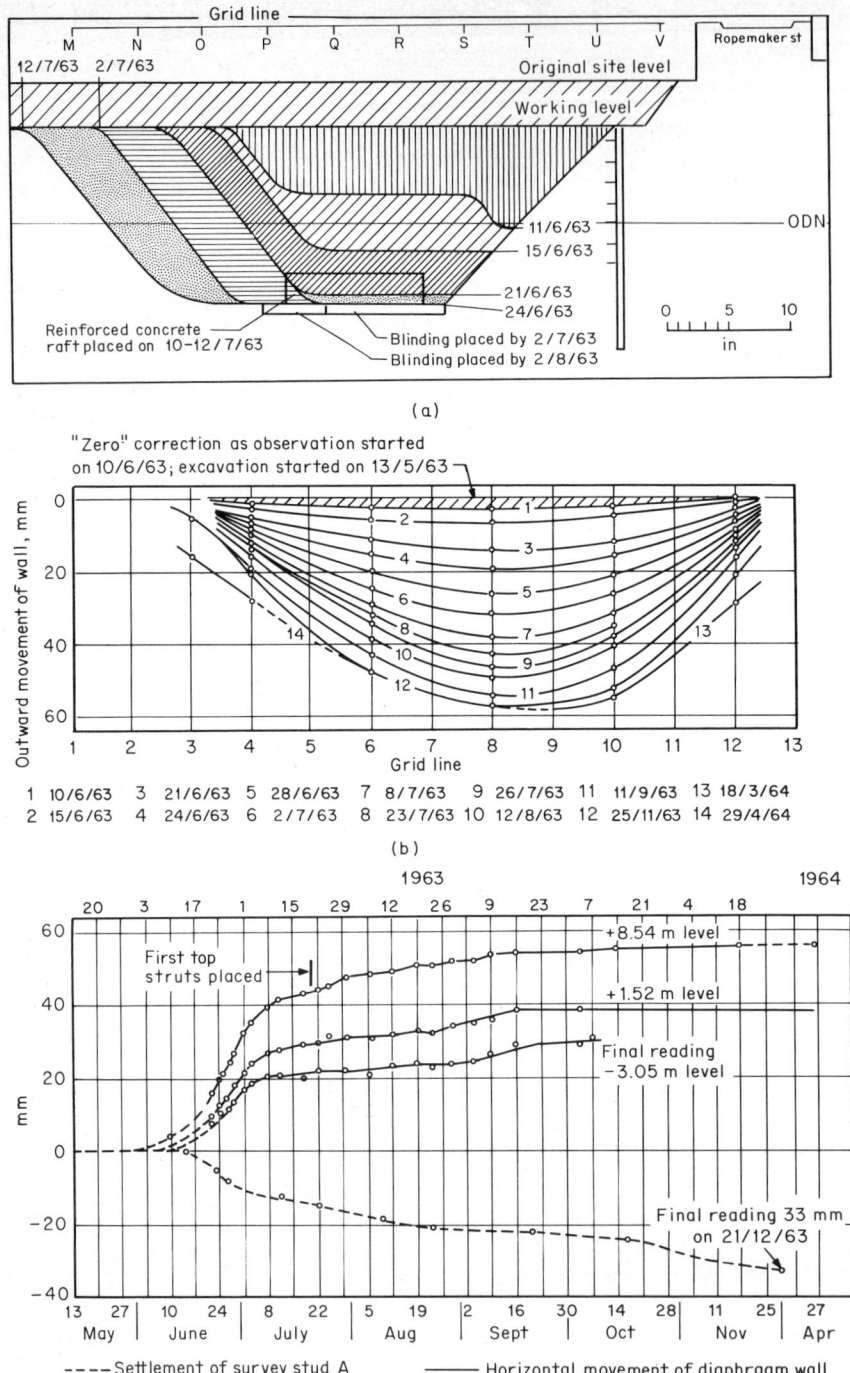

Fig. 13-52 Data from the excavation of the Brittania basement, London: (a) excavation schedule shown along a section normal to the north wall; (b) lateral movement of the top of the north wall; (c) movement of the north wall at three levels; ODN = ordinance datum Newlyn. (*From Cole and Burland, 1972.*)

Topics Relevant to Analysis and Design   555

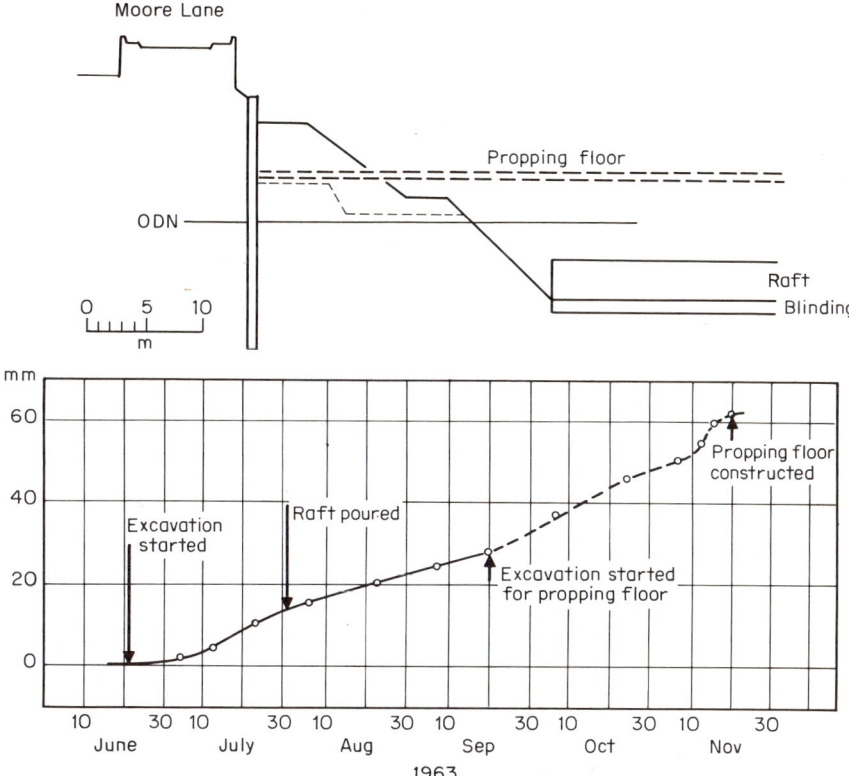

Fig. 13-53 Movement of the west wall, basement of the Brittania House, London. (*From Cole and Burland, 1972.*)

occurred during excavation was only a fraction of the total movement. The difference in the rate of movement between the north and west wall may reflect the influence of different lateral loading conditions and is consistent with the fact that the west berm was less steep than the north slope. These observations demonstrate the importance of providing a positive support in the early stage of the excavation if the movement is to be reduced to a minimum and the necessity of taking the time factor into account in making predictions of movement in plastic clays.

Southern Pacific Excavation, San Francisco Figure 13-54 shows the plan and the cross section of this excavation, together with the undrained shear strength of the surrounding soil (Clough, 1975). The excavation was carried out in very soft clay to a depth of 36 ft (11 m) using a top and bottom bracing only in conjunction with a temporary earth berm. The diaphragm wall is 2.5 ft thick (75 cm) and penetrates about 40 ft (12 m) into the ground below excavation level. The top bracing consisted of the two cross-lot struts and some diagonal bracing located as shown, whereas a group of rakers braced

556　Slurry Walls

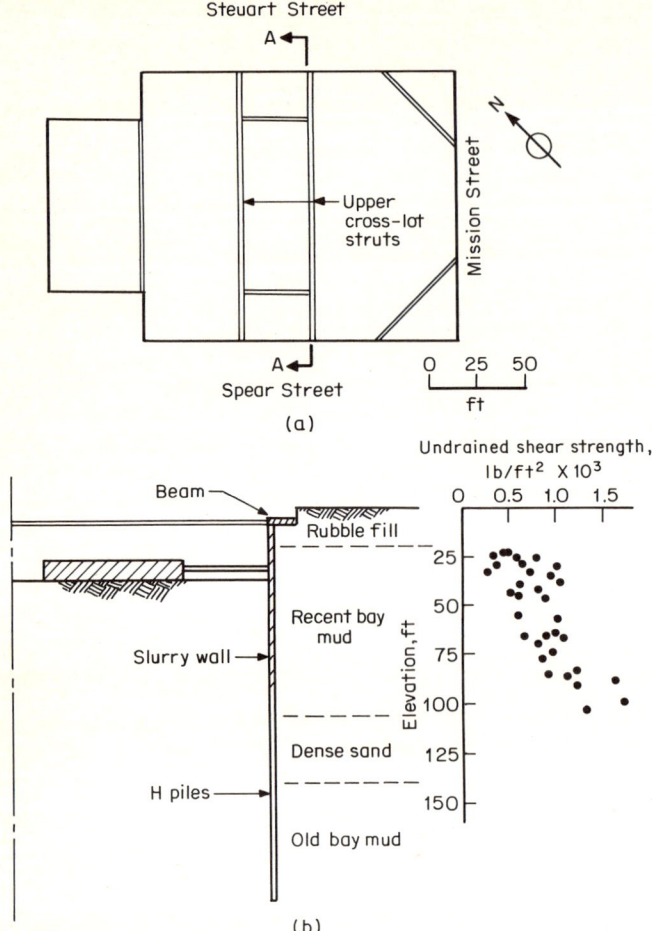

Fig. 13-54 Plan, section, and soil data for the Southern Pacific excavation, San Francisco: (*a*) plan view; (*b*) section A-A. Prestress loads: upper struts 2900 lb/ft; lower struts 6000 lb/ft.

the wall at excavation level. In order to reduce the inward bulging of the wall along its top as a result of the wide strut spacing, a thickening beam was cast on top, as shown, to stiffen the support longintudinally. The earth berm was left against the wall until the lower-level rakers were in place. Because the wall carries significant vertical loads, it is supported on piles.

The wall movement and the corresponding ground settlement are shown in Fig. 13-55. As soon as they were installed, both the top and bottom braces were prestressed to the loads shown in Fig. 13-54. The largest lateral wall movement occurred for the portion adjacent to Steuart Street, since the clay was weakest and the excavation deepest there. More than 1 in (2.5 cm) of movement was recorded at the top of the wall, in spite of the prestressing of

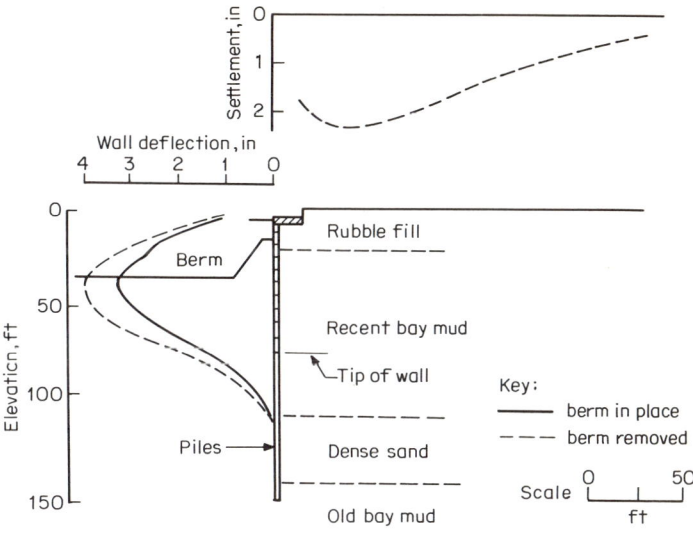

Fig. 13-55 Measured movement and settlement, Southern Pacific excavation, San Francisco. (*From Clough, 1975.*)

the struts, presumably as a result of yielding of the bracing and some bulging between supports. The wall moved 4 in (10 cm) into the excavation just below the base of the cut, whereas at the tip of the wall the inward movement was about 1.2 in (3 cm). Evidently the displacement of the wall toward the cut was largely due to movement below excavation level, and although the movement is not shown as a function of time, it can be assumed that it was time-dependent. More than 3 in of movement occurred while the earth berm was in place.

Clough (1975) has reported the results of comparative studies of the influence of wall stiffness on movement for this excavation. Initially the wall stiffness was taken as the actual EI value of the diaphragm wall (0.49×10^6 kip·ft^2), and the predicted movement and settlement agreed with the observed values. Subsequently, the wall stiffness was reduced to 0.088×10^6 kip·ft^2, which is the EI value of a PZ38 sheeting. For the sheet-pile wall the movement increased to 6 in (15 cm), and the ground settlement reached 3 in (7.5 cm). The predicted movement for both cases is shown in Fig. 13-56.

Procedure for Reducing Ground Movement and Settlement

From the foregoing analysis and discussion of movement and also from Sec. 13-2 it appears that certain useful conclusions can be drawn regarding movement and ground settlement in excavations supported by diaphragm walls, but no attempts can or should be made to quantify the results. The main problem arises from the difficulty of isolating the influence and contribution of individual system parameters, especially time-dependent effects. Some of

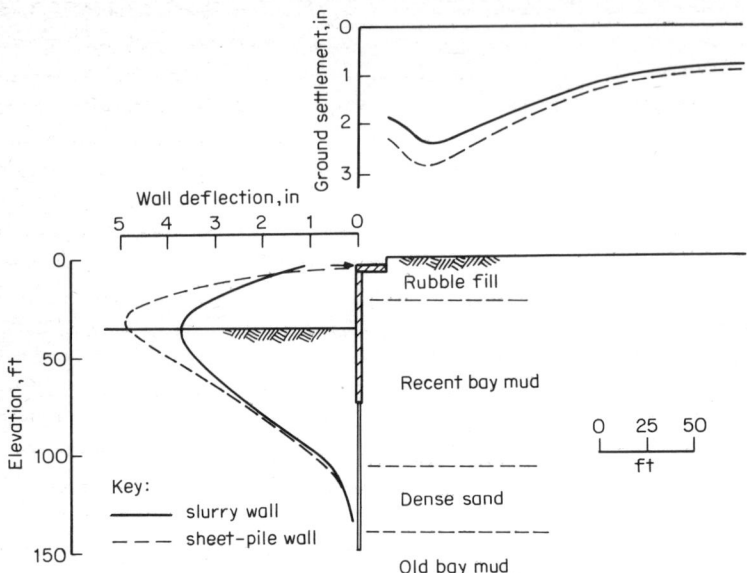

Fig. 13-56 Comparison of lateral movement for two walls of different stiffness, Southern Pacific excavation, San Francisco. Results obtained from finite-element analysis. (*From Clough, 1975.*)

the factors influencing movement can be identified better by finite-element analyses.

Braced Walls For walls braced at the top and at the bottom the construction procedure must take into account the control of base instability, which may be difficult in very large or in very deep excavations carried out in soft clay. Additionally, provisions must be made to prevent consolidation settlement due to the lowering of the water table. If the material outside the excavation is substantially incompressible, e.g., compact sand, the water table there may be allowed to respond to the dewatering inside the excavation. In relatively compressible soil the walls must isolate the construction, or the groundwater table outside the excavation must be maintained by a recharge system.

Bracing with struts offers the advantage of their prestressing for the rebound of the wall to its initial position and to compensate for further yielding of the bracing as the excavation proceeds to a lower level. However, the prestressing of struts supporting diaphragm walls does not appear to be as necessary, as it often is with flexible walls, unless the excavation is unusually wide (in which case struts would not ordinarily be used). Instead, it is better to concentrate on the construction procedure, minimize the time elapsed between bracing installations, and simplify the details. Yielding of the strut at the connection can be avoided and the workmanship can be improved if waling beams are eliminated and the struts are connected directly to bearing plates attached to the cage (see also Fig. 13-74). The extra strength of the wall, allowing greater unbraced heights, e.g., the wall of Fig. 13-54, is not initially an advantage, since in soft clay large and even excessive movement may occur as the excavation is taken down to the next bracing level if the bracing intervals are too large. This strength becomes an advantage, however, after the top and bottom braces are in place, at which time certain intermediate struts can be removed provided the remaining struts can resist the lateral loads.

Bracing with the permanent floors does not appear to contribute more to movement than bracing with struts and in fact it may even result in less movement if the top uppermot floor (near the ground surface) is cast before any excavation. The rigid connection between the concrete slab and the wall precludes the effects of poor workmanship and yielding at this location, which sometimes appear in strutted excavations, and the only component of movement associated with this bracing is initial shrinkage and elastic shortening of the concrete under the effect of lateral load. For an excavation which is 50 ft wide (15 m) the inital shrinkage of the floor slab will be of the order of $\frac{1}{4}$ in (0.6 cm), and the elastic shortening corresponding to a lateral load of 30,000 to 50,000 lb per linear foot will be of the order of 0.1 in (0.25 cm), resulting in a total change of length about 0.35 in (0.9 cm) to be distributed to both sides. If the same excavation is braced with struts stressed to 20,000 lb/in^2 (1400 kg/cm^2), the elastic shortening of the steel for a 50-ft length will be about 0.4 in (1 cm). Premature excavation under the floor before the concrete has hardened sufficiently should generally be avoided, especially if the lateral load includes full pore pressure under considerable head. A guideline in this case is to start earth moving as soon as the concrete floor can support itself in horizontal bending. It the walls are too far apart, e.g., the wall containing tube IV and the wall opposite it in Fig. 13-40, which were some 60 m (200 ft) apart, shrinkage, creep, and elastic shortening of the concrete will contribute a large component to the total movement.

Tied-back Walls The position of the top row of tiebacks is usually governed by the balance between an increase of the initial cantilever and the final bending moments on the one hand and limitations on the resulting wall movement on the other. The first row of anchors is often shown 4 to 5 m (13 to 16 ft) below ground level, anticipating that the resulting wall deflection will be reduced to some extent when the tiebacks are installed and prestressed. However, considerable movement (which sometimes may reach as much as 50 percent of the movement at full excavation depth) may occur during the cantilever stage, and in soil that becomes progressively stiffer this will take the form of rotation about the tip of the wall. Thus, if movement must be kept to a minimum, the first row of tiebacks must be located as close to the ground as possible. Littlejohn and MacFarlane (1974) suggest limiting this depth to 1.5 m (5 ft), but with this shallow anchorage there may be a risk of local ground failure behind the wall during the prestressing of the cables.

Prestressing the tiebacks to draw the wall back to its initial profile and thus compensate for movement which has already occurred may not always produce the desired results. In stiff soil considerable resistance to this displacement will be provided by the strength of the materials being compressed, or the wall may revert to its first movement profile as the tiebacks yield to mobilize the shear in the anchorage zone.

Prestressing the tiebacks to compensate for movement expected to occur is likewise governed by similar uncertainties. In stiff clay or in dense sand the prestress in the tiebacks will generally tend to push the wall toward the retained ground, but the wall stiffness will have an adverse effect on this process. In very compressible soil unusually high prestress levels may cause too much movement of the wall away from the excavation, which is likely to stay permanently if the tiebacks are anchored to rock.

Very significant is the time-dependent movement observed for excavations in clay and the fact that it extended behind the wall for a distance more than twice the depth of excavation. For the Neasden underpass (Fig. 13-44) the movement of the top of the wall during the 18 months following the excavation was almost 150 percent of the movement recorded at the end of excavation and with the four rows of tiebacks in place. For the walls of the Keybridge House (Fig. 13-46) the time effects on movement are revealed during the 3-month period following the excavation, when the differential deflection between the top and the tip of the wall more than doubled. Whereas this may be due to the consolidation of the highly stressed soil surrounding the fixed anchorage, it may also indicate a transition of the soil mass from an elastic to a plastic condition or the consolidation of the clay on the excavation side. These effects show the need for placing the permanent bracing inside the excavation, including the base slab, as soon as possible.

Certain interesting conclusions can be drawn from the performance of panel G21 of the World Trade Center (Fig. 13-48). It is conceivable that in this case the rigid subway section adjacent to the upper half of the wall acted as barrier in combination with the rock embedment of the wall and thus restrained movement of the wall. When the wall was subjected to the prestress loads, it could neither yield at the support points nor materially deflect between these points because of its considerable stiffness, so that the effect of prestressing did not materialize.

Free Cantilevers and Bracing Berms The movement of the cantilever wall shown in Fig. 13-49 is surprisingly low, and this must be attributed to its considerable embedment (almost 9 m, or 30 ft) into the very strong Woolwich and Reading beds. Almost one-half of the wall displacement occurred after completion of the excavation. The casting of the base slab for the road bed braced the walls and stopped further movement.

Berms appear to be more effective if they supplement a more rigid bracing already intalled in the upper part of the wall. Where clay berms brace the wall temporarily in its entirety, they are almost certain to result in excessive lateral movement, in spite of the wall stiffness. Several investigators (Clough and Derby, 1975) have attempted to quantify the effects of berm bracing using finite-element analyses and checking the results against case histories. The conclusion is that providing a generous berm can have a decisive effect on reducing movement in weak soils but is less important in stronger soils. Nonetheless, at least 75 percent of the theoretical passive wedge must be present in the berm for an effective reduction of movement.

Wall Stiffness The stiffness EI of diaphragm walls usually ranges from 0.5×10^6 to about 1.5×10^6 kip·ft² for the common thickness range. This means that the wall is from 7 to 20 times as stiff as the heaviest United States sheeting section and may even be stiffer.

In the examples presented in this section the stiffness of the concrete wall was used intelligently with a suitable embedment in stiff materials where they existed or in connection with the bracing procedures normally available with the diaphragm-wall method. Therefore comparison of the effect of system stiffness on movement is only relative and does not show this parameter isolated.

A true analysis of the influence of stiffness was included with the last case history, i.e., the Southern Pacific excavation, and compared the movement of a sheet-pile wall and a diaphragm wall under the same conditions. The use of the stiffer wall resulted in a decrease of movement from 6 to 4 in, or a reduction of 33 percent.

Predicted movement from finite-element analyses (Clough and Tsui, 1974b) is shown in Fig. 13-57 and demonstrates the effect of the wall stiffness as well as the tieback stiffness on wall movement. The system is a wall in medium clay having a shear strength increasing with depth from 500 to 1400 lb/ft² at 45 ft. The excavation depth is as shown. The diaphragm wall is given a stiffness EI 0.29×10^6 kip·ft² and the sheet-pile wall 0.036×10^6 kip·ft²; that is, the wall is 8 times stiffer than the sheeting. For the sheet-pile wall the tieback stiffness EA/l was varied to simulate first steel rods and then a group of high-strength steel cables. The rod system was 10 times as stiff as the cables.

From the results of Fig. 13-57 it appears that increased stiffness reduces wall movement, in this case by 40 percent. Although this required an increase in the wall stiffness by a factor of 8, it is more appropriate to point out that this increase in stiffness can readily be attained as the analysis is shifted from the sheeting to the diaphragm walls commonly used in practice. The results also show that a similar reduction in wall movement was possible by increasing the tieback stiffness.

13-12 GENERAL STABILITY OF LINEAR WALLS

Braced Walls The external-stability failure mechanism of internally braced walls is shown in Fig. 13-58. The walls can be strutted or braced with permanent floors and interior cross-lot bracing. Such walls do not act inde-

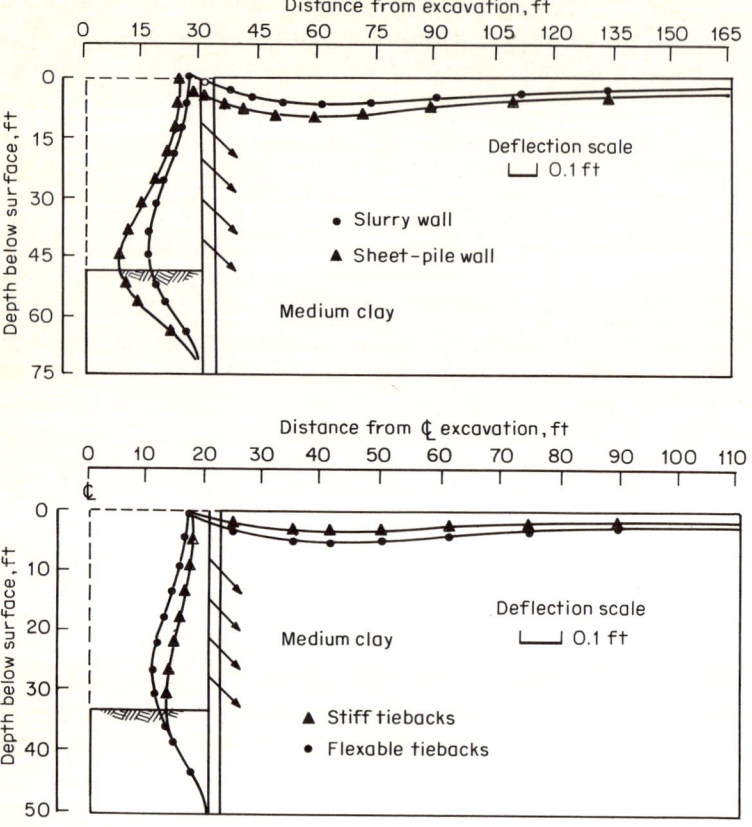

Fig. 13-57 Effect of wall stiffness and tieback stiffness on lateral wall movement. (*From Clough and Tsui, 1974a.*)

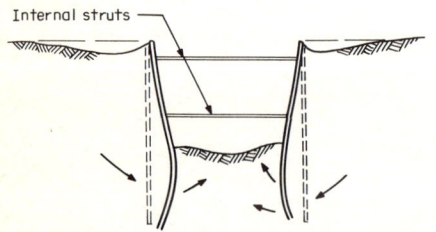

Fig. 13-58 Overall external stability of braced walls.

pendently since the lateral forces are transferred across from one wall to the other. Thus, if the bracing is adequate and sustains the imposed loads, internal failure should not occur.

The external stability is related mainly to bottom instability, which arises in weak soils and particularly in very soft clays, but may also be caused by adverse seepage conditions or if the base is left unprotected for too long. Failure can occur in the soil mass as heave, and procedures for avoiding this problem in narrow excavations are discussed in Sec. 13-3. Failure may also occur below excavation level, where sufficient passive resistance cannot be mobilized to balance the active lateral loads applied externally, and the wall begins to move inward or develops a plastic hinge at the lowest bracing point.

Walls Supported with Berms Walls braced in this manner are not uncommon but usually are constructed at sites where generous berms can be provided or where the lateral wall movement can be tolerated. The external stability is shown in Fig. 13-59. In Fig. 13-59a the excavation is made in a combined cut slope and braced cut. The unsupported slope is above the excavation, so that if slope failure occurs, it will result in a dangerous landfall into the working area. For the excavation shown in Fig. 13-59b the wall is actually braced with the berm. Slope failure in this case will lead to general failure and the collapse of the wall and the excavation. In both cases the analysis should establish the degree of slope stability.

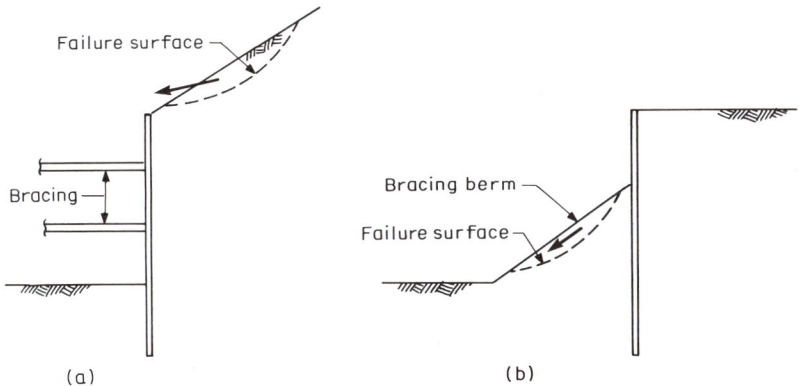

Fig. 13-59 Overall stability of walls braced with berms: (a) combined slope and braced wall; (b) braced wall supported by berm.

Tied-back Walls The general stability problem becomes more complex in this case because the tiebacks depend on the supported ground for their support. Four different failure criteria are shown in Fig. 13-60. In Fig. 13-60a failure occurs because of insufficient penetration of the wall below excavation level. In Fig. 13-60b the wall fails because of anchor pullout or tie failure. In Fig. 13-60c the problem is structural failure of the wall itself, which may occur also with braced walls. In Fig. 13-60d overall slip failure and deep-seated movement can lead eventually to the collapse of the entire system.

Further possibilities that must be investigated are bearing-capacity failure caused by the downward component of the tieback load and the special problem of deep vertical cuts, discussed in Sec. 12-13.

13-13 LATERAL EARTH STRESSES

Lateral earth stresses and loads acting on the ground-support system are generally estimated from limit theory (see also Sec. 13-9); semiempirical methods, whereby the results of measurements on full size excavations consti-

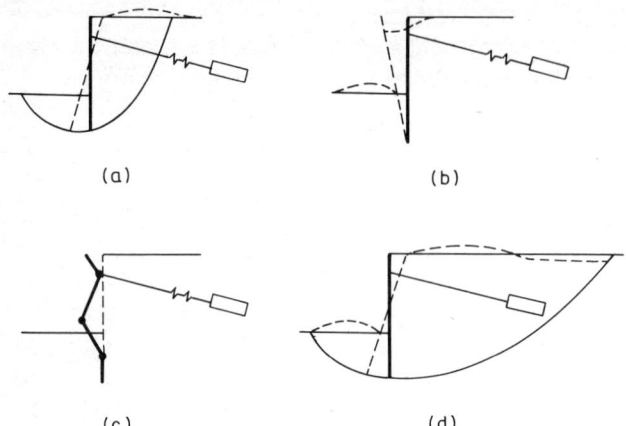

Fig. 13-60 Failure criteria of tied-back walls: (*a*) failure due to insufficient penetration; (*b*) failure of tiebacks; (*c*) structural wall failure; and (*d*) overall slip failure.

tute the basis for the design of bracing loads; and analytical methods emphasizing the importance of understanding the relation between deformations and the mobilization of lateral earth stresses and therefore based on stress-strain relationships. Each of these approaches is reviewed in this section (not necessarily in a consecutive order), and procedures are recommended for their applicability to stiff concrete diaphragm walls.

Limit Theory

The three basic states of stress, i.e., active, passive, and at rest, were mentioned briefly in Sec. 13-9. For a better description the engineer is referred to the many geotechnical sources available, and only certain principles are presented here, primarily for convenience.

Rankine States in Dry Soil The simplest case of stress occurs when

$$\sigma_h = K_a \sigma_v \qquad (13\text{-}10)$$

where σ_h and σ_v are the horizontal and vertical earth stress, respectively, and $K_a = \tan^2(45 - \phi/2)$ is the coefficient of the active state. The passive state is expressed by

$$\sigma_h = K_p \sigma_v \qquad (13\text{-}11)$$

where $K_p = \tan^2(45 + \phi/2)$ is the coefficient of the passive condition. If the difference in the friction angle ϕ is ignored, it follows that $K_p = 1/K_a$.

The strains required to produce active and passive conditions have been found empirically from the results of triaxial tests. In general, very little horizontal strain, less than -0.5 percent, is required to reach the active state. Little horizontal compression, about 0.5 percent, is necessary to reach one-half of the maximum passive resistance. However, much more horizontal compression, about 2 percent, is required to reach the full maximum passive resistance (Lambe and Whitman, 1969). These values are typical for most dense sands. For loose sand the first two statements are still valid, but full passive resistance is reached at a strain as high as 15 percent.

The foregoing conclusions are generally valid when the initial condition in the soil mass involves lateral stresses at rest; i.e., there is not lateral strain in the soil, so that $\sigma_h = K_0 \sigma_v$. If initially the soil stresses are not in a K_0 ratio, different strains are necessary to produce the active and passive conditions.

Rankine States in Soil with Water If a soil is saturated with static water, the coefficient of lateral stress is related to the effective vertical stress $\sigma'_v = \sigma_v - u$, where u is the pore-water pressure. The foregoing relations now become

$$\sigma'_h = K'_a \sigma'_v + \gamma_w h \qquad (13\text{-}10a)$$

and

$$\sigma'_h = K'_p \sigma'_v + \gamma_w h \qquad (13\text{-}11a)$$

where $\gamma_w h$ indicates the pore-water pressure at depth h and K'_a and K'_p corresponds to the friction angle ϕ' for effective stress.

A drained condition in a soil denotes a situation where pore water is allowed to flow freely in and out of the soil; i.e., excess pore pressure is permitted to dissipate by the free movement of water. In this case the pore pressure is either zero (meaning that the total stress and the effective stress are the same) or has a static value. Complex changes in pore pressure causing changes in the effective stress with time are treated in conventional soil mechanics.

The Role of Friction The relative movement of the wall with respect to the soil generally causes shear (frictional) forces at the interface. In the active zone the movement of the wall away from the soil generally produces a downward motion of the soil, as illustrated in Fig. 13-15, so that the friction is a downward force. In the passive zone the soil bulges upward, but either positive or negative friction may develop at the interface so that its sign must be determined from a consideration of the expected motion.

A complete evaluation of the active and passive state with friction becomes rather complicated due to the variation in the inclination of the slip lines. For the usual boundary conditions it is possible to obtain the orientation of the slip lines together with the stresses at each point using a numerical-integration technique (Sokolovski, 1944; Harr 1966). K_a is no longer the ratio of horizontal to vertical stress but is the ratio

$$K_a = \frac{\sqrt{\tau_w^2 + \sigma_h^2}}{\gamma h} \qquad (13\text{-}12)$$

where τ_w is the shear stress at the wall interface, and σ_v is not necessarily equal to γh owing to the curvature of the slip-line field.

Wall friction greatly reduces the horizontal thrust and especially the overturning moment against walls deriving their stability from semigravity or full-gravity action, e.g., the structures mentioned in Sec. 13-5. In this case it is appropriate to take advantage of the beneficial effect of wall friction on stability, since the downward drag will normally develop as the structure moves away from the earth.

Active and Passive State in Soil with Cohesion The effect of cohesion on the passive thrust can be considered if the thrust is assumed to be applied at a rate that does not cause excess pore pressure. Setting $N_\phi = (1 + \sin \phi)/(1 - \sin \phi) = 1/[\tan^2 (45 - \phi/2)]$, we can show that the horizontal stress σ_h and the vertical stress σ_v are related as follows:

Passive state:
$$\sigma_h = \sigma_v N_\phi + 2c\sqrt{N_\phi} \qquad (13\text{-}13)$$

so that with cohesion c present N_ϕ is not equal to the ratio of horizontal to vertical stress. Indeed the first term in Eq. (13-13) implies a linear variation of stress with depth, but the second term indicates a constant stress with depth, so that the resultant thrust is located between the midpoint and the lower-third point.

Likewise the effect of cohesion on the active thrust is included in the following relation for the horizontal and vertical stress:

Active state:
$$\sigma_h = \sigma_v \frac{1}{N_\phi} - 2c \frac{1}{\sqrt{N_\phi}} \qquad (13\text{-}14)$$

Equations (13-13) and (13-14) are easily modified for effective stress and water pressure.

It is evident from Eq. (13-14) that the presence of cohesion has a beneficial effect on the lateral earth load, since at the surface the earth stress is negative and equal to $-2c/\sqrt{N_\phi}$, indicating that the soil adheres to the wall and thus provides a negative pull toward the retained ground. Nonetheless, this adhesion often is of doubtful reliability, and there will be a tendency for the soil to crack away from the wall in the tension zone. These tension cracks may be filled with water, thus increasing the lateral pressure and changing the point of application of the resultant. The clay may shrink and pull away from the wall, which will also increase the lateral pressure. This situation can be compensated for if the theoretical benefits of cohesion are neglected ($c = 0$) and a fictitious triangular pressure is assumed for the entire wall height using an arbitrarily selected friction angle of 20 to 30°.

The strains necessary to bring a clay to the limit states are generally greater than those required for sand. For example, normally consolidated clay requires almost -4 percent horizontal strain to change from an initial $K = 1$ condition to the active stress condition. The same clay needs 5 times as much horizontal strain to go from $K = 1$ to K_p as to go from $K = 1$ to K_a. If $K_0 < 1$, less strain is required to reach the active state starting from K_0 than from $K = 1$.

Stresses for Undrained Conditions A special case will exist if $\phi = 0$, which applies, for example, to a saturated soil undergoing undrained loading. The analysis now becomes much simpler, since undrained strength is determined by the initial conditions before loading so that it is not necessary to determine the effective stress that would exist at failure. For this case $N_\phi = 1$, and c can be replaced by s_u. The corresponding critical-failure plane is inclined at 45°.

The application of the $\phi = 0$ analysis to diaphragm walls normally will indicate a temporary condition and is justified if the excavation is done quickly or if there is a rapid drawdown whereby the level of water is expected to drop rapidly. Nonetheless, consideration should be given to the temporary stages of excavation, where it is necessary to rely for stability on clay strength measured in terms of undrained strength. Even with quite small movement of the wall, the strains within the soil mass are such that unless the soil is completely homogeneous, its strength will change rapidly from that measured in terms of undrained shear to that measured in terms of effective stress. The time necessary for that usually is a few weeks, but it will be accelerated by the presence of fissures and laminations. The overall strength in terms of effective strength is commonly much lower than it is in terms of undrained strength. Hence the analysis should also be checked using effective-strength parameters, especially if there is a possibility of delaying the excavation beyond the anticipated period.

At-Rest Condition In the oedometer test (also called the *confined-compression test* and the *consolidation test*) stress is applied to the soil sample along its vertical axis, while strain in the horizontal direction is prevented. Therefore, in this test the ratio of the lateral stress to the vertical stress is the coefficient K_0.

In sedimentary soil, as the buildup of overburden continues, there is vertical compression of soil because of the increase in vertical stress, but there should be no significant horizontal compression during sedimentation. If this is true, the horizontal stress should be less than the vertical. For a sand deposit formed in this manner K_0 may vary between 0.4 and 0.5. When a granular soil is loaded for the first time, K_0 will depend on the amount of frictional resistance mobilized between particles, and so it can be related to the friction angle. Experimental values of K_0 are best represented by the (Jaky, 1944) expression

$$K_0 = 1 - \sin \phi \qquad (13\text{-}15)$$

There is evidence that the horizontal stress can exceed the vertical if a soil deposit has been heavily preloaded in the past, as the result of a process whereby this stress remained locked in and did not disappear when the loading was removed. In this case K_0 may be as high as 3.

From a series of oedometer tests, Brooker and Ireland (1965) have developed a relation between K_0, overconsolidation ratio, and plasticity index. As an average for all normally consolidated clays

$$K_0 = 0.95 - \sin \phi' \qquad (13\text{-}16)$$

Another experimental relation recommended by Alpan (1967) is

$$K_0 = 0.19 + 0.233 \log \text{PI} \qquad (13\text{-}17)$$

for normally consolidated clay, where PI = plasticity index in percent. For overconsolidated clays K_0 is in the range of 1 to 2 or even higher. Myslivec (1972) has considered the foregoing experimental relations where only a part of the shear strength of the soil is mobilized.

The importance of knowing the initial stress conditions has been demonstrated by Morgenstern and Eisenstein (1970). The initial effective stress can seldom be assumed with confidence, with the possible exception of normally consolidated clays. Thus, a history of sedimentation followed by erosion and subsequent reloading may result in a K_0 coefficient approaching K_p.

The need to determine K_0 accurately is reflected in the development of methods and equipment for in situ measurements. Bjerrum and Andersen (1972) have used a method based on the principles of hydraulic fracturing to measure the lateral earth stress at rest in normally consolidated clays. The K_0 values were found to range between 0.40 and 0.50 for the quick and 0.50 to 0.60 for the nonquick clays. Beneath a fill an increase in the horizontal stress was measured. A comparison of the K_0 values measured in the field with those obtained in the laboratory indicates that in a clay deposit K_0 increases with time. In situ measurements of soil properties became the main topic at the June 1975 Specialty Conference of the Geotechnical Division, ASCE, at North Carolina State University. The reader is referred to the proceedings of this conference.

Measured Lateral Stresses on Braced Diaphragm Walls

Walls Braced by Embedment into Rock and by Permanent Floors
For the deep basement wall shown in Fig. 13-32 lateral pressure measurements were taken for three stages: during concreting of the instrumented panel, during excavation, and for some time after completion of the structure.

The pressure exerted on the sides of the trench during concreting was the hydrostatic pressure of fresh (fluid) concrete only in the upper 5 or 6 m of the column pour and about 0.67 times that value in the deeper part of the wall, indicating the effects of the hardening process there. The total lateral force per linear meter upon completion of the pour was 235 t, adjusted to 230 t/m just before excavation.

During excavation the total thrust on the wall decreased from 230 to 170 t/m, and in the same period the point of application of the resultant did not vary by more than 1 m. The stress pattern, magnitude, and changes are shown in Fig. 13-61. Excavation to stage II proceeded before the slab at ground level was placed and was followed by a considerable reduction in earth stress along the wall. From stage II to stage III the wall was braced at the top by the A floor, but again a decrease in pressure was observed. From stage III to stage IV with two floors in place the wall rotated slightly about the B support, thereby causing an increase in lateral pressure above that level.

Long-term measurements of earth stresses are shown in Fig. 13-62 for three key stages of the construction: day 72, just before the start of excavation; day

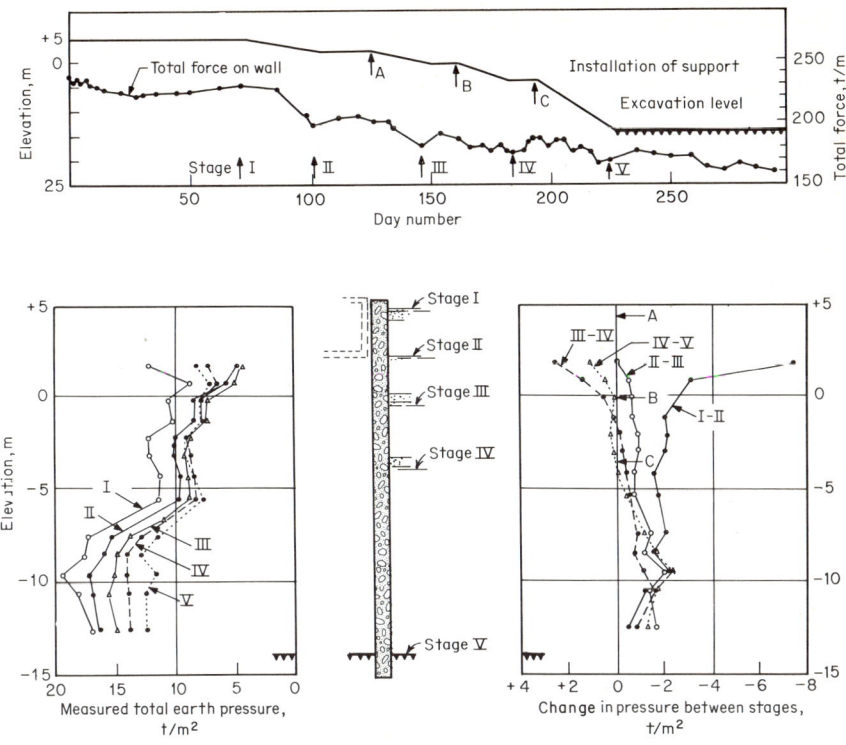

Fig. 13-61 Magnitude and changes in total lateral force acting on the wall of Fig. 13-32. (*From DiBiagio and Roti, 1972.*)

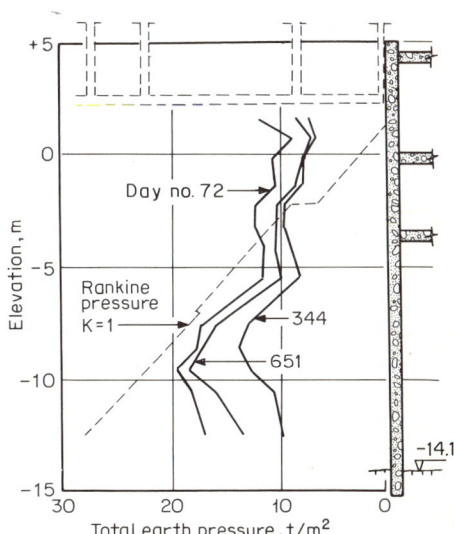

Fig. 13-62 Long-term measurements and changes in lateral earth stresses for the wall of Fig. 13-32. (*From DiBiagio and Roti, 1972.*)

570   Slurry Walls

344, when the minimum thrust was observed; and day 651. The total lateral earth thrust corresponding to these three stages is 230, 160, and 197 t/m, and the point of application is at the 0.44, 0.45, and 0.44 point of the wall height from the bottom, respectively, or nearly constant. The dashed line in Fig. 13-62 represents the earth pressure computed on the basis of $\phi = 0$ and ignoring cohesion, i.e., a pressure coefficient $K = 1$. The total thrust represented by this line is 248 t/m.

Although the resultant thrust on the wall decreased steadily during excavation, it shows a definite tendency to increase again and approach the $K = 1$ condition. However, the point of application and the distribution of the thrust show the influence of the bracing procedure and the wall movement. Thus, more thrust acts near the top and less elsewhere, raising the point of application from the 0.33 to the 0.44 point. Likewise, changes in earth stress caused by drainage of the lower part of the clay before the rock was sealed are evident in the diagrams.

Walls Braced with Struts   Figure 13-38a shows the measured lateral earth stresses on the walls of the Powell Street Station, BART, for two phases, February 1969, when the excavation was nearing completion, and February 1970, when the construction was essentially completed. Also shown are the normal water pressure, the total vertical stress, and the at-rest and water pressure $K_0 \gamma' h + \gamma_w h$. The diagrams were derived from measurements on the main braces of five panels having a load variation of 12 percent from the average. Although the development of strut loads between the two stages depicted may be subject to fluctuations due to thermal-expansion and concrete effects, it shows a tendency for the lateral pressures to increase with time after the excavation as the water level is restored.

From the pressure diagram for February 1970 it is evident that the loading in the upper part of the profile exceeds the at-rest and water pressure, whereas the lower profile indicates pressures less than the at-rest and water pressure. A tendency is evident for the two resultant pressures to be substantially equal for the long-term condition.

Figure 13-63 shows field measurements of bracing loads and total pressures for the diaphragm walls of the Civic Center Station, BART. A section through the structure, soil data, and lateral wall movement are shown in Fig. 13-43. The K_0 condition was developed by measuring values of this coefficient at the laboratory. K_0 was 0.34 for sand, 0.41 for clayey silt, and 0.38 for combined soils. For design purposes K_0 was taken as 0.45. The design combined pressure (effective stress plus hydrostatic water pressure) is shown in Fig. 13-63.

The apparent pressure (obtained from the bracing loads) is shown for two occasions: (1) during construction, January 1970, when the excavation was completed and with all the bracing in place, and (2) for a postconstruction period in December 1970. During construction the actual water level outside the structure was lower than anticipated. The significant reduction in pres-

Topics Relevant to Analysis and Design 571

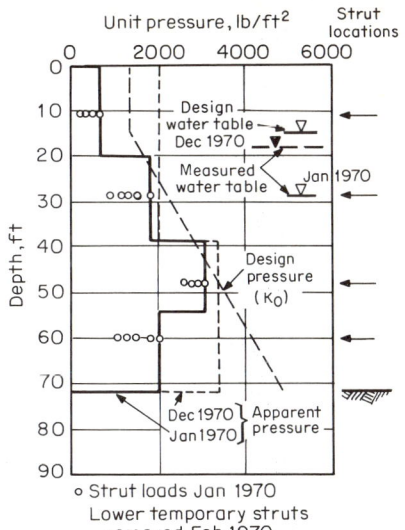

Fig. 13-63 Field measurements of bracing loads and total pressure for the diaphragm walls of the Civic Center Station, BART, shown in Fig. 13-43 (*From Thon and Harlan, 1971.*)

sure near the base of excavation shown for this stage may be due to a decrease in pore pressure in the upper clay layer, where most of the wall movement occurred. Because of the very low permeability of the upper clay, this reduction persists for some time (Thon and Harlan, 1971). A buildup of pressure is evident during the year following the completion of excavation due to the restoration of the water table and of the pore pressure in the clay layer. The distribution of the combined soil and water pressure is more nearly triangular than rectangular and approaches the design curve reasonably close, especially if the actual measured values of K_0 are used. A tendency is again evident for the loading to exceed the K_0 condition in the upper part, whereas the actual lateral pressures tend to be less than the design pressures in the lower part of the wall.

Temporary Loads for a Strutted Excavation Figure 13-64 shows a section for the diaphragm wall of the basement of the Sears Tower, Chicago (Cunningham and Carpenter, 1975). The excavation is about 45 ft deep (13.7 m) in soft Chicago clay, but the 30-in-thick (75-cm) wall is extended almost 24 ft below excavation level into stiff and tough clay. The wall was braced during excavation by three tiers of wales and inclined rakers, and horizontal diagonal bracing was used at the corners.

The design brace loads shown in Table 13-1 were obtained from the Terzaghi and Peck (Peck, 1969) trapezoidal diagram. The measured strut loads at the end of excavation are shown for three sides and one corner with all the braces in place. The brace loads for the top tiers are reasonably close to the design loads, but in some instances much lower or considerably higher. The very low raker loads measured along Wacker Drive indicate a reduced earth surcharge at this location since this is a viaduct structure. Invariably the lower tiers received loads much lower than predicted from the apparent pres-

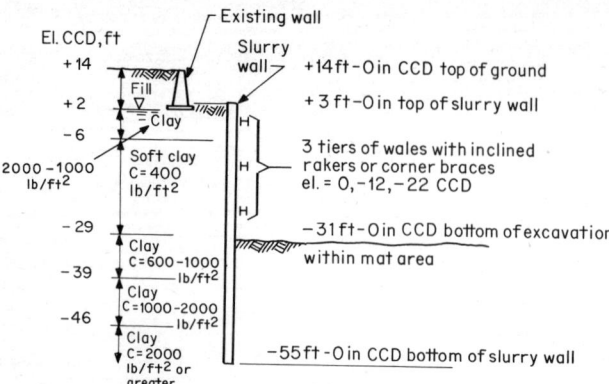

Fig. 13-64 Diaphragm wall for Sears Tower basement, Chicago: section through wall and soil data; CCD = Chicago city datum. (*From Cunningham and Carpenter, 1975.*)

TABLE 13-1 Estimated and Measured Brace Loads, Sears Tower, Chicago (from Cunningham and Carpenter, 1975)

		Load, lb × 1000	
Mark	Location	Measured	Design
Adams	No. 12 top tier	406	405
	Midtier	250	390
	Bottom tier	73	362
Franklin	No. 31 top tier	295	425
	Midtier	206	461
	Bottom tier	47	405
Adams	No. 11 top tier	461	405
	Midtier	334	390
	Bottom tier	126	362
Wacker	No. 50 top tier	117	250
	Midtier	221	400
	Bottom tier	31	381
16-27	NE first (top) tier	636	500
18-25	Second tier	480	556
19-24	Second tier	350	486
	Third tier	111	486

sure diagrams, in some cases only 10 percent of the design loads. Although this is surprising, it can be explained by the considerable wall embedment into stiff soil. It is conceivable that excavation below the intermediate brace caused more inward movement. Thus, when the lower tier was installed, the passive resistance below the base had already been mobilized and precluded the transfer of the earth load, except a small portion, to the lower tier as the excavation was carried down to the final level. A further factor explaining the measured loads is that a jacking force was applied to the upper tiers but not to the lower. Other effects are discussed in the following paragraphs.

Experimental Analysis of Strut-Load Distribution The Sears Tower example suggests the considerable effect of the wall rigidity on the load distribution in the struts, although it does not quantify it. Figure 13-65 shows diaphragm walls for an experimental enclosure, used to analyze the load distribution for a strutted excavation. Each wall panel is 60 cm thick (24 in), 5 m wide (16.5 ft), and 11 m deep (36 ft). The excavation is 7 m deep (23 ft), and each panel was braced by seven tiers of struts at 1-m vertical intervals, two sets per panel. The struts were hydraulically expandable to allow the application of jacking forces.

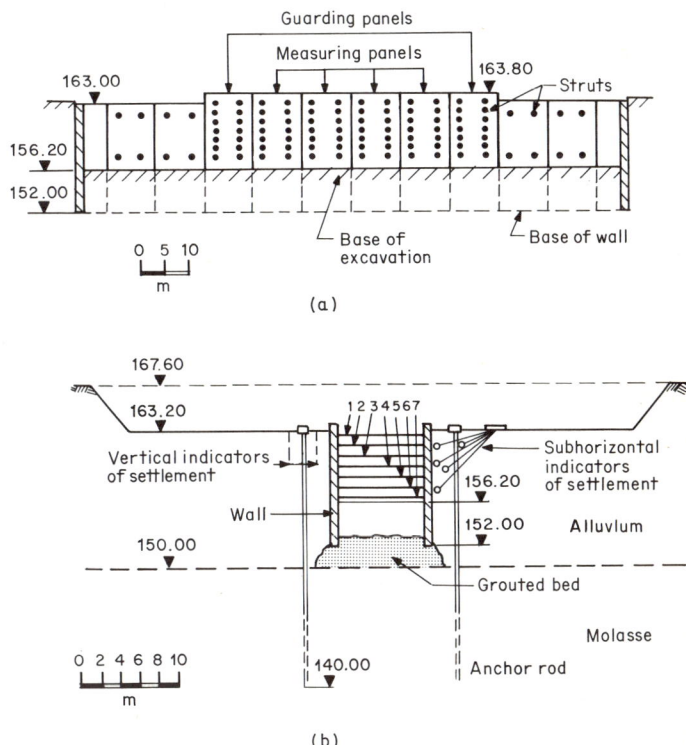

Fig. 13-65 Model walls for the measurement and distribution of strut loads: (a) wall elevation; (b) section through excavation. (*From Kastner and Lareal, 1974.*)

The soil at the test site consists of alluvial gravelly sand with bulk density 2.2 Mg/m³ and a friction angle in the range of 27 to 35°. The water table stood about 70 cm below the construction level.

Kastner and Lareal (1974) have analyzed the variation and distribution of strut loads, first during excavation and installation of the bracing and then during the process of equalization of these loads. The excavation was carried so as to minimize lateral wall movement and thus justify the K_0 condition. Each pair of struts was installed as soon as the earth moving was completed to that level and immediately brought under control for constant length. Thereafter, each strut was essentially rigid, with a maximum change in length of 0.1 mm.

The apparent pressure for each layer at the end of excavation and with all seven strut tiers in place is shown in Fig. 13-66a. Diagram A represents the average estimated pressure (earth plus water) per meter depth on each layer, based on the strut-load measurements. Diagram B depicts the apparent soil pressure (Peck, 1969), whereas diagram C represents the adjusted apparent pressure for effective stress and water pressure. Evidently there is a considerable difference between apparent and measured pressures. Furthermore, for a system of stiff walls and bracing, differences between strut loads in the same tier are quite significant. For example, diagram D of Fig. 13-66a shows the

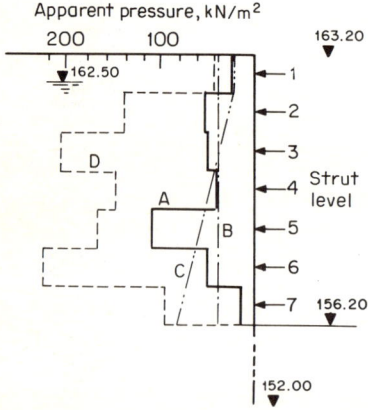

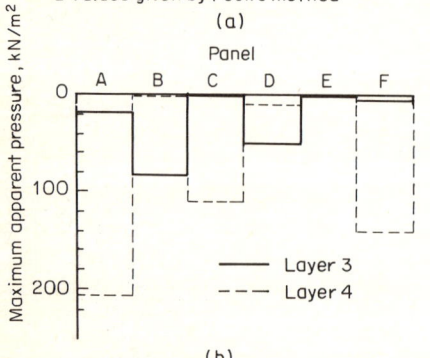

Fig. 13-66 Observed performance of the test panels of Fig. 13-65: (a) apparent pressures at the end of excavation for each layer; (b) apparent pressures at the end of excavation for each pair of struts in layers 3 and 4 (*From Kastner and Lareal, 1974.*)

maximum values measured locally on one of four pairs of struts. These values are much greater than the average stresses of diagram A and several times the values of the apparent pressures. This divergence must be attributed to the rigidity of the system.

Examination of the forces resisted by the pairs of struts shows that some of them absorb the loads from their neighbors. For example, in Fig. 13-66b the two struts in the third layer of panel B are heavily loaded, whereas the two adjoining ones and the pair immediately below (layer 4) are practically unloaded.

Although these results are not likely to be directly applicable to strutted walls because of the very close vertical strut spacing which induced a very high rigidity to the system as a whole, they nonetheless show that the variation in the strut loads is far greater when the system is stiff. The predicted strut loads may thus be significantly underestimated or overestimated. For an excavation in which the distance between walls is not fully controlled, e.g., a system of flexible walls, deformation under pressure leads to a more even load distribution. With stiff walls, the absence of any appreciable deformation may result in some struts receiving a great deal of load while others receive very little.

Measured Lateral Stresses and Loads on Tied-Back Walls

Tied-Back Walls for Guildhall Development, London The walls for this project are shown in Fig. 13-45, together with the lateral movement observed during excavation. The analysis of lateral stresses was initially based on an empirical method (Littlejohn et al., 1971) taking into account the continuous construction process and excavation stages as well as variations in the soil strata. Certain basic principles in this case rely on the assumption that the soil stress distribution is triangular rather than rectangular and that the wall yields progressively as the excavation proceeds.

The approach is based on step-by-step consolidation of the multi-tied system into a repetitive single-tied wall design of a form which allows the wall penetration to be estimated in a manner satisfying rotational as well as horizontal equilibrium (see also the last subsections of this section).

Bending moments for the wall of Fig. 13-45a, shown in Fig. 13-67, correspond to the excavation stages shown in Fig. 13-45b. Bending-moment profiles follow the pattern predicted by the design method. Particularly Fig. 13-67h and i appears to confirm the single-tied wall response, whereby the support of the two tiebacks is consolidated at a point between them so that the bending moment is negative there and positive elsewhere. Further consolidation of the bending-moment curve resulted in a lateral pressure distribution which is more triangular than trapezoidal.

An analogous analysis and monitoring of the Keybridge House wall shown in Fig. 13-46a produced the bending-moment profiles shown in Fig. 13-68 for the final excavation stage. Figure 13-68b also shows the design and the measured moments for the cantilever portion of the wall. The magnitude of

576 Slurry Walls

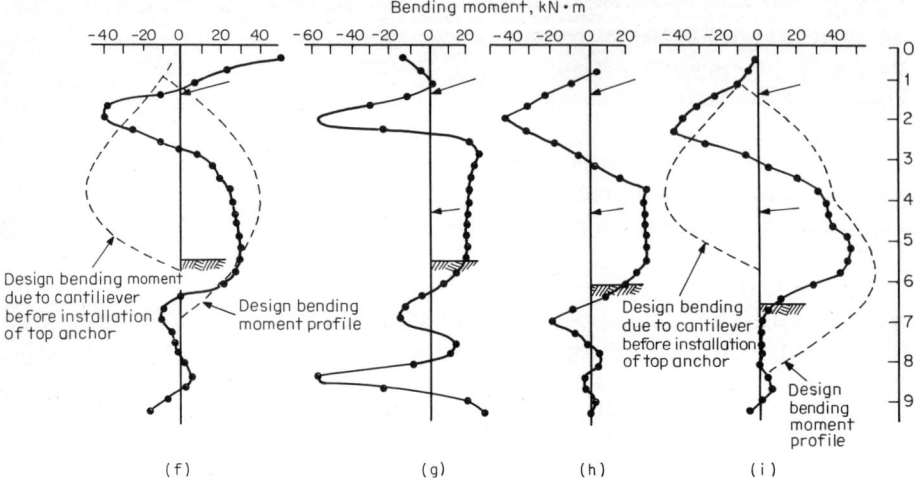

Fig. 13-67 Bending moments for the wall of Fig. 13-45 (based on a 0.3-m strip); moment diagrams (f) to (i) correspond to stages (f) to (i) of Fig. 13-45.

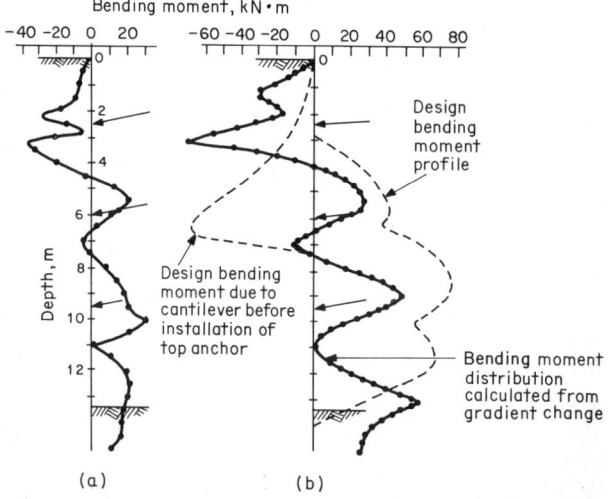

Fig. 13-68 Bending moments for the wall of Fig. 13-46 (a), based on a 0.3-m strip; final excavation stage.

the maximum bending moments is in good agreement with the design values, but a variation is evident in their distribution. The magnitude of the design cantilever moment is very similar to that produced by the stressing of the upper row of tiebacks. It should be noted that the measured moments relate to the normal seasonal groundwater level, whereas the design curves are based on the flood level (Littlejohn and MacFarlane, 1974).

Bending-moment discontinuities at the anchor locations are slightly lower (up to 1 m) than the ones predicted by the design, and this may be due to the anchor inclination and overdig before anchor installation. The shape of the moment curve above the top anchor approaches a sinusoidal curve and may indicate the influence of the guide wall at the rear of the crest. Comparison of the deformation profiles for the final excavation stage shown in Fig. 13-46b to i indicates that bending moments reached maximum values at this stage.

Regarding the overall behavior of this wall, more efficient anchoring was obtained with the gravel anchors, causing the wall to undergo some rotation about the upper anchor region. A small reduction in anchor load occurred over a period of 6 months, 2.8 percent for level 2 and 12.7 percent for level 3.

Tied-back Walls for the World Trade Center, New York Figure 13-69a and b shows bending moments and lateral stress for panel W35 of the World Trade Center diaphragm walls (see also Fig. 13-48). The moment diagrams have been correlated with the actual curvature of the wall and were obtained from slope readings following a procedure described by Saxena (1974). These values correspond closely to moments computed from dynamometers installed on the reinforcing steel (Gould, 1970), so that these two sets of observations actually compare the curvature of the wall. The general shape and magnitude of the moment diagrams appear to be correct, although slope observations are not sufficiently sensitive to reveal abrupt moment variations at the ties. The horizontal pressures at the end of excavation (August 26, 1968) shown in Fig. 13-69b have been estimated from tie-rod load cells and from data on moments. There is an apparent discrepancy between this pressure diagram and the one shown by Gould (1970) for the same panel and for the same date.

There appears to be good agreement between the moment diagrams shown by Saxena (1974) and Gould (1970) for this panel. The moment curve between depth 20 and 40 ft suggests the consolidation of several ties into a single-tie support. In the early stages of the excavation the moments are largely positive, corresponding to the behavior of an anchored bulkhead. As excavation progresses, negative moments are developed at the region of the four center ties, indicating that these ties act as single support, and thus a large positive moment remains in the lower part. The insertion of tie T_6 there causes a discontinuity in the moment curve but does not reduce its magnitude appreciably. The final large bending moments near the base of the wall should indicate a considerable horizontal subgrade reaction developed along the contact zone between the rigid wall and the hard rock and are in conformity with the steep curvature of the wall just above the rock level since it could not rotate about its tip because of the great rigidity there.

Little attention can be directed to the correlation of the actual and design pressures. Although the diagrams in Fig. 13-69b show the original (at-rest) earth and water pressures redistributed to take into account the prestressing of the tiebacks, the pressure diagram obtained from field measurements shows a substantial departure from the assumed, especially in the lower part and the center of the wall.

Bending moments and lateral stresses for panel G21 of the World Trade

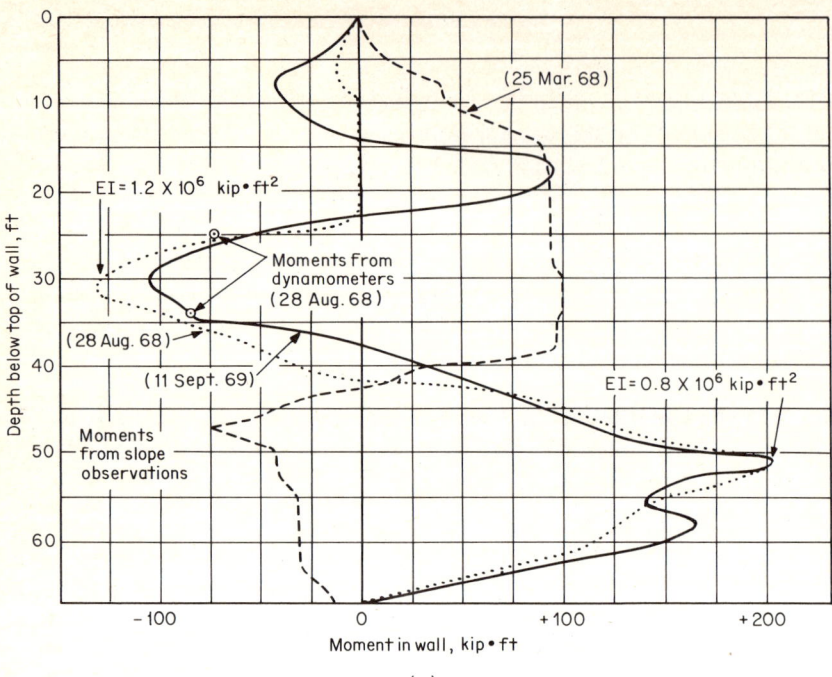
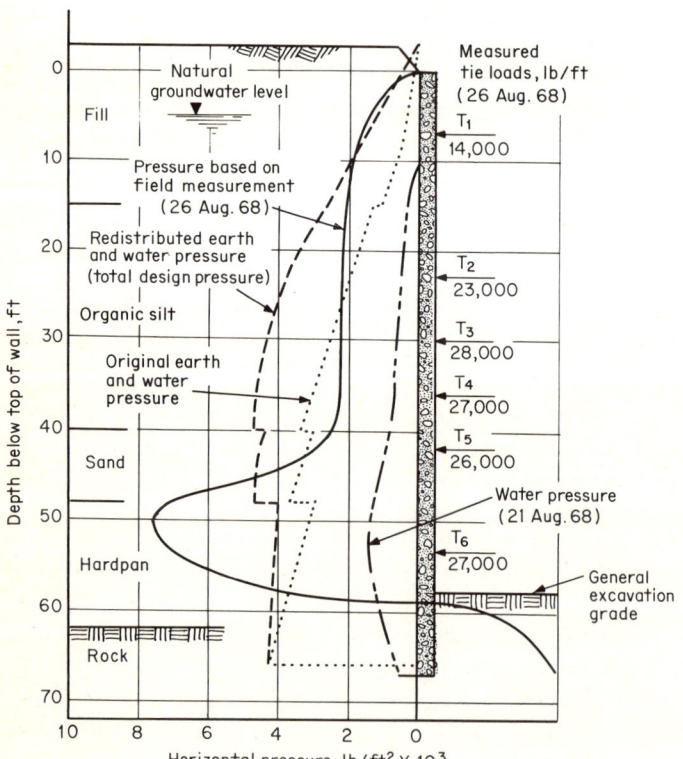

Fig. 13-69 Panel W35 of the World Trade Center, New York: (a) bending moments; (b) lateral stresses and tieback loads. (*From Saxena, 1974.*)

Center are shown in Fig. 13-70 (see also Figs. 13-47 and 13-48). The actual horizontal pressure is based on field measurements from load cells. The moment diagram suggests a concentrated reaction near the base of the existing subway section above tie T_1. The measured pressures approach the earth stress at rest plus water pressure, except in the lower part of the wall, where they are less. Gould (1970) has reported a drawdown of piezometric levels which reached about 34 ft at bedrock, following exterior dewatering near this location to accommodate a planned wall opening.

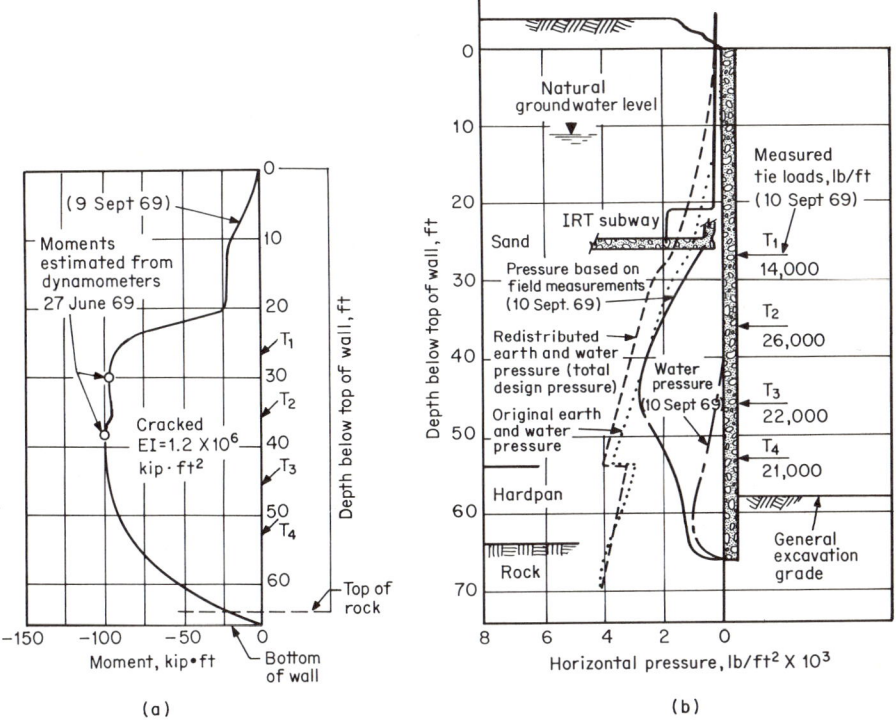

Fig. 13-70 Panel G21 of the World Trade Center, New York: (a) bending moments; (b) lateral stresses and tieback loads. (*From Saxena, 1974.*)

Factors Influencing Lateral Stresses on Diaphragm Walls

Although the application of limit theory has been useful in many practical cases, the foregoing examples clearly demonstrate its limitations. Predictions are accurate for the conditions under which they are used, but a further criticism of limit-theory solutions is that they do not evaluate stress conditions outside the failure zone. More theoretical difficulties arise for an exact solution, since for an active or passive state the resulting strains must be related to the stress through a suitable stress-strain relationship which may in itself be an uncertainty. A further consideration which complicates stress predictions is the presence of natural soil behind the wall, commonly nonhomogeneous or stratified, and with a high natural water table as opposed to a selective and uniform backfill.

Two factors become indispensable considerations in the analysis of stresses acting behind a diaphragm wall. The first is manifested by changes in the state of stress of the soil, which involve three main facets: (1) an initial alteration of the existing state of stress in the soil adjacent to the trench brought about by the slurry process; (2) a further alteration of this stress caused by the placement of the higher-density but relatively fluid concrete, which may or may not exert its full thrust, and its subsequent shrinkage; and (3) changes in the strength of the soil occurring before excavation and attributed mainly to the interaction with slurry in deep penetration and rheological blocking.

The second factor relates wholly to the methods and sequence of construction. After the wall is built, the intended excavation is carried out on one side. The wall may be left partially unsupported or allowed to cantilever; be braced internally with the permanent floors or with temporary struts; be tied back with ground or rock anchors carrying little or considerable prestressing; or be prevented from movement by a combination of bracing including earth berms. Subsequently, the wall may be incorporated in the permanent structure or abandoned.

Recommended Earth Stresses on Braced Walls

Long-Term Condition The recommended earth stresses for permanent walls have their origin in the at-rest condition. Although the influence of the system of supports on the distribution of the earth pressure is obvious, as is the influence of the excavation on the pore pressure, it can be assumed that the lateral earth stresses ultimately will return to nominal at rest values. In general, these should be linear with depth since the shear stresses mobilized during excavation as the result of ground movement and arching will eventually relax following creep, ground vibration, and groundwater fluctuations. The possibility of much larger lateral stresses in overconsolidated clays than existed before construction has been considered but discounted, the contention being that these stresses are somewhat relieved by the horizontal tensile strains accompanying the excavation.

In most instances the earth stresses at rest should be considered in conjunction with full pore pressure unless the site is known to have a permanent groundwater lowering. Good judgment is essential in the selection of at-rest coefficients, and the subsoil should be grouped on the basis of its origin and index properties.

Short-Term Condition Certain evidence exists for walls braced at the top by the permanent floor installed to minimize movement. Whereas the total resultant pressure should be unchanged from the total resultant pressure for the at-rest condition, it is conceivable that the stresses near the top will exceed the at-rest pressures while the stresses near the bottom will be less than the at-rest pressures. Consequently, walls braced at the top by the permanent floor or by other rigid bracing before excavation should be checked for the temporary stresses shown in Fig. 13-71.

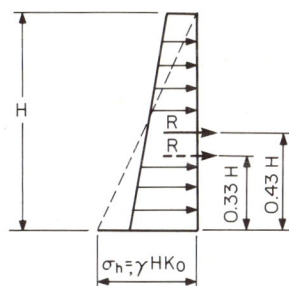

Fig. 13-71 Redistributed lateral earth stresses at rest. Short-term condition for diaphragm walls braced initially at the top. Minimum expected movement.

The purpose of this stress diagram is to recognize the results of field measurements; hence it is a semiempirical procedure with little theoretical basis. It is recommended for conditions where the soil consists of dry sand and soft to medium saturated clay. This loading may develop at the end of excavation but may persist for some time thereafter. It is not an apparent-earth-pressure diagram, but it is the original earth stress redistributed to give more load near the top and less load near the bottom of the wall. For the redistributed pressure diagram the ordinate at the top and at the base can be computed by noting that the resultant pressure is known both in magnitude and in position. This resultant has been moved from the lower third (0.33) point to the 0.43 point. For soft clay K_0 may be taken as 1.

Recommended Earth Loads for Strutted Walls

The following comments regarding strut loads in internally braced diaphragm walls are suggested only as an expedient and general guidelines, pending the development of an appropriate theoretical treatment, which may not come for some time.

The introduction of apparent-pressure diagrams (Peck, 1969) was not intended to represent the actual distribution of earth stresses but instead constitutes a method of calculating strut loads that might be approached but not exceeded during the excavation phase. The origin of these diagrams was the measurement of strut loads behind flexible walls. The behavior of these walls during excavation suggests a case of rotation about the top, as opposed to rotation about the tip or lateral translation, which is the assumed motion in the Coulomb and the Rankine theory.

Strut Loads in Sand The probable explanation of the appreciable increase of the lateral pressure near the top observed with flexible walls is the condition of true arching, which is manifested as sand grains squeeze and wedge themselves in the upper part of the failure zone. A stable condition of arching requires a major rotational movement of the wall about the top while the wall itself sustains the arch and remains unyielding. The squeezing of grains will occur as long as they are hard and nondeformable, hence it cannot develop in plastic clays. Another condition favoring arching is where the natural sand deposits are not too loose; otherwise, especially if the sand is submerged and lacks the apparent cohesion of moist sand, the tendency of the soil to wedge will initially result in densifying the materials before an arch can develop and increase the lateral thrust near the top.

Ordinarily diaphragm walls should not be expected to move, especially in dense sand. The observed inward bulging has been generally negligible and should not favor the rotational movement producing the arching effect. In sand with water, which is common with diaphragm walls where it is necessary to maintain the natural water table, the application of apparent pressures is an uncertainty by itself irrespective of the type and the behavior of ground support.

Strut loads might be considered on the basis of apparent pressure where the expected movement of the wall approaches that of flexible walls or where the struts are preloaded as they are installed, in which case the development of apparent pressures follows as a mere compatibility. It should be noted that the envelope load, obtained by integrating the apparent-pressure diagram $0.65 K_a \gamma H$, is less than the total strut load based on a K_0 condition, and the denser the soil the greater this difference.

Where doubt exists regarding the performance of the wall, many investigators have used the original earth pressure redistributed into a rectangular pressure diagram of the same total magnitude in order to obtain strut loads.

Strut Loads in Clay The evidence accumulated thus far (Peck, 1969; Flaate and Peck, 1973) appears to indicate that the behavior of the system depends, at least at present, on a consideration of the stability number $N = \gamma H/s_u$, where s_u represents the clay beside and beneath the cut and to a depth that would normally be affected by a general shear failure due to excavation. Thus, the semiempirical procedures developed for determining strut loads have been considered in terms of the transition from elastic to plastic behavior. Even in the same cut, the stability number N can change so that it may be 3, 4, or less in the shallow excavation stages and may approach or exceed 6 as the excavation is continued to deeper levels.

In stiff clays or where the dimensions of the cut result in a stability number generally less than 4, Peck (1969) has demonstrated that earth-pressure theories should not be used since they may reduce the theoretical pressure to zero, which is in conflict with reality. The same investigator has suggested that the discrepancy arises because a theory of plastic equilibrium cannot be applied to a material which is not in a plastic state. Instead, Peck proposed a trapezoidal diagram with a width corresponding to the range $0.2\gamma H$ to

$0.4\gamma H$, which has been widely used and confirmed by the majority of field observations and measurements of strut loads for cuts in stiff clay or where $N < 4$. Although a different system behavior might be expected if the bracing of a cut is changed from a flexible to a stiff wall, the absence of plastic response in this case may suggest an initial similarity in strut loads. Hence, this author suggests using the apparent-pressure diagram shown in Fig. 13-72, modified to favor as much load for the top and bottom struts as elsewhere.

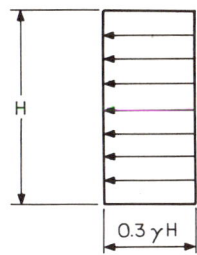

Fig. 13-72 Recommended apparent-pressure diagram for determining strut loads for diaphragm walls in stiff clay or where $N < 4$; H = excavation depth.

The value of $0.3\gamma H$ is in the author's opinion justified from a consideration of the total envelope load carried by the struts. From Fig. 13-72 this load is $0.3\gamma H^2$. The total strut load is thus seen to be equal to the total lateral pressure at rest based on a value of $K_0 = 0.6$, which is the earth coefficient for most normally consolidated clays, and satisfies the author's criterion that the capacity of any system should not be less than the earth load at rest.

The second distinction in the behavior of cuts in clay was made when the stability number reaches 7 or even higher. In the field cases monitored the soft clay extended to a considerable depth below the base of the excavation, and in some instances the clay was somewhat softer below the bottom than above. Under these conditions the inward movement of the sheeting and the settlement of the adjoining ground were extremely large. Measurements of strut loads indicated that the envelope apparent pressure would be approached if in the modified Rankine-Résal coefficient $K_a = 1 - m4s_u/\gamma H$ the value of m was taken as 0.4. This was justified noting that if $N > 6$ and the soft material extends to a great depth below the cut, the depth of the failure surface is not necessarily limited to the depth of the cut, so that much greater pressures may be mobilized followed by large movement and settlement. A further explanation was offered by Bjerrum et al. (1972), who demonstrated that wall deflection beneath the excavation leads to arching and transfer of earth pressure to the stiffer part of the system, i.e., the strutted part, with a subsequent increase in strut loads. If, on the other hand, the cut is underlain by stiff material expected to limit movement, the depth of the plastic zone will not greatly exceed the depth of excavation although the stability number will still exceed 6. The corresponding apparent earth pressure is now more likely governed by the unmodified Rankine-Résal coefficient $K_a = 1 - 4s_u/\gamma H$, so that the width of the apparent-pressure diagram is $\gamma H - 4s_u$.

The behavior of cuts in soft plastic clay supported by diaphragm walls may be influenced by the embedded portion, which usually is considerable and mobilizes earth pressures on both sides; by the ability of the system to reduce lateral ground movement even if soft saturated clay; and by the considerable wall stiffness which will often preclude a strut-load distribution according to a flexural-beam theory and thus result in erratic variations of strut loads, particularly where the struts fit loosely against the wall.

At present there is no reason to suspect that a reduction of lateral movement will significantly alter the plastic behavior of a cut. The first and third effects mentioned above are therefore the most serious since they may result in considerable overestimation or underestimation of the strut loads. Besides the examples given in the foregoing sections, this author has become aware of several instances, unfortunately not fully documented yet, where the top and bottom struts bracing diaphragm walls in soft clay received loads very different from the intermediate ones and these loads deviated considerably from apparent-pressure loads.

As long as the limitations of engineering predictions are understood, Peck's apparent-pressure diagrams developed initially for flexible walls in soft clay can be used as a method of calculating strut loads for excavations supported with diaphragm walls where the same conditions are expected to prevail. However, the author suggests using the diagram shown in Fig. 13-73 in order to induce more load in the top as well as in the bottom strut. Whereas an average load may be assumed for all intermediate struts, the top-strut load should be calculated using beam continuity, especially where the wall cantilevers above that point. Likewise, the design load for the bottom strut should recognize the actual wall embedment below the final excavation level and be consistent with the fact that this load is created as reaction to the additional lateral earth pressure mobilized as the excavation is carried from this level to the final level. However, the larger the wall embedment the higher the load at the bottom strut. Furthermore, no attempt should be made to estimate lateral ground reaction at the base from apparent-pressure diagrams.

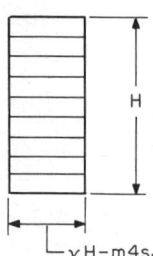

Fig. 13-73 Recommended apparent-pressure diagram for determining strut loads for diaphragm walls in soft to medium clay. H = excavation depth; $m = 1$ normally, and $m = 0.4$ where large movement is expected and the soft clay extends to a considerable distance below the excavation level. The wall does not penetrate into the stiff layer.

It is interesting to note that for soft clay ($s_u = 500$ lb/ft² or 2.5 t/m²) unless the excavation is very shallow (in which case the diagram normally should not be applicable) estimating the strut loads from Fig. 13-73 using $m = 0.4$ is much more conservative than using at-rest pressure load ($\phi = 0$ and $c = 0$) redistributed to a rectangular diagram; in many cases the difference will be greater than 50 percent. Where conditions warrant using $m = 1$, however, engineers are cautioned that if the excavation depth is less than 45 ft (15 m), the diagram of Fig. 13-73 should not be used since it is less conservative in this case and the expected wall movement may be very small. Instead using a redistributed-pressure diagram based on $K = 1$ and $c = 0$ may be more appropriate.

Bracing Details The effect of the considerable wall stiffness may be combined with poor bracing details to manifest the erratic variations in strut loads mentioned in the foregoing sections. More particularly, the connection between the strut and the wall is quite important since during excavation it is the most common cause of such difficulties as twisting, buckling, rotation, and yielding, all leading to variations in the strut load.

This author suggests avoiding the use of wales, since they are neither necessary nor beneficial. Wales will never fit right, leaving gaps and thus causing stress concentration at the contact areas. Rigid connections of wales to the concrete wall are difficult and expensive. Besides constructional benefits, serious technical advantages are derived if the struts are directly connected to the wall by means of shear plates, as in the detail shown in Fig. 13-74. In this case the wall is designed as a two-way slab, which is entirely feasible and relatively simple considering the actual thickness of the concrete section. These principles were exemplified in the design and construction aspects of the diaphragm walls for a chamber of the subway extension in New York (Xanthakos, 1976). The rigidity of the system as a whole ensured the uniform redistribution of the lateral loads, consistent with the assumption of the design, and provided structural integrity at the strut-wall connection where this is mostly needed.

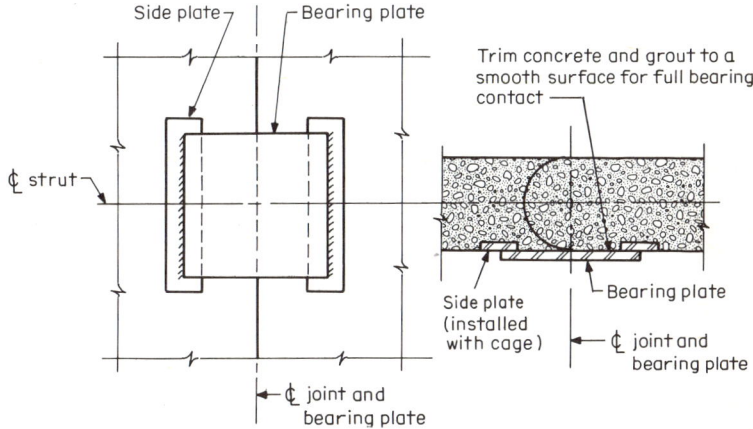

Fig. 13-74 Connection details for strutted diaphragm walls.

Stress Distribution on Tied-back Walls

Tied-back walls are generally prestressed, often to a level which exceeds the total active load or the load at rest. With the high stresses commonly used in wire or cable tendons, large strains occur throughout the length of the tieback, so that large movements of the tendons at the anchor heads are experienced during loading. Close to the junction of the free and fixed length, the movement of the bonding part relative to the surrounding soil may cause the latter to fail in shear. If the shear strength of the soil reaches a peak value beyond which further strain reduces strength, the load in the tendon will be progressively distributed along the bonded length until the load and the strength of the soil are in balance. Since cohesive soils respond slowly to stress changes, gradual progressive (creep) failure of the soil will occur, whereby the point where peak strength is mobilized in the surrounding soil will gradually move along the bonded length, leading to failure of the anchor if this peak reaches the end. The effective length of the tendon is thus gradually increased, resulting either in the movement of the anchor head as the tendon elongates or causing the anchor to fail. Progressive failure has also been reported in noncohesive soils and weak rocks. These, however, adjust more rapidly to stress changes, and the redistribution of load along the bonded length occurs within the period of loading so that creep problems are unlikely to be serious.

Tied-back walls in clay may experience a loss of load in the anchor of the order of 10 to 20 percent. Therefore, for permanent anchors in clay it is necessary to institute frequent and regular reloading for a period of several months and measure the loads thereafter until no significant loss of load in the ties is detected. On the other hand, the walls should be checked to ensure that they do not move adversely. Overstressing of the anchors, for example, may cause passive failure of the retained ground. This may be a particular hazard if the ties are very high in relation to the excavation depth since the resulting redistribution of stresses within the wall may lead to collapse.

The time-dependent movement of walls in clay means that the prestressing of anchors can be based on active pressure loads only if some movement of the wall can be tolerated. If only a small movement is allowed, the prestressing loads should be based on at-rest pressures and the forces in the ties should exceed the forces due to such pressures by some 10 to 15 percent. It appears from these considerations that the forces actually induced in a tied-back wall will often preclude the development of a particular state of earth stress. Instead the problem may approach the case of a stiff foundation mat acted upon by concentrated loads and by an appropriate soil response so that the interaction is essentially a statically indeterminate problem. The earth pressures in this case are created by the actual prestressing forces and are less dependent upon the particular state of the soil.

If the total prestress load is less than the active stress resultant, the behavior of the wall may approach the active or partially active state. This suggests that the variation of tieback loads with excavation can at first be considered to be a function of the level of prestressing. With little or no prestressing the tieback loads will normally increase. If the prestressing is very high, a decrease in the tieback load will be experienced, although exceptions have been noted.

The Creep Problem in Ground Anchors Considerable work has been carried out to trace the creep behavior of anchors in various types of soil, but further work is needed, particularly in exploring, with the help of long-term measurements, the possibility of using short-term observations to predict that behavior. Further points of interest (Ostermayer, 1974) are the change in the value and distribution of skin friction with time and the influence of prestressing on long-term behavior. Thus, for permanent tiebacks the analysis must consider long-term stability as well as creep characteristics, since the latter account for a substantial portion of the loss of prestress.

Effect of Wall Stiffness and Anchor Prestressing on Earth-Pressure Distribution The observed behavior of tied-back walls has demonstrated the influence of the wall stiffness and the amount of prestress on ground movement and therefore on the magnitude and distribution of earth pressures. The conclusions from these observations fit the practical requirements in a satisfactory way, but they lack the theoretical background.

Finite-element analyses, on the other hand, have made it possible to distinguish between the effects of different factors and in many instances isolate these effects. It is known, for example, that the influence of wall stiffness is rather insignificant with respect to the tieback reaction itself but quite important for the earth-stress distribution and therefore for the bending moments in the wall. The loads introduced into the system essentially remain there, so that the earth stresses will be compatible with the prestress diagram and the prestress level.

Situations where this analytical modeling has been used are described in the following section.

Basic Principles of Finite-Element Analysis

Advantages and Limitations It is evident that the disadvantage of the foregoing methods of analysis is that they pursue each problem independently. Earth stresses are determined by limit theory; the stability of struts and tiebacks is based on assumed stress-distribution diagrams derived from experience and semiempirical methods; and deformations are predicted by empirical data, elastic theory, and one-dimensional consolidation theory. Where the walls constitute an integral part of a more complex structure, as is often the case with permanent walls, the overall interaction is based on assumptions or, at best, experience with similar structures if this is available.

Limit theory can predict collapse loads for earth-retaining structures but does not predict deformations with the limit loads and provides no information for conditions other than those at the limit. On the other hand, a finite-element analysis gives detailed information for both stresses and deformations in the wall and the soil medium. As a predictive technique the method offers the ability to consider structures of arbitrary shape and flexibility, complex construction sequence, and heterogeneous soil conditions; the ability to analyze seepage loading and nonlinear soil and interface behavior; and the ability to predict stress changes and deformations for both the soil and the structure for conditions not at the limit.

If instrumentation is contemplated to monitor construction, the method can be employed to predict critical areas and stages of instrumentation so as to provide a logical supplement to the process. Some of the early uses of the technique in monitoring the performance of walls have been reported by Cole and Burland (1972) and Ward (1972).

Limitations in the use of the method are associated with inability to prescribe appropriate constitutive behavior and determine the parameters needed for the constitutive models. It is further conceivable that the application in soil-structure interaction involves certain special problems for which solutions are not yet complete. Other difficulties arise from the simulation of the relative movement between the soil and the structure, the special construction sequence that must be modeled, and the numerical problems which are intensified by the stress-strain pattern of the soil.

Among the various investigators who have carried out finite-element analyses are Wong (1971), Clough (1973), Egger (1972), Barla and Mascardi (1974), Tsui (1973), Clough and Tsui (1974a,b), and others.

Statement of Problem Table 13-2 shows a typical flow chart for carrying out a finite-element analysis. The procedure consists of listing the steps involved in the analysis, each step representing an idealized presentation of the actual problem, so that the work is completed by the introduction of certain assumptions.

TABLE 13-2 Typical Flow Chart and Procedure Leading to Finite-Element Analysis

Statement of problem
↓
Idealization of soil and groundwater conditions
↓
Selection of constitutive modeling techniques
↓
Selection of media properties
↓
Assumption of initial stress conditions
↓
Assumption of construction sequence
↓
Drawing of finite-element mesh to accommodate soil conditions, structural configuration, and construction sequence
↓
Analyses

It is seen that the first two steps are typical and require conversion of the soil and groundwater conditions into an idealized profile. Next, behavioral models are selected for the soil, the structure, and the soil-structure interface. These constitutive models idealize the behavior in a form that can be expressed mathematically, e.g., elastic, elastoplastic, etc. The soil model and the soil-structure-interface model are most difficult to define. Parameters indicating the media properties are selected next, and in nonhomogeneous situations they are fairly approximated. Then, initial stress conditions are defined; this can be particularly difficult in certain types of soil, e.g., overconsolidated clays and clay-shales. Finally the construction sequence is worked out, a finite-element mesh is drawn, and the analysis is carried out.

Simulation of Construction Sequence The effect of the construction sequence on wall and soil behavior has been demonstrated in real cases. Its simulation requires the division of the loading sequence into small increments, analysis of the effects of each increment in sequence, and superposition of the results to obtain the resultant stress and displacement conditions. In reality the sequence consists of trench excavation under slurry, concrete tremie placement, general excavation on one side (probably accompanied by dewatering and recharging), and installation of bracing or tiebacks. Schematically, this sequence is shown for a cut-and-cover tunnel in Fig. 13-75, where the various stages are self-explanatory.

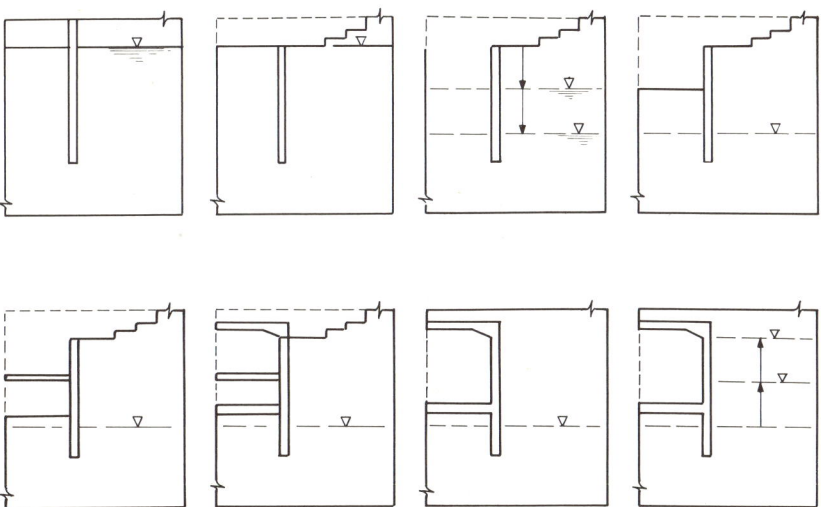

Fig. 13-75 Construction sequence for a cut-and-cover tunnel simulated in finite-element analysis.

Clough and Duncan (1971) have proposed a method for simulating excavation, which is claimed to be general and accurate in a number of situations, e.g., excavation in soil and rock and for in-place structural elements. If the soil is assumed linear elastic, results for one- and three-step excavation simulation should be comparable.

Dewatering and seepage loadings constitute pressure changes on the elements in the mesh, and they are merely special cases of the more complex loading produced by excavation.

Placement of braces and prestressing anchors is simulated in the form of a restraint or load change and is covered in the case-history studies presented in the following sections.

Among the initial conditions, the simple initial at-rest stresses are difficult to conceive. Further, the construction process of the wall prompts changes in these stresses. Initially the changes consist of replacing the soil with slurry, to be followed by the replacement of the latter by fresh concrete, so that in general the pressure at the final state exceeds the initial pressure at rest, sometimes by as much as a factor of 3. If the method of concrete placement recommended in Chap. 8 is followed, the fresh mix may exert almost its full hydrostatic head.

Among the boundary conditions, the representation of the interface is particularly pertinent to the boundary model. The soil-slurry interaction must be inferred first, and the shear resistance mobilized at the interface as the wall moves with respect to the soil must be determined next. All these will influence the shear stress-deformation behavior.

The Finite-Element Mesh The construction of the finite-element mesh is a main step in idealizing the problem. Although the details and the theoretical justification are beyond the scope of this text, it is appropriate to mention certain simple rules that should be followed for drawing the mesh. Thus, the mesh should reasonably reflect the soil conditions, the structural configuration of the problem, and the construction sequence. In areas of expected stress concentration the mesh must be refined, and a minimum number of elements are necessary to provide an appropriate flexibility for the structure and the soil.

The mesh must accommodate all the structural features entering the problem, but in the simulation process elements which eventually become structural components, e.g., the ones representing the top of the tunnel, must undergo a change in material properties to indicate the difference in the functions. Thus, these elements first enter the simulation with soil stiffness, then they are assigned very low stiffness as the soil is excavated. Finally they reenter with concrete properties to simulate the top slab of the tunnel.

For the structure just mentioned the mesh has initially two different types of finite elements, but the simulation of the bracing adds a third element, as shown in Fig. 13-76. The elements representing the media are two-dimensional, whereas the elements representing the wall-soil interface and the presence of the bracing are one-dimensional. The versions derived for two-dimensional elements have been modified. The initial versions used simple strain distribution within the element, but recently linear, quadratic, and quintic strain distribution has been suggested, giving an increased degree of flexibility. The resulting higher degree of freedom may impose a higher computer cost.

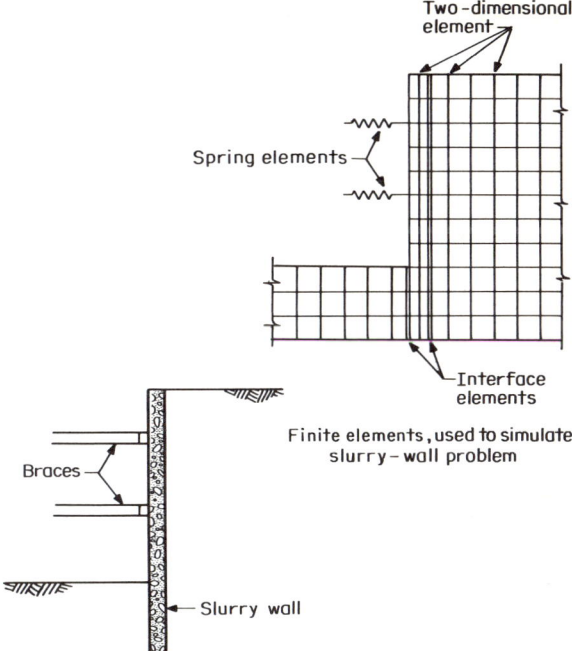

Fig. 13-76 Types of fininte elements used in cut-and-cover tunnel analysis. (*From Clough, 1973.*)

One-dimensional slip elements are essential to allow the controlled relative movement between the wall and the soil at the interface. The nature of the interface behavior to be modeled can be determined from the material presented in Chap. 11, so that a plot can be obtained relating the shear stress mobilized at the interface with the corresponding vertical displacement. In most cases this curve will be nonlinear.

Case Studies of Finite-Element Analyses

Cut-and-Cover Tunnel in Sand Clough and Tsui (1974a) have reported finite-element analyses for a cut-and-cover tunnel built with diaphragm walls. The analysis used a nonlinear elastic soil model where the tangent modulus is calculated from a hyperbolic stress-strain curve during primary loading and a straight-line unload-reload curve should the shear stresses in an element decrease during an increment of loading. The data for the soil properties are shown in Table 13-3.

Because of the complex behavior of the system no attempt was made to simulate the slurry-trench excavation and the tremie placement of concrete, so that with the walls in place a simple condition at rest was assumed with $K_0 = 0.5$. The friction at the wall-soil interface was taken to correspond to the shear strength of the soil at one-half the friction angle.

592 Slurry Walls

TABLE 13-3 Soil Data for Finite-Element Analysis, Cut-and-Cover Tunnel of Fig. 13-77* (from Clough and Tsui, 1974a)

Type of soil	= sand
Unit weight above water table γ	= 125 lb/ft^3
Friction angle ϕ	= 30°
Poisson ratio	= 0.2
Coefficient of earth pressure at rest K_0	= 0.5
Cohesion	= 0
Modulus exponent n	= 0.5
Modulus number K	= 280

*Initial tangent modulus E_i assumed to vary with minor principal stress σ_3 so that $E_i = K\sigma_3^n$.

The predicted wall movement during the construction stage is shown in Fig. 13-77. Most of the wall movement is seen to have occurred during excavation down to the first brace level. This movement was essentially a parallel translation, as shown in Fig. 13-77a, and apparently it was sufficient to reduce the earth stress at rest to a near active state for the portion of the wall above the brace level, as can be seen from Fig. 13-78a (the deflection-depth ratio for this stage was about 1:850). Excavation below the first brace level with this bracing in place caused the wall to move as shown in Fig. 13-77b, undergoing a rotation about the brace point with the bottom part moving away from the retained soil.

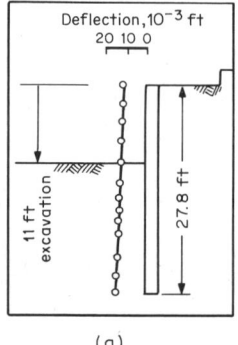
(a)
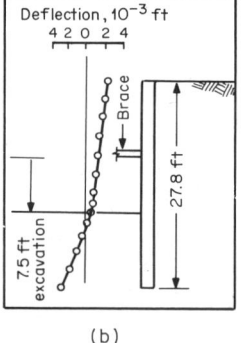
(b)
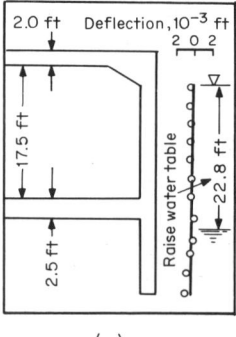
(c)

Fig. 13-77 Movement and incremental wall deflection for cut-and-cover tunnel built with diaphragm walls: (a) wall deflection due to excavation to brace level; (b) wall deflection due to excavation below brace level; (c) wall deflection due to raising water table with top and bottom slab in place. (*From Clough and Tsui, 1974a.*)

Allowing the water table to rise to its normal level following the placement of the top and bottom slab resulted, as one would expect, in a very small elastic deflection, as shown in Fig. 13-77c. From Fig. 13-78b and c the actual earth stresses are seen to approach very closely the at-rest pressures for $K_0 = 0.5$.

Of special interest is the vertical pressure distribution at the underside of the base slab. This pressure is seen to have been reduced in Fig. 13-78b to c;

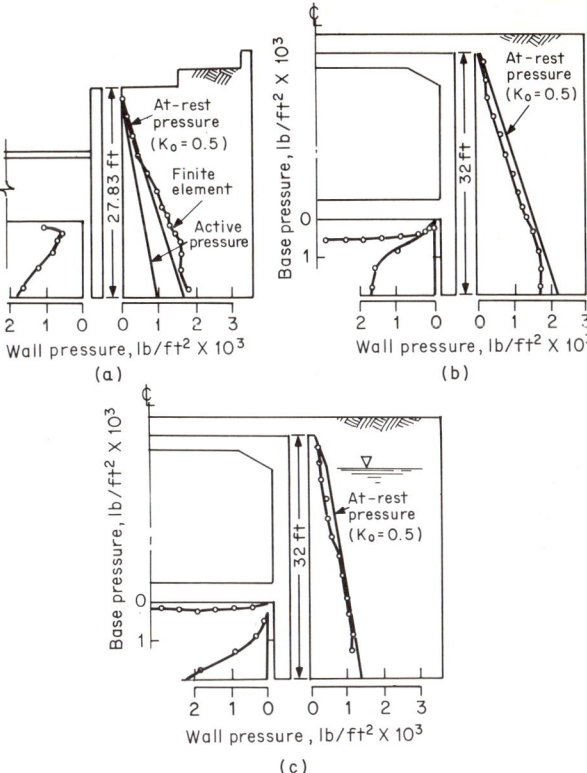

Fig. 13-78 Lateral earth stresses on diaphragm-wall cut-and-cover tunnel of Fig. 13-77: effective earth stresses (*a*) at completion of excavation; (*b*) at completion of structure and backfill above structure; (*c*) following restoration of water table. (*From Clough and Tsui, 1974a.*)

that is, after completion of construction, and the resultant is less than the weight of the overburden. The difference can be accounted for by the friction developed along the back of the walls when they were assumed to settle somewhat following installation of the concrete slabs.

Braced Excavation in Soft Clay Figure 13-79 shows the predicted earth stresses for the South Pacific excavation shown in Figs. 13-54 and 13-55. The stresses were obtained by finite-element studies and are shown for two stages of the excavation, one for the completed excavation with the lower struts not prestressed, the other for the same excavation stage with the lower struts prestressed. The analysis assumed that the prestress applied to the lower struts was held without loss (Clough, 1975).

The predicted earth pressure before prestressing was decreased to near active, corresponding to a wall movement of 0.9 percent of the excavation height. When the lower struts were prestressed, the lateral earth pressure increased to nearly the original at rest, which was anticipated since the pre-

594 Slurry Walls

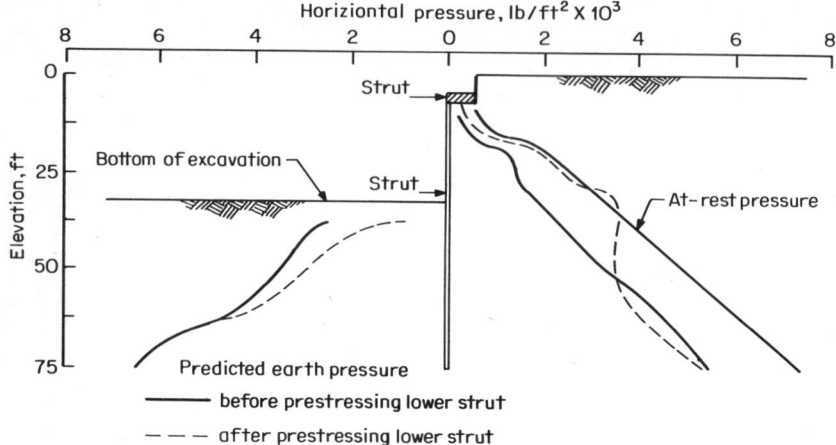

Fig. 13-79 Finite-element predictions for lateral earth stresses on diaphragm wall in soft clay. Southern Pacific excavation, San Francisco. (*From Clough, 1975.*)

stress loads were calculated from the at-rest diagram. The wall was predicted to move toward the soil by 1 in (2.5 cm) during prestressing, so that it actually remained displaced from the original position by 3 in (7.5 cm), or 0.6 percent of the excavation height. This is in agreement with the observations made in the preceding sections that applying prestress load to restore a wall to the original position will not fully produce the desired results although the lateral stresses return to the original values.

Tied-back Walls in Clays Figure 13-80 shows a section through a tied-back wall supporting an excavation 32.5 ft (10 m) deep in a homogeneous clay deposit underlain by rock (Tsui, 1973). The wall is 2 ft thick (60 cm), and the tiebacks consist of steel rods, 1 in² in area, anchored into rock. The prestress loads were estimated from the apparent-pressure diagram shown in Fig. 13-80*b* using values of width 0, $0.2\gamma H$, and $0.4\gamma H$. The soil is medium clay having an undrained shear strength increasing linearly with depth from 500 to 1400 lb/ft² (2.5 to 7.0 t/m²) at the bottom of the clay layer. K_0 is taken as 0.85, and the insertion of the wall was assumed to have no influence on the initial at-rest condition. The initial tangent modulus of the clay was taken as 400 times the undrained shear strength. The use of a plane strain condition was considered satisfactory for a wall 2 ft thick and a tieback spacing less than 10 ft (3 m).

A nonlinear elastic model was incorporated in the analysis, and tangent modulus values for the nonlinear portion of the curve were calculated from a stress-strain curve represented by a hyperbola. The interface between the wall and the soil was treated similarly on both sides of the wall, using a bilinear shear-stress–deformation relationship with an initial shear stiffness of 50,000 lb/ft³ reduced by a factor of 1000 when the yield strength of the interface was exceeded. The construction sequence was simulated by an incremental-loading process based on a nine-step modeling, as shown in Fig. 13-81.

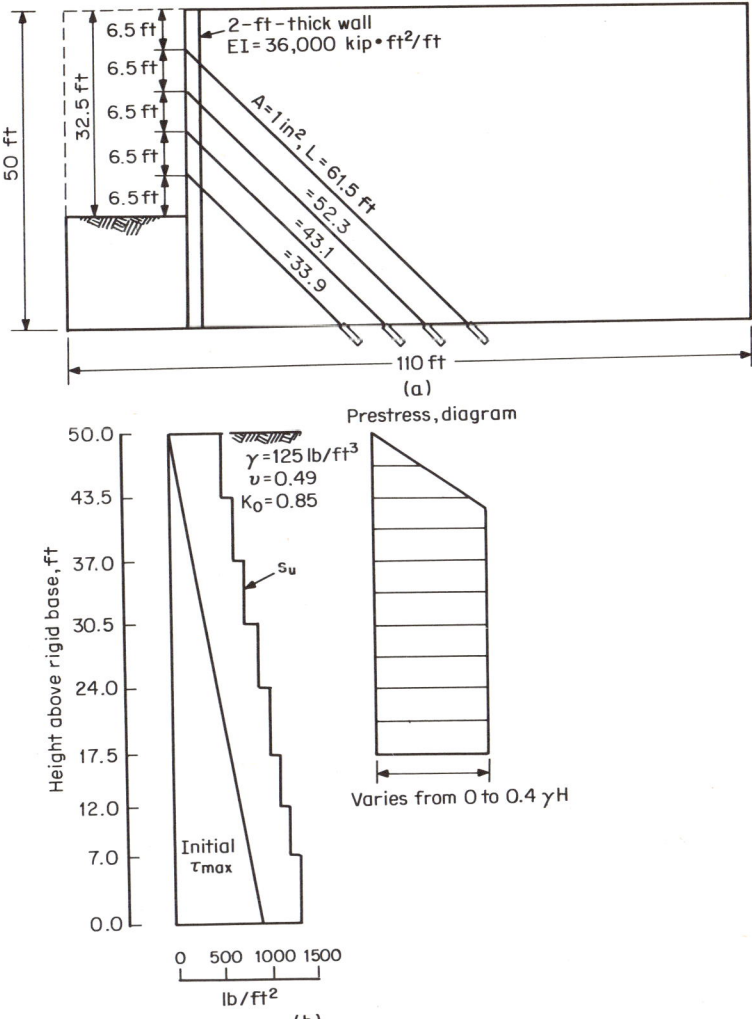

Fig. 13-80 A tied-back wall in clay: (a) section through wall; (b) soil data and prestress diagram. (*From Tsui, 1973.*)

Figures 13-82 and 13-83 show wall and ground movement and earth-pressure distribution, respectively, for the two prestress levels and with no prestress load, together with the tieback loads corresponding to the apparent-pressure diagrams. The wall movement responds consistently to the prestress load by decreasing almost linearly with the amount of prestressing. Likewise, the ground settlement behind the wall decreases as the prestress increases, but a diminishing effect is apparent as the next higher prestress load is introduced, so that the settlement is reduced more by the first increase than by subsequent prestress increments.

The predicted pressure diagrams are shown in Fig. 13-83a for the three

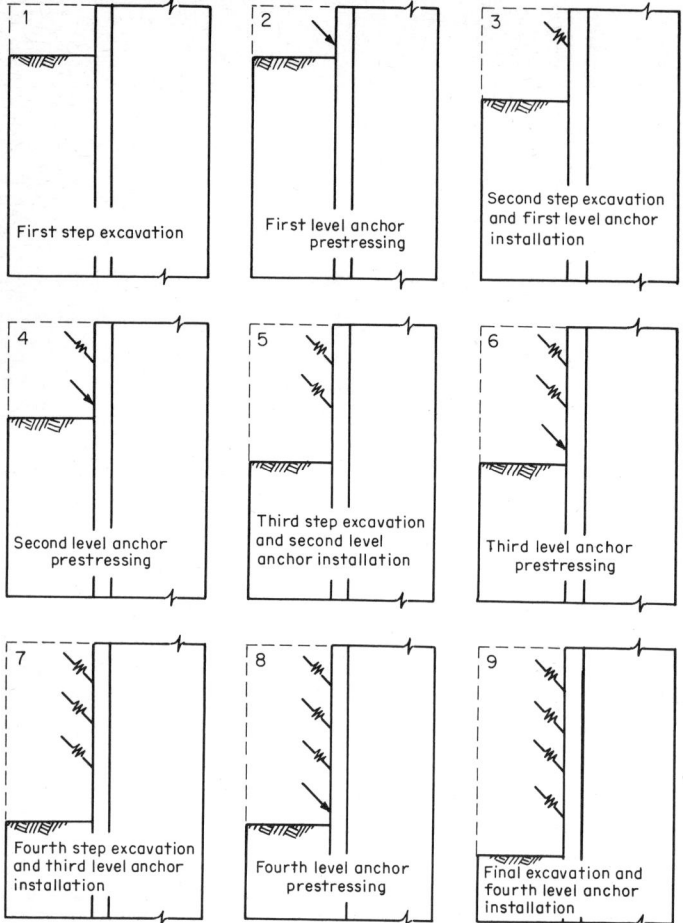

Fig. 13-81 Construction sequence. Finite-element analysis of the tied-back wall of Fig. 13-80. (*From Tsui, 1973.*)

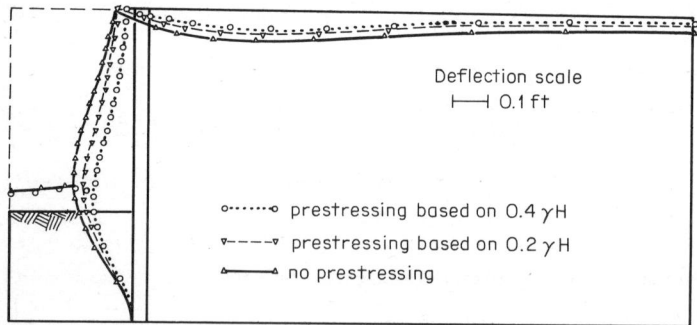

Fig. 13-82 Wall and ground movement predicted by finite-element analysis, tied-back wall of Fig. 13-80. (*From Tsui, 1973.*)

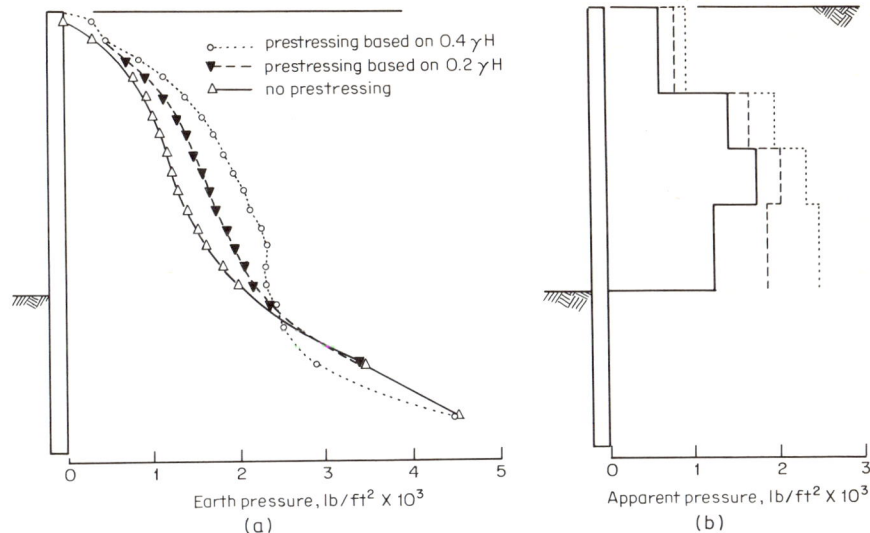

Fig. 13-83 Lateral earth pressure predicted by finite-element-analysis and apparent-pressure diagrams, tied-back wall of Fig. 13-80. (*From Tsui, 1973.*)

cases of prestress and can be compared with the apparent pressures shown in Fig. 13-83b obtained by distributing the anchor loads over the appropriate spans. The predicted lateral stresses are not very different from the original at-rest pressures and approach a definite triangular distribution. Attention is drawn to the absence of pressure bumps at the anchor points.

A second example of a tied-back wall in clay modeled by finite-element analysis is shown in Fig. 13-84 (Clough and Tsui, 1974b). The excavation is 50 ft deep (15.2 m) in clay. Two cases are demonstrated, one using four rows and the other using three rows of tiebacks. This time, the wall is of moderate stiffness equivalent to a PZ-72 sheeting; hence it can be considered flexible compared with the 2-ft thick concrete wall of the previous example. The tieback prestress load was obtained from the apparent-pressure diagram.

The predicted lateral pressures are more triangular than the design trapezoidal diagram, and this distribution should be attributed to the movement of the wall resulting from the flexible tieback supports. The pressures tend to concentrate slightly at the tieback levels. This bulging is caused not only by the application of prestress but is facilitated by the wall flexibility; hence it must be distinguished from the linear stress distribution observed with the stiff wall of the previous example. Its effect is to reduce the bending moments, and this can be done according to a procedure suggested by Hanna and Kurdi (1974).

Effect of Wall Stiffness and Anchor Prestressing Egger (1972) has reported the results of finite-element studies of flexible (sheet-pile) walls and stiff (diaphragm) walls in sand. The analysis in this case was based on a procedure developed by Malina (1969).

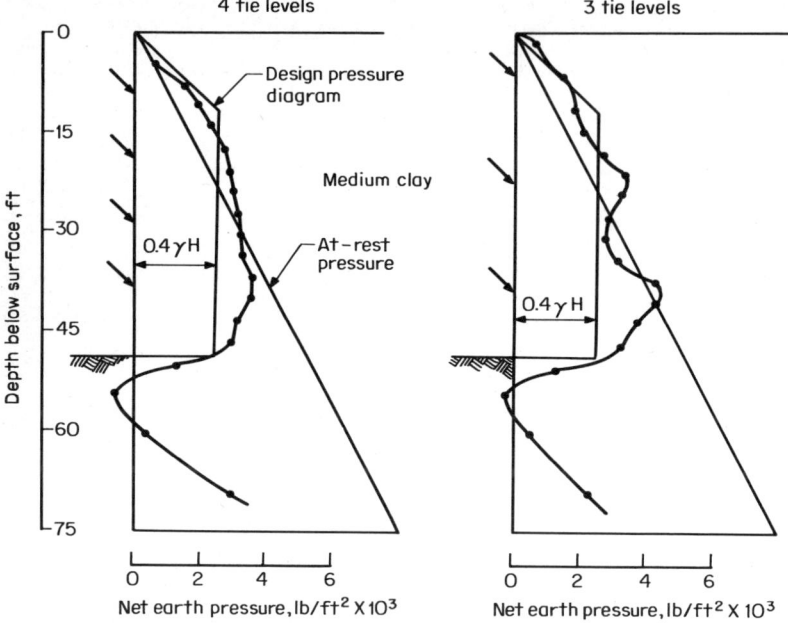

Fig. 13-84 Lateral earth pressure behind a flexible wall predicted by finite-element analysis, prestressed tied-back wall. (*From Clough and Tsui, 1974a.*)

The total wall weight, including the embedment, is 10 m (33 ft), of which 7.5 m (24.5 ft) is the excavation depth. The construction sequence involved excavation to 3.5 m depth for the installation of the first row of tiebacks, followed by a second and third excavation stage, each having a depth of 2 m for the installation of the second and third row of tiebacks. The toe of the wall is just embedded into bedrock, which precludes lateral movement at this depth, the only possible deformation thus being rotation about this point. The flexible and stiff wall have a stiffness ratio of 1:100.

Lateral earth pressures are shown in Figs. 13-85 and 13-86 for the two prestress levels. The wall movement is shown in Fig. 13-87 for the flexible and the stiff wall. During the first excavation stage the wall acts as a cantilever beam, deflecting above excavation level according to its stiffness. The lateral wall movement at the top for the flexible system is 3 times that for the rigid wall. Consequently the flexible wall mobilizes a passive earth resistance near the surface but in the highest possible degree, i.e., limit equilibrium. The stiff wall, on the other hand, mobilizes passive resistance over a greater depth, but because of the smaller movement the limit equilibrium is not reached. Likewise, the active earth pressure approaches the at-rest value near the bottom of the wall much faster in the case of the flexible wall.

As the prestress level is increased from 1 to 6 t/m, the flexible wall receives this supplementary charge in a concentrated manner, mainly between 1.5 m above and 2.5 m below the anchor level. For the stiff wall, the resulting earth-pressure diagram is substantially uniform from the top down to about a

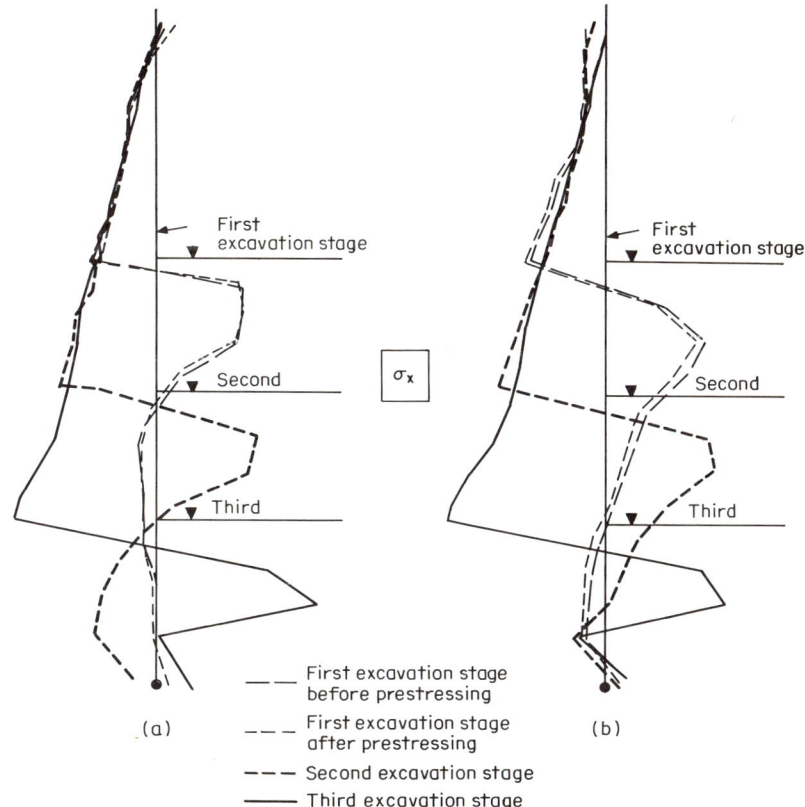

Fig. 13-85 Lateral earth pressures predicted by finite-element analysis for an excavation in sand: (a) flexible wall; (b) stiff wall. Anchor prestress 1 t/m (*From Egger, 1972.*)

depth of 6 m. These results are in good agreement with the tied-back walls described in the preceding sections, although in the last excavation stages the differences are still observable but less significant, indicating the effect of the close tieback-row spacing, i.e., the factor EI/l^4.

The effect of anchor prestressing on wall movement is seen from Fig. 13-87. The prestress of 1 t/m is far lower than the active level, whereas the value of 6 t/m corresponds roughly to the at-rest pressures existing at final excavation level. The movement of the top of the flexible wall is reduced from 2.5 to 1.7 cm (1 to 0.7 in) when the prestress is increased from 1 to 6 t/m. Likewise the movement of the stiff wall is reduced from 2.3 to 1.4 cm with the higher prestress level.

The Method of Equivalent Tie Support for Multi-tied Walls

According to this method, each stage of the excavation is analyzed assuming an equivalent single-tied wall corresponding to a unique center of rotation.

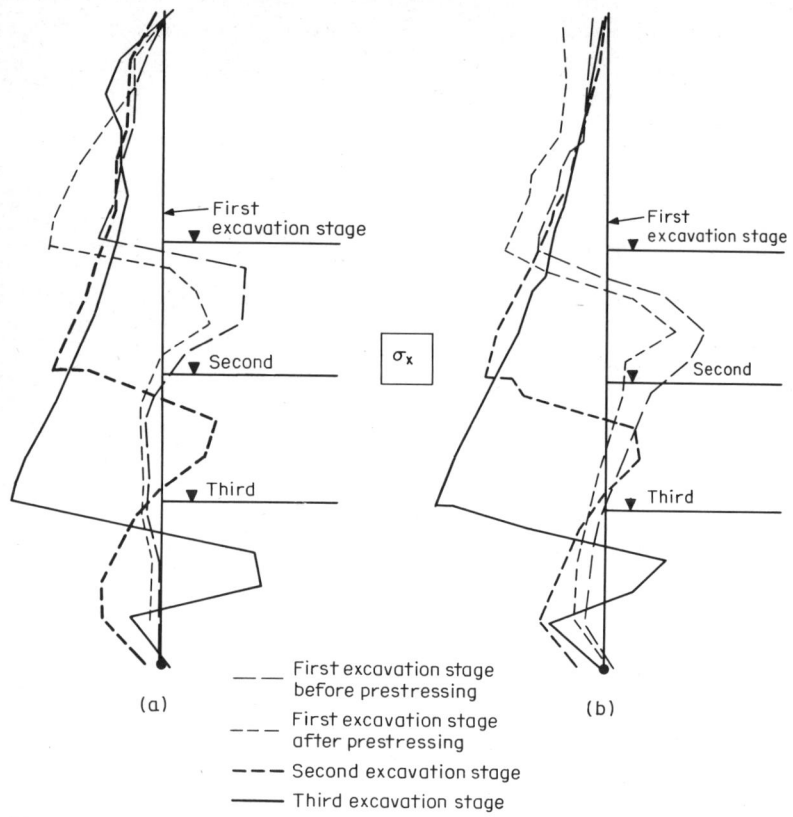

Fig. 13-86 Lateral earth pressures predicted by finite-element analysis for an excavation in sand: (a) flexible wall; (b) stiff wall. Anchor prestress 6 t/m (*From Egger, 1972.*)

This design approach has been used successfully in the last several years for multi-tied walls, and case histories were mentioned in the preceeding sections (Littlejohn and MacFarlane, 1974). The method provides an understanding of the actual behavior of multi-tied walls and can be used to determine flexibility coefficients as used in single-tie analysis (James and Jack, 1974).

Theoretical Analysis The fundamental differential equation of elasticity is

$$P = \frac{d^4y}{dx^4} EI \tag{13-18}$$

where P is the load. When P becomes a function of deformation,

$$P = \frac{d^4y}{dx^4} EI = yr \tag{13-18a}$$

where r is the equivalent spring stiffness of the soil. The simplest approach is

Topics Relevant to Analysis and Design 601

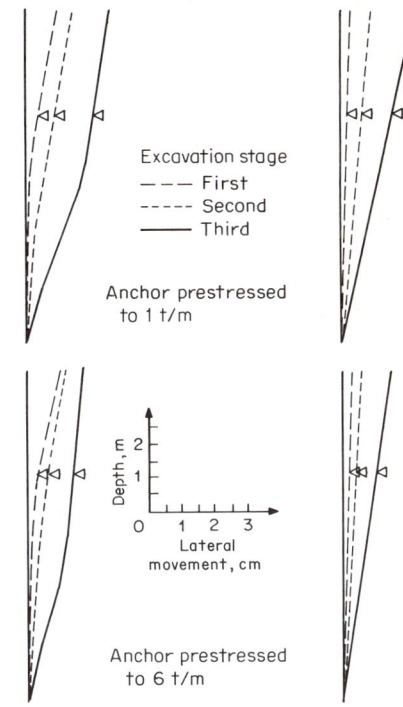

Fig. 13-87 Lateral wall movement for a wall in sand predicted by finite-element analysis: (a) flexible wall; (b) stiff wall. Anchor prestress 1 and 6 t/m (*From Egger, 1972.*)

to consider the wall as a series of members connected by nodes. At these points horizontal members simulating the soil stiffness or support points can be provided, as illustrated in Fig. 13-88. By incorporating a simple iteration routine into standard computer programs and using a simplified stress-strain relationship the elastoplastic effects of a soil can be simulated, as shown in Fig. 13-89. Estimates of P are repeatedly made for each deflection profile until convergence to a condition of equilibrium is achieved.

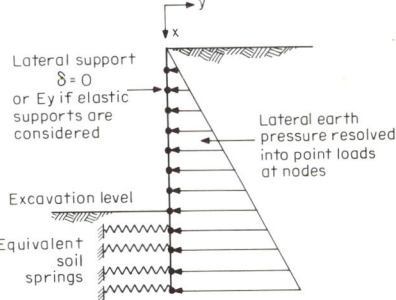

Fig. 13-88 Simulation of wall-soil interaction. (*From James and Jack, 1974.*)

The tieback forces can be predicted by considering the effects of the temporary support produced by the passive resistance at intermediate excavation stages. In this procedure the position and magnitude of a resultant tie is

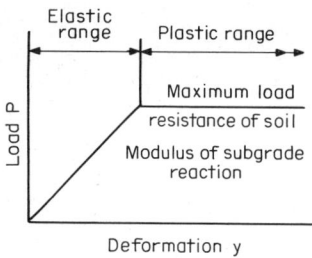

Fig. 13-89 Simplified stress-strain relationship in elastoplastic soil.

calculated by treating the wall as a single-tied structure under the following assumptions: the mobilizing and resisting soil pressures correspond to the Rankine state of stress; at failure there is rotation of the wall about a unique point in its plane; and the wall has a length which is only sufficient to mobilize a factor of safety of 1 against rotation at any excavation stage. The first assumption is made for simplicity and is justified where sufficient movement is expected. The second assumption enables one to use a simple procedure for calculating the additional tieback forces produced when the passive resistance is shifted to a different area during the next stage of the excavation. The third assumption allows the maximum resultant tie to be resolved by considering free-earth support so that the fixity produced by the continuity of the wall is neglected.

Figure 13-90 shows the equilibrium conditions of the system. From Fig. 13-90a it follows that

$$\Sigma F_H = 0 \quad \text{is satisfied when} \quad T_1 = P'_a - P'_p$$
and $\Sigma M = 0$ is satisfied when $M_p = M_a$ about point T_1 (13-19)

From Fig. 13-90b it follows that

$$\Sigma F_H = 0 \quad \text{is satisfied when} \quad T_1 + T_2 = P''_a - P''_p$$
and $\Sigma M = 0$ is satisfied when $M''_p = M''_a$ about centroid of T_1 and T_2

If R_1 is the resultant tie of T_1 and T_2, then $R_1 = T_1 + T_2 = P''_a - P''_p$ and substituting the value of T_1 from Eq. (13-19) gives $P'_a - P'_p + T_2 = P''_a - P''_p$ or

$$T_2 = P''_a - P''_p - P'_a + P'_p = P''_a - P'_a - (P''_p - P'_p) \quad (13\text{-}20)$$

from which it can be seen that the additional earth load transmitted to T_2 is temporary support offered during the previous stage. The equivalent beam loading to T_2 is shown hatched in Fig. 13-90c. As long as T_2 is unknown, so are the position and the magnitude of R_1. If one is known, the other can be estimated and the initial position checked.

The following iteration procedure is suggested to ensure a convergence to the correct value of the cantilever arm of the resultant (James and Jack, 1974). Figure 13-91 shows the stage where the excavation level has been reduced to a position for the insertion of the fourth tie. For equilibrium the following must be satisfied:

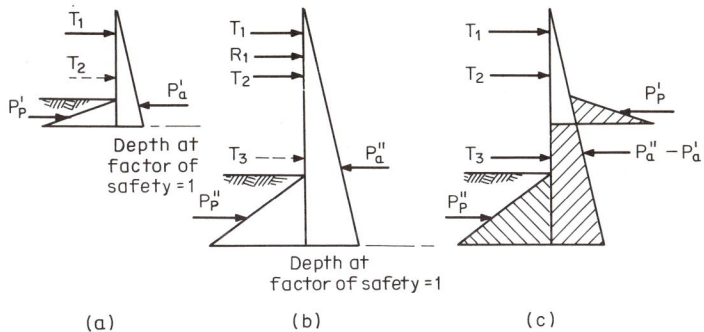

Fig. 13-90 Equilibrium conditions; multi-tied wall converted into a single-tie wall.

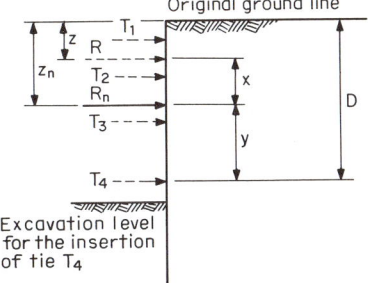

Fig. 13-91 Iteration procedure for system equilibrium, multi-tied wall. R = previous resultant tie force, R_n = new resultant tie force, z = previous resultant tie-force level, z_n = new resultant tie-force level, T_1, T_2, T_3, and T_4 = individual tie forces.

and
$$\Sigma F_H = 0 \quad \text{or} \quad T_4 = R_n - R$$
$$\Sigma M = 0 \quad \text{or} \quad Rx - T_4 y = 0 \quad (13\text{-}21)$$

where $y = D - x - z$ so that

$$f(x) = Rx - T_4 D + T_4 x + T_4 z$$

Substituting in the Newton-Raphson iteration formula gives

$$x_{n+1} = x_n - \frac{Rx - T_4 D + T_4 x + T_4 z}{R + T_4} \quad (13\text{-}22)$$

where x_{n+1} is the new estimate of x and x_n is the previous estimate of x.

Evaluation of Method James and Jack (1974) report that a comparison of the foregoing procedure with the results of published experimental work suggests a satisfactory estimate of the forces in the tiebacks. The same investigators mentioned model tests and full-scale tests performed by Tcheng and Iseux (1972). The applicability of the method was also checked in the walls described by Littlejohn and MacFarlane (1974).

At present, the behavior of tied-back walls remains a complex problem, and only approximate methods of analysis are available. The equivalent-tie-support method is useful since it considers varying soil conditions and random optimizing selection of tieback levels. The method is conservative but also shows that it may be necessary to extend the reinforcing steel, computed for the maximum moment, either for the full length or for a good distance beyond the theoretical point of zero moment.

A rational design of tied-back walls is not always possible since it must be based on a tentative interpretation of empirical findings and experimental work. A review of the limitations in the present understanding of the subject is given by Hanna and Seeton (1967), Hanna (1968, 1972, 1973, 1975), Hanna and Littlejohn (1969), Hanna and Matallana (1970), Hanna and Leonard (1969), and Hanna and Kurdi (1974).

Wall Embedment and Passive Resistance below Excavation Base

Where the reduction of wall movement and protection of the base of the excavation from groundwater are not factors influencing wall embedment, the latter is estimated considering the difference in earth pressure between the loads acting on either side of the wall. If the stability is analyzed in a general manner, the embedded portion must provide restraint against deep-seated movement and thus it controls the design.

In the usual case the stability of the toe will be affected by the development, both in magnitude and time, of active and passive earth pressure. As already mentioned, the pressures on the active face may change rapidly from those associated with the undrained condition to values associated with a fully drained state (effective stress). Likewise, the passive resistance may be expected to deteriorate at least as rapidly. If a layer of much stiffer soil or rock is found at relatively shallow depth below excavation level, it will have a marked effect on reducing toe movement, particularly if the wall is allowed to reach this layer. Finally, the possibility of overexcavation should not be discounted, although it usually is considered to be an unlikely event.

Passive Resistance in the Embedded-Wall Portion In granular soil the condition of free drainage is commonly assumed, so that changes in the pore pressure generated by strain or load changes cannot be sustained even for a short time. The analysis of passive resistance is based on drained-strength parameters and effective stresses.

In cohesive soil, because of the load decrease caused by excavation, the ground in the passive zone just below the excavation level will initially experience a decrease of pore pressure, which may even become negative. With time the pore pressure will rise again. In the immediate condition the pore pressures generated by unloading and strain have no time to dissipate so that the analysis can be based on the undrained shear strength s_u at natural water content. For the long-term condition the pore pressures generated by unloading the strain will dissipate by drainage; hence an effective stress analysis is appropriate based on static water level.

For single-anchored walls, which frequently occur with the use of posttensioning or with relatively shallow excavations, the embedment is usually estimated by the free-earth-support method (see Bowles, 1968). A partial application of this approach using modified soil-pressure diagrams is shown in Sec. 13-10.

For multibraced or multi-tied walls one conventional method is to assume an equivalent horizontal reaction at the base of the cut, to be compensated for by the passive resistance below excavation level. When the coefficient of passive resistance has been selected, the embedment depth is determined so that it satisfies force equilibrium on the horizontal plane and it is increased according to the factor of safety. A second conventional method is to assume a hinge condition at the lowest bracing and determine the embedment by equating the active and passive moments about this point. The fallacy of this approach with braced diaphragm walls is illustrated in Fig. 13-92.

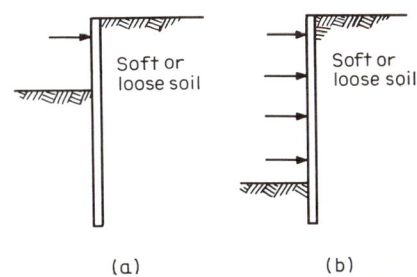

Fig. 13-92 Excavation and wall embedment of a stiff multi-braced wall in soft or loose soil: (a) initial stage (wall free to move laterally and rotate about uppermost brace point; (b) final excavation stage (wall prevented from moving laterally or rotating about any brace).

For convenience two excavation stages are shown, the initial stage, with one brace in place, and the final stage, with all the braces in place. In the first stage the wall can rotate about the uppermost brace point or move laterally toward the excavation as earth moving is continued to lower levels, both movements contributing to the mobilization of passive resistance. As the final stage is approached and with all the braces in place, the wall is expected neither to move laterally nor to rotate, undergoing only an elastic deflection below the lowest brace point. In soft or loose soil this will preclude the passive resistance from being equal to the active stress, even if the embedment is taken very deep. The wall is more likely to act as a cantilever member below the lowest brace, and thus a large horizontal reaction will develop there. If the cantilever moment is excessive, it may lead to the formation of a plastic hinge where the system possesses an inherent flexibility or cause the wall to crack if it is too stiff. Either of these conditions will cause now a redistribution of stresses, since the deformation in this case warrants the development of passive resistance.

REVIEW PROBLEMS

13-1 For the braced excavation shown in Fig. 13-4 the following data are given: $D = 22$ m, $H = 16$ m, $B = 10.5$ m, $L = 3.5$ m, $T = 1.0$ m, and $h = 6$ m. The soil properties are $s_u = 3.1$ t/m², $\gamma = 1.85$ t/m³, and $\alpha = 0.9$. A uniform surcharge $q = 1.0$ t/m² acts at ground surface. Analyze the stability against bottom heave and estimate the factor of safety (*a*) for a long excavation without diaphragm walls; (*b*) with diaphragm walls but no cross walls; (*c*) for a single opening of cross walls; (*d*) for two openings; and (*e*) for a long excavation with diaphragm walls and cross walls.

13-2 Discuss at some length the factors influencing ground movement due to excavation. Distinguish between flexible and stiff walls.

13-3 The diaphragm walls supporting an excavation rest on incompressible material precluding their vertical settlement. Analyze the problem of base failure by heave considering the shear strength along the exterior soil-concrete interface, mobilized as the wall begins to settle. Assume undrained conditions.

13-4 The two opposite faces of an excavation are supported by two walls braced against each other by interior bracing. One wall is flexible (a sheet-pile section), and the other is stiff (a concrete diaphragm wall). Explain how you would analyze the two walls.

13-5 A bored-pile wall is built as a free cantilever above excavation level. The wall consists of tangent piles with a primary and a secondary row, as shown in Fig. 10-12*a*. Explain how you would ensure the transfer of lateral load from the secondary to the primary row. Is this transfer desirable?

13-6 An underground circular shaft is constructed with diaphragm walls. The shaft has an interior diameter 10 m (33 ft) and is 30 m (98 ft) deep. The soil is soft to medium clay, $s_u = 600$ lb/ft². Design the shaft and state your assumptions. Would you investigate base stability at the end of excavation?

13-7 Investigate the structure of Prob. 13-6 if the circular enclosure now has a diameter of 50 m (165 ft) and the other parameters are the same. Outline an analytical procedure. Choose the structural details, joints, etc.

13-8 Using Eq. (13-1) as a starting point, derive Eqs. (13-2) and (13-3).

13-9 Referring to Fig. 13-4, express the factor of safety against bottom heave for three openings and estimate this factor for the problem of Example 13-1.

13-10 Using a design (working) stress 80 percent of the tensile strength of the concrete panel shown in Fig. 13-9, estimate the maximum excavation depth that can be supported by the steel-and-concrete wall. Use a soil density 1.6 t/m³ and a lateral-earth-stress coefficient 0.7.

13-11 The two composite sections shown in Fig. 13-12*a* and *b* can be assumed to respond in a different manner when acted upon by loads and stresses. Using the two methods of analysis discussed in Sec. 13-7, work out a problem to show the difference.

13-12 Design a concentrically prestressed (posttensioned) diaphragm wall using the fictitious apparent-pressure diagram shown in Fig. P13-12. The wall will carry also a permanent dead load of 50,000 lb/ft and a variable live load of 20,000 lb/ft maximum. Select a tentative wall thickness and then estimate the prestress load. Use $f'_c = 4000$ lb/in². Tension is not allowed in the concrete. Use four equal bracing intervals.

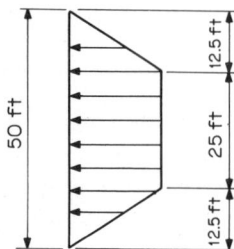

Fig. P13-12

13-13 A vertical cut in sand braced with diaphragm walls has a variable depth and width. Referring to the data and the graphs of Sec. 13-10, formulate a statement regarding the economy of a free cantilever wall vs. the economy of a wall braced near the top. Consider (*a*) a constant width of 30 ft and a depth varying from 15 to 30 ft and (*b*) a constant depth of 20 ft and a width varying from 20 to 50 ft. Base the investigation on a theoretical analysis and cost estimate for the wall and the bracing. State your assumptions. Disregard actual wall movement as a restraint to the problem.

13-14 The walls shown in Figs. 10-27 and 10-31 range from stiffened cantilevers to wall cells. Prepare a summary stating your assumptions and the factors you would consider in the analysis of stability. Discuss the pattern and magnitude of lateral stresses, wall movement, wall friction, weight of soil, and others.

13-15 For an excavation in soft clay investigate the feasibility of inserting bottom transverse cross walls (see Fig. 13-31) from a lower construction level, i.e., after the general excavation is carried down to a certain level, and in trenches without slurry support. Discuss the factors that may influence this feasibility, both technical and economical.

13-16 Estimate the lateral stress due to the continuous footing for the wall of Fig. P13-16. Use both methods shown in Fig. 13-22*a* and *b*. Repeat the problem using excavation heights 30 and 40 ft. Are there any fallacies in the methods?

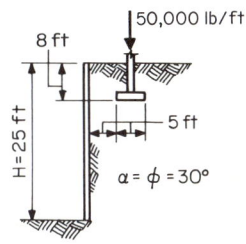

Fig. P13-16

13-17 A circular structure has an inside diameter 100 ft and is 50 ft deep. It will be used to store industrial liquids weighing 60 lb/ft³. The structure is built in soft clay with an undrained shear strength 400 lb/ft². Design the diaphragm walls for (*a*) the external lateral loads only; (*b*) loads from the liquid pressure alone; and (*c*) a combination of these loads.

13-18 For the excavation shown in Fig. 12-37*f* two procedures were mentioned, each based on a different method of analysis. Elaborate on each procedure and discuss the factors that should be considered in each analysis.

13-19 Discuss the effect of different bracing systems on movement for an excavation supported by diaphragm walls. What general guidelines would you follow in predicting movement?

13-20 Discuss the effect of strut preloading or tieback prestressing on diaphragm-wall movement. Is this procedure as effective as it often is with flexible walls? Establish the similarities and differences of movement caused by (*a*) excavation and (*b*) preloading of bracing (in this case consider the movement produced in the direction of load application).

13-21 Compute the ordinate at the top and the base of the redistributed lateral-earth-stress diagram shown in Fig. 13-71.

13-22 As a research assignment, discuss the problems associated with the analytical modeling of diaphragm walls. Consider soil-behavior models, time effects, the assumption of plane strain, and environmental factors.

13-24 Prepare a statement about the effect of wall stiffness and anchor prestressing on lateral movement and stress distribution for tied-back walls.

REFERENCES

Alpan, I., 1967: The Empirical Evaluation of the Coefficient K_0, *Soil Found., Jap. Soc. Soil Mech. Found. Eng.*, vol 7, no. 1, January.

ASCE and North Carolina State University, 1975: *Proc. Conf. in Situ Meas. Soil Prop.*, North Carolina State Univ., June.

Barla, G., and C. Mascardi, 1974: High Anchored Wall in Genoa, *Proc. Diaphragm Walls Anchorages, Inst. Civ. Eng., Lond.*

Bjerrum, L., 1963: *Discuss. Eur. Conf. Soil Mech. Found. Eng.*, Wiesbaden, vol. 2.

——— and K. H. Andersen, 1972: In-Situ Measurement of Lateral Pressures in Clay, *Proc. 5th Eur. Conf. Soil Mech. Found. Eng., Madrid*, vol. 1.

———, C. J. F. Clausen, and J. M. Duncan, 1972: Earth Pressures on Flexible Structures: A State of the Art Report, *Proc. 5th Eur. Conf. Soil Mech. Found. Eng., Madrid*, vol. 1.

——— and O. Eide, 1956: Stability of Strutted Excavations in Clay, *Geotechnique*, **6**(1):32-47.

Bowles, J. E., 1968: "Foundation Analysis and Design," McGraw-Hill, New York.

Braun, W. M., 1972: Post-tensioned Diaphragm Walls in Italy, *Ground Eng.*, March 1972.

Breth, H., and H. R. Wanoschek, 1972: The Influence of Foundation Weights upon Earth Pressure Acting on Flexible Strutted Walls, *Proc. 5th Eur. Conf. Soil Mech. Found. Eng., Madrid*, vol. 1.

Brooker, E. W., and H. O. Ireland, 1965: Earth Pressures at Rest Related to Stress History, *Can. Geotech. J.*, vol. 2 no. 1, February.

Clough, G. W., 1973: Analytical Problems in Modeling Slurry Wall Construction, *FCP Res. Rev. Conf.*, San Francisco.

———, 1975: "Deep Excavations and Retaining Structures," Stanford University Press, Stanford, Calif.

——— and G. M. Derby, 1975: "Temporary Berms in Supported Excavations," Stanford University Press, Stanford, Calif.

——— and J. M. Duncan, 1971: Finite Element Analyses of Retaining Wall Behavior, *ASCE J. Soil Mech. Found. Div.*, vol. 97, December.

——— and Y. Tsui, 1974a: Finite Element Analyses of Cut-and-Cover Tunnel Constructed with Slurry Trench Walls, *Duke Univ., Durham, N.C. Soil Mech. Ser. no. 29*.

——— and ———, 1974b: Performance of Tied-back Walls in Clay, *ASCE J. Geotech. Div.*, vol. 100, December.

Cole, K. W., and J. B. Burland, 1972: Observations of Retaining Wall Movement Associated with Large Excavations, *Proc. 5th Eur. Conf. Soil Mech. Found. Eng., Madrid*, vol. 1.

Collingridge, V. H., and R. E. Tuckwell, 1960: Underground Stations for Western District Post Office, London, *Proc. ICE Constr.* vol. 15, February.

Cunningham, J. A., and L. D. Carpenter, 1975: Monitoring of Two Braced Excavations in Chicago, *ASTM Repr. Spec. Tech. Publ.* 584.

——— and J. I. Fernandez, 1972: Performance of Two Slurry Wall Systems in Chicago, *ASCE Proc. Spec. Conf. Perform. Earth Earth-Supporting Struct., Purdue Univ.*

D'Appolonia, D. J., 1971: Effects of Foundation Construction on Nearby Structures, *4th Pam-Am Conf. Soil Mech. Found. Eng., San Juan, Puerto Rico, State-of-the-Art Volume*.

———, 1973: Cut-and-Cover Tunneling, *U.S. Dept. Transp., Fed. Highw. Admin. San Francisco Proj. Rev. Meet.* September.

DiBiagio, E., and J. A. Roti, 1972: Earth Pressure Measurements on a Braced Slurry Trench Wall in Soft Clay, *Proc. 5th Eur. Conf. Soil Mech. Found. Eng., Madrid*.

Egger, P., 1972: Influence of Wall Stiffness and Anchor Prestressing on Earth Pressure Distribution, *Proc. 5th Eur. Conf. Soil Mech. Found. Eng., Madrid*, vol. 1.

Eide, O., G. Aas, and T. Josang, 1972: Special Application of Cast-in-Place Walls for Tunnels in Soft Clay in Oslo, *Proc. 5th Eur. Conf. Soil Mech. Found. Eng., Madrid*, vol. 1.

Fisher, F. A., 1974: Diaphragm Wall Projects at Seaforth, Redcar, Bristol and Harrow, *Proc. Diaphragm Walls Anchorages, Inst. Civ. Eng., Lond.*

Flaate, K., and R. B. Peck, 1973: Braced Cuts in Sand and Clay, *Norw. Geotech. Inst. Publ.* 96, Oslo.

Gould, J. P., 1970: Lateral Pressures on Rigid Permanent Structures, *Proc. ASCE Spec. Conf., Lateral Earth Stresses Earth Retain. Struct. Cornell Univ.,* June.

Grant, R., J. T. Christian, and E. H. Vanmarcke, 1972: Differential Settlement of Buildings, *ASCE J. Geotech. Div.,* September, pp. 973-991.

Gysi, H. J., A. Linder, and R. Leoni, 1975: Prestressed Diaphragm Walls, *Proc. 1975 Eur. Conf. Soil Mech. Found. Eng., Vienna.*

Haliburton, T. A., 1968: Numerical Analysis of Flexible Retaining Structures, *ASCE J. Soil Mech. Found. Div.,* vol. SM 6, November.

Hanna, T. H., 1968: Factors Affecting the Loading Behavior of Inclined Anchors Used for the Support of Tied-back Walls, *Ground Eng.,* vol. 1.

———, 1972: Anchor Supported Walls: Research and Practice, *Ground Eng.,* March.

———, 1973: Anchored Inclined Walls: A Study of Behavior, *Ground Eng.,* November.

———, 1975: Analysis, Design and Installation of Tiebacks and Ground Anchors, *U.S. Dept. Transp., Urban Mass Trans. Admin., Proc. Semin. Underground Constr. Probl., Techniques, Solutions, Chicago.*

——— and T. I. Kurdi, 1974: Experimental Studies of Multi-anchored Flexible Retaining Walls in Sand, *Univ. Sheffield, Dept. Civ. Struct. Eng.*

———and M. W. Leonard, 1969: Some Design and Construction Considerations on the Use of Anchorage and Tiebacks, *Conf. Piling Comm., Inst. Civ. Eng., Lond.,* January.

———and G. S. Littlejohn, 1969: Design and Construction Considerations Associated with Retaining Walls Supported by Prestressed Tiebacks, *Consult. Eng., Lond.* vol. 33, nos. 5 and 6.

———and G. A. Matallana, 1970: The Behavior of Tied-back Walls, *Univ. Sheffield, Dept. Civ. Struct. Eng.*

———and J. E. Seeton, 1967: Observations on a Tied-back Soldier Pile and Timber Lagging Wall, *Ont. Hydro. Res. Q.,* vol. 19.

Harr, M. E., 1966: "Foundations of Theoretical Soil Mechanics," McGraw-Hill, New York.

Heydenrych, R. A., and B. Isaacs, 1967: The Excavations and Stabilizing of the Carlton Center Basement, *S. Afr. Inst. Civ. Eng., Symp. Deep Basements, Johannesburg,* August.

Huder, J., 1969: Deep Braced Excavations with High Ground Water Level, *Proc. 7th Int. Conf. Soil Mech. Found. Eng., Mexico City.*

Jaky, J., 1944: The Coefficient of Earth Pressure at Rest, *J. Soc. Hung. Archit. Eng.,* pp. 355-358

James, E. L., and B. J. Jack, 1974: Design Study of Diaphragm Walls, *Proc. Diaphragm Walls Anchorages, Inst. Civ. Eng., Lond.*

Kastner, R., and P. Lareal, 1974: Experimental Excavation 50 m Long Supported by Strutted Cast Diaphragm Walls: An Analysis of Stress Distribution in the Struts, *Proc. Diaphragm Walls Anchorages, Inst. Civ. Eng., Lond.*

Kitagushi, H., 1976: Observed Performance of Diaphragm Walls at Joto Site, Osaka, Aoki Construction Co., Osaka, April.

Koegler, F., and A. Scheidig, 1948: "Baugrund and Bauwerk," 5th ed., Ernst, Berlin and Munich.

Kuesel, T. R., 1969: Discussion Presented at Specialty Session, *7th Int. Conf. Soil Mech. Found. Eng., Mexico.*

Lambe, T. W., 1970: Braced Excavations, *Proc. ASCE Spec. Conf. Lateral Earth Stresses Earth Retain. Struct., Cornell Univ.,* June.

———and R. V. Whitman, 1969: "Soil Mechanics," Wiley, New York.

Lin, T. Y., 1955: "Design of Prestressed Concrete Structures," Wiley, New York.

Littlejohn, G. S., et al., 1971: Anchored Diaphragm Walls in Sand, *Ground Eng.,* September pp. 14-17, November pp. 18-21.

———and I. M. MacFarlane, 1974: A Case History of Multi-tied Diaphragm Walls, *Proc. Diaphragm Walls Anchorages, Inst. Civ. Eng., Lond.*

Malina, H. (1969): Brechnung von Spannungsumlagerungen in Fels und Boden mit Hilfe der Elementenmethode, *Veroeff. Inst. Bodenmechanik Felsmechanik Univ. Karlsruhe,* H. 40.

Martin, L. D., S. A. Gill, and N. L. Scott, 1977: Prefabricated Structural Members for Cut-and-Cover Tunnels, U.S. Department of Transportation, Federal Highway Administration, Offices of Research and Development, Washington.

Miyoshi, Y., et al., 1976: Steel and Unreinforced Concrete Panels Built in Slurry Trenches, *Constr. Techniques Mag.,* vol. 7.

Morgenstern, N. R., and Z. Eisenstein, 1970: Methods of Estimating Lateral Loads and Deformations, *Proc. ASCE Spec. Conf. Lateral Earth Stresses Earth Retain. Struct., Cornell Univ.,* June.
Myslivec, A., 1972: Pressure at Rest of Cohesive Soils, *Proc. 5th Eur. Conf. Soil Mech. Found. Eng., Madrid.*
NGI, 1962–1966: Measurements of a Strutted Excavation, *Norw. Geotech. Inst. Tech. Rep.* 1–9.
Ostermayer, H., 1974: Construction, Carrying Behavior and Creep Characteristics of Ground Anchors, *Proc. Diaphragm Walls Anchorages, Inst. Civ. Eng., Lond.*
PCI, 1972: "Design Handbook Precast and Prestressed Concrete," Prestressed Concrete Institute, Chicago.
———, 1977: "Post-tensioning Manual," Prestressed Concrete Institute, Chicago.
Peck, R. B., 1943: Earth Pressure Measurements in Open Cuts, Chicago Subway, *ASCE Trans.*
———, 1969: Deep Excavations and Tunneling in Soft Ground, *Proc. 7th Int. Conf. Soil Mech. Found. Eng., Mexico City, State-of-the-Art Volume.*
Rigden, W. J., and P. W. Rowe, 1974: Model Performance of Unreinforced Diaphragm Wall, *Proc. Diaphragm Walls Anchorages, Inst. Civ. Eng. Lond.*
Saxena, S. K., 1974: Measured Performance of Rigid Concrete Wall at the World Trade Center, *Proc. Diaphragm Walls Anchorages, Inst. Civ. Eng., Lond.*
Skempton, A. W., 1951: The Bearing Capacity of Clays, *Build. Res. Congr. Lond. pap. Div. L.,* pp. 180–189.
———and D. H. MacDonald, 1956: The Allowable Settlement of Buildings, *Proc. Inst. Civ. Eng. Lond.,* vol. 5, pt. III.
Sokolovski, J., 1944: "Statics of Granular Media," Pergamon, London.
Sowers, G. F., 1962: "Shallow Foundations," McGraw-Hill, New York.
Spangler, M. G., 1951: "Soil Engineering," art. 21.18, International Textbook, Scranton, Pa.
St. John, H. D., 1975: Recent Research into the Movement of Ground around Deep Excavations, *Build. Res. Establish. Bull.* D2/F/1.
Tcheng, Y., and J. Iseux, 1972: Full Scale Passive Pressure Tests and Stresses Induced on a Vertical Wall by a Rectangular Surcharge, *Proc. 5th Eur. Conf. Soil Mech. Found. Eng., Madrid, Sess.* II, *pap.* 13.
Terzaghi, K., 1934: Large Retaining Wall Tests, *Eng. News Rec.*
———, 1943: "Theoretical Soil Mechanics," Wiley, New York.
———, 1954: Anchored Bulkheads, *Trans. ASCE,* vol. 119.
———and R. B. Peck, 1967: "Soil Mechanics in Engineering Practice," 2d ed., Wiley, New York.
Thon, J. G., and R. C. Harlan, 1971: Slurry Walls for BART Civic Center Subway Station, *ASCE J. Soil Mech. Found. Div.,* September.
Tschebotarioff, G. P., 1973: "Foundations, Retaining and Earth Structures," McGraw-Hill, New York.
Tsui, Y., 1973: A Fundamental Study of Tied-back Wall Behavior, PhD. Thesis, Duke University, Durham, N.C.
Turabi, D. A., and A. Balla, 1968: Distribution of Earth Pressure on Sheet-Pile Walls, *ASCE Soil Mech. Found. Div.,* vol. SM 6, November.
Verdeyen, J., and J. Gillet, 1968: Evaluation de valeurs du coefficient de reaction horizontal des sols, Revue C. Genie Civil No. 9, Gand
———and ———, 1969: Rigid Diaphragm Walls, Stability after Construction, *Proc. 7th Int. Conf. Soil Mech. Found. Eng., Spec. Sess.* 14, 15, *Mexico City.*
Ward, W. H., 1963: Displacements and Strains in Tunnels beneath a Large Excavation in London, *Proc. 5th Int. Conf. Soil Mech. Found. Eng., Paris.*
———, 1972: Remarks on Performance of Braced Excavations in London Clay, *Proc. Conf. Performance Earth Earth-Supported Struct. Purdue Univ.*
———and J. B. Burland, 1972: The Measurement and Interpretation of Changes of Strains in the Earth, *Proc. R. Soc.,* London.
Wong. I. H., 1971: Analysis of Braced Excavations, Ph.D. Thesis, Massachusetts Institute of Technology, Cambridge, Mass.
Xanthakos, P. P., 1974: "Underground Construction in Fluid Trenches," Colleges of Engineering, University of Illinois, Chicago.
———, 1976: Slurry Walls for Fan Chamber No. 1, Route 133, Sec. 2, New York City Transit System, in-house rep., Petros P. Xanthakos, Ltd.
———, 1977: Post-tensioned Diaphragm Walls in Chicago, Petros P. Xanthakos, Ltd., in-house rep.

INDEX

Abandoned tunnels, 333
Access shafts, 359
Activable-tracer analysis:
   of concrete flow, 265
   of concrete strength, 273–274
   sampling of analysis of tracers,
      265–267
Active state, 20, 490, 502, 515–516,
   564–566
Active wedge, 21, 22, 48
Adhesion factor (see Coefficient, of wall
   adhesion)
Adhesion of soil to concrete (see Side
   shear at interface)
Adsorption, 65, 103, 120
Agglomerates, 98, 99
Aggregate in concrete, grading of, 258
Aggregation, 65, 99
Aggressive water:
   effect on concrete cutoffs, 247
   leaching in structures, 248
Air entraining, 259–260
Air lift (direct suction), 152, 172, 184,
   201, 378

Akasaka subway station, Tokyo,
   448–449
Angle of shear resistance, 19
Apparent pressures, 582
Arched structures, 367–371, 473, 501
Arching:
   effect on lateral loads, 582
   effect on trench stability, 30, 34–35,
      42
Arno River, Florence, Italy, 476
Association of particles, 98
Attapulgite, 117, 120, 144
Attraction, 65, 98, 100, 101
Augers, 151, 348, 352, 375

Backfill (see Earth backfill)
Backhoes, 151, 155–156
Barite (barium sulfate), 112
BART (Bay Area Rapid Transit), San
   Francisco, 10, 444, 446, 530, 533,
   540–541, 570–571
   (See also specific stations)
Base bearing, 380, 381, 386

Base bearing *(cont.)*:
  estimation of, 405–406
  as a function of allowable settlement, 406–407
  as a function of vertical displacement, 382, 384, 394
Base failure, 413, 424
  by heave, 490, 491
  (*See also* Heave, base)
Basements, deep, in buildings:
    bracing, 460–462
    plan area and shape, 460
    requirements during construction, 457
    support systems, 459–460
BBRV units, 508
Bearing-capacity factor, 41, 398, 400, 405
Bearing-capacity failure, 380
Bearing elements, load-:
    base bearing (*see* Base bearing)
    circular (*see* Piles, large-diameter)
    design for vertical load, 404–407
    diaphragm-wall panels, load tests, 396–400
    factor of safety, 407–408
    prismatic and linear: bottom cleanliness, 378, 379, 391
      defects and repairs of, 380
      effective perimeter, 385
      load tests, 380, 386
      preparation of base, 378–379
      transfer of load, 384–385, 390–392
    selection of, 400–402
    side shear at interface in, 381–383, 386
Beckton pumping station, England, 470
Benoto rig, 352–353
Bentonite, 1, 4, 97, 117, 139
  analysis of, 120–121
  concentration of, 126, 134, 207
  flow properties of, 118–120
  as main colloid, 117–121
  mixing of, 137
  resistance to contamination by cement, 140
Bentonite cake, 12
  (*See also* Filter cake)
Berms, earth, 462, 556, 560, 561, 563
Bingham bodies (fluids), 109, 112, 233
Bingham plastic flow, 109
Bituminous emulsion, 229, 251–252

Bleeding of slurries, 144–146
Bloomsbury Square parking garage, London, 469
Blowout failure of earth backfill, 207, 211
Blowout gradient, 82, 207
Blowout (pressure) tests of earth backfill, 208–211
Bond strength, 256, 278
  tests on, 279–285
  working bond stress, 285
Bored-pile walls, 2, 15, 432, 433, 448, 459
  advantages and limitations of, 251
  classification of, 345–348
  construction and installation of, 348–350
  load tests on, 386–392
Boston:
  excavations in clay, 531–532
  Sixty State Street Tower, 7, 10, 11
Boulders:
  effect on excavation, 333
  method of removal, 334–335
Bourse station utility tunnel, Brussels, Belgium, 465
Bracing:
  center-core, 462
  combined, 550–557
  cross walls, 444
  deep basements, 460–462
  design of, 485
  elastic shortening of, 559
  before excavation, 525
  long narrow cuts, 442
  with permanent floors, 534–541
  by rock socket, 527–530
  with struts, 530–534
  with tiebacks, 541–548
Bracing details, 585
Bradwell excavation, London, 23
Brittania House, London, 551–555
Brussels, Belgium, Metro, 5
Bucket excavators, 151, 375
Bucket scrapers, 160–162
Buckingham-Reiner equation, 112
Buttressed walls, 367–368, 473, 501
BW system, 177–187

Cages (*see* Reinforcing cages)
Caliper logging, 380
Cantilevered walls, 549, 560
Capillary head, 25

Cardhouse structure, 98, 99, 101, 103
Casing of holes, 348, 349, 358, 375, 376
Cavitation, 126, 183, 380
Cells, 367
   closed, 371–372
   open, 372–373, 474
   stability of, 501–502
Cement:
   cement-bentonite mixes, 224–225, 227, 228
   effect on slurry, 140
   sulfate-resisting, 247
Cement-bentonite cutoffs, 195
   (*See also* Solidified walls)
Center-core bracing, 462
Centrifuges, 94, 105
Chicago underflow system, 7, 10
Chisels:
   effect on trench stability, 50–52
   excavation with, 173
   structural details, 168
Churchill subway station, Edmonton, Canada, 446–448
Circular cuts, stability of, 40–42
Circular drills, 191–192
Circular slides, 53
Circular walls:
   construction of, 361–363
   polygonal shapes, 360–361
   stability of, 497–500
   uses of, 359
   yielding at construction joints, 359–360
Civic Center station, BART, San Francisco, 446, 540–541, 570–571
CL-CLS fluids (chrome lignite–chrome lignosulfonate systems), 94, 96–97, 122
Clamshells, 155, 159
   cable-suspended, 162, 164, 165
   excavation with pilot holes, 172
   excavation rates, 176
   hydraulic-operated, 163, 164
   power-operated, 163, 164
   rope-operated, 163
   single-stage excavation, 171
Clay:
   Boston, excavations in, 531–532
   clay-cement mixes, 222–224, 233, 235
   clay cements, time-setting, 94
   deflocculated, 226
   as earth material, 197

Clay *(cont.)*:
   properties of, 197–198
   swelling pressure of, 103
CMC (carboxymethyl cellulose), 95, 122–124, 139
CNA building, Chicago, 537–540
Coagulation of colloid particles, 86
Cobian Plaza, Puerto Rico, 7, 8
Coefficient:
   of active earth stress, 516, 564, 566, 583
   of earth stress at rest, 516, 567–568
   of fluid concrete pressure, 264
   of horizontal subgrade reaction, 246
   of passive earth stress, 516, 564, 566
   of viscosity, 108
   of wall adhesion, 386, 390, 395, 398, 400, 401, 404, 491
   of wall friction, 393, 400, 401, 405
Cohesion factor, 19, 21
Cohesion limit, 134
Colloid concentration, 111, 120, 127
Colloid particles, coagulation of, 86
Colloidal solutions, 97
Colloidal state, 98
Colloidal suspensions, 97
Colloids, 5, 97
Columns, prefounded, 418–419, 441, 506
Compacting factor test, 257–258
Composite walls:
   bored piles and linear panels, 357–358
   examples of, 444
   soldier piles and plain concrete, 355–357
   soldier piles and reinforced concrete, 353–355
   steel-and-concrete panels, 353–355, 502–505, 530–533
Concrete:
   aggregate in, grading of, 258
   compression tests on, 275–277
   contaminated, 263
   curing of fresh, 272
   defects, usual, of, 262
   elasticity of, 274
   flow motion of, 255, 263–272
   hydrostatic thrust of fresh, 568
   intermixing with bentonite, 265, 267, 269
   placement of fresh, 260, 262–263, 379
   plug motion, 270

Concrete *(cont.):*
  proportioning of mixes, 257–260
  setting time of fresh, 256
  strength of, 271–278
  usual defects of, 262
  water-cement ratio, 258–260, 272
  workability and flowability of, 256–258, 264, 271
  working stress of, 278
  (*See also* Plastic concrete)
Confined-compression (consolidation) (oedometer) test, 567
Connections, structural:
  moment, 310–311
  shear, 308–309
  wall-roof, 437
Consolidation (confined-compression) (oedometer) test, 567
Constitutive models, 588, 589
Construction joints, vertical:
  axial load, transfer of, 304–306
  casing joint by Franki, 298–299
  common joint defects, 302
  common types of, 290
  I-beam joint, 295
  interlocking-pipe (round-tube), 290, 292
  joint defects, common, 302
  locking box by Takenaka, 300–302
  modified round-tube joint, 295
  moment, transfer of, 310
  prefabricated stop ends, 341
  requirements of, 290–291
  round-tube (interlocking-pipe), 290, 292
  RPT joint, 295–296, 372
  shear, transfer of, 306, 308
  special, 297
  steel-plate-and-casing joint, 297–298
  steel plate and vinylon sheet, 292
  structural connections, 306–307
  transfer: of axial load, 304–306
    of moment, 310
    of shear, 306, 308
    of vertical load, 303–304
  vertical load, transfer of, 303–304
  water-stop joints, 296–297
Construction sequence, simulation of, 589–590, 594
Core barrels, 348
Coulis, 337, 428, 429
Counterforts, 462
Critical height of trenches, 19, 21

Critical hydraulic gradient, 61, 78, 210
Cross walls, 444, 493, 526
Crystalline state, 98
Culverts, 422–473
Cutoff concentration, 129
Cutoff walls, 2
  rigid, 244–247
  slurry-trench, 2, 194–195
Cutoffs:
  cement-bentonite, 195
    (*See also* Solidified walls)
  earth (*see* Earth cutoffs)
  noncorrosive, 247–252
  plastic-concrete, 2, 195, 237
Cycloids, 54
Cyclones, 94, 126, 172, 181, 185–186, 189
Cylindrical-surface method of trench stability analysis, 52–53

Darcy's law, 207
Dead Sea water, 144
  as base fluid in earth backfill, 211
Decking (*see* Temporary decking)
Deep filtration, 61–62, 77
  (*See also* Seal mechanism)
Deflocculants, 102
Deflocculated clays and gels, 226
Deflocculation, 101
Density of slurries:
  changes during excavation, 84
  effect on concrete placement, 130
  measurement of, 115–116
  as physical property, 107, 112
Desanding, 172
Detergents, 94
Diamond coring, 380
Diaphragm-wall panels, load tests, 396–400
Diaphragm walls:
  advantages and disadvantages of, 16–17
  construction accuracy and tolerance, 323–326
  origin of, 1
  prefabricated, 336–345
  present applications of, 2–9, 459
  on stilts, piles, subpiers, and rock, 415–418
  underpinning with, 413
  watertightness of (*see* Watertightness of diaphragm walls)
Diesel-oil fluids, 94, 96

Index 615

Dikes, 211–213, 215, 476
Dilatancy, 109
Direct circulation, 177, 193
Dispersants, 121–122
Dispersion, 65, 67
Docks, 473
Downward construction, 16, 425
　examples of, 425–427, 441, 445
Dragline buckets, 151, 156–159, 200–201
Drop shafts, 360
Dynamic factor, 48

Earth backfill, 196
　blowout failure of, 207, 211
　blowout (pressure) tests, 208–211
　case study of, 211–215
　clay mixes, 196
　consistency and water-content tests, 214
　mixing of, 201–202
　permeability and composition of, 198–199, 206–208, 210, 211
　permeability and consolidation tests, 212–213
　placement of, 202–203
　proportioning of, 208
　settlement of, 206
　slump of, 202, 208
Earth cutoffs, 195
　bottom cleaning of, 201
　cavitation of, 207
　cement-bentonite, 195
　　(See also Solidified walls)
　comparison with sheet-pile walls, 219
　connections of, 206
　efficiency of, 215
　excavation of, 200–201
　special treatment, 205
Earth stresses, lateral:
　factors affecting, 579–580
　as function of strain, 485, 515
　observed, 568, 575
　predicted by finite-element analysis, 592, 593, 595–596
　recommended, 580–581
　at rest, 516, 565, 567–568
Edmonton, Canada, subway, 446–448
Effective-stress analysis, 22, 384, 565, 604
Elastic shortening:
　of bracing, 559
　of posttensioned walls, 510, 512

Elastic theory, 587
Electrolyte concentration of slurries, 65
Electroosmotic forces, 87
Electroosmotic phenomena, 12, 86–88
ELSE (excavating machine), 160
Embarcadero station, BART, San Francisco, 530–531
Embedment of wall, 458, 513
　effect on passive resistance, 604–605
　relation to lateral displacement, 521–525, 561
Excavation, 150–151, 181
　equipment for (see specific type)

Failure surface, 20, 22, 52–54
Fann V-G meter for viscosity measurements, 114–115
FCL (ferrochrome lignosulfonate), 121, 122, 140
Ferrochrome-lignosulfonate-type materials, 94
Filter cake, 36, 59–62, 67, 86, 105, 116, 126, 207
　effect on penetration, 71–73
　effect on trench stability, 37–40
　electrochemical effects on, 85
　formation as time-dependent process, 85
　shear strength of, 69
　tests on, 86
Filter-press test, 116–117
Filtration:
　deep, 61–62, 77
　　(See also Seal mechanism)
　measurement of, 116
　rheological blocking, 62
　of slurries, 60
　surface, 61, 76
Finite-element analysis, 587–591
　case studies of, 591–599
Flocculants, 122
Flocculation, 65, 67, 86, 98–101, 103, 111
　internal mutual, 100
　resistance to, 102
Flocs, 99
Flotation, 506
Flow behavior of slurries, 101, 103, 109
Flow curve of slurries, 109
　effect of peptization or flocculation on, 111
Flow factor, 34

616 Index

Flow properties of slurries, 107
  measurement of, 112
  salt concentration, effect on, 101, 118, 141
Flow rate in pipes, 112–113
Fly ash, 228, 248–251
FPS specifications, 318, 323
Free-earth-support method for estimating embedment, 605
Freezing, 14–15
Freundlich, H., 1, 65
Friction angle, 24, 52

Gel strength of slurries:
  effect on trench stability, 27–28, 36–37
  in gravel, 83
  initial, 100, 109, 110, 115, 233, 234
  in soil pores, 61, 78
  10-min, 83, 110, 115, 132, 137
  in the trench, 67
  yield stress, 109
Gelation, thixotropic, 62, 64, 65, 67, 98–101
  (See also Thixotropy)
Gels, 99
  minimum gel structure, 101
  thixotropic, 65, 118
  transformation of, 103
  used as grouts, 226
German Embassy, London, 5
Grabs, 151
Gradation limits of earth backfills, 198–199, 201
Gravel bed:
  effect of clay on permeability of, 81–82
  effect of slurry on permeability of, 78–79
  rheological blocking of, 80
Gravity-type walls, 473, 502
Ground anchors (see Tiebacks)
Ground movement (see Movement, ground and wall)
Ground vibrations, effect of, 48
Groundwater:
  lowering, 422
  protection of excavation from, 422
Grout screens, 2
  flow properties of, 233–235
  injected, 195, 223, 231

Grout screens (cont.):
  penetrability and strength of, 235–236
Grout systems:
  displacement grout, 337–339, 343, 428
  single grout, 337, 428, 454
Grouting, 15, 236
Guide walls, 315, 320–323, 473
Guildhall Precincts Development, London, 541–544, 575–577
Gypsum fluids, 94, 95

Heave, base:
  as cause of base failure, 424–425, 506, 527, 562
  with diaphragm walls, 492–495
  in narrow excavations, 490
  pressure due to, 519
  (See also Base failure)
Holes (see Casing of holes)
Humber suspension bridge, England, 5, 476–478
Huxtable Pumping Station, Arkansas, 7, 8
Hyde Park underpass, London, 5
Hydraulic gradient, critical, 61, 78, 210
Hydrostatic thrust:
  of fresh concrete, 568
  of slurry, 21

Ideal plastic flow, 109
Impermeable layer at interface, 35
  (See also Filter cake)
Impervious blankets, 205
Initial tangent modulus of soil, 42
Injected-grout screens, 195, 223, 231
Intakes for thermoelectric power stations, 469
Interaction of particles, 98, 111
Interlocking elements, 173–176, 231–233, 236, 245
Interlocking enclosures, 470
Interparticle force, 65, 67
  (See also Interaction of particles)
Invert-emulsion muds, 94–96, 144
Ion-exchange capacity, 118, 120

Jackhammers, 348

Joosten process, 226, 236–237

Kaolinite, 101, 117
Kelly bar, 152, 164–168
Kensington Town Hall, London, 384, 399
Keybridge House, London, 544–545, 560, 575–576

Laminar tilting, 304, 370, 372
Lateral earth stresses (see Earth stresses, lateral)
Lateral loads:
   due to loads within soil, 517, 519
   due to preloading of struts, 519–520
   due to prestressing of tiebacks, 519–520
   due to shrinkage, 519–520
   due to surcharge, 516
   due to temperature changes, 519–520
   effect of arching on, 582
Leakage, 330
   (See also Waterproofing)
Lignins, 121
Lignosulfonate, 94, 95, 121
   (See also CL-CLS fluids; FCL)
Lime fluids, 94
Limit theory, 19, 20, 515, 517, 563–567, 587
Load-bearing elements (see Bearing elements, load-)
Load-distribution curves, 386
Load-settlement curves, 386, 389, 395, 396
Load tests, 380–381
   on bored-pile walls, 386–392
   on diaphragm-wall panels, 396–400
   field, large-diameter piles, 394–396
   load-bearing elements, 380, 386
Load-transfer curves, 386, 396
London clay, 23, 398, 534, 541, 544, 549, 550
Lorenz, H., 11
Loss of ground:
   in bored piles with lagging, 505
   in slurry trench excavations, 89
Lost-circulation materials, 124
Lost-circulation techniques, 94
Low-pH muds, 93, 94
Lower level, construction from, 429
   examples of, 430–431

Lubricants, pressure, 94

Manicouagan 3 dam, Canada, 169–170, 176, 177
Marsh funnel, 2, 113, 132
Mechanical diggers, 151
Mersey River underpass, Liverpool, England, 472
Mersey tunnel, England, 394
Mesh, finite-element, 590–591
Methylene blue dye, 121
Microstructure analysis, 229
Mining shafts, 360
Mixers, 137, 160
Montmorillonite, 4, 81, 101, 117, 197
   swelling of, 103
Mooring wharves, 473
Movement, ground and wall:
   bracing with permanent floors, 534–541, 559
   with combined bracing, 550–557
   with diaphragm walls, observed, 525
   differential, 486
   due to excavation, 413, 485, 487–490
   finite-element analysis method of predicting, 592
   heave in narrow excavations supported by diaphragm walls, 490–495
   observed with diaphragm walls, 525
   predicted by finite-element analysis, 592
   procedure for reducing, 557–561
   reducing, procedure for, 557–561
   with strutted walls, 530–534, 558
   with tiebacks, 541–549
   time-dependent, 541, 545, 553, 557, 560, 586
   with walls braced at the bottom, 526–530, 558
Moving loads, effect on trench stability, 49–52

Neasden underpass, London, 541, 560
New Palace Yard car park, London, 534–535
New York City subway, 7, 9
Newtonian fluids, 108, 109

Noncorrosive cutoffs, 247–252
Nonnewtonian fluids, 108, 111
Nonsalt systems, 99, 100

Oedometer (confined-compression) (consolidation) test, 567
Ohka River tunnel, Japan, 479–481
Osaka utility tunnel, Japan, 467–468
Oscillators, 348
Oslo, Norway, subway, 5, 526–527
Overtreatment, 102
Overturning of arched walls, 370, 502

Panels, wall:
    depth, 318
    length, 316–317
    sequence and arrangement, 319–320
    waterproofing, 330–331
    width, 318
Parking, underground, 359, 468–469
    (See also specific installation)
Passive state, 491, 511, 513, 516, 564–566, 604–605
Peel-off, 67, 69–71, 135
Penetration of soil by slurry, 35, 76, 78
Peptization, 101, 103, 111
Peptizers, 102
    thinning action of, 111
Percussive tools, 152, 153, 168–169
    excavation with, 171, 173, 348
Permanent floors, bracing with, 534–541
Permeability of walls (see Watertightness of diaphragm walls)
Peterfi, 1
pH of slurries, 116, 121
φ = 0 concept (total-stress analysis), 21–23
Phosphate muds, 93
Piccadilly line extension, London, 5
Pierre-Bénite power station, France, 26
Piezometer tests, 78–80
Piles:
  large-diameter, 375
    field load tests, 394–396
    installed under slurry protection, 376–377
  secant (see Secant piles)
  soldier, 13, 431, 433
Pipes, pressure drop in, 112–113

Plastic concrete:
    deformability and strength, 238–241
    examples of mixes, 241–244
    modulus of elasticity, 238–240
Plastic-concrete cutoffs, 2, 195, 237
Plasticizers, 258
Polyelectrolytes, 122
Polygonal enclosures:
    stability of, 500–501
    (See also Circular walls)
Polymer stabilizers, 124–125
Polyphosphates, 121
Pore-water pressure, 516, 565
Posttensioned walls, 5
    advantages in unbraced walls, 363
    multibraced systems, 367
    posttensioning techniques, 364–367, 508–509
    prestress loss, 512–513
    stability of, 508, 510–515
    wall types suitable for posttensioning, 513–515
Potassium aluminate, 124
Powell Street station, BART, San Francisco, 533–534, 570
Power stations, 469
Pozzolan, 249
Precast (prefabricated) walls, 5, 336, 339–341
    advantages and disadvantages of, 343–345
    installation of, 342, 343, 454–455
    uses of, 427–429, 453–455, 460
    waterproofing of, 344, 455
Prefounded columns, 418–419, 441, 506
Pressure drop in pipes, 112–113
Pressure lubricants, 94
Prestressing strands, 364, 511, 512
Progressive failure, 516, 586
Protective film, 60, 67
Pumpability of slurries, 106, 132
Pumping stations, 469, 470

Quay walls, 475
Quebracho, 121
Quick condition of sand, 48

Rankine-Résal coefficient, 583
Rankine states, 564–565
Rate of shear, 108, 112, 114

Reading bypass, London, 548, 549
Reinforcing cages:
  assembly of, 286, 317
  bar spacing, 286–288
  boxes and inserts, 288
  method of detailing, 318
  splices, 289–290
Repulsion, 65, 98, 100
Reservoirs, underground (see Tanks, underground)
Response spectra, 49
Retarders, 259
Reverse circulation, 106, 107, 169, 177
  counterflow, 190–191
Reverse-circulation machines, 152, 181–183
Reversibility, 65–67
Rheological blocking, 62, 76, 80
  (See also Seal mechanism)
Rheological properties of suspensions, 111
Rigid cutoff walls, 244–247
Rock sockets, 177, 527
  bracing by, 527–530
Rotary drilling equipment, 152, 153, 177, 348
Rotational viscometers, 114–115
Rupel tunnel, Belgium, 481–483

Salt concentration, 99
  effect on flow properties, 101, 118, 141
Salt-flocculation value, 102
Salt systems, 99
Sand, quick condition of, 48
Schneebeli method in trench-stability problems, 35
Screens, 105, 126, 131, 172, 181, 185–186, 189
Seal:
  formation of, 58, 126, 129
  in gravel bed, 78
  at interface, 35
Seal mechanism, 61–64
  (See also Rheological blocking)
Sears Tower, Chicago, 571–573
Secant piles, 15, 345, 351–352, 432, 459
Sedimentation of solids, 104, 126
Seismic coefficient, 48
Sensitization, 103, 122

Separation of solids, 105, 131, 132, 185–186
Settlement:
  control of, 489
  correlation with side and base resistance, 406–407
  detrimental, 486, 487
  with diaphragm walls, observed, 525
  differential, 486
  due to excavation, 413, 485, 487–490
  observed with diaphragm walls, 525
  procedure for reducing, 557–561
Settling tanks, 105, 185
Sewage-treatment plants, 469
Shaft resistance (see Side shear at interface)
Shafts:
  access, 359
  deep, 470–472
Shale-inhibited muds, 94, 95
Shear, rate of, 108, 112, 114
Shear breakdown, 110
Shear resistance, angle of, 19
Shear strength of slurries (see Gel strength of slurries)
Shear stress of slurries, 108, 112, 114, 120
  (See also Gel strength of slurries)
Sheet-pile walls, 14, 219, 433
Shotcrete, 15
Side shear at interface:
  in buttressed walls, 367
  correlation with friction factor, 382, 384
  correlation with undrained shear strength, 382, 384
  effect on active and passive state, 565–566
  effect of general excavation, 383–384
  estimation of, 404–405
  field tests on, 394–396
  as a function of vertical displacement, 382, 384, 386, 393, 394, 398
  laboratory tests on, 392–394
  in load-bearing elements, 381–383, 386
  as vertical load, 519
Site inspection, 313–314
Sixty State Street Tower, Boston, 7, 10, 11
Skin friction (see Side shear at interface)
Slip-caissons, 4

Sloughing, 67, 69, 135, 380
Slump of concrete, 257
Slurries:
  bleeding of, 144–146
  contamination and treatment of, 139–141
  control limits of, 127–132, 134, 200
  density of (*see* Density of slurries)
  displacement by concrete, 130
  disposal and salvage of, 141–144
  electrolyte concentration of, 65
  filtration of, 60
  first experimental work on, 11–12
  flow behavior of, 101, 103, 109
  flow curve of, 109, 111
  flow properties of, 107, 112
    effect of salt concentration on, 101, 118, 141
  functions of, 126
  gel strength of (*see* Gel strength of slurries)
  grouts, similarity to, 236–237
  hydrostatic thrust of, 21
  interaction with soil, 58–90
  large-diameter piles, use in, 375–377
  loss of, 61, 88–89, 124, 129, 229, 316
  mixing of, 137–138
  origin and first uses of, 1
  penetration of soil by, 35, 76, 78
  pH of, 116, 121
  preparation and control of, 136–141
  prismatic elements, use in, 377–380
  proportioning of, 135–136
  pumpability of, 106–107, 132
  shear stress of, 108, 112, 114, 120
    (*See also* Gel strength of slurries)
  similarity to grouts, 236–237
  soil-slurry interaction, 58–90
  specific gravity of, 107, 112
    (*See also* Density of slurries)
  stiffening of, 67
  testing of, 138–139
  use in large-diameter piles, 375–377
  use in prismatic elements, 377–380
  yield stress of (*see* Yield stress of slurries)
Slurry face shields, 4
Slurry-loss control agents, 124, 134
Slurry plant, 159, 315
Slurry-trench-cutoff walls:
  classification of, 194–195
  origin of, 2
Sodium humate, 121

Soldier piles, 13, 431, 433
Solidified walls (cement-bentonite), 2, 5, 195, 227–230, 252
Solids:
  sedimentation of, 104, 126
  separation of (*see* Separation of solids)
Solutions, colloidal, 97
Southern Pacific excavation, San Francisco, 555–557, 593–594
Specific gravity of slurries, 107, 112
  (*See also* Density of slurries)
Specific surface, 81, 83
Splices of reinforcing bars, 289–290
  (*See also* Reinforcing cages)
Stability number, 488, 491, 582, 583
Stability of trenches:
  arching, effect on, 30, 34–35, 42
  chisels, effect on, 50–52
  circular excavations, 31
  clay, special considerations for excavations in, 40–43
  during concreting, 263–264
  construction equipment, effect of, 50
  cylindrical-surface method of analysis, 52–53
  under dynamic loading, 47–52
  effect of arching on, 30, 34–35, 42
  effect of chisels on, 50–52
  effect of construction equipment, 50
  effect on panel length, 316
  effective stress and, 22
  equal-stress method, 29–30
  under external concentrated loads, 47
  without filter cake, 37–38
  gel strength of slurry and, 27–28
  gelation of soil, 36–37
  with horizontal curvature, 44–46
  impermeable layer at interface, 35
    (*See also* Filter cake)
  lateral displacement of face, 43, 45–46
  loss of, 78
  under moving loads, 49–52
  panel length, effect on, 316
  with permeable filter cake, 38–39, 72–73
  rotational failure, 52–54
  sand, special considerations for excavations in, 29–31
  slurry-filled trenches in clay, 20–21
  slurry-filled trenches in sand, 24–27
  in terms of effective stress, 22
  in terms of gel strength of slurry, 27–28

Stability of trenches (cont.):
  in terms of total stress, 21
  unsupported trenches, 18–20
  validity of the $\phi = 0$ concept, 21–22
Stagnation gradient, 37, 39, 71–74, 85
Starch, 124
Sticky limit, 134
Stiffening of slurries, 67
Stiffness, wall:
  effect on load distribution, 584, 587, 597–599
  effect on wall movement, 557, 561
Stilts, 416
Stormer viscometer, 2
Stress-strain relationship of soil, 43, 564, 579, 591, 601
Strip panels, 419–422, 432, 453
Structural connections (see Connections, structural)
Strut loads:
  distribution of, 573–575
  measured, 570–573
  recommended, 581–585
Struts, 462, 478, 530–534, 558, 561, 570–575
Subbasements, existing, 333
Subway stations, 441
  (See also specific station)
Subways:
  composite sections, 439, 449, 507
  construction at crossings, 440–441
  construction under the roof, 433–437
  examples of construction, 444–449
  transfer of load, 438
  ventilation during construction, 438
Suction, direct (see Air lift)
Suction pumps, 153, 181, 184
Surface filtration, 61, 76
  (See also Filtration)
Surface finish of walls, 326
Surfactant fluids, 94, 95
Suspension(s):
  colloidal, 97
  of excavated materials in slurries, 129
  rheological properties of, 111
Swelling pressure of clay, 103

Tangent walls, 348
Tanks, underground, 359, 469–471
Tannins, 121
TBW system, 187–189
Telmarch, 125, 140, 144

Temporary decking, 431, 433, 435, 439–441
Tension cracks, 20, 21, 46, 566
Thames gravel, 144
Thinners, 121–122, 139
Thinning action of peptizers, 111
Thixotropic gelation (see Gelation, thixotropic; Thixotropy)
Thixotropic gels, 65, 118
Thixotropy, 2, 58, 64–65, 67, 110, 234, 237
  (See also Gelation, thixotropic)
Tieback loads:
  distribution of, 586–587
  effect of prestressing, 587, 597–599
  loss of, 586
  single-tie response, 575, 599–604
Tiebacks, 414, 423, 460, 462, 541–549, 559–560, 563, 587
Time-setting clay cements, 94
Tolerance:
  allowable on verticality, 323
  angular deviations in plan, 325
  of bored piles, 348, 349
  control of, 163
  devices for measuring deviations, 325
  in ground with boulders, 335
  irregularities and protrusions, 323, 325–326
  walls terminated below guide walls, 327
Top concrete layer, 327
Tortuosity, 82, 83
Total-stress analysis ($\phi = 0$ concept), 21–23
Traffic, effect on construction, 314, 413
Traffic maintenance, 313, 433
Traffic underpasses, 449–456
Tremie pipes, 260–262, 264, 377
  spacing of, 263
Trenches (see Stability of trenches)
Tunnels (see specific tunnel; Utility tunnels)

Ultimate vertical load, 384
Ultrasonic devices, 189
Unconfined compressive strength, 22
Under-the-roof construction method, 16, 86, 433, 441
Underpinning:
  with diaphragm walls, 413

Underpinning (cont.):
  as ground strengthening, 413
  as lateral protection, 413, 495
Undrained condition analysis, 566–567
Undrained shear strength, 22
UNO building, Vienna, 396–398
Uplift, 413, 422, 423, 506, 519
Utilities:
  availability of, 314
  handling of, 313, 433
  interference with construction, 314, 331, 413, 432, 440
Utility tunnels, 464–465
  examples of, 465–468, 532–533

Veder, C., 11–12
Velocity gradient, 108
Venice, Italy, aquifer recharge in, 5
Vertical construction joints (see Construction joints, vertical)
Vertical load(s), 519
  ultimate, 384
Viale Monte Grappa, Milan, Italy, 511
Vibrations, ground, effect of, 48
Viscometers, 113–115
Viscosity, 99, 100, 107, 111
  apparent, 108, 109, 115, 132
  coefficient of viscosity, 108
  correlation with soil type, 135–136
  effect on concrete placement, 130
  funnel, 113, 136
  measurement of, 113–115
  plastic, 108, 109, 111, 115, 118, 130
  relative, 111

Wall embedment (see Embedment of wall)
Wall movement (see Movement, ground and wall)
Wall panels (see Panels, wall)
Wall stiffness (see Stiffness, wall)
Wall surface finish, 326
Wall systems:
  arched (see buttressed, below)
  bored-pile (see Bored-pile walls)

Wall systems (cont.):
  buttressed, 367–368, 473, 501
  cantilevered, 549, 560
  circular (see Circular walls)
  composite (see Composite walls)
  cross, 444, 493, 526
  cutoff (see Cutoff walls)
  diaphragm (see Diaphragm walls)
  gravity-type, 473, 502
  guide, 315, 320–323, 473
  polygonal, 500–501
  posttensioned (see Posttensioned walls)
  precast (see Precast walls)
  quay, 475
  secant-pile, 351–353
  sheet-pile, 14, 219, 433
  slurry-trench-cutoff, 2, 194–195
  solidified (see Solidified walls)
  tangent, 348
Wall thickness:
  actual, 326
  effective, 319
Water:
  aggressive, in alkali regions, 247–248
  as basic fluid, 123–124, 136
Water jetting, 184
Water Tower Place, Chicago, 7, 8
Waterproofing:
  of bored-pile walls, 350
  of concrete boxes, 437–438
  of construction joints, 330
  of precast walls, 344, 455
  of wall panels, 330–331
Watertightness of diaphragm walls:
  of concrete, 328, 329
  construction joints, leakage of, 329–330
  effect of curing conditions, 329
  of precast panels, 344, 455
Weighting agents, 128
World Trade Center, New York City, 7, 463, 546–549, 560, 577–579

Yield stress of slurries, 99, 102, 106, 111–113, 118, 120, 132
YMCA building, London, 550–551
Young's modulus, 239